Lecture Notes in Civil Engineering

Volume 183

Lecture Notes in Civil Engineering (LNCE) publishes the latest developments in Civil Engineering - quickly, informally and in top quality. Though original research reported in proceedings and post-proceedings represents the core of LNCE, edited volumes of exceptionally high quality and interest may also be considered for publication. Volumes published in LNCE embrace all aspects and subfields of, as well as new challenges in, Civil Engineering. Topics in the series include:

- Construction and Structural Mechanics
- Building Materials
- Concrete, Steel and Timber Structures
- Geotechnical Engineering
- Earthquake Engineering
- Coastal Engineering
- Ocean and Offshore Engineering; Ships and Floating Structures
- Hydraulics, Hydrology and Water Resources Engineering
- Environmental Engineering and Sustainability
- Structural Health and Monitoring
- Surveying and Geographical Information Systems
- Indoor Environments
- Transportation and Traffic
- Risk Analysis
- Safety and Security

To submit a proposal or request further information, please contact the appropriate Springer Editor:

- Pierpaolo Riva at pierpaolo.riva@springer.com (Europe and Americas);
- Swati Meherishi at swati.meherishi@springer.com (Asia - except China, and Australia, New Zealand);
- Wayne Hu at wayne.hu@springer.com (China).

All books in the series now indexed by Scopus and EI Compendex database!

More information about this series at http://www.springer.com/series/15087

Indrajit Pal · Sreevalsa Kolathayar
Editors

Sustainable Cities and Resilience

Select Proceedings of VCDRR 2021

Springer

Editors
Indrajit Pal
Disaster Preparedness, Mitigation
and Management
Asian Institute of Technology
Klong Luang, Pathum Thani, Thailand

Sreevalsa Kolathayar
Department of Civil Engineering
National Institute of Technology Karnataka
Mangalore, Karnataka, India

ISSN 2366-2557 ISSN 2366-2565 (electronic)
Lecture Notes in Civil Engineering
ISBN 978-981-16-5542-5 ISBN 978-981-16-5543-2 (eBook)
https://doi.org/10.1007/978-981-16-5543-2

This Springer imprint is published by the registered company Springer Nature Singapore Pte Ltd.
The registered company address is: 152 Beach Road, #21-01/04 Gateway East, Singapore 189721, Singapore

Contents

Cross-Cutting Issues

About the Editors

Dr. Indrajit Pal presently working as Assistant Professor and Chair at Disaster Preparedness, Mitigation, and Management program at the Asian Institute of Technology, Thailand. Prior to joining at AIT, Dr. Pal served as a faculty member at Centre for Disaster Management at Lal Bahadur Shastri National Academy of Administration, Mussoorie, India (Premier National Institute for training Indian Administrative Services Officers). Dr. Pal has done extensive work on capacity development of decision-makers, risk assessment and disaster risk governance. Dr. Pal has written 6 books and more than 55 articles in international and national peer-reviewed journals apart from supervising masters and doctoral research. Dr. Pal having about 18 years of experience in research and capacity development on Disaster Risk Management and Governance and Disaster Risk Science and Education. RESEARCH EXPERTISE: Disaster Risk Governance, Disaster Risk Reduction (DRR) and Development, Incident Command System (ICS), GIS/RS in disaster risk management, Climate Change Adaptation, Community Based Disaster Risk Management, Disaster Risk Science and Education, Sustainability, Public Health Risk assessment.

Dr. Sreevalsa Kolathayar pursued M.Tech. from IIT Kanpur, Ph.D. from Indian Institute of Science (IISc) and served as International Research Staff at UPC BarcelonaTech Spain. He is presently Assistant Professor in the Department of Civil Engineering, National Institute of Technology, Surathkal, Karnataka, India. Dr. Sreevalsa has authored six books and over 75 research papers. He is Associate Editor of two International Journals and Executive Guest Editor for one. His research interests include Seismic Hazard Assessment, Geohazards, Earthquake Preparedness, Landslides, and Water geotechnics. The New Indian Express honoured Dr. Sreevalsa with South India's Most Inspiring Young Teachers Award. He is in the roster of two technical committees of ASCE Geo-Institute. He received "IEI Young Engineers Award" by The Institution of Engineers (India), in recognition of his contributions in the field of Civil Engineering.

City Resilience and Sustainable Infrastructure—An Introduction

Indrajit Pal, Sreevalsa Kolathayar, and Satya Venkata Sai Aditya Bharadwaz Ganni

1 Introduction and Background

The cities are vulnerable to adverse effects of natural and manmade hazards and pose great challenges due to rapid urbanization and climate change. More than half of the population of the world lives in cities, and it is likely to increase in future. Cities also account for about 70% of global carbon emissions [43]. The urban sustainability focuses on managing resources in a way that ensures welfare and promotes equity for current and future generations [29]. Resilience is the capacity of a system to absorb disturbance, such as a natural hazard or disaster, essentially retaining the same function, structure, feedbacks and identity [29]. Sustainability should the goal of society while resilience signifies a characteristic of the urban system. Both sustainability and resilience come together as a powerful tool in the context of urban development. OECD [29] defines resilient cities as "those which have the ability to absorb, recover and prepare for future shocks (environmental, economic, social and institutional). Resilient cities promote sustainable development, well-being and inclusive growth."

The guiding principles are based on the predefined ten essentials helping the cities and local governments to share learning, access information, develop indicators, and performance measures and track progress. The Ten Essentials for Making Cities Resilient were proposed to hasten execution of the Sendai Framework for Disaster Risk Reduction (2015–2030) [31, 44].

Essential 1: Institutional and Administrative Framework.
Essential 2: Financing and Resources.
Essential 3: Multihazard Risk Assessment—Know Your Risk.

I. Pal (✉) · S. V. S. A. B. Ganni
Asian Institute of Technology, Pathum Thani, Thailand
e-mail: indrajit-pal@ait.ac.th

S. Kolathayar
National Institute of Technology Karnataka, Surathkal, India

I. Pal and S. Kolathayar (eds.), *Sustainable Cities and Resilience*, Lecture Notes in Civil Engineering 183, https://doi.org/10.1007/978-981-16-5543-2_1

Essential 4: Infrastructure Protection, Upgrading, and Resilience.
Essential 5: Protect Vital Facilities: Education and Health.
Essential 6: Building Regulations and Land Use Planning.
Essential 7: Training, Education and Public Awareness.
Essential 8: Environmental Protection and Strengthening of Ecosystems.
Essential 9: Effective Preparedness, Early Warning and Response.
Essential 10: Recovery and Rebuilding Communities.

Sustainability and resilience need important attentions in planning, design, construction and operation of the cities' infrastructure. Thanvisitthpon et al. [42] describe flood risk management in urban areas relies on infrastructure development for flood prevention and management for resilience building and also through flood adaptive capacity of urban residents (i.e., non-structural strategy). Climate variability contributed to the unpredictability of precipitation in many parts of the world and significantly impacted the city development, livelihoods, the environment and the economy [41]. Similarly, seismic risk can be better addressed though the micro-zonation mapping to emphasis in any infrastructure development planning [37].

The chapters in this book volume present diverse insights on urban resilience and sustainable infrastructure. This article is an attempt to summarize the contents of the book volume on City Resilience and Sustainable Infrastructure. All the chapters in this volume are segregated in five clusters, e.g., resilient infrastructure in construction, innovative construction interventions, waste management and disaster risk reduction, urban development and sustainability and cross-cutting issues.

2 Resilient Infrastructure in Construction

Rungskunroch et al., in their exploratory study on *Risk and resilience of railway infrastructure: An assessment on uncertainties of rail accidents to improve risk and resilience through long-term data analysis,* aimed to analyze uncertainties of railway accidents and evaluating risk and resilience of rail's infrastructure after occurring an accident. The datasets are analyzed by using Bayes' and decision tree methods through Python programming. The model uses long-term data to measure the severity level of an accident by infrastructure failures. The result shows the severity level is scored at 18 of 32, which can be interpreted at 'high risk' [12]. Alok Rathore et al., Paper *Assessing Resilience of Transportation Networks under Multi-Hazards: A Review,* present a review of various methodologies for assessing and improving the resilience of the bridge network system against disasters. There are broadly two assessment approaches used by the authors in this paper, namely resilience aspect (quantitative) and resilience assessment (qualitative approaches). Resilience optimizes the maintenance method for faster recovery when the system's life cycle is considered. Therefore, a structural system's resilience is considered as a critical performance indicator for infrastructure [13]. S. Divya Sankar et al., in their *Risk*

Management In Construction Industry, present how the people in construction organizations need to manage the risks and need to know how to manage the risks by using different techniques of construction management. The analysis of risk and also techniques of management are applied rarely by the Visakhapatnam construction organizations because of the absence of knowledge and also expertise. Therefore, the construction industry is not confident relating to the techniques used and also its applicability in the construction projects [2]. Poornima Ramesh and Bharani Alagirisamy in their paper Achieving Sustainability Goals through infrastructure modifications: Lessons learnt from COVID-19 pandemic present how the world produced by antivirus looks on the basis of the lessons learned and the value of a stable and safe environment. This paper looks at the future COVID-19 steps to gradual and systemic improvements in varying time frames and sizes, which enhance air quality and less energy use, or the use of materials that eventually fulfill the sustainability objectives. The pandemic caused decision makers, designers and architects to reflect more, attempt to mold our physical areas and reset the current built environment, or to create more ideas to confront potential attacks on the virus [4]. Katopodis et al., in their exploratory study, explains the vulnerability and impact assessment of extreme climate events in the Greek oil industry by identifying the vulnerability status of employees, processes and structures through a three-level analysis. The proposed approach is built on the use of the high resolution WRF regional climate model to determine the climate hazards with the most severe impacts to oil infrastructure, the most exposed structures and processes, and the changes of the trends in extreme events, under the future analysis. The study highlighted the need for the re-design or improvement of the defenses of the oil infrastructure, taking into consideration the prospect of climate change, to withstand extreme hazards and loads expected in 50 years ahead [14]. Elena Dolgopolova in the paper Catastrophic events at the river basins due to permafrost thawing: review and examples discussed about how thawing of permafrost results in intensive erosion of river banks, breaking of constructions and pipes, thermokarst subsidence of landscape, etc. Most of the river discharges under consideration show increased trends, the rate of increase of water discharge decreases with a rise of air temperature. At the same time, correlation between the rate of river discharge increase and growth of ground temperature is found to be positive. It is found that a river bend is a potential source of hazard connected with the cemetery erosion. Growth of water discharges increases water levels which result in intensification of river bank erosion and may be a source of catastrophic pollution of water due to wash out of burial wastes [10]. Smaranika Panda in the paper Effect of Land Uses on Personal Exposure of Street Vendors at a Metropolitan City presents the impact of contrasting land uses on personal exposure of street vendors in one of the metropolitan cities of India (Bengaluru). Study results indicated heavy particulate exposure at both the locations. The PM5 personal exposure was observed to be exceeding the PM10 and PM2.5 standards by several folds at both the land uses. However, at the traffic intersection the exposure concentration of PM5 was observed three times higher than the residential area [32]. Sumeet Moktan et al., in their paper Social Vulnerability Assessment to Natural Hazards in Western Nepal, explained about the social vulnerability to natural hazards in western Nepal and have aimed to assess social vulnerability in 40 districts of the Western regions of Nepal. The overall vulnerability map was prepared using the indicator-specific outputs obtained in this research. The indicators considered in this study were population, gender, age, education, health, poverty, ethnicity, economic and housing unit. The results of this study have the potential in contributing to policy making and preparedness, emergency planning, public awareness and altogether in creating a more successful and focused crisis response program [7]. The paper Sustainable Urban Drainage System to avoid flooding of rain origin and improving green areas, Lima, Peru by López Amaro et al. aims to provide an engineering solution to the problem with a sustainable urban drainage system (SUDS). It consists of a set of elements of the drainage network that will allow the collection, transport, decontamination, retention, infiltration and rainwater evacuation sustainably. It is recommended to implement SUDS alternatives to reduce the risk of floods due to the urbanization process and improvement of green areas and to avoid their occurrence in the urban development process in this and other areas of the city [39].

3 Innovative Construction Interventions

The explorative study on *Replacement of river sand with coal bottom ash as fine aggregate in cement mortar* by Wasnik et al. explained about utilization of coal bottom ash as fine aggregate in cement mortars. Compression tests are conducted on cement-mortar cubes (with different proportions of coal bottom ash) to determine compressive strength. Further, a compression test is conducted on cement-mortar cylinders in a displacement controlled Universal Testing Machine (UTM) to obtain the stress–strain curve and modulus of elasticity. The study found that river sand replaced with up to 50% coal bottom ash exhibited satisfactory performance as fine aggregate in cement mortar [45]. Vasavi G S et al., in their paper *Strength and Durability properties of Alkali activated Flyash Earth bricks,* discussed the strength and durability characteristics of alkali activated Flyash Earth bricks. Two kinds of bricks were produced, one set of bricks with the use of manufactured sand or M-sand and another set without M-sand. Soil Fly Ash alkali activated bricks can be used as an alternative to traditional burnt clay bricks as the strength parameters of these bricks are in the same range or even slightly higher than that of burnt clay bricks. It was found that the dry compressive strength of the bricks is in the range of 8 to 10 MPa, and wet compressive strength is around 70% of dry compressive strength [26]. Roy et al., in their paper *Experimental optimization of GGBS Fly ash-based geopolymer concrete paver blocks,* discussed about an extensive experimental investigation is carried out using various mix proportions of Geopolymer concrete (GPC), to identify the design mix considering the compression testing results and other key parameters of GPC with low-cost technology. From the compression testing results, GPC can be considered as the right material for construction from both strength and deformation considerations. Apart from its high strength, this material requires low production cost at high durability. Practical implementation of GPC technology in precast concrete products such as paver block has been developed in Indian scenarios under ambient curing [11]. The explorative study on An Experimental Study of Creep Behavior for Disturbance of Unsaturated Expansive Clay Soil by Balaganesh P et al.. explained about how the recent trends in technological innovations involved in the design and implementation of a smart composter enabled with IoT. The control and accuracy of results mainly depend on the selection of the major processing units having suitable output-related hardware such as Arduino UNO, Node MCU, Raspberry Pi model 3, track recorder, Siretta antenna and wireless communication module. It is also observed that selection of proper internet gateway is essential to address the challenges in data transmission, data safety and power consumption as in the case of a community-based cluster of smart composters. The study makes design-oriented prospects toward development of a smart-rapid composter involving provisions for advanced process control and quality checking (Expansive Soils). Kok Shien Ng and Yee Ming Chew in their paper *Practical Design of Stone Column in Predicting Settlement Performance* present a simple yet practical design method for both the end bearing and floating stone columns for large and small column groups. Hence, the methods introduced here focus mainly on the estimation of the settlements of stone columns and improved ground. They were developed based on a series of numerical studies using 2D and 3D finite element analyses. Three methods have been introduced for large stone columns groups and two methods for small stone columns groups. The methods are derived from numerical analyses where the effect of influencing factors on the settlement performance is taken into accounts [28]. Thanh Quang Khai LAM et al., in their paper Using prestressed reinforced concrete piles as basement walls for high-rise buildings discussed the latest construction techniques. There are currently some construction methods for building houses with basements, such as: pile construction method is done, then dig the soil to the bottom of the foundation and then build the house from the bottom up, or build according to the soil wall method (Barret wall), or the top down method, semi top down methods, etc. The authors present in this paper: "Using pre-stressed reinforced concrete piles for high-rise buildings as a basement wall." The use of pre-stressed reinforced concrete piles saves and lowers the project's cost, bringing the construction to high quality [20]. Prakash et al., in their paper *Analytical Study on Single Span Reinforced Concrete Beam with Continuous Spiral Reinforcement Under Pure Torsion* discussed the construction methods. In this paper, four inclination angles of stirrups and traditional stirrups are adopted. The torsion behavior in beams is studied by monitoring the load deflection curves, ultimate load values, vertical deflections measurements and crack propagation during static tests. Test results indicate that using rectangular spiral shear reinforcement improved the torsion capacity of beams compared with traditional individual closed stirrup beams. Using rectangular spiral shear reinforcement is recommended because it enhances the stiffness in beams [33]. Vigneshwaran R and Prabavathy S in their paper *Investigation of wind loads on setback building using Computational Fluid Dynamics* explained about the actual wind loads acting on the setback tall building. The simulation is carried using computational fluid dynamics (CFD) with the help of turbulence model realizable k-ε. The analysis is carried for a particular wind angle 0° at a scaled wind velocity of 10 m/s.

From the results, the mean pressure coefficient (Cp) is obtained for each face of the setback building and further pictorial representation the pressure contour on each face of the building is presented. Additionally, the physics of the wind flow behavior around the setback is studied [34]. Mahesh Mungulea and Kannan K. R Iyer in their paper *A Review on Role of Pavement Materials on Urban Heat Island Effects* have done the comparison of thermal performance of pavement materials, namely asphalt and concrete. Despite the obvious advantage with concrete, little attention has been paid to improve albedo values of concrete. Limited experimental investigations on modified concrete pavement highlight the potential gain that can be achieved. The utilization of admixtures for enhancing the albedo value of concrete pavement can be seen as an effective way to reduce the UHI effects, especially in urban areas [21]. Anjali A et al., in their paper Studies on Strip footings resting on lateritic slopes, had investigated the settlement behavior of strip footings resting at various levels on lateritic slopes, by carrying out a series of laboratory scale load tests on model footings resting along the slope surface. The parameters varied are distance of footing from edge of slope and slope angles. Finite element analyses are carried out with the FE software PLAXIS 2D, and the results are compared with those obtained from laboratory scale load tests for validation. Eccentricity and slope angle are major factors which affect the load settlement behavior of footing resting on slopes [1].

4 Waste Management and Disaster Risk Reduction

Rohan Deshmukh et al.., in their paper *Effective Utilization of Waste Plastic Bottle as Geocell in Road Pavement: A Numerical Study,* describe about an approach to use cellular reinforcement made up from Waste Plastic Bottle Geocell (WPBG) for the improvement of pavement. In this study, unreinforced and cellular reinforced pavement was analyzed by numerical modeling in PLAXIS 2D and 3D. This study gives an effective result and proves to be a boon for road construction as it improves the subgrade and also for environmental conservation as it provides effective utilization of waste plastic bottles in the form of geocell [40]. The explorative study on *Impacts of Temple waste on the environment and its mitigation* by Jahagirdar S. et al. reviewed about the temple waste management. India is land consisting of millions of temples which in turn generate thousands of tons of waste every year. This study and analysis indicated that proper management of temple waste has emerging benefits like excellent quality manure and reusability of water, making the temples more sustainable and smarter, rather than disposing of the temple waste with domestic waste [22]. Dani Irwan et al., in their paper *Age and Household Solid Waste Arising in Suburban Malaysia: A Statistical Approach,* aim to examine the relationship between age and the rate of household solid waste (HSW) generation per capita from households in selected residential areas of the suburban Malaysian townships of Bandar Baru Bangi, Putrajaya and Kajang. Subsequent data refinement resulted in a final sample of 219 households consisting of 4623 discrete measurements being used for statistical analysis in the IBM SPSS software package. Results of statistical analyses show that weighted average age has a small and positive but statistically insignificant correlation to average daily per capita HSW arising. However, evidence from this study shows that the relationship between the two is insignificant from a statistical perspective [19]. The paper Feasibility study on Municipal Solid Waste (MSW) as sustainable engineering material using suction characteristics by M. V. Shah and A. J. discussed about the study which attempts to observe the suction characteristics of the waste for different moisture content and to establish the relation between suction parameters and strength of the waste. To check its suitability as the alternate material for construction purpose, the same study is carried out on silty sand so that the influence of the volatile matters present in the MSW on suction can be observed. The average pore size of the compressed mass is carried out using the relation of matric suction to pore size and the void formation is observed in a high magnification microscope to visually check and corelate the variation in pore size. The suction has an unavoidable impact on strength, compressibility and pore size of the compressed mass. The compaction on the dry side of optimum is preferable for achieving higher strength and more continuous structure of compressed solid waste [24].

5 Urban Development and Sustainability

Bharani Alagirisamy and Poornima Ramesh in their paper *Smart Sustainable cities: Principles and Future Trends* discussed new sustainable development concepts and intelligent city approaches with respect to the planning policy. This further addresses the successes and shortcomings with special focus on the degree to which it is related to in incorporating the principles of sustainable growth. As the examined papers revealed, many elements are needed to create sustainable cities, but the main emphasis is on the threefold approach to sustainability (i.e., environment, economics and equity). Each aspect is linked with these principles of sustainability either directly or indirectly. In the end, the potential of our future cities to imitate the carbon and water cycles of nature is critical for regeneration and reuse, rather than the present mitigation conditions that prioritize recycling [8]. Saif Al Zabeeb in the paper *Tire derived aggregate as a sustainable technique to mitigate transient seismic effect on buried concrete pipes* explains how the tier derived aggregate (TDA) is used as sustainable lightweight fill during the last few years. This research has been conducted to investigate the static and seismic performance of buried concrete pipe that is supported using TDA. The finite element method has been utilized in the analyses using earthquake records that cover a predominant frequency range and a peak ground acceleration (PGA) range. The results showed that the TDA reduces the maximum bending moments induced in the pipe wall by 27% for static condition and by 22 to 39% for seismic conditions. Based on the obtained results, the TDA can be considered as one of the feasible options to be used as material supporting buried concrete pipes subjected to transient seismic shake [3]. Ebne Alam et al., in their paper *Temporal Variation of Land Surface Temperature in Response to Changes in Vegetation Index of Bhawal National Park, Bangladesh,* discussed about the generality of the Normalized Difference of Vegetation Index (NDVI) and Land Surface Temperature (LST) correlations encountered over a wide range of vegetation coverage areas of the Bhawal National Park during the winter season. Information on LST and NDVI was obtained from long-term (30 years) datasets acquired from Landsat 5 TM and Landsat 8 OLI after atmospheric correction. Regression analysis is used to find the correlation between LST and NDVI. The rise in temperature in the recent year and the correlation coefficient found in this study support the progressive declination of vegetation coverage in the Bhawal National Park of Bangladesh. The findings also indicate the negative effects of deforestation and the importance of the forest areas in the city areas which are altered by the rapid human settlement, urbanization day by day [46]. Daniel Rios in the paper Use of stream power as a tool for the detection of critical reaches in channeled streams has explained how we can use stream power for detection of critical reach in channeled streams. The need of a tool for prioritizing intervention reach in the Aburrá River arises because there is a canalized reach of approximately 24 km along the valley, and additionally, there is a metro line and main roads parallel to the canal to great extent of its length. Finally, the classification obtained with the criterion of the Task Force Stream Power was compared against a personal photographic record that the author built over four years in the different analysis sites. Finding that the active process in the six sites corresponds to predicted with the criterion of unit stream power, therefore, the use of stream power as a tool for the detection of critical reaches and schedule your preventive maintenance in channeled streams is recommended [6]. Dasandara et al., in their paper *Urban Facilities Management: A Way of Attaining Sustainable Cities in Sri Lanka Dynamics,* intended to analyze the drivers and barriers to the successful adaptation of UFM into Sri Lankan context. A qualitative research approach with qualitative interview survey strategy was followed to accomplish the aim of this study. The empirical findings revealed that UFM adaptation into Sri Lankan context is still at its nascent stage in Sri Lanka. The findings would therefore enable relevant stakeholders, who engaged in management and development of cities, to make informed decisions and thereby promote modern, innovative and sustainable ways for ensuring the well-being of the whole society [35]. Shubham Pandeyet et al., in their paper *Fuzzy-AHP Based Design and Performance Indexing Model for Tall Buildings,* aim an effort to encompass all such parameters into a simple, easy to understand hierarchical unit, which is very simple and can highlight the crucial parameters and make it possible to assess the comparative need and requirement of one factor and criterion over the other. In the end, the prevailing design practices and norms are reviewed, and some suggestions have been summarized considerations of which can further enhance and improve the likely performance of the tall buildings [36]. The paper *Carbon Neutral Communities: Model for Integrating Climate Action into Development Planning* by Sajan et al., explains that Climate change is one of the major discussions pertaining to economic development across the globe. The concept of 'carbon neutrality' puts forth the notions of zero carbon development, food and energy self-sufficiency at local government level and falls in line with the sustainable development goals put forward by the United Nations. This pilot project paves way for low carbon development in Kerala and serves as a model for the rest of the country to follow. Achieving carbon neutral status is an arduous but worthy task as the resulting sustainable economy will benefit all. By adopting policies and schemes and setting short-term and long-term projects, a community can easily become carbon neutral with the support of all actors in the community [27].

6 Cross-Cutting Issues

Akheel K. et al., in their paper *Review of Free Vibration Response of Spar Supported Wind Turbine with Tuned Mass Damper*, have reviewed the necessity and advantages of offshore floating-type wind turbines compared to the onshore wind turbines are carried out. Successively a floating-type spar-supported OC3-Hywind wind turbine is selected for the investigation. The structural modeling details are studied, and OpenFAST software framework is used for the analysis. The corresponding changes in the considered structural degrees of freedom vibrations are evaluated. Corresponding to this instant, the fore-aft, surge and pitch responses had reductions of 80%, 53% and 99%, respectively. It is found that the response control improves (further) positively with the increase of Tuned Mass Damper mass ratio [25]. Abitha Babu and Sitaram Nayak in their paper *A Review on Methods for Analysis of Laterally Loaded Piles* have reviewed different methods for the analysis of laterally loaded piles. The applicability, advantages, limitations and comparison of various methods are also included in the study. With the increased computational power of the computers and the availability of appropriate software, numerical simulation and analysis of the problem has become the most cost effective and accurate method [23]. Sumathi A and Saravana Raja Mohan K in their paper *Effect of Silica fume and Steel fiber on Mechanical Characteristics of High Strength Concrete reviewed* about characteristics of high strength concrete. In the present work, the effect of silica fume (SF) as a partial substitute to cement on mechanical properties of steel fiber reinforced concrete with a characteristic compressive strength of 60 MPa was considered. ANOVA statistical method was used to evaluate the substantial effect of SF and fiber in concrete strength. The additions of steel fibers to HSC containing silica fume change the complete basic trend of stress–strain response. The descending branch slope enhances based on the increase of fiber in volume fractions. The strength model was developed using experimental findings on the basis of regression analysis. In estimating the different intensity at 28, 90 and 180 days on the basis of estimation more or less similar with measurements, the proposed regression linear model was found to be accurate [38]. Nurhanani A et al., in their paper *Particle Deposition Analysis using DPM-DEM,* have explained about coastal erosion. The effect of regular strengths, such as wind and waves, can alter the shoreline structure and rapid coastal erosion. The numerical study showed that the number of particles changes as time increases due to the generation of waves. The findings have been validated with the smooth-particle hydrodynamics method to investigate the effectiveness and accuracy of the numerical analysis of DPM-DEM. It is also proved that the length of time would generally increase the number of particles erosion and change the formation of sand that would lead to erosion [5]. Balaganesh

P et al., in their paper *Recent Trends in IOT enabled composter for Organic Wastes,* reviewed the recent trends in technological innovations involved in the design and implementation of a smart composter enabled with IoT. The control and accuracy of results mainly depend on the selection of the major processing units having suitable output-related hardware such as Arduino UNO, Node MCU, Raspberry Pi model 3, track recorder, Siretta antenna and wireless communication module. It is also observed that selection of proper internet gateway is essential to address the challenges in data transmission, data safety and power consumption as in the case of a community-based cluster of smart composters. The study makes design-oriented prospects toward development of a smart-rapid composter involving provisions for advanced process control and quality checking [9]. The paper *Study on the Effects of CNT and Nano Graphene in Clayey Soil of Aligarh City of Northern India* by Jibran Qadri et al. explains and analyzes the possible advantages of nanotechnology for revolutionary applications in the soil enhancement area. The aim of the study is to investigate the effect of the addition of different nano-CNT and nano-Graphene on the geotechnical properties of clayey soil samples from Aligarh city in university campus. When the nanoparticles apply the liquid limit, the plastic limit and the plasticity index decreases. With the inclusion of nanoparticles exceeding the optimal quantity, the state of a mass cluster is associated, thereby influencing the mechanical features of soils. Thus, even in very tiny quantities, the inclusion of nanoparticles will exquisitely increase strength and strengthen properties. The overall cohesion and the angle of friction have increased, which suggest an increase in shear strength [16]. Site Onyejekwe et al.., in the paper *Highway Development-Related Gully Erosion: The Case of the Okigwe-Isuikwuato Highway, Southeastern Nigeria,* discussed about the anthropogenic, hydraulic and hydrological factors that led to the development of these gullies. It also proffered suggestions for the mitigation of the effect of these gullies. This paper reviewed the causes of these gullies and found that there was interplay of a number of factors at work in this respect on the inception and progression of gully development and erosion generally was demonstrated. Considering the scale of the problem, it is clearly evident that strategies for the control and remediation of erosion ravaged areas should involve a serious reforestation program and all other strategies that reduce the rate of flow of surface runoff [17]. Site Onyejekwe et al.., in their paper *Assessment of the Causes of Erosion and Gullying Along the Leru—Nkwoagu, Amuda (Isuochi)—Mbala (Isuochi) Highway, Southeastern Nigeria: A Case Study,* reviewed the causes of erosion and gullying on the Leru–Nkwoagu–Mbala highway, a highway in the semi-tropical environment of Southeastern Nigeria. The failure to implement safeguards, poor design considerations, particularly in relation to the aspects of design relating to hydraulics, hydrology and geotechnics, and the general lack of enforcement of environmental regulations on this highway have been found not to be an outlier. Hence, a river basin-wide, intergovernmental, multisectoral, multidisciplinary, and all-stakeholder approach to the solution of highway-related soil erosion is strongly recommended for adoption [18].

7 Summary

UN Sustainable Development Goal 11: "Make cities inclusive, safe, resilient and sustainable," target 11 focuses on resilience and sustainability in urban development and human settlement. It suggests to substantially increase the number of cities and human settlements adopting and implementing integrated policies and plans toward inclusion, resource efficiency, mitigation and adaptation to climate change, resilience to disasters, and develop and implement, in line with the Sendai Framework for Disaster Risk Reduction 2015–2030, holistic disaster risk management at all levels, and to support least developed countries, including through financial and technical assistance, in building sustainable and resilient buildings utilizing local materials. System approach for flood vulnerability and community resilience assessment at the local level is one of the major components for resilience building [30].

This chapter summarized the contents of the book volume on City Resilience and Sustainable Infrastructure. All the chapters in this volume are segregated in five sections such as resilient infrastructure in construction, innovative construction interventions, waste management and disaster risk reduction, urban development and sustainability and cross-cutting issues.

References

1. Acharyya, R., Dey, A.: Assessment of bearing capacity for strip footing located near sloping surface considering ANN model. Neural Comput. Appl. **31**(11), 8087–8100 (2019). https://doi.org/10.1007/s00521-018-3661-4
2. Al-Bahar, J.F., Crandall, K.C.: Systematic risk management approach for construction projects. J. Constr. Eng. Manag. **116**(3), 533–546 (1990). https://doi.org/10.1061/(asce)0733-9364(1990)116:3(533)
3. Alzabeebee, S.: Seismic settlement of a strip foundation resting on a dry sand. Nat. Hazards **103**(2), 2395–2425 (2020). https://doi.org/10.1007/s11069-020-04090-w
4. Appolloni, C., D'Alessandro: The pleasure of walking: an innovative methodology to assess appropriate walkable performance in urban areas to support transport planning. Sustainability **11**(12), 3467. https://doi.org/10.3390/su11123467
5. Aziz, N.A., Zawawi, M.H., Zahari, N.M., Abas, A., Azman, A.: Simulation of Homogeneous Particle Size in Fluid Flow by Using DPM-DEM. In Water Resources Development and Management. Springer, pp. 549–555. https://doi.org/10.1007/978-981-15-1971-0_53
6. Barker, D.M., Lawler, D.M., Knight, D.W., Morris, D.G., Davies, H.N., Stewart, E.J.: Longitudinal distributions of river flood power: the combined automated flood, elevation and stream power (CAFES) methodology. Earth Surf. Proc. Land. **34**(2), 280–290 (2009). https://doi.org/10.1002/esp.1723
7. Berkowitz, L.: Frustration-aggression hypothesis: examination and reformulation. Psychol. Bull. **106** (1989)
8. Bibri, S.E., Krogstie, J.: Smart sustainable cities of the future: an extensive interdisciplinary literature review. Sustain. Cities Soc. Elsevier Ltd (2017, May 1). https://doi.org/10.1016/j.scs.2017.02.016
9. Boniecki, P., Idzior-Haufa, M., Pilarska, A.A., Pilarski, K., Kolasa-Wiecek, A.: Neural classification of compost maturity by means of the self-organising feature map artificial neural

network and learning vector quantization algorithm. Int. J. Environ. Res. Public Health **16**(18) (2019). https://doi.org/10.3390/ijerph16183294
10. Burn, C.R., Kokelj, S.V.: The environment and permafrost of the Mackenzie delta area. Permafrost Periglac. Process. **20**(2), 83–105 (2009). https://doi.org/10.1002/ppp.655
11. Deb, P.S., Nath, P., Sarker, P.K.: The effects of ground granulated blast-furnace slag blending with fly ash and activator content on the workability and strength properties of geopolymer concrete cured at ambient temperature. Mater. Des. **62**, 32–39 (2014). https://doi.org/10.1016/j.matdes.2014.05.001
12. Dindar, S., Kaewunruen, S., An, M., Sussman, J.M.: Bayesian network-based probability analysis of train derailments caused by various extreme weather patterns on railway turnouts. Saf. Sci. **110**, 20–30 (2018). https://doi.org/10.1016/j.ssci.2017.12.028
13. Dong, Y., Frangopol, D.M.: Risk and resilience assessment of bridges under mainshock and aftershocks incorporating uncertainties. Eng. Struct. **83**, 198–208 (2015). https://doi.org/10.1016/j.engstruct.2014.10.050
14. Emmanouil, G., Vlachogiannis, D., Sfetsos, A.: Exploring the ability of the WRF-ARW atmospheric model to simulate different meteorological conditions in Greece. Atmos. Res. **247**,(2021). https://doi.org/10.1016/j.atmosres.2020.105226
15. Expansive Soils: Problems and Practice in Foundation and Pavement Engineering | Wiley. (n.d.). Retrieved May 4, 2021, from https://www.wiley.com/en-ad/Expansive+Soils%3A+Problems+and+Practice+in+Foundation+and+Pavement+Engineering-p-9780471181149
16. Firoozi, A.A., Guney Olgun, C., Firoozi, A.A., Baghini, M.S.: Fundamentals of soil stabilization. Int. J. Geo-Eng. **8**(1), 1–16 (2017). https://doi.org/10.1186/s40703-017-0064-9
17. Floyd, B.: Soil erosion and deterioration in Eastern Nigeria. Geogr. J. 117, 291–306 (1965). Retrieved May 4, 2021, from http://www.sciepub.com/reference/109323
18. Gobin, A.M., Campling, P., Deckers, J.A., Poesen, J., Feyen, J.: Soil erosion assessment at the Udi-Nsukka Cuesta (southeastern Nigeria). Land Degrad. Dev. **10**(2), 141–160 (1999). https://doi.org/10.1002/(SICI)1099-145X(199903/04)10:2%3c141::AID-LDR325%3e3.0.CO;2-N
19. Grossman, D., Hudson, J.F., Marks, D.H., (n.d.).: Waste Generation Models for Solid Waste Collection. Retrieved May 4, 2021, from https://cedb.asce.org/CEDBsearch/record.jsp?dockey=0021718
20. Holland, R.B. (2012). Durability of Precast Prestressed Concrete Piles in Marine Environments Durability of Precast Prestressed Concrete Piles in (August)
21. Ikechukwu, E.E.: The effects of road and other pavement materials on urban Heat Island (a case study of Port Harcourt City). J. Environ. Prot. **06**(04), 328–340 (2015). https://doi.org/10.4236/jep.2015.64033
22. Kaushik, S., Joshi, B.D., Vishwavidyalaya, S.D.: Solid waste management at Mansa Devi and Chandi Devi temples in the Shiwalik foothills, during Kumbh Mela at Haridwar (Uttarakhand). Report and Opinion, vol. 4 (2012). Retrieved from http://www.sciencepub.net/report
23. Kavitha, P.E., Beena, K.S., Narayanan, K.P.: A review on soil–structure interaction analysis of laterally loaded piles. Innovative Infrastructure Solutions. Springer (2016, December 1). https://doi.org/10.1007/s41062-016-0015-x
24. Marinho, F.A.M., Stuermer, M.M.: The influence of the compaction energy on the SWCC of a residual soil. In: Proceedings of Sessions of Geo-Denver 2000 - Advances in Unsaturated Geotechnics, GSP 99, vol. 287, pp. 125–141 (2000). American Society of Civil Engineers. https://doi.org/10.1061/40510(287)8
25. Mechanical Vibrations : Jacob P.Den Hartog : 9780486647852. (n.d.). Retrieved May 4, 2021, from https://www.bookdepository.com/Mechanical-Vibrations-Jacob-P-Den-Hartog/9780486647852
26. Mujahid, A., Zaidi, A., Riza, F.V., Rahman, I.A.: Preliminary study of compressed stabilized earth brick (CSEB). Aust. J. Basic Appl. Sci. **5**(9), 6–12 (2011)
27. Nagar, J. (n.d.).: Published by : Thanal. Retrieved from www.thanal.co.in
28. Ng, K.S., Tan, S.A.: Design and analyses of floating stone columns. In: Soils and Foundations, vol. 54, pp. 478–487. Japanese Geotechnical Society (2014). https://doi.org/10.1016/j.sandf.2014.04.013

29. Organisation for Economic Co-operation and Development (OECD) (2018) Resilient cities
30. Pal, I., Doydee, P., Utarasakul, T., Jaikaew, P., Razak, K.A.B., Fernandez, G., Huang, T., Chen, C.S.: System approach for flood vulnerability and community resilience assessment at the local level – a case study of Sakon Nakhon province, Thailand. Kasetsart J. Soc. Sci. **42**(1), pp 67–76. https://doi.org/10.34044/j.kjss.2021.42.1.xx
31. Pal, I., Bhatia, S.: Disaster risk governance and city resilience in Asia-pacific region. In: Science and Technology in Disaster Risk Reduction in Asia: Potentials and Challenges, pp. 137–159 (2017). https://doi.org/10.1016/B978-0-12-812711-7.00009-2
32. Panda, S., Shiva Nagendra, S.M.: Chemical and morphological characterization of respirable suspended particulate matter (PM10) and associated heath risk at a critically polluted industrial cluster. Atmos. Pollut. Res. **9**(5), 791–803 (2018). https://doi.org/10.1016/j.apr.2018.01.011
33. Park, R., Paulay, T.: Reinforced Concrete Structures. Wiley (1975). https://doi.org/10.1002/9780470172834
34. Paul, R., Dalui, S.K.: Wind effects on 'Z' plan-shaped tall building: a case study. Int. J. Adv. Structural Eng. **8**(3), 319–335 (2016). https://doi.org/10.1007/s40091-016-0134-9
35. Pearce, A.R.: Sustainable urban facilities management. In: Encyclopedia of Sustainable Technologies, pp. 351–363. Elsevier (2017). https://doi.org/10.1016/B978-0-12-409548-9.10183-6
36. Roshan, P., Pal, S., Kumar, R.: Performance assessment indexing of buildings through fuzzy AHP methodology. In: Lecture Notes in Civil Engineering, vol. 58, pp. 503–519. Springer (2020). https://doi.org/10.1007/978-981-15-2545-2_42
37. Sekac, T., Jana, S.K., Pal. I., Pal, D.K.: Earthquake hazard assessment in the Momase region of Papua New Guinea. In: Spatial Information Research, Springer (2016). ISSN: 2366–3286
38. Selvaraj, R., Muralidharan, S., Srinivasan, S.: The influence of silica fume on the factors affecting the corrosion of reinforcement in concrete—a review. Struct. Concr. **4**(1), 19–23 (2003). https://doi.org/10.1680/stco.2003.4.1.19
39. Son, R., Wang, S.Y.S., Tseng, W.L., Barreto Schuler, C.W., Becker, E., Yoon, J.H.: Climate diagnostics of the extreme floods in Peru during early 2017. Clim. Dyn. **54**(1–2), 935–945 (2020). https://doi.org/10.1007/s00382-019-05038-y
40. Thakare, S.W., Sonule, S.K.: Performance of plastic bottle reinforced soil. IJEIR **2**(3), 207–210 (2013)
41. Thanvisitthpon, N., Shrestha, S., Pal, I.: Urban flooding and climate change: a case study of Bangkok Thailand. Environ. Urbanization ASIA **9**(1), 86–100 (2018). https://doi.org/10.1177/0975425317748532
42. Thanvisitthpon, N., Shrestha, S., Pal, I., Ninsawat, S., Chaowiwat, W.: Assessment of flood adaptive capacity of urban areas in Thailand. Environ. Impact Assess. Rev. **81**,(2020). https://doi.org/10.1016/j.eiar.2019.106363
43. UN Habitat: Hot Cities: The Battle-Ground for Climate Change. UN Sustainable Development Goals (2011)
44. UNISDR: How To to Make Cities More Resilient: A Handbook for Local Government Leaders. UN Office for Disaster Risk Reduction, Geneva (2017)
45. Venkatarama Reddy, B.V., Gupta, A.: Influence of sand grading on the characteristics of mortars and soil-cement block masonry. Constr. Build. Mater. **22**(8), 1614–1623 (2008). https://doi.org/10.1016/j.conbuildmat.2007.06.014
46. Zhang, X.X., Wu, P.F., Chen, B.: Relationship between vegetation greenness and urban heat island effect in Beijing City of China. In: Procedia Environmental Sciences, vol. 2. Elsevier, pp. 1438–1450 (2010). https://doi.org/10.1016/j.proenv.2010.10.157

Resilient Infrastructure in Construction

Risk and Resilience of Railway Infrastructure: An Assessment on Uncertainties of Rail Accidents to Improve Risk and Resilience Through Long-Term Data Analysis

Panrawee Rungskunroch, Anson Jack, and Sakdirat Kaewunruen

1 Introduction

Recently, the reduction of railway accident has become a critical development for railway authorities. It leads to provide sustainability development on the rail network. Based on the UIC's report, the railway accident has been slightly decreased due to high technologies applied to the system [1]. However, an attempt to make zero accident is a challenge to rail authorities.

The railway infrastructure failures have shown an increasing trend to a railway accident based on the collected long-term accident data set. The fraction of infrastructure failures is changed from 4.19% to 5.14% during 2000–2019; moreover, the severity level of an accident by infrastructure failures have illustrated at a high level. Also, more than 70% of all accident shows as a train derailment.

The research has created two novelty models through the Python programming language. Firstly, the main problem is that railway risk and resilience contain various uncertainties; therefore, the Bayesian method is taken to understand the uncertainties of railway accident and predict probabilities on each type of accident in the future. Then, the risk-based DT model has been provided to evaluate the risk level. This model offers precisely the risk level based on fatalities and injuries rates. This is a direct benefit for railway authorities to provide an effective plan to combat avoidable accidents.

P. Rungskunroch (✉) · A. Jack · S. Kaewunruen
School of Engineering, University of Birmingham, Birmingham B15 2TT, UK
e-mail: pxr615@student.bham.ac.uk

I. Pal and S. Kolathayar (eds.), *Sustainable Cities and Resilience*, Lecture Notes in Civil Engineering 183, https://doi.org/10.1007/978-981-16-5543-2_2

2 Literature Reviews

An attempt to reduce railway accidents has occurred across the railway operators and other related sectors. High technologies and new policies have been applied to the railway operation to eliminate all network risks. As a result, the number of railway accident illustrated slightly decreased in some areas. The European Union agency for railways (ERA) reveals an overall number of Europe's railway accidents that it has been reduced by almost one-third within five years [2]. The evidence can imply that the safety level on the rail network has been successfully improvement. The research deeply studies the causes of railway accidents. We classified them into seven groups including driver's error, signalmen's error, infrastructure failures, improper maintenance, human's error, natural causes, and contributing factors. Nevertheless, the cause of infrastructure failures has slightly increased. It has changed from 4.19% between 2000 and 2010 to 5.14% during 2011–2019.

Regarding the railway accident from infrastructure failures, using high technologies and other monitoring techniques can prevent railway accidents [3, 4]. Previous studies state that installing technologies can increase safety infrastructure, such as using sensors on the rail network to prevent hazardous events [5–7]. Also, some studies suggest providing effective maintenance to the network [8, 9]. However, the failure in railway infrastructure has occurred from climate change. And, it may lead to an unexpected railway accident. Some scholars study on the cold weather impacts on the railway infrastructure in Sweden [10]. The research aims to provide high quality and secure service for winter climate. As a result, the study finds that the weather condition impacts the safety level and suggests improving maintenance condition. Another research also provides a solution to reduce the accident from infrastructure failures. The study stated that installing a thermoelectric heater to heat rail pads is purposed to maintain railway infrastructure's condition [11].

With the aims at evaluating risk among railway network, various methods have been taken to develop risk models. Several studies focus on increasing railway performance by using decision tree (DT) methods [12]. Also, the fuzzy logic and Bayes methods are taken to improve the reliability across the railway industry [13, 14]. On the other hand, some authors plan to avoid railway accident by providing safety policies [15]. Similarity, other research applies the analytical hierarchy process (AHP), maximum absolute weighted residual (MAWR), maximum entropy method (MEM), fault tree analytic (FTA) and Petri-net (PT) methods [16–19]. Several studies also focus on accident analysis to predict the accident rate from long-term accident data. Researchers provide methods to reduce the accident rate on a freight train for dangerous products [18, 20].

These are rarely analysed in the literature about the different damage size on each accident. Also, only a few studies intend with a number of fatalities and injuries passenger. To fill those research gaps, this study examines two novelty models including;

1. The prediction model based on Bayesian theorem to understanding uncertainty and forecasting future accident.

Table 1 The summary of the infrastructure failures' sub-cause of railway accident

Cause of accident	Sub-cause of accident
Infrastructure failures	Track geometry
	Frogs, switches and track appliances
	Other ways and structure (bridge/design construction)
	Rail joint bar
	Roadbed

2. The risk-based DT model to evaluate the risk level from the long-term railway accident data sets.

By analysing through both novelty models, the study's outcomes show a high accuracy risk score; moreover, the research's prediction model can update real-time information of railway accidents. Those advantages lead to a high-level evaluation of the risk and resilience of the railway infrastructure.

3 Methodology

3.1 Data Availability

This study collected railway accidents from official companies, government, and rail authorisations' reports worldwide. Also, the research focuses on passenger train accidents that occurred during 2011–2019 [2, 21, 22]. The 1005 appropriately data sets of infrastructure failures, including injuries and fatalities, are provided in this study.

There are five main sub-causes of infrastructure failures that are frequently investigated after an accident, as shown in Table 1. The research also classified an effect with train after an accident into three groups including collision, derailment, and other effects (such as fire, bomb, and vandalism).

3.2 An Application on the Bayesian Network

The Bayesian statistic is a frequently used analytic tool explaining the probability of two events, which relates to prior knowledge. The Bayesian outcome shows a term of conditional probability. It also can be converted to the likelihood of a single event, as illustrated in Eqs. 1 and 2.

$$(A|B) = \frac{P(B|A) * P(A)}{P(B)} \tag{1}$$

$$P(B) = \frac{P(B|A) * P(A)}{P(A|B)} \tag{2}$$

Given ‘A’ is an effect with train after an accident consists of A1 (collision), A2 (derailment), and A3 (other causes). Given ‘B’ is an infrastructure failure that is one of the causes of a railway accident. Therefore, the posterior probability of train derailment (TD) from railway infrastructure failure (IF) is shown in Eq. 3, and the probability of infrastructure failure is shown in 4 as follows;

$$P(TD|IF) = \frac{P(IF|TD) \times P(TD)}{P(IF)} \tag{3}$$

$$P(IF) = \frac{P(IF|TD) \times P(TD)}{P(TD|IF)} \tag{4}$$

The prediction model has created based on the Bayesian concept through the Python programming language. The causes and effects with a train after an accident are predicted based on the conditional probability through prediction model. The result leads to estimate damage’s size of railway accident by infrastructure failures.

3.3 Research Framework

Figure 1 shows a whole research framework. The research collected long-term secondary accident data sets from railway companies, authorities, and official report. All the recorded data sets are cleaned and verified. Only the accident data set that occurs from infrastructure failures accidents is taken in the pre-processing stage. Then, the development of the prediction model is provided. In this part, there are two involved with data sets, including the prior belief and likelihood. At the end of this section, the prediction model at 95% efficiency level has been created.

Next, the prediction model has taken to estimate railway accident rate that happens from infrastructure failures. After that, the novelty risk-based decision tree model has been provided to evaluate the risk level. The model involves injuries and fatalities numbers, and the outcome shows in the range 1–32 that score 1–8 means low risk, score 9–16 means moderate risk, 17–24 means high risk, and 25–32 means too high risk.

3.4 Risk Prediction Model

The risk and resilience of train accident are hard to predict because it relates to many external factors and contains with uncertainties. This study characterise different aspects of a train after accident into three groups including collision, derailment,

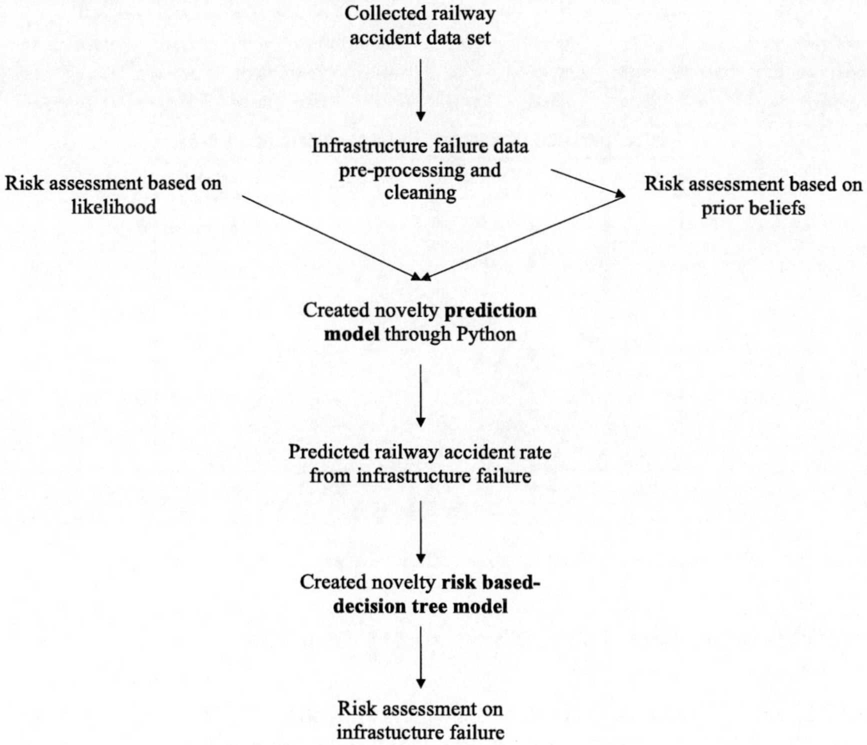

Fig. 1 A whole research framework

and other effects. And, each effect with train after an accident causes difference in damage's size.

Therefore, the risk prediction model has been created to evaluate the future's railway accident risks. The outcome leads to precisely prevent accidents. Within this case, the prediction model adopts the Bayes' theory by using two data sets, including prior knowledge and collected data. The novelty prediction model has been developed through Python. This study found that using posterior distribution at 4:4:1 has provided the highest efficiency prediction result. Also, it has been qualified with FRA's data set at 95% effective level to verify this model's effect.

Figure 2 shows the comparison on the posterior distribution among train collision, derailment, and other effects with train, and the outcome shows probability at 0.279, 0.651, and 0.070, respectively. The result can interpret that the rail's infrastructure failures have a high effect on train derailment.

	mean	sd	hpd_2.5%	hpd_97.5%	mcse_mean	mcse_sd	ess_mean	ess_sd	ess_bulk	ess_tail	r_hat
Collision	0.279	0.074	0.142	0.423	0.002	0.002	1301.0	1179.0	1331.0	1161.0	1.0
Derailment	0.651	0.076	0.508	0.795	0.002	0.001	1302.0	1302.0	1321.0	1133.0	1.0
Other	0.070	0.039	0.010	0.145	0.001	0.001	1938.0	1898.0	1819.0	1225.0	1.0

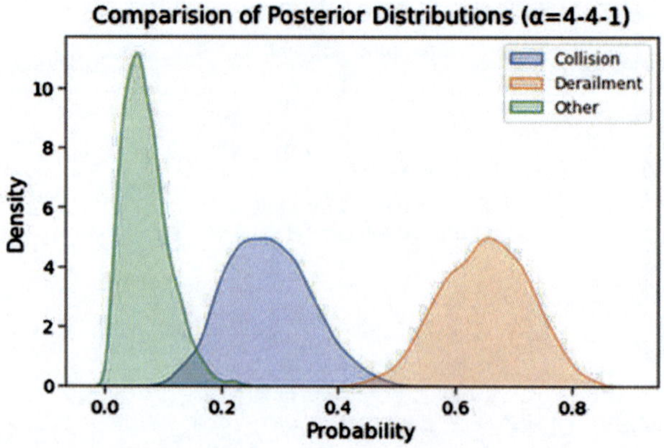

Fig. 2 An analysis of railway risk from infrastructure failures

4 Evaluating Risk Level Through DT Models

The DT, which is a non-supervised tool, is widely applied to classify complex decision rules [23]. Within this study, the risk-based DT model is provided to evaluate the risk level. Five essential factors of this research consist of fatalities and injuries rate, probabilities of collision (A1), derailment (A2), and other causes (A3). All factors have been placed as a decision node to design a useful decision tree into the complex decision rules.

Regarding the decision nodes, the research places fatalities rate at 12 people per accident, and injuries rate at 66 people per accident as the main decision nodes. With that, threshold number has been provided by the average number of global railway accident between 2000 and 2019. Next, A1, A2, and A3's probabilities are placed as a threshold by comparing with the global average A1, A2, and A3 values. As shown in Fig. 3, the outcomes are classified at DT's leaves into 32 scales of risk levels, which the small number means low risk and large number means high risk.

5 Result and Discussions

The analysis result through the DT model shows that the severity level of infrastructure failures equal to 18, which is at a high-risk level, as shown in Fig. 4. Based on the collected data, the accident by infrastructure failures bring an average number of fatalities and injuries at 12 and 66 people per accident. The severity of an accident

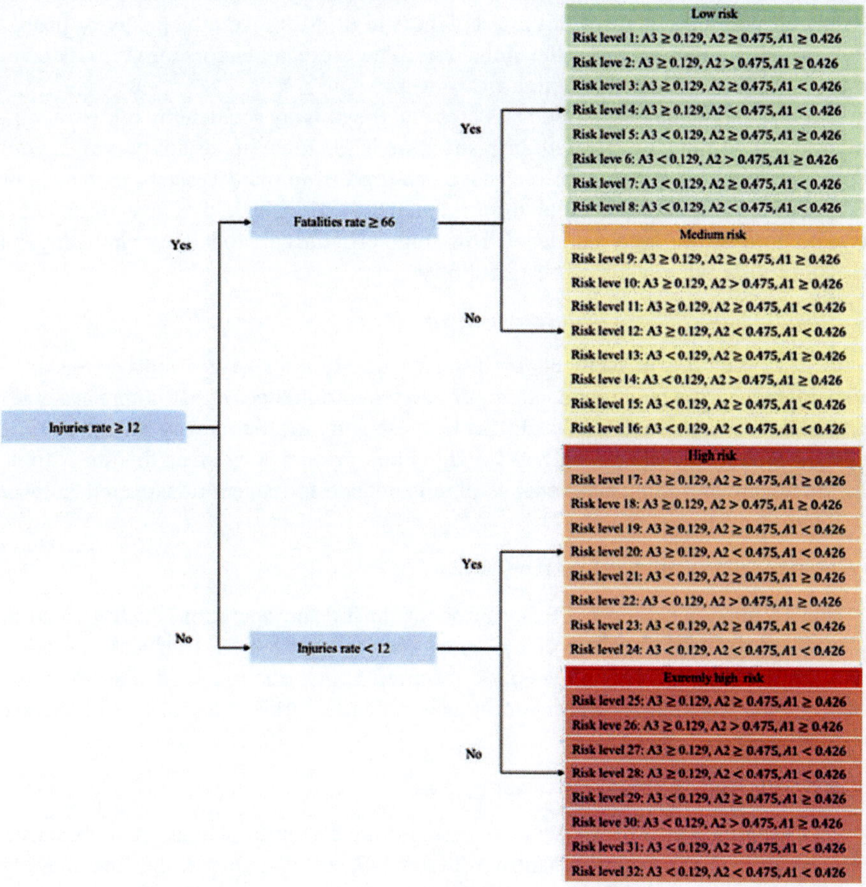

Fig. 3 The created risk-based decision tree framework

Fig. 4 Risk interface's result analysis through the risk-based DT model

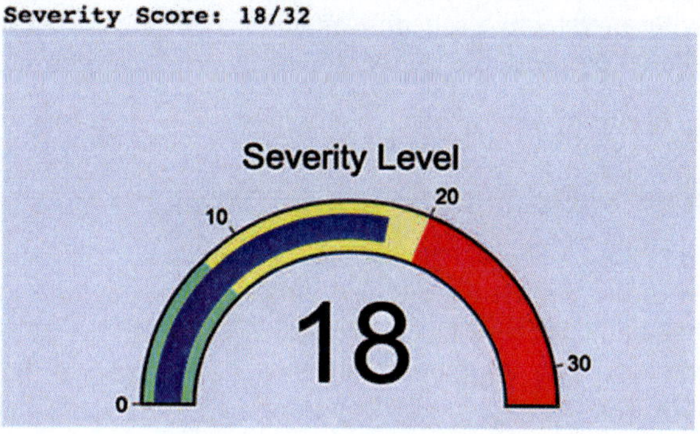

compared with other causes of the accident is high. Moreover, the posterior probability of other effects is above the global rate. Therefore, all factors turn the risk level of infrastructure failures into 'high risk' level.

Our finding on the risk-based DT states that the railway accident by infrastructure failures should not be neglected. Infrastructure is the most important part of the rail network. The infrastructure failures have occurred from many reasons such as poor design, lack of maintenance, global warming, natural disaster, etc. Hence, its defect is possible to show in 'high risk' level. This study provides practical recommendations to avoid future railway accidents as follows;

- *Providing an effective maintenance plan*:

Most of the railway accidents by infrastructure failures have occurred from lack of maintenance. As mentioned, attempting to repair outdated infrastructure is a challenge for a civil engineer [24–26]. It is because some of the railway infrastructures have been built since 1800s such as London underground. Moreover, this study finds that an adequate proper maintenance plan is needed to prevent unexpected railway accident.

- *Increase the safety level on the rail track*:

Due to uncertainty events that can occur during the operation, such as natural disaster, the rail track's increasing safety level is needed as a critical operation plan. Some issues should be deeply concerned about, such as landslide, drainage flow. Adequate safety and earthwork plans can lead to preventing long-term railway accident.

- *Maintain operational performance:*

As mentioned, the railway accident contains uncertainties, which can be occurred by external factors such as a natural disaster [27, 28]. Therefore, the fundamental improvement is to maintain operational performance into normal conditions. In some cases, installing new equipment on rail's infrastructure is required to re-operate the system. The research recommends the rail authorities to combines high-technologies with an effective schedule maintenance plan. By following these solutions, they should decrease the accident by infrastructure failures.

6 Conclusion and Future Work

With the dramatic growth of railway infrastructure, the evaluation of the railway's risk and resilience has played an essential role in maintaining the safety level. Rail authorities have provided various policies to reduce the number of railway accidents. Also, the new technologies, equipment, and strategies have been addressed along with the network. However, railway accident contains uncertainties, especially from external factors. Hence, this research generates two new novelty models including (i) the prediction model that uses for estimating future accident. The model adopted long-term accident data sets and combined with Bayes' theorem. The prediction rate is more than 95%. It becomes a direct benefit to rail authorities to precisely prevent an accident. (ii) The risk-based DT model to evaluate risk level that shows severity level of accident. The model uses long-term data to measure the severity level of an accident by infrastructure failures. The result shows the severity level is scored at 18 of 32, which can be interpreted at 'high risk'. The main conclusion that can be drawn is that the rail accident by infrastructure failures are harmful to a passenger, and it should be eliminated to make the network reach safety level. This study's prediction model illustrates high accuracy outcomes. The model can also be up-to-date, based on real-time railway accidents. Future research should consider the effect of railway accident on infrastructure more carefully by maintaining all equipment in normal conditions. Also, the research relates to high-technologies with rail's infrastructure to prevent severe natural disasters. These can lead to sustainable development on a rail network.

Acknowledgements The first author gratefully acknowledges the Royal Thai Government for the Ph.D. scholarship at the University Of Birmingham, United Kingdom and the RISEN funding for one year at University of California, Berkeley. The first author also thanks to the second and third authors for giving recommendation during studying Ph.D. at UOB. The third author acknowledges the Australian Academy of Science (AAS) and the Japan Society for the Promotion of Sciences (JSPS), for the JSPS Invitation Fellowship for Research (Long-term), Grant No. JSPS-L15701, at the Railway Technical Research Institute (RTRI) and the University of Tokyo, Japan. The authors are sincerely grateful to the European Commission for the financial sponsorship of the H2020-RISEN Project No. 691135 "RISEN: Rail Infrastructure Systems Engineering Network", which enables a global research network that tackles the grand challenge of railway infrastructure resilience and advanced sensing in extreme environments (www.risen2rail.eu).

References

1. UIC: UIC Safety report 2019—Significant Accident Public report (2019). ISBN 978-2-7461-2863-7
2. ERA (European union agency for railway): ERAIL database (2018). Available at: https://www.era.europa.eu/activities/rail-accident-investigation_en. Accessed on 20 Dec 2020
3. Ngamkhanong, C., Kaewunruen, S., Costa, B.J.A.: State-of-the-Art review of railway track resilience monitoring. Infrastructures **3**, 3 (2018)
4. Kaewunruen, S., Wu, L., Goto, K., Najih, Y.M.: Vulnerability of structural concrete to extreme climate variances. Climate **6**, 40 (2018)
5. El Miloudi, El Koursi, Bruyelle, J.L: Railway accident prevention and infrastructure protection. J. Civil Eng. Architecture, David Publishing Company, pp. 96–107 (2016). https://doi.org/10.17265/1934-7359/2016.01.010
6. Krezo, S., et al.: Field investigation and parametric study of greenhouse gas emissions from railway plain-line renewals. Transp. Res. Part D: Transp. Environ. (2016). https://doi.org/10.1016/j.trd.2015.10.021
7. Kaewunruen, S., Sussman, J.M., Matsumoto, A.: Grand challenges in transportation and transit systems. Front. Built Environ. (2016). https://doi.org/10.3389/fbuil.2016.00004
8. Al-Douri, Y.K., Tretten, P., Karim, R.: Improvement of railway performance: a study of Swedish railway infrastructure. J. Mod. Transport. **24**, 22–37 (2016). https://doi.org/10.1007/s40534-015-0092-0

9. Lu, C., Cai, C.: Overview on safety management and maintenance of high-speed railway in China. J. Transp. Geotechnics **25** (2020). https://doi.org/10.1016/j.trgeo.2020.100397
10. Stenström, C., Famurewa, S.M., Parida, A., Galar, D.: Impact of cold climate on failures in railway infrastructure. In: The Second International Congress on Maintenance Performance Measurement & Management Conference, University of Sunderland, Sunderland, UK, pp. 1–9, 12–13 September 2012
11. Yang, F., Gao, M., Cong, J., Wang, P.: System dynamics modelling and experimental study of railway track with thermoelectric heater/generator in extreme weather conditions. J. Cleaner Prod. **249**,(2020). https://doi.org/10.1016/j.jclepro.2019.119367
12. Zhou, J.L., Lei, Y.: A slim integrated with empirical study and network analysis for human error assessment in the railway driving process. Reliab. Eng. Syst. Saf. (2020). https://doi.org/10.1016/j.ress.2020.107148
13. Dindar, S., et al.: Bayesian Network-based probability analysis of train derailments caused by various extreme weather patterns on railway turnouts. Saf. Sci. (2018). https://doi.org/10.1016/j.ssci.2017.12.028
14. Jia, C., Xu, W., Wang, H.: Study of management information system of railway permanent way safety risks and comprehensive evaluation. In: Procedia Engineering (2011). https://doi.org/10.1016/j.proeng.2011.08.239
15. Dindar, S., Kaewunruen, S., An, M.: Bayesian network-based human error reliability assessment of derailments. Reliab. Eng. Syst. Saf. (2020). https://doi.org/10.1016/j.ress.2020.106825
16. Song, H., Schnieder, E.: Evaluating fault tree by means of colored petri nets to analyse the railway system dependability. Saf. Sci. (2018). https://doi.org/10.1016/j.ssci.2018.08.017
17. Liu, C., et al.: An improved risk assessment method based on a comprehensive weighting algorithm in railway signaling safety analysis. Saf. Sci. (2020). https://doi.org/10.1016/j.ssci.2020.104768
18. Huang, W., Liu, Y., et al.: Fault tree and fuzzy D-S evidential reasoning combined approach: an application in railway dangerous goods transportation system accident analysis. Inf. Sci. (2020). https://doi.org/10.1016/j.ins.2019.12.089
19. Vileiniskis, M., Remenyte-Prescott, R.: Quantitative risk prognostics framework based on Petri Net and Bow-Tie models. Reliab. Eng. Syst. Saf. (2017). https://doi.org/10.1016/j.ress.2017.03.026
20. Harris, N., Ramsey, J.: Assessing the effects of railway infrastructure failure. J. Oper. Res. Soc. **45**, 635–640 (1994). https://doi.org/10.1057/jors.1994.101
21. ETSC (European Transport Safety Council): Transport Safety Performance in the EU a Statistical Overview (2003). Available at: https://etsc.eu/wp-content/uploads/2003_transport_safety_stats_eu_overview.pdf. Accessed on: 28 Feb 2021
22. Eurostat (EC): Rail accident fatalities in the EU (2020). Available at: https://ec.europa.eu/eurostat/statistics-explained/index.php/Rail_accident_fatalities_in_the_EU. Accessed on: 28 Feb 2021
23. Zheng, Z., Lu, P., Tolliver, D.: Decision tree approach to accident prediction for highway-rail grade crossings: empirical analysis. Transp. Res. Rec. (2016). https://doi.org/10.3141/2545-12
24. Kaewunruen, S., Sussman, J.M., Einstein, H.H.: Strategic framework to achieve carbon-efficient construction and maintenance of railway infrastructure systems. Front. Environ. Sci. (2015). https://doi.org/10.3389/fenvs.2015.00006
25. Rungskunroch, P., Kaewunruen, S., Shen, Z.-J.: An improvement on the end-of-life of high-speed rail rolling stocks considering cfrp composite material replacement. Front. Built Environ. **5**,(2019). https://doi.org/10.3389/fbuil.2019.00089
26. Binti Sa'adin, S.L., Kaewunruen, S., Jaroszweski, D.: Heavy rainfall and flood vulnerability of Singapore-Malaysia high speed rail system. Austr. J. Civil Eng. (2016). https://doi.org/10.1080/14488353.2017.1336895

27. Sáadin, S.L.B., Kaewunruen, S., Jaroszweski, D.: Operational readiness for climate change of Malaysia high-speed rail. In: Proceedings of the Institution of Civil Engineers: Transport (2016). https://doi.org/10.1680/jtran.16.00031
28. Binti Sa'adin, S.L., Kaewunruen, S., Jaroszweski, D.: Risks of climate change with respect to the Singapore-Malaysia high speed rail system. Climate **4**, 65 (2016)

Assessing Resilience of Transportation Networks Under Multi-hazards: A Review

Alok Rathore and Rajeev Kumar Garg

1 Introduction

The quality and quantity of the infrastructure are essential criteria for a smart city's development, and maintenance of its functional level is equally important. The infrastructure consists of physical network (railways, bridge, roads, to name a few) and institutions that develop the economic, health, environmental and cultural excellence. Bridges are an essential part of a transportation system of the civil infrastructure. Recent studies in India and worldwide [19] have highlighted vulnerability of infrastructure in hazards. The occurrence of a hazard event may not cause collapse or visible damages, but still these damages may get enhanced at a subsequent hazard event.

Disruption to the functionality of a transportation network may impact societal and economic losses; for example, a bridge collapse may cause indirect losses as traffic delay and network downtime. Apart from the natural hazards, some environmental or materialistic behaviour such as corrosion, thermal stresses, large soil settlement, fatigue, etc. affects the functionality of the structure which can be prevented by proper planning, and losses can be avoided by retrofitting appropriately. A bridge system's capacity to sustain against the different level of damage states and steps required for improvised functionality over the service life is termed as resilience. Important aspects of analysing and improving resilience for bridges against multi-hazards are presented in this paper.

A. Rathore
Academy of Scientific and Innovative Research (AcSIR), New Delhi, India

R. K. Garg (✉)
CSIR-Central Road Research Institute, New Delhi, India

I. Pal and S. Kolathayar (eds.), *Sustainable Cities and Resilience*, Lecture Notes in Civil Engineering 183, https://doi.org/10.1007/978-981-16-5543-2_3

2 Infrastructure and Hazards

2.1 Infrastructure and Hazards

Hazards are natural disaster events or human-made phenomena and may occur without warning. Hazard events usually adversely affect the performance of infrastructures and social life. The hazards are categorised by their origins, such as a geological process that is continuously undergoing process but is noticeable when it causes loss of infrastructure and life. Common types of hazards are geological hazards (that cause due to geological changes such as earthquake, flood, landslides, tsunami, volcanic eruptions), atmospheric hazards (occur in the atmosphere such as tropical cyclones, tornadoes and thunder storms) and hydrological hazards (such as drought, flood, rain).

It has been observed that bridges behave differently leading to different states of damage level during natural hazard conditions like floods or earthquakes. A damage scenario may not always result in a bridge's sudden failure but may leave some minor damages that get highlighted and are disastrous on subsequence disaster event. Apart from the extreme events, some hazards are time dependent (corrosion, fatigue cycles and excessive settlement) that can accelerate the capacity degradation process as time progresses [22] like the scenario of failure of Surajbari Bridge in Bhuj earthquake [16]. The bridge was having various degrees of corrosion in reinforcement due to adverse environment, which degraded the capacity and led to the failure during the Bhuj earthquake.

Flood hazards result in significant disruption in the built environment and damage to infrastructure, particularly bridge structures. During a flood, soil inundation may cause instability (large settlement) to foundations, including its collapse (Fig. 1a). Highway Bridge can be affected by flood and depending on its intensity (peak discharge). Flood damages the infrastructure in several ways such as increased debris deposition around the substructure or culverts, removing the foundation support and

(a) (b)

Fig. 1 Bridge collapse, **a** Pier failure in Ramtek, **b** Shear failure of pier wall (*Courtesy* www.ndtv.com) (*Courtesy* Sung, 2009 [33]) dislodging the superstructure of bridges. It is expected that extreme climatic change may cause more flood resulting in more damages to infrastructure.

Earthquake is a geological phenomenon that occurs due to rock masses' relative movement along faults. The movements of faults are the result of changes in the Earth's crust. The behaviour of infrastructure during an earthquake depends on two basic parameters, namely, the quality of infrastructure and the intensity of the seismic event. The performance of the bridge structure against earthquake depends on the structural configuration, design procedure, the ductile detailing of the reinforcement and good construction practices. Earthquakes may cause permanent ground settlement and may induce unseating of supported superstructures, hinge formation at the pier support, etc. (see Fig. 1b). Failures of expansion joint due to pounding effect and bearing failure are secondary damages.

Occurrence of some hazards may coincide and are termed as multi-hazard events. Multi-hazard process may result varying degree of primary, secondary and tertiary effects on the infrastructure. Primary effects result from the process itself as structural damages may be induced during an earthquake or a collapse takes place due to flood or landslides. Secondary effects are some of the direct effects related to services such as the disruption of the road network, downtime of traffic, water service damage under the event. Tertiary effects are time-dependent effects that depend on primary events' impact as economic losses towards strengthening infrastructure, societal issues, etc.

3 Research in Bridge Resilience Assessment Under Multi-hazards

3.1 Modelling Resilience

The performance of the bridge and its components is dependent on several factors, including the material and structural form used, type of hazard events and its intensity which can be assessed using appropriate methodologies. Further, the restoration of infrastructure to its pre-hazard level can be achieved using appropriate retrofitting technology. Still these issues need to be practised in an integrated manner so that the infrastructural functionality is restored as early as possible, depending on the desired performance goal. This process is best implemented using the resilience approach and is reviewed in the next sections.

Resilience is defined by the US National Academies as "ability to prepare and plan for, absorb, recover from, and more successfully adapt to adverse events" [8]. The concept of resilience was first introduced in ecology [25] that is gradually introduced in various disciplines, including environmental studies, social studies, engineering science and economics. Historical development of the concept of resilience provides eleven aspects that include four dimensions (social, technical, organisational and economical), four properties (resourcefulness, redundancy, rapidity and robustness)

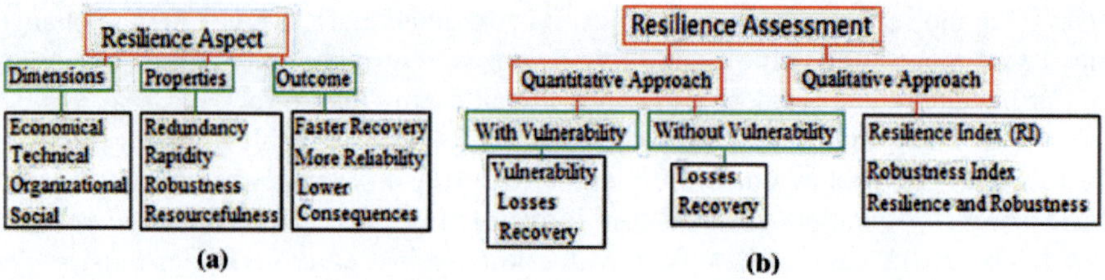

Fig. 2 **a** Resilience Aspect, **b** Resilience Assessment of Bridges

and three outcomes (more reliability, faster recovery and lower consequence) of resilience. It is shown in Fig. 2a.

Analytically, vulnerability function in the form of fragility curve corresponding to the damage state, loss model and recovery function is adopted that provides vital information to estimate the mathematical model of the resilience. In recent years, several researchers have carried out such studies using different type of bridges and different type of hazards. There are broadly two assessment approaches, namely quantitative and qualitative approaches, summarised in the flow chart (Fig. 2b).

3.2 Recent Research Studies

Several articles have discussed bridge vulnerability assessment and their components failure subjected to hazards resulting from natural and other events. Most of these studies discuss the displacement response and damage modes of the bridge components subjected to seismic ground motions and flood hazards, including the retrofitting types. A brief description of the studies considering several multi-hazards is presented in Table 1. Typical multi-hazards are earthquake, earthquake and flood, earthquake and scour, earthquake and corrosion, traffic and corrosion, earthquake and liquefaction and earthquake main shock and aftershocks.

4 Modelling Resilience and Post-Event Recovery

4.1 Operational Level

Resilience is defined as a dimensionless parameter representing the system's quickness to restore from different damage states to pre-damaged operational level or functionality level. Post-event recovery of the system and loss due to collapse or damage state caused by hazard event are two crucial components in measuring the infrastructure system's resilience. Resilience is represented as Eq. 1.

Table 1 Studies on bridge assessment under multi-hazard conditions

Authors (Year)	Hazards	Description of modelling and outcome
Zhang et al. [40]	Earthquake and liquefaction	The study considered the RC bridge with six different end conditions and derived the fragility curve based on the damage states and different soil strata during the seismic event. The effect of lateral spreading due to liquefaction shows varying bridge responses related to the damage index
Choe et al. [6]	Earthquake and corrosion	They developed probabilistic models for seismic demand in the deformation and shear of RC bridge systems subject to different corrosion stages. They noted that the sensitivity and importance measures change over time. Influence of different chloride exposure conditions, environmental oxygen availability, water-to-cement ratios, curing conditions and structural geometry are studied
Franchin and Pinto [18]	Earthquake sequences (main shocks and aftershocks)	They have suggested some criterion for opening the bridge for traffic after the main shock event considering the residual stiffness and strength in different response directions. The collapsing risk of the main shock-damaged structure and the intact structure's pre-main-shock risk are compared based on the transition probabilities for the damaged structure under different damage states. They showed that the aftershock-risk decreases with time lapse and the bridge could not be opened until the risk level reach a threshold
Ghosh and Padgett [21]	Earthquake and corrosion	The author developed the fragility curves showing ageing and deterioration due to corrosion of pier reinforcement and steel bearing in the steel bridges. The study highlighted the vulnerability of steel bearing assemblies (plate and anchor bolts) due to corrosion. The increased seismic demand in columns is influenced by cracking and spalling with ageing
Simon et al. [32]	Earthquake and corrosion	The authors considered deteriorated RC bridge (columns and piles) due to corrosion (under splash zone). They observed that spalling of the concrete cover has a more significant influence on strength than the reduction in reinforcing steel until the reduction in reinforcing bars reaches 10%
Akiyama et al. [1]	Earthquake and corrosion	The study is carried out on RC Bridge showing cumulative failure probability of corroded pier under seismic events. The study considered marine environmental condition and corrosion due to airborne chlorides. The life cycle assessment is accomplished considering the model for bucking of reinforcing bars and seismic vulnerability under cyclic loading

(continued)

Table 1 (continued)

Authors (Year)	Hazards	Description of modelling and outcome
Deco and Frangopol [9]	Earthquake, scour, corrosion and traffic	Estimated the running cost and time lost considering traffic (different vehicles) and uncertainties in structural parameters. They also considered annual hazard events (seismic, corrosion and scouring) and evaluation of annual probabilities of different limit states' correspondence to the structural failure
Aygun et al. [3]	Earthquake and liquefaction	The study is carried out on continuous steel bridges resting on liquefiable soil, including the uncertainties in geotechnical parameters. They suggested a computationally economical approach towards a one-dimensional set of force and stiffness of springs for a coupled bridge-soil-foundation system model compared to a three-dimensional model. They noted that columns' fragility depends on the type of soil overlying the liquefiable sands while the vulnerability of rocker bearings, piles, embankment soil and the probability of unseating increase with liquefaction
Zhong et al. [41]	Earthquake and corrosion	They have considered PSC bridges under corrosion due to chloride and carbonation. Also, considered stiffness degradation due to spalling of cover for developing probabilistic capacity and demand model. They found that the cracking of cover has a minor effect on high seismic fragility curve of column, bent and bridge levels
Prasad and Banerjee [30]	Earthquake and scour	The authors studied the effect of seismic activities on bridge piles subjected to flood-induced scouring and concluded that the larger diameter piles are more resilient to scouring than the piles with the smaller diameter
Ou et al. [27]	Earthquake and corrosion	They have studied RC bridges near the coastal region by modelling the crack in concrete cover (compressive stress–strain curve) with softening coefficient. They noted the effect of corrosion on longitudinal and shear reinforcing bars, indicating a marginal difference in their performance. The shear rebar got corroded earlier compared to longitudinal reinforcing bars
Padgett et al. [29]	Earthquake and liquefaction	Fragility analysis on steel bridges by modelling uncertainties in geotechnical and structure is performed considering the ANOVA method, and a wide variation of bridge seismic fragility is observed
Wang et al. [35]	Earthquake and liquefaction	Steel bridges and soil foundation are considered for the study, including liquefaction corresponding to horizontal and vertical ground motions. They analysed the influence of the frequency of modes and the vertical force component; and showed that the latter has a significant impact

(continued)

Table 1 (continued)

Authors (Year)	Hazards	Description of modelling and outcome
Alessandri et al. [2]	Earthquake sequences–MS & AS	They studied RC bridge with unequal pier heights under MS-AS seismic events. They used a Latin Hypercube sampling technique and a regression analysis to eliminate the errors during data analysis
Wang et al. [36]	Earthquake &scour	They studied the seismic performance of continuous RC box girder bridge degrading due to flood (discharge parameter and scour depth). They found the relation of stiffness and depth of foundation in fragility curve. The small flood may affect minimal scour but increases the chances of unseating failure. Also, an enlarged shaft foundation is observed to provide better performance in seismic and flood events
Biondini et al. [4]	Earthquake and corrosion	They have considered continuous box girder bridge exposed to aggressive environments considering chloride transportation in concrete by diffusion process (corrosion initiation and corrosion propagation). The study on seismic performance includes time-variant behaviour for several variables such as moment and curvature and degradation of concrete and steel
Wang et al. [37]	Earthquake and liquefaction	They assessed the reliability of multi-span continuous steel girder bridges using first-order and second-order methods considering fragility in terms of peak ground velocity (PGV). The effect on the pile cap and column's displacements under stiff soil and liquefaction of soil are studied
Dong and Frangopol [13, 14]	Earthquake sequence–(MS and AS); Earthquake and scouring	This paper proposes a framework for assessing risk, including the probabilistic direct loss, indirect loss, and resilience of bridges under main shock (MS) and aftershocks (AS). The authors observed an increase in uncertainties associated with residual functionality and repair loss during MS–AS sequences
Ghosh et al. [25]	Earthquake sequence—MS and AS	They developed the damage index based on multiple main shocks using individual ground motions and accumulation of damage states in the bridge due to a series of aftershocks
Fakharifar et al. [17]	Earthquake sequence (MS and AS)	The authors performed the analysis for enhancing the seismic performance of repaired bridge pier with FRP and steel jacketing after main shock. The authors recommend bi-directional fibre wraps (FRP jacketing) for plastic hinge confinement of MS-damaged bridge columns subjected to aftershocks

(continued)

Table 1 (continued)

Authors (Year)	Hazards	Description of modelling and outcome
Yilmaz et al. [38, 39]	Earthquake and scour	The authors carried out a study on the RC bridge in multi-hazards. Fragility surface is obtained considering annual peak flow discharge and PGA. They observed that the ductile pier and pile shaft with larger diameter could minimise seismic vulnerability during a flood event. Also, observed variation in risk due to parameter uncertainty and varied flood hazard levels
Guo and Chen [23]	Earthquake and scour	The authors performed a time-sensitive assessment of bridge failure. They have observed that scour significantly affects seismic vulnerability even after 65 years of service life but the insignificant effect in the first 45 years. They suggested performing seismic analysis that includes a scouring effect for evaluating ageing of the bridge in high flood regions
Gehl and D'Ayala [20]	Earthquake and scour	The study incorporates component fragility curves considering the sequence of failure modes and the Bayesian network considering inter-relation between failures of components in a bridge. Hazards due to earthquake, flood and ground failure events are considered, and the significance of flood hazard on bridge seismic fragility is observed at the bridges' collapse stage

$$R = \int_{t_O}^{t_O+T_C} \frac{F(t)}{T_C}\,\mathrm{d}t \tag{1}$$

where t_O represents the time when the hazard event E occurs, and T_C is a controlled time set to evaluate resilience. Also, $F(t)$ represents the operational level or system functionality, which can be represented as a dimensionless function of time t (in days). Figure 3a, represents the system functionality schematically before and after a hazard event and during the post-event recovery process. Therefore, resilience is the area of this functionality curve between t_O and $t_O + T_C$ averaged over T_C. The analytical expression of $F(t)$ given in the Eq. 2, constitutes a loss function and a recovery function of system performance during the period of system interruption because of the extreme event [13].

$$F(t) = 1 - [L_S(I, T_R) \times \{M(t - t_O) - M(t - (t_O + T_R))\} \times frec(t, t_O, T_R)] \tag{2}$$

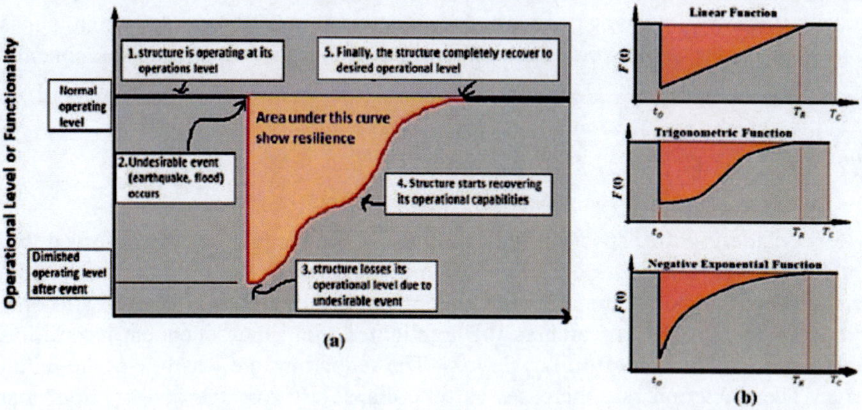

Fig. 3 **a** Represents the components of Resilience, **b** Recovery function used in resilience estimation

Here, $L_S(I, T_R)$ is the loss function, f_{rec} is the recovery function, I is seismic intensity, and T_R is the recovery time for an event E. Figure 4b represents schematically three recovery functions, namely linear, trigonometric and negative exponential functions. Here, $M(x)$ is the Heaviside step function; this discontinuous function takes value equal to either one or zero based on positive and negative arguments. Calculation of loss requires the information on system vulnerability under natural disaster.

In the bridge network's qualitative assessment, simplified procedures are considered by Ikpong et al. 2015 and Domaneschi et al. 2016 [12, 26]. This process is useful for rapid estimation of needs of resilience for a large number of bridges and their network. Proposed methodology [12, 26] adopts a resilience index (RI) applicable for bridge networks. The term resilience index is used to assess the effects of climate change to extreme events based on some weighting factor to account for the bridge's replacement cost as a percentage of its initial construction cost and post-event cost. RI is also estimated based on bridge robustness index. The researchers [12] studied

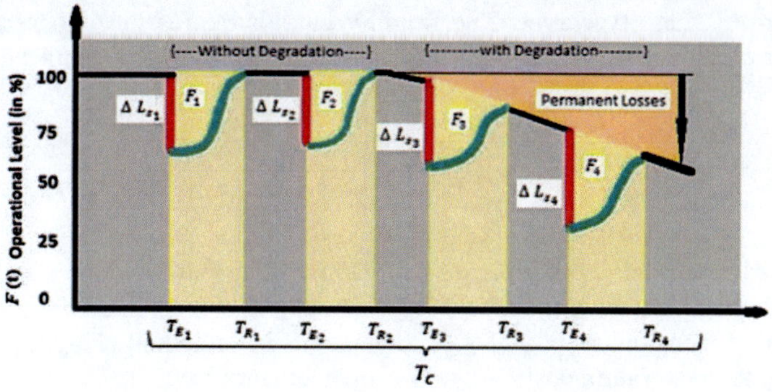

Fig. 4 Functionality function with degradation and without degradation

the robustness by observing passive and semi-active control algorithms' behaviour as applied on bridges and recorded the response of various critical bridges components.

4.2 Role of Gradual Degradation

After considering the aspect of multi-hazard on bridges [15], it was observed that corrosion changes the life cycle of bridge structure with the time that means the seismic performance of the bridge (and components) degrades gradually with time due to corrosion. The researchers [4] highlighted that time-dependent hazard has significant effects on total life cycle loss. The reduction rate was low at the initial stage. The reduction rate increases as time elapse after bridge construction; that makes a non-linear variation of bridge resilience over its life cycle. In a study [5], the seismic resilience of RC bridge is assessed under corrosion of reinforcing bars. The time-variant functionality indicator was developed to evaluate the post-event residual functionality and recovery of the deteriorating system as a function of time of occurrence of a seismic event.

Figure 4 shows resilience estimation in terms of the system's functionality without considering its gradual degradation. At the time of the extreme event, the functionality may drop due to system damages.

4.3 Losses and Recovery Time

The studies show that loss and recovery are the two significant components of system resilience. To estimate the loss, the models consider direct loss and indirect loss proposed in HAZUS MH 2005 [24]. Further, the direct loss is related to the cost associated with the post-event system restoration capability. Loss can be estimated considering the state of the structural damage due to event [24, 34]. Vulnerability model (Eq. 3) of the bridge is developed, which defines the different damage states in the form of fragility curves. The fragility curve's expression is based on two parameters c_j (median value) and ζ_j (log-standard deviation) for a damaged state, j.

$$F(PGA_i, c_j, \zeta_j) = \phi\left[\frac{\ln\left(PGA_i/c_j\right)}{\zeta_j}\right] \tag{3}$$

where PGA_i represents peak ground acceleration of a ground motion i and j show different damage (minor, moderate, major damage and complete collapse) of the bridge. The dimensionless cost, D_L of the event E is computed using Eq. 4 and is the ratio of the restoration cost C_{DL} to the replacement cost C_R.

$$D_L = \frac{C_{DL}}{C_R} = \sum_j P_E(DS = j) \times r_j \tag{4}$$

Here, $P_E(DS = j)$ is the probability of bridge failure at damage state j during the seismic event E, and r_j is the damage ratio corresponding to damage state j. The values of $P_E(DS = j)$ are obtained from fragility curves (Eq. 3) developed for various damage states. Indirect loss is related to the socio-economic losses that include casualties and delay in travelling time, increased vehicle operating costs and loss of economic activities due to inability to travel. Indirect loss due to increased travelling time resulting from bridge and highway network failure is typically 10–20% times larger than bridge repairing cost [7, 11].

The bridge restoration process depends on the extent of damage and the community's preparedness to recover. Accurate simulation of the path to recovery becomes challenging in the absence of specific event or location information. Two types of seismic restoration models were developed for bridges to solve this challenge. Firstly, the bridge's probability will be recovered entirely (i.e. it will gain 100% functionality) given the bridge's damaged state and the time after the seismic event [31]. In contrast, the other quantifies the percentage of the bridge's functionality that is conditional on the damage state and the time after the earthquake occurrence [24, 28]. Therefore, resilience is measured as a percentage of functionality. A fully serviceable system without any loss in functionality after an extreme event is defined as $R = 100\%$. For partial functionality, it is expressed as $0 < R < 100\%$. The researchers [7] have proposed a simplified mathematical model for the recovery process considering linear, exponential and trigonometric functions (Figure 3b). Recent studies have used different recovery patterns for different seismic bridge damage states [10, 42].

5 Case Study

The considered bridge is a multi-cell RC box girder (2.06 m wide and 1.58 m deep with a total span of 140.4 m) supported on five circular 1.219 m diameter piers in Los Angeles and was damaged during the Northridge Earthquake [34]. The non-linear rotation springs are provided at the hinges location on the pier, and the rigid links are provided at the girder connecting with the pier. Expansion joints are modelled as hook and gap elements. The bridge is analysed for Northridge Earthquake event and developed the fragility curves under time history of 60 ground motions at PGA of 0.883 g. The damage states are based on the ductility drift ratio. The uncertainties considered are control time, recovery time and fragility parameter and modelled using FOSM. The authors used mainly two modes of failure. The shear failure indicated the complete failure and the flexure failure based on different damage states (namely slightly, moderate, extensive and complete collapse). During analysis, the pier got failed when shear demand exceeded the shear capacity and is corroborated as was damaged in the seismic event.

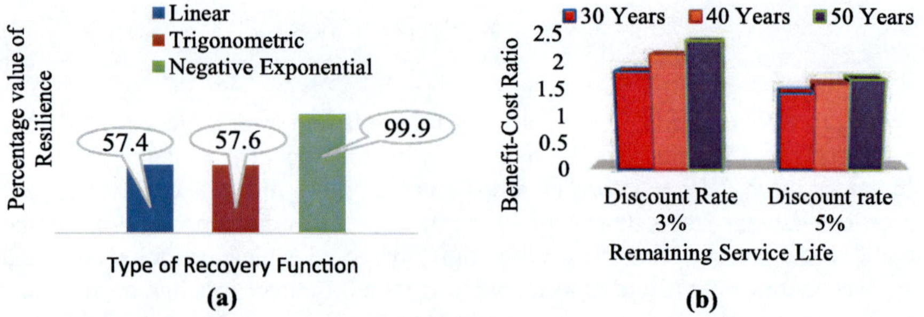

Fig. 5 **a** Influence of recovery functions on resilience, **b** Summary of benefit cost ratio

Steel jacketing was used to retrofit against seismic vulnerability to enhance the performance in flexural ductility and shear strength while reducing shear demand in the bridge pier. Seismic resilience is assessed using Equations 1 and 2 based on fragility curves combining with loss function and recovery function (linear, trigonometric and negative exponential functions). The assessment of resilience is based on an evaluation of failure and losses reduced by proposed steel jacketing. The resilience is observed to increase by 74% due to the seismic retrofitting. They observed that uncertainties related to control time and recovery time have significant influence, but fragility and indirect to direct loss ratio have very low influence on resilience sensitivity. The authors estimated the total cost of retrofitting as USD 168,800 with enhanced service life by another 45 years had it been retrofitted before the Northridge Earthquake. The study highlighted the importance of loss function and recovery time for assessing seismic resilience and provided cost-effectiveness of the retrofitting based on the benefit–cost ratio (Fig. 5) which is high for the expected higher service life of the bridge.

6 Conclusions

The current review presents methodologies and influences of several associated factors for studying the resilience of bridges and bridge networks under different hazard conditions. Bridge capacity and its performance degrade with time due to various events in a multi-hazard scenario while sudden damages are observed with extreme events. Resilience assessment of the bridges provides developing the recovery model and loss function model as per performance goals. These models help in making judgments and realistic decision for bridge management. Resilience optimises the maintenance method for faster recovery when the system's life cycle is considered. Therefore, a structural system's resilience is considered as a critical performance indicator for infrastructure.

Acknowledgements The authors acknowledge the support provided by the Director, CSIR-CRRI and AcSIR, New Delhi.

References

1. Akiyama, M., Frangopol, D.M., Matsuzaki, H.: Life-cycle reliability of RC bridge piers under seismic and airborne chloride hazards. Earthq. Eng. Struct. Dynam. **40**(15), 1671–1687 (2011). https://doi.org/10.1002/eqe.1108
2. Alessandri, S., Giannini, R., Paolacci, F.: Aftershock risk assessment and the decision to open traffic on bridges. Earthq. Eng. Struct. Dynam. **42**(15), 2255–2275 (2013). https://doi.org/10.1002/eqe.2324
3. Aygun, B., Duenas-Osorio, L., Padgett, J.E., DesRoches, R.: Efficient longitudinal seismic fragility assessment of a multi span continuous steel bridge on liquefiable soils. J. Bridg. Eng. **16**(1), 93–107 (2011). https://doi.org/10.1061/(ASCE)BE.1943-5592.0000131
4. Biondini, F., Camnasio, E., Palermo, A.: Lifetime seismic performance of concrete bridges exposed to corrosion. Struct. Infrastruct. Eng. **10**(7), 880–900 (2013). https://doi.org/10.1080/15732479.2012.761248
5. Biondini, F., Camnasio, E., Titi, A.: Seismic resilience of concrete structure under corrosion. Earthq. Eng. Struct. Dynam. **44**(14), 2445–2466 (2015). https://doi.org/10.1002/eqe.2591
6. Choe, D.E., Gardoni, P., Rosowsky, D., Haukaas, T.: Seismic fragility estimates for reinforced concrete bridges subject to corrosion. Struct. Saf. **31**(4), 275–283 (2009). https://doi.org/10.1016/j.strusafe.2008.10.001
7. Cimellaro, G.P., Reinhorn, A.M., Bruneau, M.: Framework for analytical quantification of disaster resilience. Eng. Struct. **32**, 3639–3649 (2010)
8. Cutter, S.L., Ahearn, J.A., Amadei, B., Crawford, P., Eide, E.A., Galloway, G.E., Scrimshaw, S.C.: Disaster resilience: a national imperative. Environ.: Sci. Policy Sustain. Dev. **55**(2), 25–29 (2013). https://doi.org/10.1080/00139157.2013.768076
9. Deco, A., Frangopol, D.M.: Risk assessment of highway bridges under multiple hazards. J. Risk Res. **14**(9), 1057–1089 (2011). https://doi.org/10.1080/13669877.2011.571789
10. Deco, A., Bocchini, P., Frangopol, D.M.: A probabilistic approach for the prediction of seismic resilience of bridges. Earthq. Eng. Struct. Dyn. **42**(10), 1469–1487 (2013)
11. Dennemann, L.K.: Life-Cycle Cost Benefit (LCC-B) Analysis for Bridge Seismic Retrofit. MS Thesis, Rice University, TX (2009)
12. Domaneschi, M., Martinelli, L.: Earthquake-resilience-based control solutions for the extended benchmark cable-stayed bridge. J. Struct. Eng. **142**(8), C4015009 (2016). https://doi.org/10.1061/(ASCE)ST.1943-541X.0001392
13. Dong, Y., Frangopol, D.M.: Risk and resilience assessment of bridges under mainshock and aftershocks incorporating uncertainties. Eng. Struct. **83**, 198–208 (2015). https://doi.org/10.1016/j.engstruct.2014.10.050
14. Dong, Y., Frangopol, D.M.: Probabilistic time-dependent multi-hazard life-cycle assessment and resilience of bridges considering climate change. J. Performance Constr. Facilities **30**(5) (2016). https://doi.org/10.1061/(ASCE)CF.1943-5509.0000883
15. Dong, Y., Frangopol, D.M., Saydam, D.: Time-variant sustainability assessment of seismically vulnerable bridges subjected to multiple hazards. Earthq. Eng. Struct. Dynam. **42**(10), 1451–1467 (2013). https://doi.org/10.1002/eqe.2281
16. EEFIT: The Bhuj, India Earthquake of 26th January 2001, Earthquake Engineering Field Investigation Team, Institution of Structural Engineers, London (2005). https://www.istructe.org/downloads/resources-centre/technical-topic-area/eefit/eefit-reports/bhuj-india.pdf
17. Fakharifar, M., Chen, G., Sneed, L., Dalvand, A.: Seismic performance of post-mainshock FRP/steel repaired RC bridge columns subjected to aftershocks. Compos. B Eng. **72**, 183–198 (2015). https://doi.org/10.1016/j.compositesb.2014.12.010

18. Franchin, P., Pinto, P.E.: Allowing traffic over main-shock damaged bridges. J. Earthq. Eng. **13**(5), 585–599 (2009). https://doi.org/10.1080/13632460802421326
19. Garg, R.K., Chandra. S., Kumar. A.: Analysis of bridge failures in India from 1977 to 2017. Struct. Infrastructure Eng. (2020). https://doi.org/10.1080/15732479.2020.1832539
20. Gehl, P., D'Ayala, D.: Development of Bayesian Networks forthe multi-hazard fragility assessment of bridge systems. Structural Safety **60**, 37–46 (2016). https://doi.org/10.1016/j.strusafe.2016.01.006
21. Ghosh, J., Padgett, J.E.: Aging considerations in the development of time-dependent seismic fragility curves. J. Struct. Eng. **136**(12), 1497–1511 (2010). https://doi.org/10.1061/(ASCE)ST.1943-541X.0000260
22. Ghosh, J., Padgett, J.E., Sanchez-Silva, M.: Seismic damage accumulation in highway bridges in earthquake-prone regions. Earthq. Spectra **31**(1), 115–135 (2015). https://doi.org/10.1193/120812EQS347M
23. Guo, X., Chen, Z.: Lifecycle multi-hazard framework for assessing flood scour and earthquake effects on bridge failure. ASCE-ASME J. Risk Uncertainty Eng. Syst. Part A: Civil Eng. **2**(2), C4015004 (2016). https://doi.org/10.1061/AJRUA6.0000844
24. Hazus, M.H.: Multi-hazard Loss Estimation Methodology-Earthquake Model, Federal Emergency Management Agency, USA (2005). https://www.fema.gov/sites/default/files/2020-09/fema_hazus_earthquake-model_user-manual_2.1.pdf
25. Holling, C.S.: Resilience and stability of ecological systems. Annu. Rev. Ecol. Syst. **4**(1), 1–23 (1973). https://doi.org/10.1146/annurev.es.04.110173.000245
26. Ikpong, A., Bagchi, A.: New method for climate change resilience rating of highway bridges. J. Cold Regions Eng. **29**(3), 04014013 (2015). https://doi.org/10.1061/(ASCE)CR.1943-5495.0000079
27. Ou, Y.C., Fan, H.D., Nguyen, N.D.: Long-term seismic performance of reinforced concrete bridges under steel reinforcement corrosion due to chloride attack. Earthq. Eng. Struct. Dynam. **42**(14), 2113–2127 (2013). https://doi.org/10.1002/eqe.2316
28. Padgett, J.E., DesRoches: Bridge Functionality relationships for improved seismic risk assessment of bridge network. Earthq. Spectra. https://doi.org/10.1193/1.2431209(2007)
29. Padgett, J.E., Ghosh, J., Duenas-Osorio, L.: Effects of liquefiable soil and bridge modelling parameters on the seismic reliability of critical structural components. Struct. Infrastructure Eng. **9**(1), 59–77 (2013). https://doi.org/10.1080/15732479.2010.524654
30. Prasad, G.G., Banerjee, S.: The impact of flood-induced scour on seismic fragility characteristics of bridges. J. Earthq. Eng. **17**(6), 803–828 (2013). https://doi.org/10.1080/13632469.2013.771593
31. Shinozuka, et al.: Effect of seismic retrofit of bridges on transportation networks. Earthq. Eng. Eng. Vibration **2**, 169–179 (2003)
32. Simon, J., Bracci, J.M., Gardoni, P.: Seismic response and fragility of deteriorated reinforced concrete bridges. J. Struct. Eng. **136**(10), 1273–1281 (2010). https://doi.org/10.1061/(ASCE)ST.1943-541X.0000220
33. Sung, Y., Liu, K.: Enhancing the structural Longevity of the bridges with insufficient seismic capacity by retrofitting. Tech Science Press, SL, vol. 1, no.1, pp. 1–16 (2009)
34. Venkittaraman, A., Banerjee, S.: Enhancing resilience of highway bridges through seismic retrofit. Earthq. Eng. Structural Dyn. Earthq. Eng Struct. Dyn. **2014**(43), 1173–1191 (2013)
35. Wang, Z., Duenas-Osorio, L., Padgett, J.E.: Seismic response of a bridge–soil–foundation system under the combined effect of vertical and horizontal ground motions. Earthq. Eng. Structural Dyn. **42**(4), 545–564 (2013). https://doi.org/10.1002/eqe.2226
36. Wang, Z., Duenas-Osorio, L., Padgett, J.E.: Influence of soil structure interaction and liquefaction on the isolation efficiency of a typical multi-span continuous steel girder bridge. J. Bridg. Eng. **19**(8), A4014001 (2014). https://doi.org/10.1061/(ASCE)BE.1943-5592.0000526
37. Wang, Z., Padgett, J.E., Duenas-Osorio, L.: Risk-consistent calibration of load factors for the design of reinforced concrete bridges under the combined effects of earthquake and scour hazards. Eng. Struct. **79**, 86–95 (2014). https://doi.org/10.1016/j.engstruct.2014.07.005

38. Yilmaz, T., Banerjee, S., Johnson, P.A.: Performance of two real-life California bridges under regional natural hazards. J. Bridg. Eng. **21**(3), 04015063 (2016). https://doi.org/10.1061/(ASCE)BE.1943-5592.0000827
39. Yilmaz, T., Banerjee, S., Johnson, P.A.: Uncertainty in risk of highway bridges assessed for integrated seismic and flood hazards. Struct. Infrastructure Eng. **14**(9), 1182–1196 (2018). https://doi.org/10.1080/15732479.2017.1402065
40. Zhang, J., Huo, Y., Brandenberg, S.J., Kashighandi, P.: Effects of structural characterizations on fragility functions of bridges subject to seismic shaking and lateral spreading. Earthq. Eng. Eng. Vib. **7**(4), 369–382 (2008). https://doi.org/10.1007/s11803-008-1009-2
41. Zhong, J., Gardoni, P., Rosowsky, D.: Seismic fragility estimates for corroding reinforced concrete bridges. Struct. Infrastruct. Eng. **8**(1), 55–69 (2012). https://doi.org/10.1080/15732470903241881
42. Zhou, H., Wang, J., Wan, J., Jia, H.: Resilience to natural hazards: a geographic perspective. Nat. Hazards **53**(1), 21–41 (2010). https://doi.org/10.1007/s11069-009-9407

Risk Management in Construction Industry

Surisetti Divya Sankar, Kulkarni Shashikanth, and Sangamalla Mahender

1 Introduction

Every construction project is exposed to various risks, and the risk is unpredictable. This construction industry needs a large number of individuals having different talents and also need a good coordination of people performing different activities in the construction organization. The construction project risks are of two different types. They are known risks and unknown risks. Risk is expected when a selected activity is leading towards loss. Risks can emerge out of unpredictable budgetary business sectors, failure of projects, lawful accountabilities, and hazards due to accidents, and also hazards occurring naturally. In ISO 31000, risk management is defined as the risk identification; risk assessment and also risk prioritization are followed by team work coordination and also affordable use of resources towards limiting, monitoring, and also controlling the likelihood effect of unpredicted occurrences. The risk is assessed by means of functioning of following three variables. They are possibility if hazard is present; possibility if they are susceptible, and final possibility is impact towards the construction industry business. The hazards idea in statistics is frequently demonstrated as the normal estimated value results are not favorable. This merges the probabilities of different likely occurrences and also particular evaluation

S. Divya Sankar (✉)
Research Scholar, Department of Civil Engineering, Lincoln University College, Selangor, Malaysia
e-mail: divyasankar@lincoln.edu.my

K. Shashikanth
Supervisor, Lincoln University College, Selangor, Malaysia

Associate Professor, Department of Civil Engineering, University College of Engineering, Osmania University, Hyderabad, TS, India

S. Mahender
Master of Engineering, Department of Civil Engineering, University College of Engineering, Osmania University, Hyderabad, TS, India

I. Pal and S. Kolathayar (eds.), *Sustainable Cities and Resilience*, Lecture Notes in Civil Engineering 183, https://doi.org/10.1007/978-981-16-5543-2_4

of comparing sufferings into a single price. A straightforward formula is utilized for a chance of either occurrence or non-occurrence of an accident. Therefore, the risk formula is below:

Risk = (Probability of an accident occurrence) × (Expected loss in case of an accident occurred).

Risk is possible when an activity or action is not carefully chosen, and then, it leads towards loss. An idea concludes that a decision has an impact on the existing result (or else existed). Possible losses itself may perhaps recognized as "risk." According to PMBOK guide, risk is defined as "an event occurring uncertainly and has a positive effect or negative effect on project objectives".

1.1 Research Significance

The reason for not finishing the project and delivering the project within an agreed low-priced budget and also within the indicated time, may possibly remain for not observing towards the specifications which are established previously and are unable to see the hazards and also the steps are not been towards cutting off the hazard, and this is also causing loss towards all the parties which are concerned.

1.2 Objectives

The objective of this analysis is assessing the acceptability of various management risks practices utilized by the contractual workers in the urban areas of Vijayawada and Visakhapatnam. Furthermore, discovering the best reasonable and suitable method of managing the hazard location in India, thus, the projects possibly will bring better-quality, completion of project within given time, and reducing the conflicts and moreover, getting profit.

2 Literature Review

Hertz and Thomas [1] stated that the construction project activities are built on the open area, and therefore, the risk complexity is added by different uncontrollable factors externally.

Chapman [2] explained that risk analysis assists in calculating possible effects and also decision making of risks towards retaining of hazards and transferring of risks to different parties. For risk analysis, both qualitative and also quantitative techniques are available. The quantitative strategy depends on distributing the hazards probability and also gives precise outcomes than the qualitative strategy. On individual judgments and earlier experiences of the experts, the results can be different from individual to individual in this qualitative method. Consequently, the quantitative methods are significant if the two decisions are available.

Al-Bahar [3] stated that risks in construction recognizes the occurrences, which impacts the project aims of costs, time, and quality. In construction process, few risks are predicted and also identified. Risks in construction have been classified as follows:

- Acts of God: for example, floods, storms, earthquakes, etc.
- Physical risks: for example, work wounds, fire, and equipment damage;
- Financial and monetary risks: for example, price increasing, funds are not being available;
- Political and environment risks: for example, changing instructions and guidelines, political indecisiveness;
- Design risks: for example, defective plans.
- Changes of risks occurring in Bill of Quantities (BOQ), for example, changing orders, productivity of labor, etc.

Thompson and Perry [4] explained the motivation behind the analysis of risk and the management is towards assisting stakeholders avoiding the failures.

Edwards [5] stated that large projects which are undertaken are failing to meet the targeted given time and budget. The customers, contractual workers, the general population, and others also had undergone suffering.

Nummedal [6] explained risk management definition as a procedure of controlling risk levels and mitigating its effects and an organized methodology towards identification, assessing, and reacting to risks faced in a construction project.

In a construction project, the four particular ways responding towards risks are:

(1) Risk elimination (for example, by putting a high bid), (2) Risk transfer (for example, employing subcontractors), (3) Risk retention (for example, by means of insurance) and (4) Risk reduction (by giving training to employees about the risk awareness and also its management).

Zou et al. [7] explained the importance of being conscious of the risks allotted for every party, whichever is involving, and then, the preparation must be successful aiming through the hazards dealing and also finally contributing towards the successful project.

PMI [8] stated that the risks are measured as the occurrences influencing the goals of a specific project in construction. Risk management can be defined as a structured and also comprehensive process designed to analyze, identify, and also respond to the factors of risk towards attaining the goals of project.

Perera et al. [9] stated that various options and also actions are developing for the promotion of opportunities and decreasing threats towards the project aims. Few levels of risks need to be accepted by the party which is involved in the contract and also need to have full knowledge of sharing their own risks and later putting losses into consideration.

Rostami [10] stated that when risks are unidentified, then it may result into shortage in the complete procedure. In the organization, critical effects can be available on the resources when it deals with risk. Identification of risks helps firms in involving risk management by identifying best and also most important input information. Better knowledge in risk management need to be present and can identify risk and their impacts by providing information for the people who are taking decisions in the field of construction projects.

3 Research Methodology

The data was gathered by means of questionnaire survey from the construction organizations in Vijayawada and Visakhapatnam for this research study. The questionnaire is containing the proofs of identity of serious hazards and also their influence on costs (Table 1).

The following details gathered from the construction firms are the data of the respondents and construction company's time and quality, the plan towards dealing with the hazards identification, having knowledge about the ease of understanding analysis of risk and mitigation of risks response techniques. The data collection was gathered through questionnaires with professional experts in the organizations which are carefully chosen. Therefore, the analysis was done from the collected data, and the outcomes also discussed. The following are the definitions of the terms which are included in this research paper.

- Contractor: Towards obtaining civil engineering contracts, this type of classification is for the business people, who will place tenders additionally organizes and also finishes the work. The designing and surveying work, drawing plans, and also managing the contracts for the clients must be placed in category civil engineers.
- Subcontractor: The architect or engineer is chosen by a subcontractor towards carrying out work of a specialist. A provisional sum or else prime cost sum remains included in the bill of quantities are appointed on the expenditure.
- Expert systems: Technical and economic issues are being associated with the influence on healthiness of a human and environment. This one is involving experts from various fields towards solving problems. Expert systems are very interesting and also complex solutions for problem solving.
- Probability analysis: On behalf of quantifying hazards or else uncertainty in engineering problems, an official source is provided. Through engineering judgments, the qualitative approach can also be dealt with.

Table1 General data of the respondent

District	Vijayawada	Visakhapatnam
No. of companies participating	40	40
No. of companies that responded	29	29
Response rate (%)	72.5	72.5

- Sensitivity analysis: The set of assumptions with changed values of an independent variables affecting particularly with a dependent variable will be determined.
- Simulation analysis: In computer-aided design, the process of emerging a mathematical illustration of an actual computer model is simulation analysis.
- Direct judgment: The judgment should be compulsorily based on exact standards or else on professional expert and also should be managed in an application part and in an engineering industrial area.
- Comparing analysis: This one implements the comparative analysis which is critical, and also abundant practical methods are opening their chances which are proposed in the civil engineering for further educations.

4 Analysis of Results and Discussion of Vijayawada and Visakhapatnam

Questionnaire response rate

From topmost management till the site offices in the construction organizations, the respondent's qualification, years of experience additionally, and their professional background details are gathered and which indicate that all of them are having good knowledge about hazards in the construction industry.

The response rate of Vijayawada is as per the following:

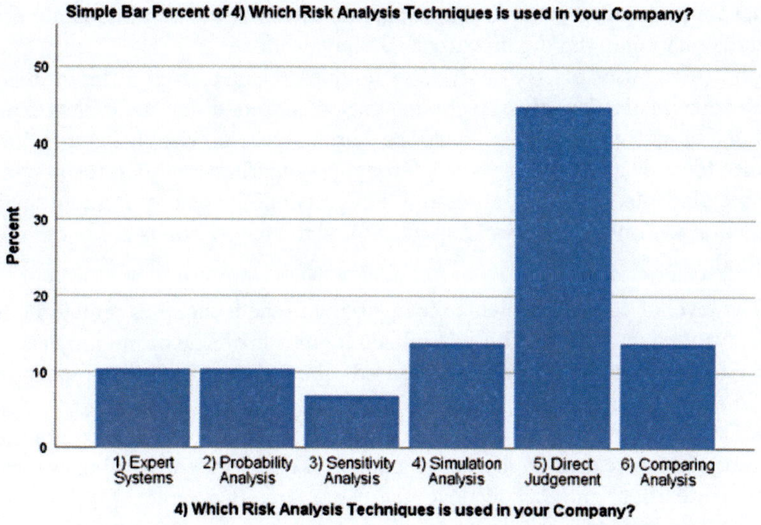

Graph 1 Risk analysis techniques evaluation in Vijayawada, *X*-axis indicates risk analysis techniques; *Y*-axis indicates % of respondents

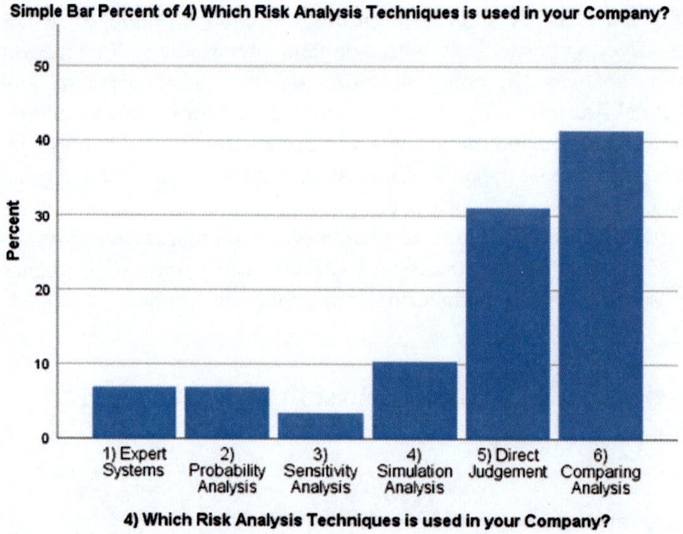

Graph 2 Evaluation of risk analysis techniques in Visakhapatnam. *X*-axis indicates risk analysis techniques; *Y*-axis indicates % of respondents

Graph 1. The responses from selected districts about different risk analysis techniques

Graph 1 explains the results of the most of the construction organizations (about 45%) have a preference towards direct judgment techniques towards dealing with the hazards relating to the organizations in Vijayawada.

Graph 2 Responses of risk analysis techniques selected from different districts Fig. 2. Responses of risk analysis techniques selected from different districts explains the results of construction organizations about 42% in Visakhapatnam having preference towards analyzing the risks through Comparing Analysis Technique.

The respondents are cooperating and comprehending for not utilizing analysis of simulation within area of Visakhapatnam, and the below points are gathered and understood:

(a) The level of complexity which is involving in the techniques remaining there is not justified or authorized if it is related by means of size of the project.
(b) The organizations are not assured about usage of the management administration techniques.
(c) Most of the risks are related to the construction process and also contractually. Therefore, they are dealing in a better manner, and also they are having experience from their earlier contracts whatever they have already completed.
(d) The management techniques relating to risk requires the ease of accessing information, which remains very hard towards assuring confidence.
(e) All the projects are dissimilar and are not comparable.

(f) It is discussed from the above conversation that within given period of time, organizations will finish the construction projects and will also handover for about one year or two years or three years. Therefore, for collecting necessary information and analyzing the information requires few months, but the construction organizations are not willing to spare time in recognizing the hazards and also not understanding the mitigation of risks.

The following are the definitions of different techniques relating to the risk elimination, risk transfer, risk retention, risk reduction, technical risks, and financial risks.

(a) Risk Elimination: When it comes to risk control, the first step is towards considering elimination or avoidance of risk.
(b) Risk Transfer is the risk management technique used for transferring risk.
(c) Risk retention is the cost of avoiding the risk which remains greater than the cost of damaging. Therefore, it is better to avoid risk if it is frequently maintained.
(d) Risk reduction can be frequently reduced by means of taking different precautions against it and also when exposing the company towards risk remains unavoidable.
(e) Technical risk includes uncertainty of resources, availability of construction materials, incomplete designs, and inadequate site exploration.
(f) Financial risks on a construction project is an extensive area and also consist of difficulties by means of under financed, misusing of projects funds, contractor nonpayment issues, etc.

Risk responses practice evaluation: In this risk response practices questionnaire survey, the respondents were asked which risk methods are used by their particular organization. The responses towards the four principal strategies, i.e., risk elimination, risk transfer; risk retention, and risk reduction is as follows:

Response rate of Vijayawada is as per the following:

Note: *X*-axis indicates risk response techniques; Y-axis indicates percentage of respondents used technique %.

In Graph 3, the results reveal that risk elimination and risk transfer are the utmost valued strategies utilized by contractors in Vijayawada. Risk transfer towards a specialty subcontractor or else through monetary methods, such as insurance, also was examined. In Vijayawada construction, contractors utilize the two techniques by transferring the hazard to subcontractors once the loss predicted is more. The response rate of Visakhapatnam is as per the following:

Note: Percent of respondents utilized that technique % along *Y*-axis.

In Graph 4, the results revealing risk elimination, risk transfer, and risk reduction are three valued strategies utilized by the general contractors in Visakhapatnam.

Risk transfer to a subcontractor, such as an insurance, was also examined. These general contractors now in Visakhapatnam construction industry utilize the three techniques; however, transfers hazard to a subcontractor once the loss is expected.

At the point when the construction firms are not in a position to do research and development (R&D), at that point, they move the risk of R&D to the external party

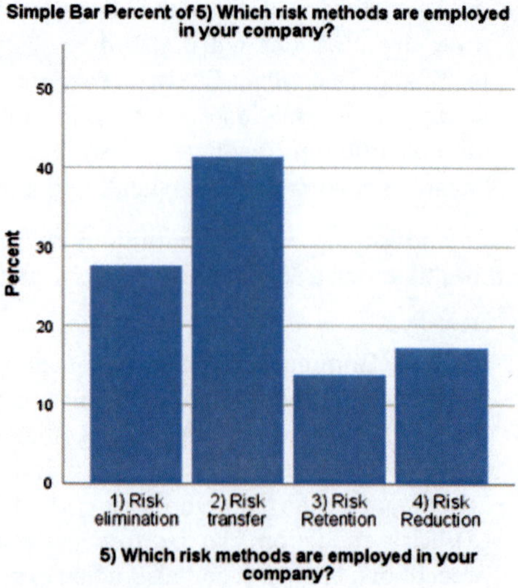

Graph 3 Vijayawada: Responses from chosen regions about various risk response techniques. X-axis indicates risk methods; Y-axis indicates % of respondents

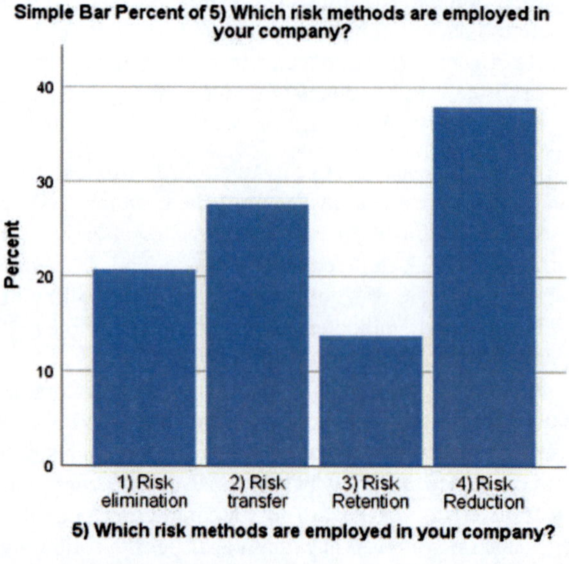

Graph 4 Visakhapatnam responses from chosen regions about various risk response techniques

or some other subcontractor; furthermore, it is gotten that if this R&D is finished by people, they may.

Whenever the construction organizations unable to do research and development, then transferring of risk of research and development takes place to the party which is outside or else to any subcontractor.

Graph 5 Risk transferring method used in Vijayawada

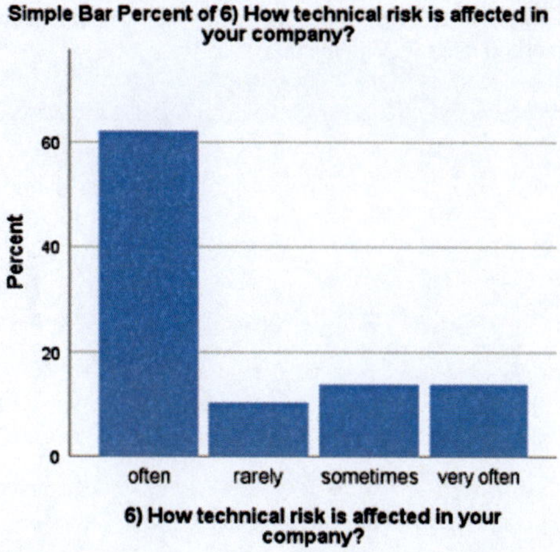

Note: Response rate in percent along the *Y*-axis. From the above Graph 5, it shall be interpreted that these construction companies transfer technology risk often about 61%.

Note: Response rate in percent along the *Y*-axis. From the above Graph 6, it is understood that these construction organizations transfers technology risk often about 45%.

Note: Response rate in percent along the *Y*-axis. From the above Graph 7, it is demonstrated that, these construction companies transfer financial risk often about 65%.

Note: Response rate in percent along the *Y*-axis. From the above Graph 8, it is observed that rarely (about 62%) to the contractors or else subcontractors lose

Graph 6 Risk transferring method used in Visakhapatnam

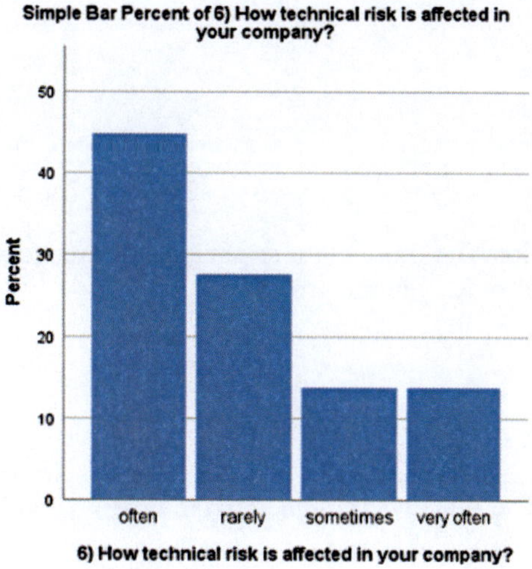

Graph 7 Risk transferring method used in Vijayawada

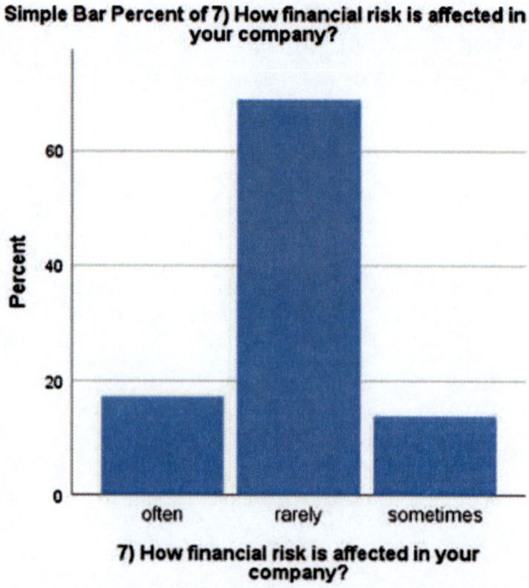

Graph 8 Risk transferring method used in Visakhapatnam

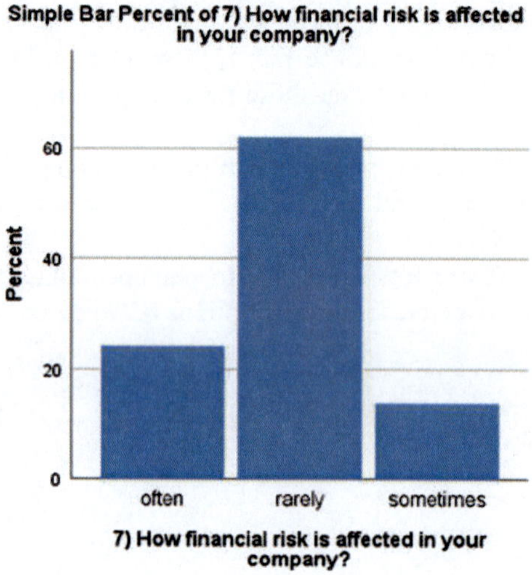

finishing time, and it might lead towards lowering productivity, reduced quality, and also delay in the projects.

5 Limitations of the Study

The study was confined to cities of Vijayawada and Visakhapatnam only.

6 Recommendations

In the construction industry, the professionals need to know about the techniques of management used in the construction organizations by conducting risk alertness programs and also by conducting training sections on construction project management. It is important to recruit construction engineers having knowledge in using management techniques.

7 Conclusion

The analysis of risk and also techniques of management are applied rarely by the Visakhapatnam construction organizations because of the absence of knowledge and also expertise. Therefore, the construction industry is not confident relating to the techniques used and also its applicability in the construction projects. The contractors recognized that risk is based on their years of experience and also in their decision making or judgment.

References

1. Hertz, D., Thomas, H.: Risk analysis: important new tool for business planning. J. Bus. Strategy **3**(3) (1983) (Cambridge)
2. Chapman, C.B.: A risk engineering approach to project management. Int. J. Project Manage. **8**(1), 5–16 (1990)
3. Jamal, F., Al-Bahar, K.C.: Crandall: systematic risk management approach for construction projects. J. Constr. Eng. Manag. **116**(3), 533–546 (1990)
4. Thompson, P.A., Perry, J.G. (eds.): Engineering Construction Risks—A Guide to Project Risk Analysis And Risk Management. Thomas Telford, London (1992)
5. Edwards, L.: Practical Risk Management In The Construction Industry. Thomas Telford Services Ltd, London (1995)
6. Nummedal: Proceedings of the International Conference on Health, Safety and Environmen, vol. 2, Texas Univ, Austin, pp 112–117 (1996)
7. Zou, P.X.W., Zhang, G., Wang, J.: Understanding the key risks in construction projects in China. Int. J. Proj. Manag. **25**(6), 601–614 (2007)
8. PMI : A Guide to The Project Management Body of Knowledge (PMBOK guide) (Newtown Square: Project Management Institute) (2013)
9. Perera, B., Rameezdeen, R., Chileshe, N., Hosseini, M.R.: Enhancing the effectiveness of risk management practices in Sri Lankan road construction projects: a delphi approach. Int. J. Constr. Manag. **14**, 1–14 (2014)
10. Rostami, A.: Tools and techniques in risk identification: a research within SMEs in the UK construction industry. Universal J. Manag. **4**(4) 203–210 (2016)

Achieving Sustainability Goals Through Infrastructure Modifications: Lessons Learnt from COVID-19 Pandemic

Poornima Ramesh and Bharani Alagirisamy

1 Introduction

Any infrastructure investments results in detrimental impacts, such as erosion, air and water emissions and low climate resilience. Calls for sustainable infrastructure have risen in recent years. Concerns about barriers to exploiting small public services were increased by the COVID-19 pandemic. The broad economic effects of the pandemic made it imperative that expenditure be focused on the required and well-designed. GDP growth decreased and was gradually negative during the first quarter of 2020, even within the global economies [1]. The world's per capita GDP is predicted to decrease by 4.2% overall by 2020 compared to a 1.6 per cent drop after the 2009 global financial crisis. COVID-19 also provides a chance in the form of recovery kits. When nations are trying to rebound as soon as possible, all recovery package-supported spending needs to be sustainable. There can be hardly any second chance of more disasters, considering climate change and the grim global macroeconomic outlook. While measures for solving COVID-19 socio-economic challenges, including economic lockdowns, have led to reduced carbon emissions, the pandemic will not slow down climate change and associated risks in the long run [2]. In addition, if new projects are not directed to low-carbon projects, this leads to lock-in, making it much harder for countries to fulfil their Paris Climate Agreement obligations as set out in their Nationally Determined Contributions (NDCs). To promote a green recovery, countries must use their financial stimulus. Sustainable investment, especially in infrastructure, would offer an opportunity to improve economic growth while increasing climate resilience and reducing projected losses.

P. Ramesh · B. Alagirisamy (✉)
Department of Environmental Sciences, Tamil Nadu Agricultural University, Coimbatore, Tamil Nadu 641003, India

I. Pal and S. Kolathayar (eds.), *Sustainable Cities and Resilience*, Lecture Notes in Civil Engineering 183, https://doi.org/10.1007/978-981-16-5543-2_5

2 Impact of COVID-19

The global reported COVID-19 cases exceeded 60 million infections and 1.4 million deaths by the end of November 2020 and covered 191 countries and regions (https://coronavirus.jhu.edu/map.html). However, COVID-19 is not only a health problem, but also a major economic catastrophe. GDP statistics obtained during the first four months of 2020 from global economies indicate that the ongoing economic downturn is worse than the 2008–2009 financial crisis, rivalling even the Great Depression of 1930 [1]. The International Monetary Fund (IMF) forecasts that per capita global GDP will fall by 4.2% by the end of 2020 [1]. Forecast forecasts show in every country this year a negative GDP rise and by the end of 2021 per capita GDP will already be smaller than in most countries before the pandemic levels [1]. In developed nations with less money and social security systems, this would hit more profoundly. Studies have found that the pandemic has contributed to more than 850 million persons—most of them in the poorer economies—collapsing into poverty [3]. At the global level, the pandemic caused both supply disruptions and lower demand. For example, 72.3% of domestic workers in 137 countries were decreased or lost their employment during the pandemic in June 2020 [4]. For instance, the average price shift of stocks was massively: −40% and −36%, respectively, in both the aviation and holiday services sectors [5]. Projections suggest that 14% of consumer spending will be at risk in the USA alone by 2020 because of physical-distancing requirements [6]. In general, the number of customer damages will be 84 billion dollars by 2023. The pandemic also threatens growing business costs, the danger of the nation and public costs, particularly in developed countries [7]. However, the magnitude and length of the COVID-19 crisis will depend on these estimates. In order to promote economic recovery after a crisis, fiscal stimulus—for example tax relief, currency exchanges and credits to vulnerable companies—is required [1]. Macroeconomic policies are intended to mitigate the magnitude to prepandemic peaks of the present economic crisis. The economic recovery is nevertheless threatened due to the fundamental unpredictability; in particular, the major factors that affect medium-term situation have been described as five potential uncertainties [8] which includes the actual number of mortality; means of herd immunity; transmission seasonality; the efficacy of action in public health; and commitment to indicators of public health.

3 Consequences of Infrastructure

In reaction to the COVID-19 [9] pandemic, the global infrastructure has become critical—from communication networks to water facilities that reduce the effect of good hygienic practises on businesses. Global economies, especially the developing economies, already had large infrastructural gaps and were unable to effectively respond to the pandemic [10]. Yet the infrastructure of the countries that had recently undergone big outbreaks allowed the COVID-19 pandemic to be controlled. South

Korea for instance has developed a successful contactability mechanism for delaying the dissemination of COVID-19, which was an epidemic of MERS for 2015 [11]. Government must determine if capital investment can be expanded in order to boost national economies or invest in other industries. Public investments decline as history has shown, as economic growth declines [10]. Moreover, the amount of public and private spending in OECD countries was well below the precrisis baseline of 2008 before the COVID-19 crisis [12]. The quality of local infrastructure has deteriorated in some countries and existing infrastructure deficits can impede the socio-economic recovery and the build-up of resilience in entire regions. Around the same time, before the pandemic, infrastructure needs were high, not only for new building programmes, but also to sustain current facilities.

Investing in modern technology, health care, transportation access, welfare and public sector adaptation to the health crisis has become a big pandemic. Nevertheless, states and foreign organizations are mobilizing to heavily spend and stimulate the economy after the pandemic—investing in industrial recovery and not creating resilient facilities. The USA and the EU have declared US$2 trillion, while the IMF Rapid Credit Facility has authorized debt relief and disbursement for more than 30 countries [13]. The concern is whether this can fix the system's inherent susceptibility to climate change and ensure that infrastructure systems are robust and more effective. In this background, policymakers will be asked to stabilize their post-crisis economies and to invest in the required stimulation field of resilience infrastructure [14]. If stabilization is not a core priority, medium-to long-term cost rises, exacerbating the vicious cycle of deficit and debt growth.

With respect to existing promises and actions, Tracker Energy Policy reports that 54% of all investments announced are carbon intensive. The meaning "clean" can only be 35%. For those countries and organizations, financial funding for renewable infrastructure has also been announced [11]. For example, GBP 1,3 billion in housing and infrastructure investments are approved in the UK to support green economic recovery while EUR 225 million from the Electricities and Growth Program of Egypt (EGGSP) have been approved by the African Development Bank Group [15].

4 Reverting Back to Better Infrastructure

The pandemic of COVID-19 supported the concept of what would be "new normal". Consumers, for example, are speeding up new channels; 75% of end-users after this pandemic are preferring to choose digital platforms. The health problem has also resulted in the widespread use of remote work; such a transition created tremendous benefit from improved organizational productivity to greater satisfaction for employees. In addition, the COVID-19 crisis has reduced global energy consumption. Movement controls have resulted in a decline in transport demand and changing usage habits, resulting in a predicted reduction of −2 to −7% in carbon emissions by 2020 [2]. However, such a reduction is estimated to be only transient and still insufficient in order to meet the Paris Climate Agreement goals, requiring a reduction of some 7 per cent per year in global carbon emissions to reduce global temperature change to 1.5 °C [16].

The COVID-19 crisis offers the potential to counter climate change while promoting economic recovery; climate-resilient infrastructure projects would play an important part in its realization. Overall, where 1% of national GDP is spent in infrastructure, the profit increases in the same year by 0.4% and in four years by 1.5% [17]. Thus, renewable infrastructure projects will foster creativity, efficiency and equity funding in key industries, such as energy and transport [18]. Sustainable infrastructures will reduce emission of GHGs and foster climate resilience. They are related to major sources of climate pollution, from electricity to transport. It is a mutually beneficial option, especially for those countries which are continuing to develop most of their basic infrastructure. It should be remembered that productive structural capability and strategies ought to maintain a sustainable system.

Many countries are expected to help COVID-19 development programmes aimed at improving economic recovery. Sustainable infrastructure in particular will produce major medium-term gains while maintaining the long-term effects of climate resilience. For example, low- and middle-income nations will benefit from investment in climate sensitive infrastructure by over US dollars in addition to US$4 trillion [17, 19]. In other words, any $1 dollar expended on green infrastructure is connected to a return of US $4. Although early capital spending for sustainable development and equivalent cost of operating can be significantly higher, studies have found the profit rises indirectly in public expense savings. For example, decreases in air emissions from transition to green energies save on healthcare costs and improvements in facilities for sanitation and waste management. Improved water drainage and climactic road networks have been shown to minimize repair costs and infrastructure distribution disruptions, including flooded highways or accidents, creating induced social savings. Studies also demonstrate that the protection of environmental sustainability requires sustainable facilities, for instance by eliminating waste rushes from roads to surface waters, which in turn increases water quality while reducing the risk of threats to human and livestock health. Sustainable infrastructure thus creates gains across markets and actors, which imply a fairer allocation of benefits that goes beyond the primary objective of delivering utilities, which is the central asset of traditional infrastructure.

Infrastructure expenditure would play an important role in helping the stabilization of economies and will also be important for climate change targets in ensuring environmental sustainability. As the OECD points out, a weak carbon economy would be related to the efficiency of post-COVID-19 investments in infrastructure projects. This allows for the long-term economic and environmental purpose of international obligations such as the Paris Convention or the Sustainable Development Goals to be balanced with short-term emergency responses. For instance, urban planning and transport pricing on the new infrastructure will reduce urban energy usage by a quarter compared to the typical business scenario, especially in Asia and Africa [20]. In combination with current infrastructure alternatives, such initiatives could reduce infrastructure-related pollution by 45–68%. The research gaps in post-COVID 19 architecture and urbanism is presented in Fig. 1.

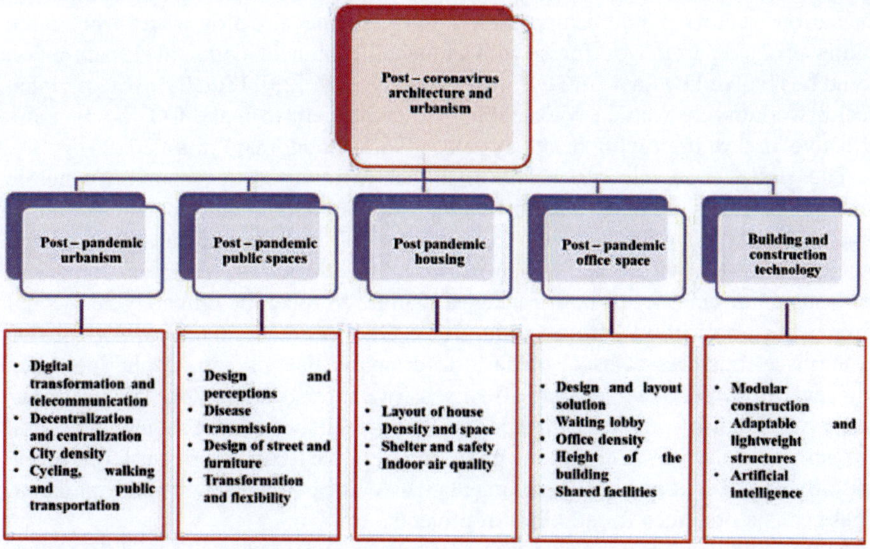

Fig. 1 Research gaps in post-COVID-19 architecture and urbanism

Integrating emerging technology into projects in infrastructure will also greatly minimize costs and increase functionality. Climate information and early warning system (CIEWS) is used to modernize national meteorological and hydrological systems to mitigate the damages incurred by disasters; for example they would yield a return on the investments that could lower costs and alleviate environmental harm, while growing revenues like 5G, artificial intelligence, cloud storage, renewables and 3D printing [21]. Innovations in the current structural and governance structures would definitely be followed by technical innovations. This is already apparent when it comes to sustainable development planning or the SDGs. Instead of formulating plans and investments to facilitate total framework optimization, the present structure of organizations contributes to sectoral/thematic/departmental optimization.

Ecosystem-based adaptation is an example of a modern approach that will involve a reflexion on governance structures (EbA). EbA, even in conjunction with infrastructure, will reduce the effect on industries of climate change [22] and needs strong coordination between local plant engineers and national bodies, and incorporation of ecological, climatic, infrastructure and local growth expertise. Hydropower producers, for example, will rely on upstream cloud forests to manufacture water while mangrove habitats guard against increasing sea levels. EbA interventions will also provide co-benefits such as carbon sequestration, water quality and better livelihoods for the environment.

However, despite their proven commitment to reducing costs of responding to climate change, even in infrastructures, EbA are seldom included in climate risk strategies. Women's advancement will also benefit from the efficiency and climatic stability of resources provided by infrastructure: the value of women as partners in the various sectors should be understood. In the perfect situation where women are taking an equal part in workforces with women, the financial deficit in climate action could be resolved by 26% of GDP worldwide by 2025 [23]. Initially, 70% of global health workers are women whose expertise could help to make COVID-19 more effective, in turn improving health systems (World Economic Forum 2020d).

The global economic and health crisis has just started to respond. Economic solution packages are being established by national governments and supranational organizations such as the European Commission. It is necessary to balance economic recovery priorities with climatic goals in order to ensure both the socio-economic and environmental viability of both packages. Failure to meet the climate goals would have major implications, from yearly global GDP decreases by up to 3% relative to the usual business-as-usual scenario to dramatic rises in catastrophe incidence and severity, which may severely affect sensitive infrastructure [24, 25]. In formulating sustainability projects, strategies must be established to assist construction and infrastructure planners. The starting point is to improve EIAs for project-level review and Strategic Environmental Assessments (SEA) for policy growth and evaluation. These can be practiced for all kinds of projects.

More initiatives are needed at the same time to incorporate EIA and SEA findings entirely into analyses of project financing. The same refers to measures of sex and fitness. This is a step forward for organizations such as the Green Climate Fund (GCF), which allows project backers to measure their project according to multiple parameters addressing social, economic, environmental and financial metrics. Instead, organizations such as IISD create frameworks and models for such evaluations to be carried out. The process and models of Sustainable Value Asset Appraisal are an example [26–28].

The solution packages and quantitative easing schemes do not involve fossil energy and carbon-intensive industries [29]. Moreover, in order to accelerate investments, environmental laws need not be relaxed. There would still have to be clarity in recovery kits. Since capital investment needs to be included in recovery efforts, the promotion of climate change and adaptation plans must also be environmentally friendly. The potential preventive measures that can be taken in the built environment are presented in Table 1.

5 Green Alternatives

Supporting renewables in particular, India has indicated consistently that it has a commitment to climate action. India was also proactive in carbon taxes, with an INR 400 per tonne coal levy and the world's largest motor fuel tariff. In recent years, subsidies for fossil fuel have been gradually reduced. However, any contribution to combating climate change needs immense financial capital. In order to follow a green transformation, the pandemic aggravated India's already limited financial place. The task is to concentrate efforts on leading industries such as storage and solar, helping achieve economic growth, social regeneration and carbon transfer objectives.

Table 1 Modifications in the built environment after the lesson learnt from COVID-19

Built environment system	Measure
Street path	Pedestrian paths can be made unidirectional, wide by narrowing the road width and increasing the area of the pedestrian path
Public equipment	Distance (safety) markers on benches. Distributing public disinfectants, washing infrastructures and local services like water and food
Public transport facilities	Protective and safety barriers can be kept for public transport drivers. Establishment of automatic doors and different doors can be kept for entry and exit. Safety distance markers can be kept between seats, at stops and in waiting rooms
Green areas and nature-based solutions	Establishing natural environment in urban regions
Information technology	Internet of things (IoT) approaches integrates the risk of new disease transmission
Exterior enclosure	Sunlight can be provided by glazed opening, orientation and open blinds. Windows can also be provided with a pleasant view
Roofing	Sufficient openings can be given for light to promote natural lighting
Interior construction and finishing	Hand—free, easy to clean (non-porous) doors and selection of materials with less virus residence time (e.g. copper)
Conveying	Safety markers in lift, unidirectional stair case with direction and distance markers can be kept
Water system	Regular disinfection of wastewater, testing of wastewater to ensure hygienic conditions
HVAC	Heat recovery unit to ensure complete separation of air with 40–60% relative humidity to reduce the residence time of virus
Waste	Specific segregation of wastes from quarantined house and properly incinerating them with strict disinfection of collecting vehicles
Furniture	Use of rugs and carpets should be avoided to reduce the accumulation of pathogens in building
Terraces, balconies and flat roofs	Minimum one outdoor area with sufficient space should be provided for fresh air and sunlight
Immediate surroundings	Natural green landscapes and health promoting landscape should be kept to facilitate the mental wellness and reduce stress

5.1 *Energy Storage Systems (ESS)*

Energy storage systems (ESS) are crucial for the Indian transition to greener technologies, whether stationary or mobile. By 2035 [30] the developed countries will make up 80% of the global market for electricity and more than two thirds of that demand will come from renewable energy sources. India's energy consumption per household is much less than the average worldwide. It gives the chance to place them on a lower carbon footprint scheme. The transformation of a rural economy will lead to green energy. With a battery storage facility, micro-mini-grid systems for remote areas will provide a viable energy source. It will be an inexpensive means of providing electricity to agriculture, schools and whole societies. It would also have employment and livelihood.

5.2 *Green Investment*

A Centre for Politics Research (CPR) report on state finances [31] shows that, also prior to the pandemic, the fiscal room in countries was already limited by structural reforms such as a GST, a central funded (CSS) consolidation system and the Ujwal DISCOM Guarantee Yojana (UDAY) scheme. Thus, their capacity to make additional investment in the midst of the pandemic's economic effect will depend on how well independent source of income they organize. The same refers to state and subregional bodies. It will be important for them to tap into independent financial outlets.

Municipal green bonds may be a way to borrow market funds. However, municipal bonds in India have only been partially effective and reflect a small share of overall municipal revenue [32]. Municipal bonds in developing countries are the primary source of funding for municipal administrations. India would need both financial and structural changes to make the municipalities and local councils more credible if these resources were to be used. The second tranche of the economic stimulus connects states' spending caps to their success to boost the budgets of their municipalities. This could provide states with an extra impetus for long-term metropolitan local authority reforms [33].

6 Strategies Required for Both Existing and for New Buildings

Many experiments have demonstrated how physical and mental well-being is affected by sensitivity levels to natural conditions. The involvement of the green elements leads to a wide array of health advantages for all age ranges as well as a significant contribution in alleviating the effects of a developed environment on the environment and enhancing the eco-climate conditions [34, 35]. One of their primary impacts is mitigating the urban atmosphere, leading to minimizing the health effects of heat waves, rising air humidity and decreasing the temperature over the hottest time cycle. Various foreign experiences have offered a potential solution to this issue and the construction of green roofs as an avenue for the growth of green spaces in urban contexts. The function of the garden, both in terms of education and well-being, is equally interesting. This practise occurs in particular in some ages to improve happiness with life, vigour and psychologic well-being, healthy interactions among individuals, a sense of community and a cognitive function [36]. It also helps to alleviate fatigue, cold, tiredness, depression and anxiety [37]. As a result, gardening is increasingly recognized both as an intervention to benefit well-being [38] and as a medicinal treatment for persons suffering from mental health issues, so-called horticultural therapy [39]. Another big construction technique is the redevelopment of the core values and archetypes of sustainable design, thermal convenience and indoor air quality (IAQ). This requires the required orientation for the optimum use of sunshine and natural heat and light.

It is also critical that natural illumination by the glass surfaces is taken into account for an adequately high degree of lighting in the room, that external views are taken into consideration and at the same time light colours used by the interior surfaces are used in order to improve reflexion [40]. In addition, home automation systems (smart homes) with suitable and programmed measures, assisted even by in situ detectors, can play a strategic role to ensure proper thermohygrometric, natural and/or mechanical ventilation and adequate IAQ, which can ensure optimal living conditions by automatic window opening for frequent air changes. The co-presence of the disease during home isolation was the greatest obstacle, since it was important to protect the other family members: it is also crucial that homes be reprocessed in compliance with safety guidelines. While it is, on the one hand, decisive to ensure a supply source of healthy for the consumed, the provision of adequately handled and delivered drinking water, on the other hand, the handling of both liquid and solid industrial waste becomes a further critical stage. Fluid waste in particular, including its potential function as a vehicle for infection, is currently being investigated by the international scientific community. As far as waste water is concerned, it should be beneficial to have isolated bathrooms cleansed and disinfected at least once a day by means of personal safety devices for persons who have suspected (or confirmed) COVID-19. If the toilets in the home are not connected to the drains, the treatment systems on the property should be assisted.

7 Conclusions

Since the financial crash of 2008, the movement for a green recovery was noted. However, green initiatives are not implemented into their stimulus packages in many nations. Many who have done so today lead the advancement of such technologies. It is time for India to focus on its incentive to cope with both the economic slowdown triggered by COVID-19 and the economic direction towards a balanced growth direction. As the countries continue to rebound from the economic effects of the COVID-19 pandemic, they are in danger of developing recovery plans with no focus on low-carbon policies. While these activities may yield short-term economic development, they may aggravate climate change and contribute to continuing problems in public finances and environment. In the global economic recovery after COVID-19, sustainable development investments will play an integral role. These ventures are intended to improve economic activity, encourage GDP growth and build jobs. Investing in infrastructure environmental protection would give countries the potential to adapt to the current economic downturn effectively while still ensuring their long-term climate commitments.

The global pandemic has exposed the restrictions on how to handle our urban environment in terms of how to plan, build and operate our built environment, but has presented us with a chance to learn. However, several concerns persist such as can these particular lessons be taken into account? If so, we can devote further attention to the advantages and further innovations of these compulsory trials so that we can choose them from a revolutionary perspective as long-term changes. In this sense, the pandemic caused decision-makers, designers and architects to reflect more, attempt to mould our physical areas and reset the current built environment, or to create more ideas to confront potential attacks on the virus. This development gives us an insight into the long-term and better direction in which our towns will change. However, it is too early to determine the effect of COVID-19 responses on the theories of architecture and urbanism. These outputs include urgently investigating our urban environment and not hoping for another pandemic to be a reminder. This strategy would go hand in hand with other sustainable practises that do not compete with natural resources or harm our climate. Our new architecture and cities continue to serve us well if we are able to handle that. However, there will be many problems in the post-pandemic period involving deeper awareness of COVID-19 and its demographic and social consequences. There is no question about the future and thus future multidisciplinary studies are expected.

References

1. Economonitor. The Impact of Coronavirus on the Global Economy. June. https://www.thestreet.com/economonitor/emerging-markets/coronavirus-global-economy (2020)
2. Le Quéré, C., Jackson, R.B., Jones, M.W., Smith, A.J., Abernethy, S., Andrew, R.M., Friedlingstein, P., et al.: Temporary reduction in daily global CO_2 emissions during the COVID-19 forced confinement. Nature Climate Change 1–7 (2020)
3. UN News: Impacts of COVID-19 Disproportionately Affect Poor and Vulnerable: UN chief. June 30. https://news.un.org/en/story/2020/06/1067502 (2020)
4. ILO: Impact of the COVID-19 crisis on loss of jobs and hours among domestic workers. International Labour Organization. June 15. https://www.ilo.org/global/topics/domestic-workers/publications/factsheets/WCMS_747961/lang--en/index.htm (2020)
5. McKinsey and Company: COVID-19 Briefing Materials: Global Health and Crisis Response, Updated March 3, 2020 (2020a). https://www.mckinsey.com//media/mckinsey/business%20functions/risk/our%20insights/covid%2019%20implications%20for%20business/covid%

2019%20march%2030/covid-19-facts-and-insights-april-3.ashx. Accessed December 20, 2020
6. S&P Global Ratings: Economic Research: A U.S. Recession Takes Hold As Fallout From The Coronavirus Spreads. https://www.spglobal.com/ratings/en/research/articles/200317-economic-research-a-u-s-recession-takes-hold-as-fallout-from-the-coronavirus-spreads-11392127 (2020)
7. McKibbin, W.J., Fernando, R.: The Global Macroeconomic Impacts of COVID-19: Seven Scenarios (2020)
8. McKinsey & Company: Safeguarding our Lives and our Livelihoods: The imperative of our time (2020b)
9. Contreras, C.C.: Sustainable Infrastructure in a Post Covid Era. June 24. https://www.ispionline.it/it/pubblicazione/sustainable-infrastructure-post-covid-era-26585 (2020)
10. Abadie, R.: COVID-19 and Infrastructure: A Very Tricky Opportunity. May 15. https://blogs.worldbank.org/ppps/covid-19-and-infrastructure-very-tricky-opportunity (2020)
11. Sharma, A., Borah, S.B., Moses, A.C.: Responses to COVID-19: the role of governance, healthcare infrastructure, and learning from past pandemics. J. Bus. Res. **122**, 597–607 (2020)
12. OECD: OECD Policy Responses to Coronavirus (COVID-19)—The territorial impact of COVID-19: Managing the crisis across levels of government. June 16. http://www.oecd.org/coronavirus/policy-responses/the-territorial-impact-of-covid-19-managing-the-crisis-across-levels-of-government-d3e314e1/ (2020)
13. IISD: Environmental, Social, and Governance Targets of the March 2020 Economic Stimulus. June 4. https://www.iisd.org/sustainable-recovery/march-2020-stimulus/ (2020a)
14. Nelson-Jones, M., Wilson, C., Bassford, H.: Partners from DLA Piper's global infrastructure team consider the future of infrastructure after the COVID-19 crisi. April 28. https://www.dlapiper.com/en/uk/insights/publications/2020/04/infrastructure-in-a-post-covid-19-world/ (2020)
15. IISD: Covid-19 Financial Response Tracker. https://docs.google.com/spreadsheets/d/1s6EgMa4KGDfFzcsZJKqwiH7yqkhnCQtW7gI7eHpZuqg/edit#gid=0 (2020b)
16. World Economic Forum: The COVID-19 Recovery can be the Vaccine for Climate Change. June 09. https://www.weforum.org/agenda/2020/06/covid-recovery-climate-and-health-hand-in-hand/ (2020a)
17. World Economic Forum: How Sustainable Infrastructure can Aid the Post-COVID Recovery. April 28. https://www.weforum.org/agenda/2020/04/coronavirus-covid-19-sustainable-infrastructure-investments-aid-recovery/ (2020b)
18. New Climate Economy: The sustainable infrastructure imperative: financing for better growth and development. In: The 2016 New Climate Economy Report (2016)
19. Hallegatte, S., Rentschler, J., Rozenberg, J.: Lifelines: The Resilient Infrastructure Opportunity. Sustainable Infrastructure. World Bank, Washington, DC (2019)
20. Creutzig, F., Agoston, P., Minx, J.C., Canadell, J.G., Andrew, R.M., Le Quéré, C., Peters, G.P., Sharifi, A., Yamagata, Y., Dhakal, S.: Urban infrastructure choices structure climate solutions. Nature Climate Change **6**(12), 1054–1056 (2016)
21. WMO: Valuing Weather and Climate: Economic Assessment of Meteorological and Hydrological Services. Geneva (2015)
22. Goldstein, A., Turner, W.R., Gladstone, J., Hole, D.G.: The private sector's climate change risk and adaptation blind spots. Nature Climate Change **9**(1), 18–25 (2019)
23. World Economic Forum: We Can Solve Climate Change—If We Involve Women. September 16. https://www.weforum.org/agenda/2019/09/why-women-cannot-be-spectators-in-the-climate-change-battle/ (2019)
24. Dellink, R., Lanzi, E., Chateau, J.: The sectoral and regional economic consequences of climate change to 2060. Environ. Resource Econ. **72**(2), 309–363 (2019)
25. IPCC: Summary for Policymakers. In: Global Warming of 1.5°C. Intergovernmental Panel on Climate Change (2018)
26. Bassi, A.M., Pallaske, G., Stanley, M.: An Application of the Sustainable Asset Valuation (SAVi) Methodology to Pelly's Lake and Stephenfield Reservoir, Manitoba, Canada. IISD, Geneva (2019)

27. Bassi, A.M., Casier, L., Pallaske, G., Perera, O., Uzsoki, D.: The Sustainable Asset Valuation of the Southern Agricultural Growth Corridor of Tanzania (SAGCOT) Initiative: A focus on irrigation infrastructure. Geneva: International Institute for Sustainable Development (IISD). https://www.iisd.org/sites/default/files/publications/savi-tanzania-sagcot-initiative.pdf (2018)
28. Bassi, A.M., Perera, O., Wuennenberg, L., Pallaske, G.: Lake Dal in Srinagar, India: Application of the Sustainable Asset Valuation (SAVi) methodology for the analysis of conservation options. Geneva: International Institute for Sustainable Development with support of the MAVA foundation, EMSD and GIZ (2018)
29. Greenpeace: Coronavirus Recover: A Free Ride for Polluters (2020). https://www.google.com/search?q=coronavirus+recover+a+free+reide+for+polluters&oq=coronavirus+recover+a+free+reide+for+polluters&aqs=chrome..69i57j33.5658j0j7&sourceid=chrome&ie=UTF-8
30. Sawhney, A.: Battery energy storage systems in India: New kid on the block. In: Energy World, 22 March https://energy.economictimes.indiatimes.com/energy-speak/battery-energy-storage-systems-in-india-new-kid-on-the-block/3487#:~:text=On%208%20May%202020%2C%20ReNew,%2Dthe%2Dclock%20power%20supply (2019)
31. Kapur, A., Pandey, S., Ranjan, U., Irava, V.: Study of State Finances 2020–21 (Provisional), Centre for Policy Research, 10 May 2020. https://www.cprindia.org/research/papers/study-state-finances-2020-21-provisional-0 (2020)
32. Ahluwalia, I.J., Mohanty, P.K., Mathur, O., Roy, D., Khare, A., Mangla, S.: State of Municipal Finances in India. ICRIER, March 2019, https://fincomindia.nic.in/writereaddata/html_en_files/fincom15/StudyReports/State%20of%20Municipal%20Finances%20in%20India.pdf (2019)
33. Sarma, N:. Green Recovery: Opportunities for India (2020)
34. Appolloni, L., Corazza, M.V., D'Alessandro, D.: The pleasure of walking: an innovative methodology to assess appropriate walkable performance in urban areas to support transport planning. Sustainability **11**(12), 3467 (2019)
35. Engemann, K., Pedersen, C.B., Arge, L., Tsirogiannis, C., Mortensen, P.B., Svenning, J.C.: Residential green space in childhood is associated with lower risk of psychiatric disorders from adolescence into adulthood. Proc. Natl. Acad. Sci. **116**(11), 5188–5193 (2019)
36. Soga, M., Gaston, K.J., Yamaura, Y.: Gardening is beneficial for health: a meta-analysis. Preventive Med. Rep. **5**, 92–99 (2017)
37. Wood, C.J., Pretty, J., Griffin, M.: A case–control study of the health and well-being benefits of allotment gardening. J. Public Health **38**(3), e336–e344 (2016)
38. Clatworthy, J., Hinds, J., Camic, P.M.: Gardening as a mental health intervention: a review. Mental Health Rev. J. (2013)
39. Gonzalez, M.T., Hartig, T., Patil, G.G., Martinsen, E.W., Kirkevold, M.: Therapeutic horticulture in clinical depression: a prospective study of active components. J. Adv. Nurs. **66**(9), 2002–2013 (2010)
40. Berto, R., Barbiero, G., Pasini, M., Pieter, U.: Biophilic design triggers fascination and enhances psychological restoration in the urban environment. J. Biourbanism **1**, 27–34 (2015)
41. Pinheiro, M.D., Luís, N.C.: COVID-19 could leverage a sustainable built environment. Sustainability **12**(14), 5863 (2020)

Vulnerability and Impact Assessment of Extreme Climate Events in the Greek Oil Industry

Theodoros Katopodis, Athanasios Sfetsos, and Emmanuel D. Adamides

1 Introduction

Extreme weather events are expected to increase in frequency and intensity under the prospect of climate change [1], exposing oil infrastructure and their operations at risk, leading to more devastating Natech (natural disasters triggering technological accidents) events in the future. Midstream to downstream oil services have been designed to withstand climate change pressures over the facility's design working life (DWL), up to specific design thresholds. However, frequently extreme events may exceed these thresholds, and thus adaptation strategies are necessary to enhance their resilience. Climate change is anticipated to modify the distribution of the extreme values of climatic actions, and hence to impact the design values of climatic parameters, leading to the increase of the mean or maximum values [2]. Structures and infrastructures in Europe have already set their DWL at 50 years, based on an annual probability of being exceeded of 0.02, according to the Eurocodes methodology [3]. However, recent approaches highlight the need for the redesign or improvement of the defenses of the infrastructures and structures to withstand the extreme hazards and loads expected in 50 years ahead, taking into account the climate change effect [2]. The concept of return period is commonly used to carry information about the likelihood of rare and extreme events, and it is the main element for the development of a risk assessment tool as it overall includes the collection and processing of all the associated climate information [4]. Hence, adaptation of critical infrastructures operations under the future conditions is necessary to incorporate the establishment of design thresholds and risks to the impacts of climate change over the life span of assets. The exposure of the oil infrastructure to climate change at a global level has

T. Katopodis (✉) · A. Sfetsos
National Center of Scientific Research NCSR "Demokritos", Athens, Greece
e-mail: tkatopo@ipta.demokritos.gr

T. Katopodis · E. D. Adamides
Department of Mechanical Engineering and Aeronautics, University of Patras, Patras, Greece

I. Pal and S. Kolathayar (eds.), *Sustainable Cities and Resilience*, Lecture Notes in Civil Engineering 183, https://doi.org/10.1007/978-981-16-5543-2_6

recently been studied in [5] through the collection of the critical event parameters in a hazard threshold matrix. However, due to the idiosyncratic characteristics of local environments, specific regional climate change analysis with the use of very high-resolution climate data is required in order to analyze the local conditions, opportunities, and challenges for each facility [6].

This work highlights the need for the development of a comprehensive approach to assess the vulnerability of the Hellenic Petroleum (HELPE) facilities, located in the western part of Attica, near the Megara Gulf, in Greece, for the future non-stationary climate change conditions. In the context of this study, the regional analysis focused on the Megara installations and Elefsina/Aspropyrgos refineries (hereafter installations), which are broadly located in a region that experienced extreme events in the past [7]. It is the first time that a systematic investigation of the climate change impacts to the Greek oil industry is undertaken and constitutes the first step in identifying the vulnerability impact of the most exposed "assets" (employees, processes, and structures), in a three-level analysis, which can provide further guidance for the resilience capacities of oil infrastructure.

This study is organized as follows. Sect. 2 suggests the recommended research methodology for the vulnerability and impact analysis. Sect. 3 presents the validation of the high-resolution WRF climate model simulations, the projections of future climate conditions at HELPE facilities and the return periods. Sect. 4 performs the vulnerability analysis of the most exposed processes, including structures and employees of the Greek oil industry. Finally, conclusions are introduced in Sect. 5.

2 Methodology

The overall methodology of the vulnerability and impact assessment of extreme climate events in the Greek oil industry, under the prospect of future climate change impacts, is presented in Fig. 1. The proposed approach is built on the use of the high-resolution Advanced Weather Research and Forecasting (WRF-ARW) (v3.6.1) climate model [8], forced by EC-EARTH (a European community Earth-System Model), which has been dynamically downscaled to the domain of Greece, with grid spacing of $5 \times 5\ \text{km}^2$ [9, 10]. Climate simulations with the WRF model were performed for the historic period (1980–2004), and were used for model validation purposes, and the future (2025–2049), for the two RCPs (4.5 W/m^2 and 8.5 W/m^2).

Most of the exposed critical services of the upstream, midstream, and downstream services, in a global analysis, can been found in [5] (Table 1) as well in publications of relevant projects in the field [11, 12]. Under the prospect of the present study, we focused on the most exposed assets/processes of the Greek oil downstream and midstream sector (HELPE facilities), which are located in the western part of Attica, as shown in Fig. 2. Additionally, the most characteristic site-specific climate hazards that are considered in this study are presented in Fig. 1. These are used to provide an estimation of the likelihood of the climate-induced risks to infrastructures and contribute to the identification of the operational or structural thresholds.

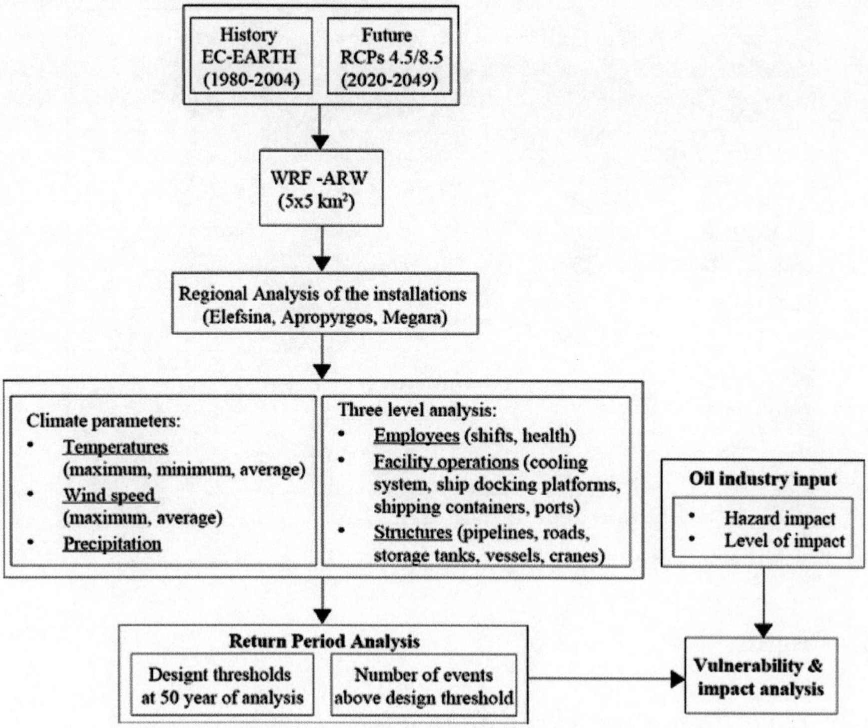

Fig. 1 Methodology of vulnerability assessment of extreme climate events in the Greek oil industry

Table 1 Statistical indices between model simulations and observations from the HNMS meteorological station for 300 months during the 25 years comparison

	Bias	RMSE	r	r_p-value	MAE	IOA
TX (°C)	−0.96	2.57	0.95	0	2.03	0.97
TN (°C)	−0.25	2.01	0.952	0	1.58	0.974
TG (°C)	−1.85	2.81	0.958	0	2.36	0.96
Precipitation (mm)	27.7	94.5	0.3	0	41.2	0.33
Wind speed (m/s)	0.3	1.89	0.088	0.126	1.46	0.435

In principle, these are the climate hazards that impact directly and indirectly the employees/processes/structures of the oil sector. In addition, a summary of the most exposed critical processes to the representative climate parameters, with new quantitative critical values (thresholds) is presented, based on a 50 year of return period analysis derived from the examined scenarios. Then, the climate change vulnerability analysis of the most exposed assets of the HELPE facilities has been proposed, by identifying the vulnerability of employees, processes, structures in a three-level analysis.

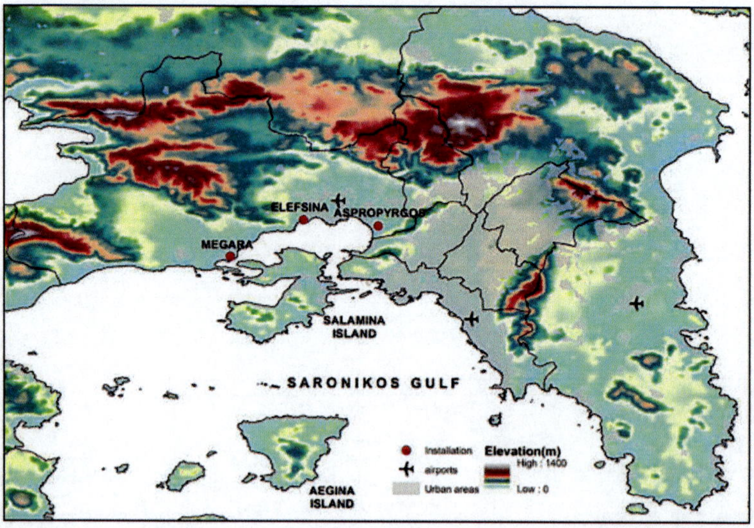

Fig. 2 Indicative map of the studied installations over the simulation domain of 5 km

3 Results

3.1 High-Resolution Model Simulations Against Observations

For the purpose of this study, the implemented WRF regional climate model was assessed with the use of the "nearest neighbor" to the position of the monitoring stations method, by comparing the WRF simulations for the historical period 1980–2004, with the most recent available meteorological dataset (1980–2004) validated by the Hellenic National Meteorological Service (HNMS) of the nearest station at Elefsina's airport. The capability of the WRF to identify satisfactorily the present climate conditions in Greece has already been evaluated in the studies of [9, 10, 13, 14]. The statistical analysis for the validation procedure was performed using monthly values (300 over the 25 years period) of temperatures—maximum (TX), minimum (TN), average (TG), monthly mean wind speeds, and monthly accumulated precipitation values. In Table 1, we present the performance of these simulations when compared to the surface observations from the nearest HNMS station. The statistical significance of the Pearson correlation (r) using t-Test at the 95% confidence interval is also depicted in the table.

The analysis yields that having greater correlation coefficients (r) with high significance of results (above 0.95), temperatures (TX/TN/TG) are better simulated with the WRF model compared to the other climate parameters. As a general comment, it could be said that the WRF model underestimates the temperatures of the historic period. Within them, the minimum temperatures dataset has the lowest errors (mean bias and root mean square error (RMSE)) in (°C) and the highest correlation. Also, the index of agreement (IOA), as a measure of the degree of model prediction error based on [15], indicates an almost perfect match, with values in the range of above 0.974. The statistical analysis of the mean monthly wind speeds and precipitation values yielded a low correlation associated with changes of low statistical significance. The results agree with results of other similar studies [9, 10, 13], considering that Greece is a country with a very complex terrain, thus climate simulations still remain a challenging task [16]. However, analysis yielded that the mean monthly precipitation for the specific domain cannot be simulated well with model, indicating low correlation and very high bias errors.

A graphical representation of the probability density functions (PDF) of the WRF model simulations and HNMS observational dataset is presented in Fig. 3. Regarding wind speed (Fig. 3a), WRF simulations correspond to wind speeds with smaller standard deviation, overestimating the frequency of appearance of lower wind speeds values (5–10 m/s) and underestimating the higher ones (23–27 m/s). For precipitation (Fig. 3a), the model has performance close to observations when filtering the lowest values (<0.5 mm). Also, WRF has a tendency to predict more intense precipitation events in comparison to observations in the particular area. Both findings are in accordance to [13], which states that precipitation still remains a very difficult meteorological parameter to estimate locally.

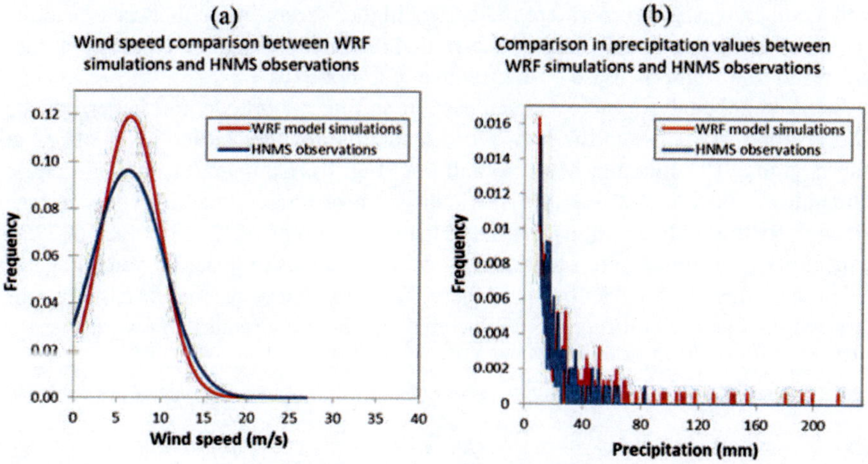

Fig. 3 Comparison of WRF model simulations and HNMS observations in the monthly wind speeds and precipitation values due to changes in the probability density functions (PDF)

3.2 Projections of Future Climate Conditions at HELPE Facilities

This section presents the projected changes in temperature characteristic values (TX, TN, TG), in mean and maximum winds (WSavg and WSmax) and in precipitation daily accumulated values, as a result of climate change, for the studied region in Greece (2025–2049). For every yearly dataset of the climate parameters and for each installation (at Elfsina, Aspropyrgos, and Megara), which corresponds to different grid-cell points, daily values of the maximum, minimum, and average air temperatures, mean and maximum wind speeds, and daily accumulated precipitation values were extracted for the historic and future periods. A continuous probability distribution developed with extreme value theory (GEV) [17] was fitted to the 25 year datasets. At the end, the differences between the present and future threshold values with a mean return period of 50 years were computed (Sect. 3.3). The method for calculating return periods and the probability of exceeding a value is based on the extreme value theory (EVT), as in [18]. The variations in the characteristic values of TX, TN, and TG between future and historic period are presented in Fig. 4 and illustrates the expected variations on average temperature distribution, showing how climate change cannot only induce a shift in the distribution but also a change in its shape. Dashed lines in Fig. 4 represent the probability density of the historic conditions that will be shifted to the continuous lines in the future. A flattened distribution with a higher average temperature indicates a higher variability with increases in the frequency of extremes. It is, hence, shown that the increase in the mean temperature will result in a slight increase of the frequency of hot days.

On average, analysis indicated increase of similar magnitude and pattern of the TX, TN, and TG between the two future scenarios, for the studied installations at Aspropyrgos, Elefsina, and Megara facilities (Fig. 4). Compared to mean historic climatology for TX and TG, the WRF model near-term climatology percentage changes, for the studied regions, are found to be in the range of 12–13% and 16–22%, respectively, for both future scenarios. However, analysis indicated a percentage of TN in the range of 25–28% for the Megara, Elefsina, and Aspropyrgos installations between the historic and future RCPs scenarios. The most remarkable variations of

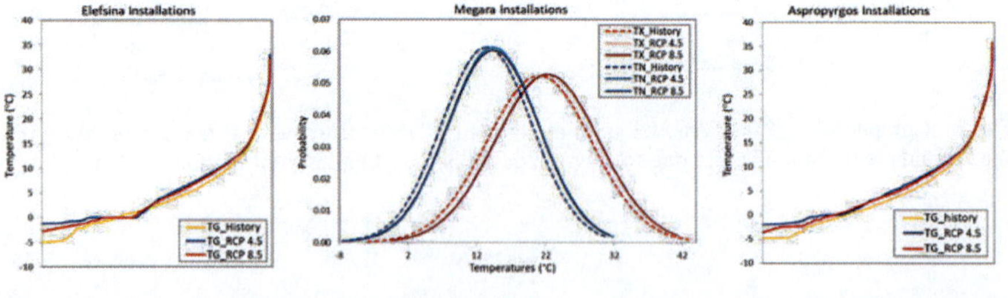

Fig. 4 Expected changes in the mean value and variance in maximum and minimum temperatures under the RCP 4.5 and RCP 8.5 future scenarios due to changes in the probability density functions (PDF)

TX (>1.3 °C), for RCP 4.5 and RCP 8.5, respectively, are estimated in all the studied regions, and of TG in Aspropyrgos and Megara (Table 2).

Future analysis showed that increase in the wind speeds for the future scenarios are predicted to be in the range of 5%, most of them occurring in the Aspropyrgos installation. Moreover, analysis indicated that RCP 8.5 is considered to be the scenario with the highest expected maximum and average winds in comparison to RCP4.5 and historic scenarios. Concerning the daily accumulated precipitation values, future analysis projected the decrease to be in the range of 10–25%, for all the installations over the future scenarios. The highest decrease is projected for the Elefsina facilities under the RCP 8.5 future scenario. However, the highest daily accumulated precipitation (PR) values in the range of 150 mm have been recorded only during the historic period, in the installations of Aspropyrgos and Elefsina.

3.3 Analysis of 50 Years Return Period

In this study, climate analysis determined the trends in extreme events and the return period of the most severe impacts to HELPE facilities. As it was already mentioned, climate change might change the design values of climatic and environmental actions, resulting in increase in the mean or max values. For instance, [19] found that projected changes due to greenhouse effect, might bring about significant changes in mean temperatures, especially on the extreme tails of the PDF. As the need for the redesign or improvement of the defenses of the infrastructure to withstand extreme hazards and loads expected in 50 years ahead it is of high priority [2], in the context of climate change, it is considered noteworthy to calculate the design thresholds based on a return period analysis of 50 years. The determination of the return periods necessitates the choice and determination of the underlying distribution function. The so-called block maximum method has been used, where the sample of extreme values is obtained by selecting the maximum or minimum of the daily values observed during a year based on [20]. Threshold analysis indicated a significant increase in TG and TX design thresholds for the installations studied, under the RCP 4.5 scenario, while major increases are expected to occur concerning the TN values (Table 3). Moreover, an increase in the WSmax thresholds of about 1.0 m/s is projected to occur in the installations of Elefsina, under the RCP 8.5 scenario. Besides, Heat index (HT) [21] threshold analysis indicated an increase during the return period of 50 years for the installations of Elefsina and Aspropyrgos. On the contrary, Wind chill [22] indicator analysis, recorded a significant increase for the installations of Aspropyrgos, under the future scenarios, which might be related to the projected changes in the variables of temperatures, relative humidity, and wind speeds over the future period.

Table 2 Statistical comparison between WRF climatology for the particular region

	History (1980–2004)						RCP 4.5 (2025–2049)						RCP 8.5 (2025–2049)					
	TX		TN		TG		TX		TN		TG		TX		TN		TG	
	m	St.dev	m	St.dev	m	St.dev	m	St.dev	m	St.dev	m	St. dev	m	St. dev	m	St.dev	m	St. dev
Aspropyrgos	21.26	7.69	14.33	6.66	17.08	7.07	22.55	7.62	15.48	6.60	18.3	7.03	22.49	7.61	15.41	6.66	18.2	7.05
Elefsina	19.34	6.77	13.71	6.01	15.94	6.29	20.59	6.58	14.84	5.92	17.1	6.24	20.56	6.59	14.80	5.98	17.1	6.26
Megara	21.10	7.60	13.47	6.51	16.51	7.0	22.45	7.58	14.60	6.56	17.7	7.08	22.42	7.56	14.58	6.61	17.7	7.09

Table 3 Thresholds at a return period of 50 years analysis, based on the maximum or minimum of the daily values observed during a year, between historic and future periods for the studied installations

Precipitation (mm)		Threshold at return period of 50 (y)			TG (°C)		Threshold at return period of 50 (y)		
Period	Scenarios	Megara	Elefsina	Aspropyrgos	Period	Scenarios	Megara	Elefsina	Aspropyrgos
History (1980–2004)	Historic	135.4	179.4	169.2	History (1980–2004)	Historic	37.0	33.8	36.2
Future (2025–2049)	RCP 4.5	95.5	122	124.8	Future (2025–2049)	RCP 4.5	37.3	35.0	37.3
	RCP 8.5	95.4	127.7	110.7		RCP 8.5	37.1	33.8	37.5
TX (°C)		Threshold at return period of 50 (y)			**TN (°C)**		Threshold at return period of 50 (y)		
History (1980–2004)	Historic	45.4	39.3	43.6	History (1980–2004)	Historic	−10.6	−9.4	−10.9
Future (2025–2049)	RCP 4.5	44.6	39.7	44.3	Future (2025–2049)	RCP 4.5	−5.4	−5.1	−6.4
	RCP 8.5	43.6	39.2	43.7		RCP 8.5	−9.9	−9.1	−9
WSmax (m/s)		Threshold at return period of 50 (y)			**Wind chill (°C)**		Threshold at return period of 50 (y)		
History (1980–2004)	Historic	30.5	29.9	25.4	History (1980–2004)	Historic	−25.3	−24.1	−12
Future (2025–2049)	RCP 4.5	29.6	27.8	22	Future (2025–2049)	RCP 4.5	−15.3	−15.3	−16
	RCP 8.5	30	30.8	25.4		RCP 8.5	−21.9	−21.8	−21.6

4 Vulnerability and Impact Analysis

As it is already mentioned, recently a first step to identify the exposure and to evaluate the vulnerability of the oil industry to extreme weather events, on a global scale, has been performed by [5]. In this section, we try to address the impacts of the climate hazards in the installations studied as found from the literature and as validated and contributed by the HELPE organization (Table 4). The proposed approach performs an expansion of the climate change vulnerability assessment of the most exposed assets of the HELPE installations, by identifying the vulnerability of employees, processes, and structures, in a three-level analysis. With the objective to quantify the changing frequency of extremes for the future period of 2025–2049, we also present in Table 4 the return periods of the climatological parameters the number of events above their critical design threshold and their impacts in case of exceedances, as found in literature (scientific journals, international guidelines, etc.,) and as provided by HELPE installations via internal discussions and technical documentations.

5 Conclusions

This study carried out an improved climate change vulnerability and impact assessment due to climate change of the most exposed assets of the HELPE facilities located in the western part of Attica, by identifying the vulnerability of employees, processes, and structures in a three-level analysis. In addition, the study highlighted the need for the redesign or improvement of the defenses of the oil infrastructure, taking into consideration the prospect of climate change to withstand extreme hazards and loads expected in 50 years ahead. In the study, the implemented regional climate model WRF was assessed with the most recent available meteorological dataset (1980–2004) validated by the Hellenic National Meteorological Service (HNMS) of the nearest station of Elefsina's airport. The high-resolution regional climate downscaled WRF model indicated increase in the values of temperatures (TX, TN, TG), slight increase in the wind speeds (max, average), and decrease for the daily accumulated precipitation values in the mid-twenty-first century (2025–2049). Moreover, threshold analysis during the return period of 50 years indicated a slight increase in TX design thresholds for the installations of Aspropyrgos and Elefsina under the RCP 4.5 future scenario, while major increases are expected to occur to all the studied locations, concerning the TN values. Also, TG design threshold for the installation of Aspropyrgos is projected to increase, by more than 1.0 °C, under the future scenarios. Additionally, an increase in the WSmax thresholds of about 1.0 m/s is projected to occur in the installations of Elefsina, under the RCP 8.5 scenario, while on the contrary, a significant increase in the Wind chill [22] threshold analysis by the order of -10 °C is projected to occur in the installations of Aspropyrgos, under the future scenarios. It is the first time a systematic investigation of the climate change impacts to the Greek oil industry was carried out, and it is a first step to identify

Table 4 Employees, processes, and structures exposure in a three-level analysis under the climate change context for the historic (1980–2004) and future (2025–2049) periods

Exposure	Sub-asset	Design threshold	Return period (years)			Number of events above threshold			Impact
			History	RCP4.5	RCP8.5	History	RCP4.5	RCP8.5	
Employees	Shifts and health	HT = 41 °C	7	5.5	5.9	2	8	10	Possible stroke, heat cramps, or heat exhaustion [23]
		WSavg = 17 m/s	0.9	0.91	1.1	88	91	76	Review any work at height and outdoor working when the wind speed reaches >17 m/s to ensure it is safe to continue [24]
Processes	Street traffic	PR = 100 mm	5.49	14.23	17.22	6	1	1	Reducing visibility and causing flooding of passageways [25]
	Shipping ports	WSmax = 26 m/s	11.02	21.1	10.91	1	1	2	Limit tanker approaches, interrupt fuel deliveries, postpone pumping, or interrupt construction and maintenance [26]

(continued)

Table 4 (continued)

Exposure	Sub-asset	Design threshold	Return period (years)			Number of events above threshold			Impact
			History	RCP4.5	RCP8.5	History	RCP4.5	RCP8.5	
	Transit operations	WSmax = 22 m/s	1.69	1.58	2.33	33	45	29	Severe risk and significant impact to safety and transit operations [26, 27]
	Ship docking platform	WSavg = 12 m/s	0.65	0.71	1.00	99	100	83	Ships might not be able to approach[a]
	Water/ wastewater biological treatment	TX = 45 °C	41.89	60.66	113.72	0	0	0	Treatment efficiency decreases by reducing bacterial floc formulation[a]
	Cooling system	TG = 32 °C	2.71	1.58	1.83	34	63	60	The cooling capacity to cool down the cooling water is decreased[a]. In the past, HELPE installations decreased their activity level[a]
		TX = 38 °C	2.22	1.52	1.81	29	57	60	
	Refinery processes	TN = −5 °C	6.09	26.38	12.08	7	0	2	Extreme low temperatures might impact the refinery processes. In the past, HELPE installations decreased their activity level[a]

(continued)

Table 4 (continued)

Exposure	Sub-asset	Design threshold	Return period (years)			Number of events above threshold			Impact
			History	RCP4.5	RCP8.5	History	RCP4.5	RCP8.5	
Structures	Cooling tower	RH = 90%	1.13	1.54	1.74	65	33	31	The cooling capacity of the tower might be decreased[a]
	Roads	PR = 150 mm	22.06	169.1	117.8	1	0	0	Road constructions might be a failure, and roads might be inundated. In 2007, refinery faced an extreme precipitation event (with associated effects in its activity levels), which caused irregular increases in the water level with the inundation depth in the facilities > 1.5 m[a]

(continued)

Table 4 (continued)

Exposure	Sub-asset	Design threshold	Return period (years)			Number of events above threshold			Impact
			History	RCP4.5	RCP8.5	History	RCP4.5	RCP8.5	
	Cranes	WSmax = 20 m/s	1.10	1.21	1.65	68	65	46	High risk might occur with a loss of asset [28]
	Storage tanks								Failure of the insulation of the storage tank roof might occur[a] [29]
	Vessel	TN = 0 °C	1.02	2.72	2.29	84	41	63	Low temperatures events below 0 °C might be responsible for freezing [30]

HELPE installations' info and technical contribution

the risk of the Greek oil industry, which can provide further guidance related to the resilience capacities of the HELPE organization, so that in the long term, it adapts to future climate conditions.

Acknowledgements This work was supported by computational time granted from the Greek Research &Technology Network (GRNET) in the National HPC facility—ARIS—under projects ID HRCOG (pr004020) and HRPOG (pr006028). We thank also the Hellenic National Meteorological Service for providing all the necessary observational data sets.

Funding This publication is supported by the program of Industrial Scholarships of Stavros Niarchos Foundation and Hellenic Petroleum Group Grant.

References

1. Forzieri, G., Bianchi, A., Silva, F.B., Marin Herrera, M.A., Leblois, A., Lavalle, C., Aerts, J.C.J.H., Feyen, L.: Escalating impacts of climate extremes on critical infrastructures in Europe. Glob. Environ. Change. **48**, 97–107 (2018). https://doi.org/10.1016/j.gloenvcha.2017.11.007
2. Athanasopoulou, A., Sousa, M.L., Dimova, S., Rianna, G., Mercogliano, P., Villani, V., Croce, P., Landi, F., Formichi, P., Markova, J.: Thermal design of structures and the changing climate. Publications Office, Publications Office of the European Union, Luxembourg (2020)
3. CEN: Eurocode—Basis of structural design (1990)
4. EU-Circle: D3.4 Holistic CI Climate Hazard Risk Assessment Framework (2016)
5. Katopodis, T., Sfetsos, A.: A review of climate change impacts to oil sector critical services and suggested recommendations for industry uptake. Infrastructures. **4**, 74 (2019). https://doi.org/10.3390/infrastructures4040074
6. Gething, B.: Design for future climate. Opportunities for Adaptation in the Built Environment. Technology Strategy Board, UK (2010)
7. Arizona State University: WMO Region VI (Europe, Continent only): Highest Temperature. https://wmo.asu.edu/content/europe-highest-temperature
8. Skamarock, W.C., Klemp, J.B., Dudhia, J., Gill, D.O., Barker, D.M., Duda, M.G., Huang, X.-Y., Wang, W., Powers, J.G.: A Description of the Advanced Research WRF. , National Center for Atmospheric Research Boulder, Colorado, USA (2008)
9. Politi, N., Nastos, P.T., Sfetsos, A., Vlachogiannis, D., Dalezios, N.R.: Evaluation of the AWR-WRF model configuration at high resolution over the domain of Greece. Atmos. Res. **208**, 229–245 (2018). https://doi.org/10.1016/j.atmosres.2017.10.019
10. Politi, N., Sfetsos, A., Vlachogiannis, D., Nastos, P.T., Karozis, S.: A sensitivity study of high-resolution climate simulations for Greece. Climate. **8**, 44 (2020). https://doi.org/10.3390/cli8030044
11. Cruz, A.M., Krausmann, E.: Vulnerability of the oil and gas sector to climate change and extreme weather events. Clim. Change **121**, 41–53 (2013)
12. Parry, M.L., O.F. Canziani, J.P. Palutikof, P.J. van der Linden, C.E. Hanson eds: Contribution of Working Group II to the 4th Assessment Report of the Intergovernmental Panel on Climate Change. Cambridge University Press (2007)
13. Emmanouil, G., Vlachogiannis, D., Sfetsos, A.: Exploring the ability of the WRF-ARW atmospheric model to simulate different meteorological conditions in Greece. Atmos. Res. **247**, 105226 (2021). https://doi.org/10.1016/j.atmosres.2020.105226
14. Katopodis, T., Markantonis, I., Politi, N., Vlachogiannis, D., Sfetsos, A.: High-resolution solar climate atlas for Greece under climate change using the weather research and forecasting (WRF) model. Atmosphere **11**, 761 (2020). https://doi.org/10.3390/atmos11070761

15. Willmott, C.J., Robeson, S.M., Matsuura, K.: A refined index of model performance. Int. J. Climatol. **32**, 2088–2094 (2012). https://doi.org/10.1002/joc.2419
16. Wang, Y., Yang, K., Zhou, X., Chen, D., Lu, H., Ouyang, L., Chen, Y., Lazhu, Wang, B.: Synergy of orographic drag parameterization and high resolution greatly reduces biases of WRF-simulated precipitation in central Himalaya. Clim Dyn. **54**, 1729–1740 (2020). https://doi.org/10.1007/s00382-019-05080-w.
17. Coles, S.: An Introduction to Statistical Modeling of Extreme Values. Springer London, London (2001). https://doi.org/10.1007/978-1-4471-3675-0
18. Holmes, J.D.: Wind Loading of Structures (2017)
19. AghaKouchak, A., Easterling, D., Hsu, K., Schubert, S., Sorooshian, S. eds: Extremes in a Changing Climate. Springer Netherlands, Dordrecht (2013). https://doi.org/10.1007/978-94-007-4479-0
20. Jenkinson, A.F.: The frequency distribution of the annual maximum (or minimum) values of meteorological elements (1955)
21. National Weather Service: The Heat Index Equation. https://www.wpc.ncep.noaa.gov/html/heatindex_equation.shtml
22. National Weather Service: Wind Chill. https://www.weather.gov/epz/wxcalc_windchill
23. Opitz-Stapleton, S., Sabbag, L., Hawley, K., Tran, P., Hoang, L., Nguyen, P.H.: Heat index trends and climate change implications for occupational heat exposure in Da Nang. Vietnam. Climate Serv. **2–3**, 41–51 (2016). https://doi.org/10.1016/j.cliser.2016.08.001
24. OSHAD: Dealing with adverse weather conditions **24** (2017)
25. Chinowsky, P.S., Price, J.C., Neumann, J.E.: Assessment of climate change adaptation costs for the U.S. road network. Glob. Environ. Change. **23**, 764–773 (2013). https://doi.org/10.1016/j.gloenvcha.2013.03.004
26. OFCM: Office of the Federal Coordinator for Meteorological Services and Weather Information for Surface Transportation: National Needs Assessment Report. U.S. Department of Commerce, National Oceanic and Atmospheric Administration (NOAA). Office of the Federal Coordinator for Meteorological Services and Supporting Research (2002)
27. Garg, A., Naswa, P., Shukla, P.R.: Energy infrastructure in India: Profile and risks under climate change. Energy Policy **81**, 226–238 (2015). https://doi.org/10.1016/j.enpol.2014.12.007
28. HSE: The effect of wind loading on the jib of a luffing tower crane. Prepared by the Health and Safety Laboratory for the Health and Safety Executive. Health and Safety Executive (2012)
29. Olivar, O.J.R., Mayorga, S.Z., Giraldo, F.M., Sánchez-Silva, M., Pinelli, J.-P., Salzano, E.: The effects of extreme winds on atmospheric storage tanks. Reliab. Eng. Syst. Saf. **195**, 106686 (2020). https://doi.org/10.1016/j.ress.2019.106686
30. IFC: Climate Risk and Business Practical Methods for Assessing Risk (2010).

Catastrophic Events at the River Basins Due to Permafrost Thawing: Review and Examples

Elena Dolgopolova

1 Introduction

Arctic coastal development and climate warming demand research of rapidly changing properties of permafrost under modern conditions. A considerable portion of the Arctic Ocean coast in Russia, Canada and the USA (Alaska Peninsula) lies in the permafrost zone. Permafrost occupies about 24% of the land (an area of 22.8 million km^2), including more than 60% of the Russian territory (Fig. 1). Increase of the global air temperature [1] is followed by climate changes, leading to changes in precipitation, melting of sea ices and glaciers, permafrost thaw and increase of water flux of pan-Arctic rivers. The territories of watersheds and mouths of the rivers of Siberia, Alaska and Canada are located in permafrost of different properties (continuous, discontinuous and sporadic). Air and frozen grounds temperature monitoring in Russia shows increase in 1990–2000 [2] in comparison with normal values in 1961–1990. Approximate forecast gives increase of ground temperature in continuous permafrost zone, which is still staying negative, and 100 km shift of the south permafrost boundary to the North about by 2025. About 80% of Alaska is underlain by permafrost [3]. Thaw of permafrost causes numerous thermokarst terrain. Thin permafrost layer thawing from the bottom to the surface results in permafrost disappearance. Investigation of permafrost degradation in central Alaska (Tanana Flats) shows the area of no permafrost increases from 39 to 47% during the period 1949–1995 [3]. Frozen grounds of sporadic and discontinuous permafrost zones at the eastern coast of Hudson Bay in Canada are also rapidly disappear [4]. At the Mackenzie Delta which is mostly underlain by continuous permafrost of the thickness less than 100 m, the annual air temperature increases from 1970 by about 2.5°C [5]. For years 1983–2008, mean active layer depth measured at 12 stations along the delta showed the increase from 42 to 50 cm.

E. Dolgopolova (✉)
Water Problems Institute, Russian Academy of Sciences, Gubkina str. 3, 119333 Moscow, Russia

I. Pal and S. Kolathayar (eds.), *Sustainable Cities and Resilience*, Lecture Notes in Civil Engineering 183, https://doi.org/10.1007/978-981-16-5543-2_7

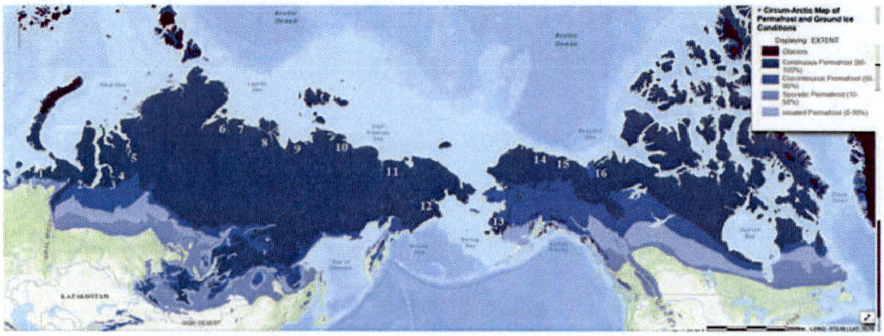

Fig. 1 Circum-Arctic map of permafrost and ground ice conditions from national snow and ice data (Last modified May 12, 2011) [6] with river mouths listed in Table 1

Permafrost thawing and growth of active layer depth assisted by increase of surface temperature contribute to water river fluxes. The dependence of water discharge change at the Arctic river mouths on permafrost properties demands to be clarified. In this study, we consider the changes of river discharges in cryosphere, river deltas morphology (meandering, braided branches and score holes) and abrupt erosion of the channel banks of a bend creating reformation of river channel structure and pollution of water.

2 Discharges, Air and Ground Temperature at Pan-Arctic River Mouths

2.1 Data, Methods and Location

To analyze the long-term trends of annual river discharges Q at the mouth heads of the rivers listed in Table 1, we used the data on the temporal series of the mean month river discharges available in the sources [7, 8]. Calculated long-term discharges $\langle Q \rangle$ were compared with the values from modern-day papers [9–14]. The rate of increase/decrease of Q during the period presented in Table 1 was estimated as a ratio of the difference of Q at the last and first years, ΔQ, to $\langle Q \rangle$ in %. The increase in air temperature T_a leads to a rise in ground temperature T_g and changes properties of frozen soils which finally results in permafrost shrinking. In northern regions, permafrost temperature in the period of seasonal thawing varies because of phase transitions in the frozen soil independently of the external heat exchange factors. In cold months, after complete freezing of the active layer, the permafrost mass starts actively interacting with the atmosphere through the snow cover. Therefore, the effect of winter season on soil stability in the cryosphere is greater than that of the summer season. Changes of T_a, T_g were taken from the sources [15–21].

Table 1 Pan-Arctic rivers in order of decreasing mean annual water discharge <*Q*>, changes of T_a, T_g per year with periods of observation. *D* is delta, *E* is estuary and *E–D* is estuarine-deltaic river mouths, the numbers are shown in Fig. 1

River	<*Q*> (m^3/s)	*Q*/<*Q*> (%/year)	ΔT_a (°C/year)	ΔT_g (°C/year)
Yenisei	17,917	0.12	0.04	0.03
E–D, 5	1936–08		1966–05	1966–2005
Ob (Irtysh)	12,817	0.10	0.05	0.02–0.03
E–D, 2	1936–08		1966–05	1966–2005
Lena	17,178	0.21	0.02	0.01–0.02
D, 8	1936–08		1966–20	1966–2005
Mackenzie	9261	0.21	0.08	0.06
D, 16	1973–11		1969–06	1962–08
Yukon	6557	0.67	0.03	0.03
D, 13	1976–19	1976–19	1949–16	
Pechora	4231	0.24	0.02	0.00–0.02
D, 1	1936–05		1966–05	1966–2005
Kolyma	3162	– 1.74	0.02	0.02–0.03
D, 11	1978–93		1966–05	1966–2005
Anadyr	993	– 0.09	0.03	0.03
E, 12	1958–88		1966–05	1966–2005
Indigirka	1603	0.18	0.2–0.3	0.02
D, 10	1937–98		1966–05	1966–2005
Taz	1047	– 0.11	0.04	0.03
D, 4	1962–96		1966–05	1966–05
Olenek	1167	1.10	0.02	0.02
D, 7	1936–99	1964–99	1966–05	1966–05
Yana	1063	0.61	0.02–0.03	0.01–0.02
D, 9	1973–07		1966–05	1966–05
Pur	897	– 0.34	0.04	0.03
D, 3	1939–91		1966–05	1966–05
Anabar	450	0.81	0.03	0.02
E, 6	1954–99		1966–05	1966–05
Colville	289	1.00	0.05	0.06
D, 14	2003–19	2003–19	1949–16	1976–03
Sagavanirktok	48	1.20	0.05	0.06
D, 15	1983–19	1983–19	1949–16	1976–03

2.2 Variations of Water Discharges

The temporal trends in Q of large rivers in the cryosphere show a tendency toward an increase under the climate warming. At the mouths of the *Ob*, Yenisei and Lena, for which long enough observation series are available (60–70 years), the annual increase in water discharge Q varies from 0.1 to 0.2% per year [22]. The most western river in cryosphere—Pechora River also having a long series of Q—shows the largest increase of water discharge reaching ~0.24% per year. At the eastern mouth of the Indigirka River during 1937–1998, Q increases by 0.18%. The analysis of shorter Q observation series at the mouths of the nonregulated rivers of Yana, Anabar and Olenek mouth (Fig. 2) shows a larger increase in Q—from 0.6% to 1.1% per year. Temporal trends of Q for the rivers Yukon, Colville and Sagavanirktok having data till 2019 show increase in the range 0.67–1.2% per year.

The period of data series for the Kolyma River 1978–1993 includes the year 1983, when the Kolymskaya HPP was brought into operation, which results in decreasing of Q. Investigation of long-term river flux for the Ob, Yenisei, Lena and Kolyma in 1978–2016 shows increase of the total flux of four rivers, but the shear of the Kolyma is unknown [23]. Decreasing Q for the rivers Anadyr, Pur and Taz can be explained by local precipitation tendencies.

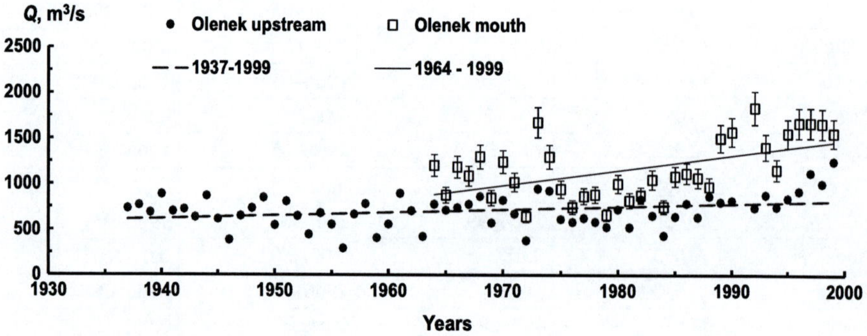

Fig. 2 Increase of Q at the Olenek R. upstream gauge (1937–1999) and at the mouth

Fig. 3 Dependence $T_g = f(T_a)$ with 20% error

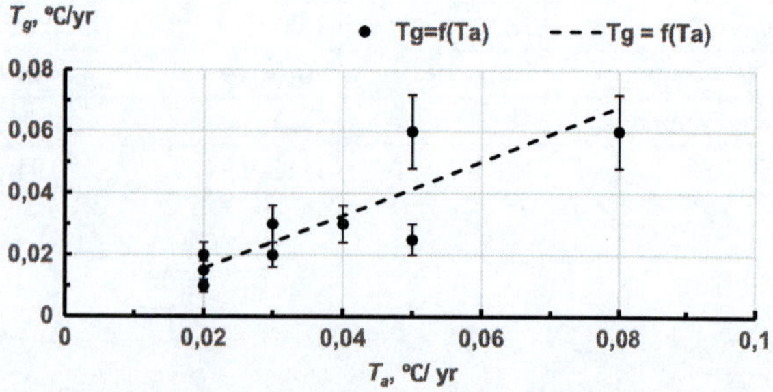

2.3 Impact of Changes of T_a and T_g on the Pan-Arctic River Discharges

Data of observation of T_a and T_g averaged through 1966–2005 [18] at the river mouths in Russia show the rate of temperature increases in the ranges 0.02–0.05 and 0.01–0.03 °C/year correspondingly. At the Yukon Delta, air temperature growth is just the same as it is at the Anadyr—0.03 °C/year. At the coast of Beaufort Sea, T_a increases more rapidly from 0.05 °C/year at the Colville and Sagavanirktok rivers [21] to 0.08 °C/year in Inuvik at the Mackenzie Delta [5]. The smallest rate of T_g increase was observed at the Yukon Delta 0.03 °C/year 0.03, and at the Beaufort Sea coast, it reaches 0.06 °C/year [17, 24]. Dependence of frozen ground temperature on T_a is linear with $R^2 \cong 0.8$.

Many factors influence the fresh water flux of all pan-Arctic rivers under climate warming such as: air temperature, changes in precipitation and permafrost thawing [3, 25, 26]. The rate of water discharge change for the pan-Arctic rivers decreases with growth of T_a (Fig. 4), whereas it increases with growth of T_g. Ground temperature increase provokes active layer growth, and permafrost thawing enhances infiltration which results in significant contribution of groundwater to the base flow [25, 27]. So, growth of T_g results in increase of $Q/ <Q>$. Air temperature increase causes decrease of snow cover season and areas covered with snow which lowers rate of water discharge increase [23].

3 Catastrophic Events in Permafrost Zone

The slopes of pan-Arctic river mouth plains are much smaller than medium river reaches causing meandering and ice jams formation. The coastal tundra in Russia is 50–80 m above the ocean level; therefore, river falls vary as follows: $i = 2–5 \times 10^{-4}$ from the Pechora to the Kolyma. The slope of the Mackenzie Delta plain is $i = 5 \times 10^{-5}$, for the Yukon Delta $i = 10^{-4}$, for the Colville Delta $i = 7 \times 10^{-5}$ and

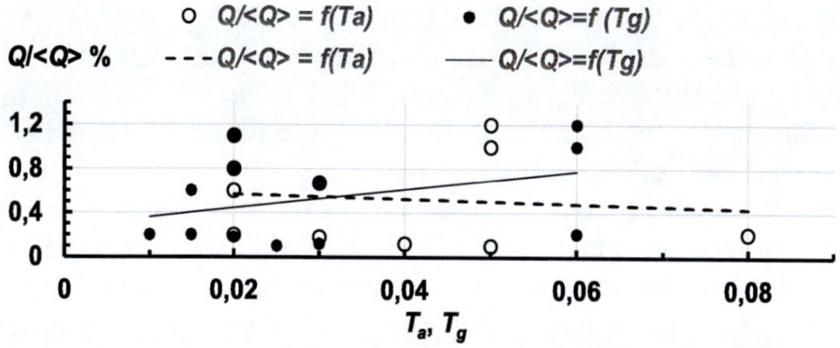

Fig. 4 Dependence rate of water discharge increase on air and ground temperatures

for Sagavanirktok Delta $i = 5.3 \times 10^{-4}$ [28–31]. The degree of sinuosity of a river is estimated using the sinuosity coefficient $k = L/l$ (L is the channel length; l is the straight-line distance from the source to the mouth), varying from 1.1 to 3 and more.

3.1 Ice Jams

The ice melting beginning upstream the pan-Arctic rivers generates a flow of water with ice floats moving downstream to the ice-covered flows at the river mouths causing conditions for ice jam formation especially near the channel sharp bend, abrupt narrowing and bifurcation. In case an ice jam with high water level is formed, the flow moves over the jam and further over the ice-covered reach, entraining ice blocks and sediment toward the sea edge of the delta which is typical for the delta branches of the Yukon and Colville [32, 33]. Growth of the river discharges mentioned above enhancing water levels during freshet or ice jam results in strong flooding over the vast territory. Due to climate warming, ice cover thickness becomes smaller and winter base flow increases resulting in spring the process of break up turns to be "thermal break up" instead of "mechanical" one. Further growth of air temperature will diminish frequency of ice jams which will be strengthen with higher water levels [34]. At high levels during ice jams, water can reach frozen soil at the banks promoting their intensive erosion and retreating.

3.2 Erosion of Banks in River Bends and Land Sliding

Meandering is a characteristic feature for pan-Arctic rivers, especially for delta branches of the rivers at the Arctic coast which form many bends. Investigations of 3D structure of the flows in a river bend show considerable transversal velocity v_y directed to the outer bank (Fig. 5) and vertical velocity v_z directed to the bottom which cause clockwise cross-sectional circulation initiating scouring of concave bank and deepen the flow at this bank [35, 36].

If the ratio $R/B \geq 5$, the river bend is considered as a sharp one (R is the curvature radius, B is the width of the flow). For the sharp bend, the outer-bank cell is described mathematically and is found in experiments [35, 37]. The circulation in this cell is opposite in direction to the central cell slightly reduce the activity of the central cell protecting outer bank against erosion.

High water levels during ice jams cause niche creation in a concave bank especially in presence of snow cover (Fig. 6a) which moves inside the bank until the upper layer of soil with snow becomes too heavy and fall into the flow [38]. The thermo-erosion process at high water levels during spring freshets and ice jams is characteristic for different rivers, among them for the middle stream of the Lena River [39] and the Colville River near the Nuiqsut, where maximal erosion rate reaches 2.2 m/year during 1980–2015 [40].

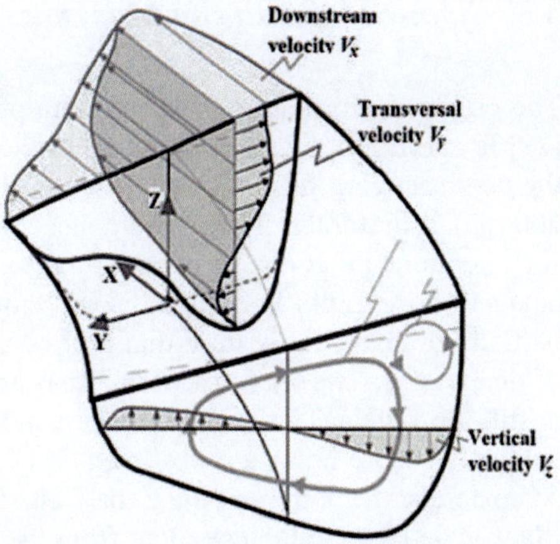

Fig. 5 Scheme of water movement in river meander after [35]

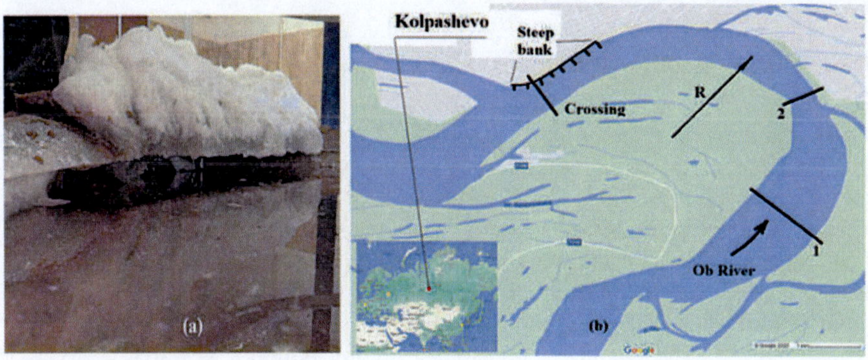

Fig. 6 River bend in floom model (**a**) and map of the Ob River at the Kolpashevo city (**b**)

Erosion of steep concave bank is observed at the Ob River in Kolpashevo city (58° 19′ N, 82° 55′ E, 65 m above sea level) (Fig. 6b) with the rate 3–4 m/year in 2007 2009 [11]. Upstream the steep bank and crossing ($B = 500$ m), there is a narrow reach, where $B = 700$ m (Sect. 2) and $B = 1530$ m (Sect. 1). The ratio R/B for the bend at Kolpashevo changes in the range 4–5 at high water level and at dry season correspondingly. Catastrophic washout of the cemetery inside the bank happened at the beginning of May (1979) at high freshet. At this time, minimal magnitude of R/B resulted in minimal protective effect of the outer-bank cell. The right bank of the Ob River continues to retreat and the whole streets vanish in the river.

3.3 *Scour Holes in the Mackenzie River*

The origin of score holes in the bottom of Mackenzie Delta branches described in [28] is another puzzle of nature. In 1974, there were observed 7 score holes with the depth ranging from 13 to 37 m, and then in 1985–1986, there were 13, and in 2004–31 score holes. Their emergence could not be explained by ice jam release. It is mentioned that one of the main reasons of these score holes formation is delta channel erosion into floodplain lakes. Some of these captured lakes appeared to be partially drained, suggesting that permafrost may also be a factor in the formation of deep holes. The share of permafrost ground in originating of these score holes is still a problem. Until now, Mackenzie Delta is the sole river mouth where such dangerous score holes are observed. Our investigation [42] reveals one spasmodic score hole at the bottom at the Lena Delta branch of depth comparable with those in Mackenzie Delta, which resulted from the strong ice jam release.

4 Discussion and Concluding Remarks

It was found that the rate of increasing trend of pan-Arctic river discharges lowers with the air temperature increases despite of the long-term magnitudes of discharges continue to increase; see also [23, 27]. Temperature of soil increasing with T_a causes increase of active layer thickness and permafrost thawing which enhances base flux of the rivers. Table increase trend of water discharges of pan-Arctic rivers can be used as a gross criterion of permafrost shrinking.

Growth of water discharges increases water levels which result in intensification of river bank erosion, may be a source of catastrophic pollution of water due to wash out of burial wastes. Thawing of permafrost provokes subsidence of soil all over the cryosphere [17, 24, 25]. The icy structure of landscapes in the permafrost zone is a stabilizing factor during natural evolution, while under urbanization and mining operations, it aggravates the decay of natural landscapes. Destruction of surface vegetation due to urbanization and oil/gas constructions results in accelerating of ground thawing and subsidence which causes destruction of construction and catastrophic pollution of the surface [43].

Due to the climate warming and growth of active layer thickness, discontinuous permafrost at the south boundary turns into sporadic, then into isolated. Frozen grounds disappear, and the soil becomes excessively moist. At such terrain, easily eroded soils are formed which cause landslides, as it happened at the Bureya River, the left tributary of the Amur River, where the landslide occurred in December 2018 [44]. The landslide of volume about 22.5 ml m^3 (more 55 ml. ton) completely blocked the river channel and formed tsunami wave in the opposite tributary of the Bureya. The landslide occurred in sparsely populated area, that is why there were no victims, but it was necessary to blow up the blocking dam to avoid a catastrophic flood upstream the landslide in spring.

Acknowledgements The work was supported by state program № 0147-2019-0001, registration AAAA-A18-118022090056-0.

References

1. Hansen, J., Sato M., Ruedy R. et al.: Global Temperature 2017, pp. 1–5 (2018)
2. Pavlov, A.V.: Permafrost Monitoring. Academic Publishing House "Geo", Novosibirsk (2008) (in Russian)
3. Osterkamp, T.E., Jorgenson, M.T.: Permafrost conditions and processes. In: Young, R., Norby, L. (eds.) Geological Society of America, Boulder, Colorado, pp. 205–227 (2009)
4. Jolivel, M., Allard, M. Impact of permafrost thaw on the turbidity regime of a subarctic river: the Sheldrake River, Nunavik, Quebec. Arctic Science Special issue entitled "Arctic permafrost systems", pp. 1–24 (2016)
5. Burn, C.R., Kokelj, S.V.: The environmental and permafrost of the Mackenzie delta area. Permafrost Periglac. Process. **80**, 83–105 (2009)
6. Conservation Biology Institute (CBI) Data Basin Homepage, https://databasin.org/datasets.
7. NOAA's Arctic Program Homepage. http://www.arctic.noaa.gov NOAA's
8. USGS Current Water Data for USA Homepage. https://waterdata.usgs.gov/nwis/rt
9. ATLAS: morphodynamics of mouth systems of large rivers at Arctic coast of Russia. Geographic Department MSU im. M. V. Lomonosova: Shirshov Institute of Oceanology of Russian Academy of Sciences. APR, Moscow (2017) (In Russian)
10. Doxaran, D., Devred, E., Babin M.: 50% increase in the mass of terrestrial particles delivered by the Mackenzie River into the Beaufort Sea (Canadian Arctic Ocean) over the last 10 years. Biogeosciences, **12**, 3551–3565 (2015)
11. Benke, A.C. Cushing, C.E.: Field Guide to Rivers of North America. Elsevier Academic Press. 459p. (2010)
12. Magritsky, D., et al.: Changes in hydrological regime and morphology of river deltas in the Russian Arctic. In: Proceedings of HP1, IAHS-IAPSO-IASPEL Assembly. Gothenburg, Sweden. IAHS Press. pp. 67–79 (2013)
13. Rood, S.B., Kaluthota, S., Philipsen, L.J., et al.: Increasing discharge from the Mackenzie River system to the Arctic Ocean. Hydrol. Process. **31**, 150–160 (2017)
14. Wilson, N.J., Walter, M.T., Waterhouse, J.: Indigenous knowledge of hydrologic change in the Yukon River basin: a case study of Ruby. Alaska. ARCTIC. **1**(68), 93–106 (2015)
15. Alaska Climate Research Center Homepage. http://climate.gi.alaska.edu/ClimTrends/Change/TempChange.html
16. Government of Canada Homepage. https://www.canada.ca/en/environment-climate-change/services/environmental-indicators/temperature-change.html
17. Jorgenson, T., Shur, Y., Osterkamp, T., Ping C.-L., Kanevskiy M.: Environment of the beaufort coastal plain. In: Jorgenson, M. T. (ed.) Coastal region of Northern Alaska guidebook to permafrost and related features. Guidebook vol. 10, pp. 1–39, Division of Geological & Geophysical Surveys, 3354 College Rd., Fairbanks, Alaska pp. 99709–3707 (2011)
18. Pavlov, A.V., Malkova, G.V.: Small-scale mapping of trends of the contemporary ground temperature changes in the Russian North. Earth's Cryosphere. **4**(13), 32–39 (2009) (In Russian)
19. Report on climate features on the territory of the Russian federation in 2016 //Russian federal service for hydrometeorology and environmental monitoring (roshydromet), Moscow, 2017 (in Russian)
20. Walsh, J.E. Overland, J.E., Groisman, P.Y. et al.: Ongoing climate change in the Arctic. AMBIO **40**, 6–16 (2011)

21. Wendler, G., Gordon, T., Stuefer, M.: On the precipitation and precipitation change in Alaska. Atmosphere **8**(253), 1–10 (2017)
22. Dolgopolova, E.N., Kotlyakov, A.V.: Permafrost in estuarine areas of the arctic rivers of Russia. Ice and Snow. Russian Acad. Sci. Assoc. Glaciolog. Stud. **1**(113), 81–92 (2011) (In Russian)
23. Suzuki, K., Hiyama, T., Matsuo, K., et al.: Accelerated continental-scale snowmelt and ecohydrological impacts in the four largest Siberian river basins in response to spring warming. Hydrol. Process. **34**, 3867–3881 (2020)
24. Coleman, K.A., et al.: Tracking the impacts of recent warming and thaw of permafrost peatlands on aquatic ecosystems: a multi-proxy approach using remote sensing and lake sediments. Boreal Environ. Res. **20**, 363–377 (2015)
25. Jacques, St. J-M., Sauchyn, D.J.: Increasing winter base flow and mean annual stream flow from possible permafrost thawing in the Northwest Territories, Canada. Geophysi. Res. Lett. **36**, L01401 (2009)
26. Suzuki, K., Hiyama, T., Yamazaki, D., et al.: Hydrological variability and changes in the arctic circumpolar tundra and the three largest pan-arctic river basins from 2002 to 2016. Remote Sens. **10**(402), 1–20 (2018)
27. Xu, M., et al.: Climate and hydrological changes in the Ob River Basin during 1936–2017. Hydrological Proc. 1–16 (2020).
28. Beltaos, S., Carter, T., Prowse, T.: Morphology and genesis of deep scour holes in the Mackenzie Delta. Can. J. Civil Eng. **38**, 638–649 (2011)
29. Dolgopolova, E.N.: Flow structure in a river bend in the cryolithozone. Power Technol. Eng. **3**(54), 309–312 (2020)
30. Hodel, K.L.: The Sagavanirktok River, North Slope Alaska: Characterization of an Arctic Stream. Interior Geol. Survey, Report 86–267 United States Depart. California, pp. 1–28 (1986)
31. Scott, K.M.: Effects of permafrost on stream channel behavior in arctic Alaska. Geological Survey Professional Paper 1068, US Gov. Print. Office, Washington, pp. 1–19 (1978)
32. Brabets, T.P., Wang, B., Meade, R.H.: Environmental and hydrologic overview of the Yukon River Basin, Alaska and Canada. US Geological Survey, Water-Resources Investigations Report, 106 p. (2000)
33. Walker, H.J., Hadden, L.: Placing river colville research on the internet in a digital library format. In: 7th International Proceedings of Permafrost Conference, pp. 1103 – 1107. Collection Nordicana, Yellowknife, Canada (1998)
34. Beltaos, S., Prowse, T.: River-ice hydrology in a shrinking cryosphere. Hydrol. Process. **23**, 122–144 (2009)
35. Blanckaert, K., De Vriend, H.J.: Secondary flow in sharp open-channel bends. J. Fluid Mech. **498**, 353–380 (2004)
36. Rozovskii, I.L.: Motion of Water in a Bend of an Open Channel. Izd. AN UkrSSR, Kiev (1957) (in Russian).
37. Graf, W.H., Blanckaert, K.: Flow around bends in rivers. Proceedings 2nd international conference on new trends in water and environmental engineering for safety and life, pp. 1–9. Aquatic Environments, Capri (2002)
38. Dolgopolova, E.N., Maslikova, O.Ya., Gritsuk, I.I., Ionov, D.N.: Laboratory investigation of flow and bank interactions of the river in cryosphere. In: Proceedings of international conference "Solving the puzzles from Cryosphere". "Okabiolab" Ltd., Pushchino, Russia, pp 83–84 (2019)
39. Tananaev, N.I.: Morphology and dynamics of banks of large rivers of the cryolithozone (for the Middle Lena near Yakutsk as an example). Geomorfologiya **1**, 81–91 (2014)
40. Payne, C., Panda, S., Prakash, A.: Remote sensing of river erosion on the colville river. North Slope Alaska. Remote Sens. **397**(10), 1–20 (2018)
41. Zavadskii, A.S., Il'yasov, A.K., Ruleva, S.N., et al.: The evolution of the Kolpashevskaya meaner of the Ob' and ongoing hazardous manifestations of channel processes. Geogr. Nat. Resour. **34**, 26–33 (2013)
42. Dolgopolova, E.N., Isupova, M.V.: Influence of permafrost soil on the hydrological and morphological processes in the mouths of the Lena and Mackenzie Rivers. Eng. Ecology **4**(118), 10–26 (2014). (In Russian)

43. Gorelik, J.B., Soldatov, P.V., Seleznev, A.A.: Thermal stability of frozen ground at sites of well clusters in the Yamburg gas condensate field. Earth's Cryosphere **1**(XIX), 53–60 (2015)
44. Makhinov, A.N., Kim, V.I., Ostroukhov, A.V., Matveenko, D.V.: Large landslide in the valley of the Bureya River and tsunami in the reservoir of the Bureya hydroelectric power station. Vestnik of Far Eastern Branch of Russian Acad. Sci. **2**, 35–44 (2019). (in Russian)

Effect of Land Uses on Personal Exposure of Street Vendors at a Metropolitan City

Smaranika Panda

1 Introduction

Air pollution in India is one of the major issues of concern due to rapid growth in transportation, industrial, power generation sectors coupled with planned and unplanned urbanization, infrastructure development, expansion of cities, industrial and population growth. According to the recent reports, 50% of global deaths attributed to ambient air pollution. The National ambient monitoring program (NAMP) revealed routine exceedance of $PM_{2.5}$ and PM_{10} concentration from the prescribed standards in all major cities in India [1]. In fact, 27 cities in 2011 and 37 cities in 2014 from India were listed in top 100 cities with worst $PM_{2.5}$ pollution [2]. In urban areas, large number of population are exposed to elevated air pollutant concentrations. Short- and long-term exposure to high and low pollutant concentrations at different urban locations can result in various diseases such as asthma, bronchitis, lung and heart-related diseases, allergies and cancer [3]. The total deaths attributed to air pollution were reported to be more than 0.5 million per year in India [4]. Among the major cities in India, Bangalore has witnessed a rapid growth and become the silicon valley of India in a very short span. The multi-dimensional growths in infrastructure, road network, vehicular population, industries and street shops in and around the city center have increased the pollutant concentration beyond the permissible limits. The exposure of the pollutant concentration varies considerably with time and land use pattern of the location [5]. In general, the work group population such as traffic polices, street vendors, etc., get exposed to high pollutant concentrations for a considerable amount of time. However, the amount of exposure for street vendors and traffic polices also vary notably in different land uses. In India, studies on personal exposures for the most exposed group of people in metropolitan cities are very limited. In this regard, the present study focused on developing PM exposure profiles and chemical characterization of PM_5 samples for street vendors in two contrasting land uses in order to

S. Panda (✉)
CMR Institute of Technology, Bengaluru, India

I. Pal and S. Kolathayar (eds.), *Sustainable Cities and Resilience*, Lecture Notes in Civil Engineering 183, https://doi.org/10.1007/978-981-16-5543-2_8

understand the impact of land uses on personal exposure. Also, health risk assessment of street vendors in different land uses were carried out.

2 Methodology

2.1 Study Area Description

The personal exposure monitoring of street vendors were carried out for Bangalore city at two contrasting land uses such as traffic intersection and residential area. Kundanahalli traffic intersection (latitude: 12° 57′ 22″ N; longitude: 77° 42′ 54″ E) which is one of the congested traffic locations was considered for monitoring. Personal exposure monitoring for a street vendor sitting in a font open shop was carried out. The four-legged traffic junction receives traffic from three major four lane roads and one two-lane minor road. Banaswadi residential area (13° 00¢ 24″ N and 77° 38¢ 13″ E) was considered as the other monitoring location. The area is dominated with individual houses and apartments with few numbers of street shops for grocery and vegetables. However, movements of private vehicles were observed in the location. A major road at a distance of 300 m for movement of city buses and other vehicles were located near the monitoring location. Figure 1 presents the map of Bangalore city with the monitoring locations.

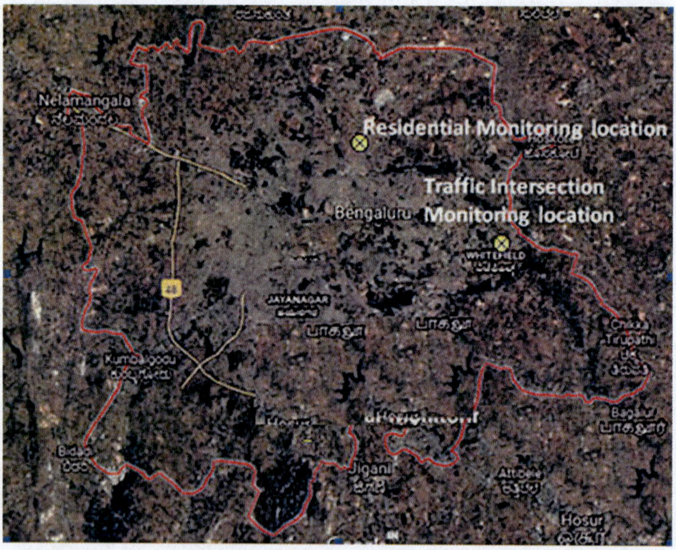

Fig. 1 Map of Bangalore city showing the monitoring locations at Kundanahalli (traffic intersection) and Banaswadi (residential area)

2.2 *Monitoring and Instrumentation*

PM5 personal exposure monitoring for 8 h was carried out at the traffic intersection and residential area for 7 days using personal sampler (EnviroTech APM 801) from Feb 12, 2019 to Mar 6, 2019. Monitoring was carried out for weekdays and weekends. The deposition of PM5was obtained by weighing the filter papers before and after the monitoring and the concentration was calculated by utilizing the weight of deposition and total amount of air flow during the entire sampling duration.

2.3 *Chemical Characterization of Particulate Samples*

Chemical characterizations of PM5 for 15 elements (Al, Ba, Ca, Cd, Co, Cr, Fe, K, Mg, Mn, Na, Ni, Pb, Sr and Zn) were carried out using inductive coupled plasma with optical emission spectroscopy (ICP-OES, make: Optima 5300-DV). The elements were extracted using hot acid digestion procedure (IO-3.1) and were analyzed (IO-3.5) based on U.S. EPA methods.

2.4 *Health Risk Assessment*

Excess cancer risk was performed for the carcinogenic heavy metals (Cr, Ni, Pb and Cd). Inhalation unit risks (IUR) for the metals were taken from USEPA. The excess cancer risks were obtained by multiplying the exposure metal concentration with the respective IUR.

3 Results and Discussion

3.1 *Variation of Personal Exposure Profiles at Different Land Uses*

The 8-h average PM5 personal exposure concentration at the traffic intersection during weekdays and weekends was observed to be 585.2 ± 107 $\mu g/m^3$ and 431.4 ± 51 $\mu g/m^3$, respectively. Similarly at the residential area, the concentrations were in the range of 203.4 ± 48 $\mu g/m^3$ and 124.3 ± 50.9 $\mu g/m^3$ during weekdays and weekends, respectively. The week end concentration at the traffic location was observed 26% lower than the weekday exposure concentration. However, the weekend exposure concentration was observed considerably low, i.e., 39% reduction from weekdays at the residential area which may be due to non-working days for offices and

Fig. 2 PM5 exposure concentration at Banaswadi (residential area) and Kundanahalli (traffic intersection)

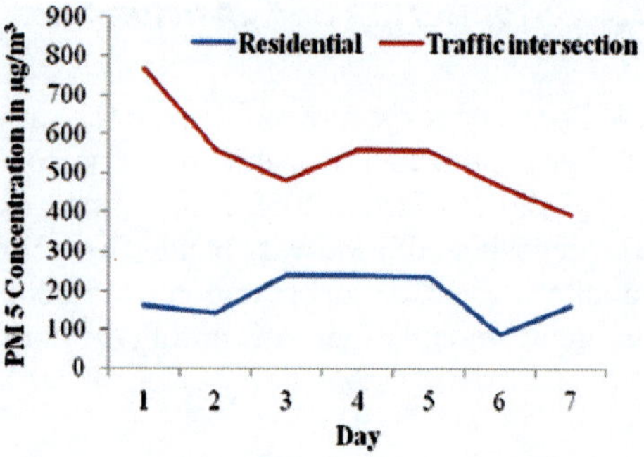

schools which lower the vehicular activities in the vicinity. Figure 2 presents the variation of PM5 exposure concentration in traffic intersection and residential area.

3.2 *Chemical Characterization of Particulates*

Analyzed 15 elements contribute 15% of total PM_5 exposure concentration at the traffic intersection. The elevated metal concentration of Mn, Ba, Cu, Zn and Pb at the traffic location could be from the vehicular emissions and wear/tear of tires. The heavy metals such as Ni, Cd and Pb concentrations were observed 2, 4 and 5 times higher than the national ambient standard concentration at the intersection. At the residential area the contribution of heavy metals was observed to be 12%. Ni and Cd were observed below detection limit for all the analyzed samples. The Pb concentration was observed below the ambient air quality standard. Figure 3 presents the elemental concentration of abundance and trace elements at traffic intersection and residential area.

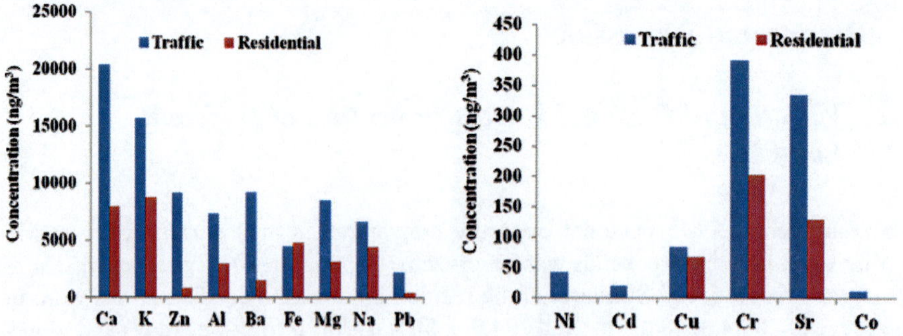

Fig. 3 PM5 bound abundant and trace elemental exposure concentration at Banaswadi (residential area) and Kundanahalli (traffic intersection)

Table 1 Excess cancer risk for heavy metals at the traffic intersection and residential area

Pollutant	IUR ($\mu g/m^3)^{-1}$	Excess cancer risk (10^{-6} population)	
		Traffic	Residential
Cd	1.8×10^{-3}	33.8	0.0
Cr	1.2×10^{-2}	130.8	68
Ni	2.4×10^{-4}	10.1	0.0
Pb	1.2×10^{-5}	26.4	5.8

3.3 Health Risk Assessment

The excess cancer risk for the residential and traffic intersection is provided in Table 1. Cr showed highest cancer risk at both the land uses.

4 Conclusion

Personal exposure monitoring for the street vendors were carried out at a busy traffic intersection and residential area of Bangalore. Study results indicated heavy particulate exposure at both the locations. The PM5 personal exposure was observed to be exceeding the PM10 and PM2.5 standards by several folds at both the land uses. However, at the traffic intersection, the exposure concentration of PM5 was observed three times higher than the residential area. Weekend concentrations were observed to be lower at both the land uses. The elemental analysis indicated presence of elevated concentration of carcinogenic heavy metals at the traffic intersection. Further, at the residential area, presence of two carcinogenic elements were observed which was exceeding the permissible WHO standard.

References

1. Central Pollution Control Board, New Delhi. https://app.cpcbccr.com/ccr/#/caaqm-dashboard-all/caaqm-landing/data
2. World health organization. https://www.who.int/airpollution/data/cities/en/
3. Panda, S., Nagendra, S.S.: Chemical and morphological characterization of respirable suspended particulate matter (PM10) and associated heath risk at a critically polluted industrial cluster. Atmos. Pollut. Res. **9**(5), 791–803 (2018)
4. Silva, R.A., West, J.J., Zhang, Y., Anenberg, S.C., Lamarque, J.F., Shindell, D.T., Collins, W.J., Dalsoren, S., Faluvegi, G., Folberth, G., Horowitz, L.W.: Global premature mortality due to anthropogenic outdoor air pollution and the contribution of past climate change. Environ. Res. Lett. **8**(3), 034005 (2013)
5. Babadjouni, R.M., Hodis, D.M., Radwanski, R., Durazo, R., Patel, A., Liu, Q., Mack, W.J.: Clinical effects of air pollution on the central nervous system; a review. J. Clin. Neurosci. **43**, 16–24 (2017)

Social Vulnerability Assessment to Natural Hazards in Western Nepal

Sumeet Moktan, Kumud R. Kafle, and Suman Chapagain

1 Introduction

The complicated geophysical structure makes Nepal vulnerable to various kinds of disasters [1]. Natural hazards including floods, epidemics, landslides, wildfires, wind storms, GLOF, cold and heat waves, landmines, road accidents, drowning incidents, and earthquakes are frequent in Nepal [2]. Floods, landslides, and fires often occur frequently, and the nation is vulnerable to the earthquake as it sits at an edge of Tibetan and Indian Plates alongside the Himalayan Region [3]. The mountainous and hilly regions are more prone to the occurrence of landslides and GLOF events [4] whereas Terai plains are more prone to seasonal flooding as the result of the monsoon rains and the complex river systems [5]. The country's DesInventar data shows that more than 30,000 people were killed by any one of the disasters, with more than 60,000 people injured creating a great loss of human lives and properties. Nepal is situated in the tectonically active central part of the Himalayas, being one of the global earthquake hazard hotspots [6]. In the context of global vulnerability, Nepal ranks as the 20th most multi-hazard-prone country and 11th in the case of earthquake hazards [7].

Underdeveloped nations lacking resilience and mechanisms to respond to disasters suffer the most. Disasters claim human lives as well as destroy properties and infrastructures, raise the risk of outbreaks of infectious disease, threaten food security, cause disruptions of social and economic aspects contributing to population displacement [8, 9]. Disaster vulnerability is highly dynamic and diverse throughout Nepal. Factors such as weak governance, population and rapid urban growth, relatively weak land use planning, irregular settlement, poor construction techniques,

S. Moktan (✉) · K. R. Kafle
Department of Environmental Science and Engineering School of Science, Kathmandu University, Dhulikhel, Nepal

S. Chapagain
Himalayan Risk Research Institute, Bhaktapur, Nepal

I. Pal and S. Kolathayar (eds.), *Sustainable Cities and Resilience*, Lecture Notes in Civil Engineering 183, https://doi.org/10.1007/978-981-16-5543-2_9

steep agricultural practices and settlements invading forest and plains, and the deterioration of the environment have contributed to the rise in the occurrence of disasters in Nepal [10].

Quite a few studies have explored social vulnerability to natural hazards amid the subsequent rise in public understanding of natural hazards across substantially all areas of Nepal. Social vulnerability analysis and mapping are of urgent need in Nepal as most urban as well as rural settlements are exposed to multiple risks which can have a significant effect on planning, policies, preparedness, and the allocation of resources [4].

2 Social Vulnerability

Cutter et al. [11] has defined social vulnerability as a factor that influences or shapes the susceptibility of society to harm and the ability to respond. Such factors are further characterized by types of communities and built-environment and level of urbanization, growth rates, and economic vitality. The substantial variability of disaster depends on the local circumstances and its likelihood of severity of impacts compared to different places depends on the local vulnerability components of the affected society, i.e., cultural, social, and economic. Consequentially, there is a significant connection between the potential hazard risk and the social resistance and resilience of a particular place. In this manner, the disaster response differs based on the social structure. Hence, the evaluation of social vulnerability is a vital step for understanding the ability of a society to envision, resist, cope, and recover from the impacts of natural disaster events in the future [12]. Socially vulnerable groups are more likely to be adversely affected by disaster events. Hence, addressing social vulnerability constructively reduces both human casualties and properties. This is guided by the provision of social services and public assistance in a post-disaster scenario [13].

Yoon [14] examined and compared the developing methodologies for the assessment of social vulnerability to natural disasters. Current literature on vulnerability indicates that two methods have been applied in creating social vulnerability indexes: (a) a deduction method using standardization methods, for example, z scores [15, 16] and (b an inductive method using data reduction techniques such as factor analysis [11]. The contrast between the deductive and inductive indicator integration and index development methods in evaluating social vulnerability to natural hazards is experimentally depicted in this study. The findings showed that although there are no variations in the measurements of the overall social vulnerability scores using three separate standardization methods, the overall social vulnerability scores determined by factor analysis and by standardization are different.

Gautam [4] focused on the assessment of the social vulnerability on a local scale of all 75 districts. Altogether, 13 variables were selected, and the social vulnerability index was constructed by the maximum value transformation method and cumulative indexes were mapped using ArcGIS. It was observed that only four districts were having a very low social vulnerability index and 46 districts having moderate to high social vulnerability levels. A study done by Aksha et al. [17] looks into the social vulnerability in Nepal by a modified Social Vulnerability Index (SoVI). 39 variables were reduced to seven factors using the principal component analysis which explained 63.02% of the variance in the data. The factor scores were then summarized to calculate final SoVI scores. Higher social vulnerability levels were observed in the central and western hill, Western Mountain, and central and eastern Terai regions, while the central and eastern hilly regions are less vulnerable.

3 Study Area

Western Nepal includes western, mid-western, and far-western from the old administrative division and recently has been divided into a province which includes provinces 4, 5, 6, and 7. It has an alpine environment of mid-latitude, extending from the altitude of 150 m covering Terai plains to the 8172 m including the Himalayas where slope varies from flat to steep indicating the possibilities of slope failures. The climate is warm temperate. Heavy monsoon rains between 1000 and 2000 mm can be observed between the month of June and September and slightly varying depending on altitude and topography. Top hills and steep areas are covered with highly degraded forests. The majority of people make their livelihood from agriculture. In Western Nepal, possible mega earthquakes, famines, and epidemics have long been known as possible threats [4]. Considering the scenario of western Nepal, there is a need for the development of effective prevention, mitigation, and contingency plans for a multi-hazard scenario.

4 Methodology

4.1 Data

The national census report published by the Government of Nepal in 2011 was used as a primary source of data. The indicators and variables for the study were identified and selected from the literature review. The deductive approach was applied for the selection of the variables. Altogether, 9 indicators and their 17 variables were identified for the research (Fig. 1).

4.2 Calculation of Social Vulnerability Index

The maximum value transformation approach by Cutter et al. [18] was used to convert each variable was to a common scale from 0 to 1 for the calculation of SoVI. The

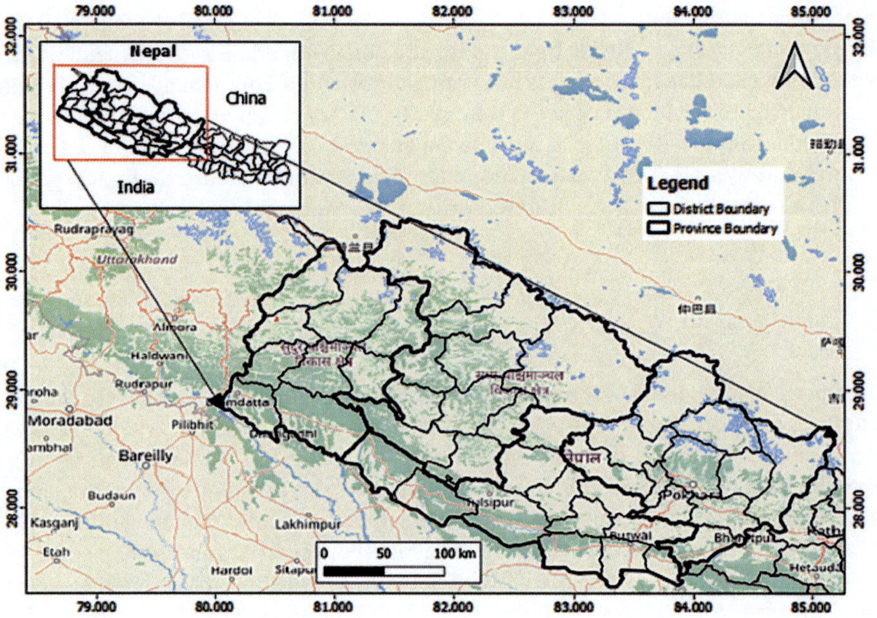

Fig. 1 The location of the study area

ratio of the value of that variable (X_i) to the maximum value ($X_{\max}$) for the variable was calculated as given by the equation:

$$\text{Score}(N_i) = \frac{\text{value of variable, } i}{\text{maximum value}} \tag{1}$$

As considered by Cutter et al. [18], higher index values indicate higher vulnerability. The social vulnerability index was calculated for each district by integrating the scores of each variable as the overall vulnerability score given by the following equation (Table 1):

$$\text{Total vulnerability score } = \sum_{i=1}^{13} N_i \tag{2}$$

4.3 Development of Social Vulnerability Map

The social vulnerability indexes were classified into five different classes: very low, low, moderate, high, and very high based on standard deviation, as shown in Table 4. QGIS was used to generate the social vulnerability map in terms of vulnerability

Table 1 Indicators and Variables used for SoVI analysis

Indicators	Variables
Population	(N1) Population Density (N2) % of Absentee population (N3) Average household size
Gender	(N4) % of female-headed households (N5) % of the female population
Age	(N6) % of children of age below 14 years of age (N7) % of elders above 60 years of age
Education	(N8) % of illiterate people above 5 years of age
Health	(N9) % of disabled people above 5 years of age (N10) % of malnourished children under 5 years of age (N11) % Household without toilet (N12) % of the population without safe water
Poverty	(N13) Human poverty index
Ethnicity	(N14) % of low caste ethnic group
Economic	(N15) % of agriculture holding with livestock (N16) % without economic provisioning
Housing unit	(N17) % of house foundation with mud-bonded bricks/ stone

Table 2 Vulnerability level classification based on the standard deviation approach

Standard deviation	Level of vulnerability
>12.49	Very high
11.18–12.49	High
9.87–11.18	Moderate
8.57–9.87	Low
<8.57	Very low

level which shows the distribution of social vulnerability to multi-hazards in Western Nepal (Table 2).

5 Result and Discussion

The overall vulnerability map was prepared using the indicator-specific outputs obtained in this research. The indicators considered in this study were population, gender, age, education, health, poverty, ethnicity, and economic and housing unit. The direction of these indicators was determined with regards to their known influences on vulnerability, which are identified with the help of existing literature. Social vulnerability assessment of all the 40 districts of Western Nepal shows us that most of the vulnerable areas fall under mid-western and far-western parts which include hilly and mountain regions of provinces 6 and 7. The sum of results shows that 5% of the districts fall under a very high social vulnerability region, whereas 25% of districts fall under the high socially vulnerable level. The least vulnerable areas are located mostly in the western region and extend along the Terai region as listed in Table 4).

Table 3 Descriptive statistics of the selected variables

Variables	Standard deviation	Mean	Max	Min
N1	111.51	138.23	600.95	2.79
N2	15.19	12.89	64.15	0.17
N3	0.64	4.95	6.26	3.92
N4	11.28	28.47	49.50	7.71
N5	2.75	52.47	56.81	44.00
N6	5.54	37.39	46.23	21.58
N7	2.14	8.33	12.74	4.99
N8	7.53	31.83	49.37	16.04
N9	0.95	2.83	5.39	1.12
N10	10.30	46.09	65.70	22.90
N11	19.02	40.88	78.35	0.85
N12	13.68	22.06	48.12	2.14
N13	7.79	34.31	49.26	16.50
N14	7.40	20.28	47.68	5.14
N15	15.47	82.28	96.03	14.41
N16	9.99	34.07	48.14	15.22
N17	27.44	73.98	99.16	12.53

Table 4 District frequency related to social vulnerability level

Level of social vulnerability	Number of districts	Districts
Very high	2	Achham, Doti
High	10	Baitadi, Bajhang, Bajura, Dailekh, Humla, Jajarkot, Kalikot, Rolpa, Rukum, Salyan,
Moderate	15	Arghakhanchi, Baglung, Dadeldgura, Darchula, Dolpa, Gorkha, Gulmi, Jumla, Mugu, Mustang, Myagdi, Kapilbastu, Nawalparasi, Pyuthan, Tanahu,
Low	11	Banke, Bardiya, Dang, Kailali, Kanchanpur, Lamjung, Mustang, Palpa, Parbat, Rupandehi, Syangja,
Very Low	2	Kaski, Manang

5.1 *Social Vulnerability of All Indicators*

5.1.1 Population

Higher population density indicates the high chances of the existence of socially vulnerable populations [19]. Higher population density increases the risk of losses and makes the evacuation process harder [11]. The Post Disaster Need Assessment of the 2015 Gorkha earthquake published by the National Planning Commission addressed that the 14 most heavily affected districts had a high absentee population migrating overseas for job opportunities, and their household income depends on remittances. According to [11], a large number of dependents in the families obtain limited finance to support the dependents so there is a need for switching in between work responsibilities and family care.

5.1.2 Gender

Women tend to have more harsh times than men during recovery the reason being family responsibilities, lower wages, and sector-specific employment. A study was done by [20], which explains that women are most affected by earthquakes than men in the case of every major earthquake in Nepal, for instance, the Gorkha earthquake of April 25, 2015 killed more women than men. The reason behind this is that most women in Nepal are mainly involved in household activities and were inside their houses during the disasters [4]. In relatively poor families, the loss of properties along with the lack of family security, and stress of finding alternate livelihoods could lead to dire consequences like sexual abuse, human trafficking, child marriage, and labor on particularly women, girls, and children.

5.1.3 Age

According to [11], caring for children requires a huge amount of time and money and while the elderly may face limitations in mobility. According to NPC [21], children are at risk of dropping out of school and get engaged in activities that generate income to support their household. Several effects on children have been addressed by various child-focused varying from issues such as food, hygiene, shelter, education, and access to drinking water.

The elderly people may suffer from "mobility, cognitive, sensory, social, and economic restrictions that can hamper their adaptability and ability to cope during disasters" [8].

5.1.4 Education

Education is associated with socioeconomic status [11]. Higher education results in greater job opportunities and lifetime earnings, whereas lower education limits the ability to understand the warning and recovery information. So, in this regard, higher education decreases and lower education increases social vulnerability. The developmental potential of a society can be depicted by its average education level. A higher level of education indicates a better capacity to respond cope and recover from natural disasters [22].

5.1.5 Health

Disabled people have been immensely affected as many do not possess the means for the reconstruction of the houses [21]. People, especially with sensory disabilities such as speech, hearing, vision are at greater risk of not being able to receive information before, during, and after a hazardous event [23]. From the assessment of post-earthquake, it was observed that food consumption practices have deteriorated in the affected districts which resulted in the decline below the levels as compared to the pre-earthquake assessment data. Sanitation is a great concern in Nepal. Most people do not have access to safe drinking water and toilets. Risk factors like high population displacement with limited access to the toilet, safe drinking water, and lack of sanitary facilities may lead to epidemics in the disaster-affected areas [21].

5.1.6 Poverty

Many have to emigrate to neighboring countries and the Middle East in search of employment opportunities due to poverty which provides much-needed remittances [17]. The poor communities are the most vulnerable than other groups in terms of economic, health, and natural hazards. These groups are deprived of access to material, economic, social, political, or cultural resources and struggle to meet their basic needs [24, 25]. This in turn will lead to decreased potential in coping at the time of disaster.

5.1.7 Ethnicity

Low caste groups have always been the victim of disparity. Dalits are considered as "untouchable" and face social and economic and systemic hardships [17]. In Nepal, the problem of untouchability still exists to this day and age. People including Dalits, marginalized ethnic minorities residing in remote geographical regions are underprivileged and lack access to social services.

5.1.8 Economic

According to NPC [21], mostly rural areas in the central and western regions have been affected by the earthquake. A large population there was highly dependent on agriculture for livelihood, which the earthquakes and the subsequent landslides have affected. Besides, these districts have a relatively higher per unit of livestock than the national average which is regarded as one of the key sources of income for rural households. Loss of livestock at a massive level would potentially trigger a significant loss of income. Economic deprivation means the absence of material assets which are regarded as the basic necessities in society. Economic deprivation generates mental distresses like anxiety and depression because of the stress of not having sufficient resources available to support their everyday living [26].

5.1.9 Housing Unit

Housing foundations and wall materials have a significant impact on vulnerability since Nepal is situated in a highly active seismic belt. The case of Gorkha Earthquake 2015 claimed many lives of the people living in poor quality houses. Especially, the traditional mud and stone houses in rural areas suffered the most in terms of damage [17].

5.2 *Overall Social Vulnerability*

The result shows that 5% of the districts fall under the very high vulnerability regions, whereas 25% of districts fall under the highly vulnerable level. The least vulnerable areas are located mostly in the western region and extend along the Terai region encompassing some parts of province 4, 5, 6, and 7 of the western regions with very low to moderate levels of social vulnerability. 2 out of 40 (5%) districts are at a very high social vulnerability level, 10 districts (25%) are at a high vulnerability level, 15 districts (37.5%) are at a moderate social vulnerability level, 25 districts (27.5%) are at a low vulnerability level, and 2 districts (5%) are at a very low vulnerability level in Nepal. Social vulnerability assessment of all the 9 indicators shows us that most of the vulnerable areas fall under mid-western and far-western parts which include the hilly and mountain region of provinces 6 and 7. The reason is well interpreted regarding all the 9 selected indicators. It can be observed that there are ample reasons such as backward development, gender and caste disparity, geological features, insufficient road, and transport facilities, illiteracy, and dependency in only agriculture that makes up the overall social vulnerability. Out of all the districts, the study has shown that in Doti and Achham the social vulnerability is very high. According to the Nepal Hazard Risk Assessment of 2010, Doti and Achham are prone to three types of Geological, Hydro-meteorological hazards. Doti is one of the five districts with the highest percentage of permanent houses exposed to a very

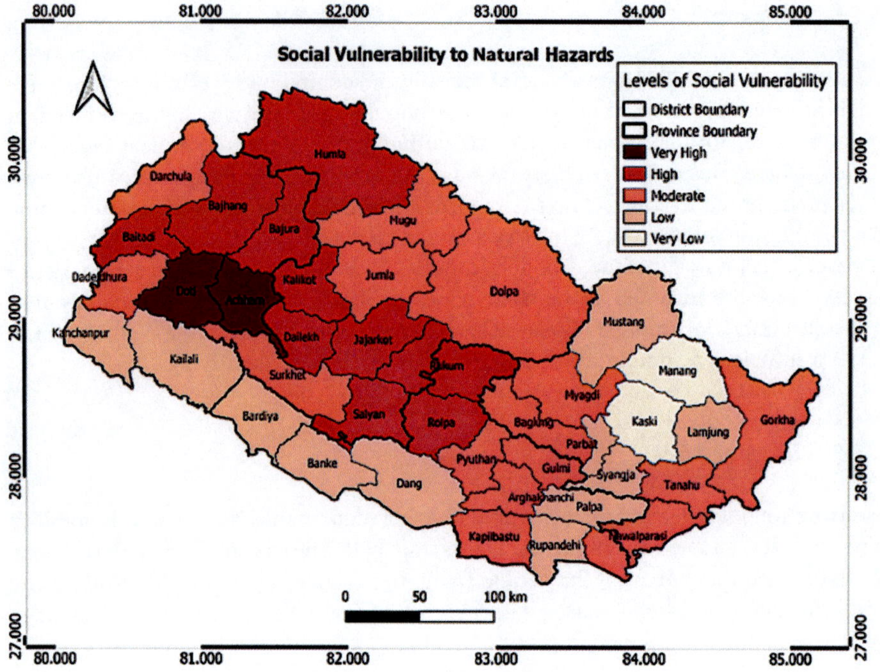

Fig. 2 District-wise social vulnerability to multi-hazard in Western Nepal

high hazard zone. Also, the district area percentage distribution of earthquake hazard assessment for a 500-year return period is very high in both districts (Fig. 2).

6 Conclusion

This study attempts to illustrate district-level social vulnerability to multi-hazard in Western regions of Nepal with the help of the SoVI method. Multi-hazards are beyond human control. The only way to counter this natural phenomenon is to be prepared. Nepal lies in a seismically active zone. Considering the scenario of western Nepal, there is a need for the development of effective prevention, mitigation, and contingency plans for different types of hazards. The results illustrate that social vulnerability ranges from mid-hills and mountains of mid-western to the mid-hills of far-western districts. These results have the potential to help policymakers and emergency planners in creating a more successful and focused crisis response program. Damages and losses due to several hazards in western districts would be very high and there is a high need for preparation. Social vulnerability due to natural hazards in various districts of Nepal varies, so there is also a need for decentralization in regards to preparedness, response, and recovery.

This study is based on a deductive approach. The results may vary depending on the number of considered variables. It is important to consider furthermore variables for the precision of the work. So, to reduce such issues, the municipality-wise data is required for further uniformity in the variation of the variable; an inductive approach like principal component analysis is recommended for more reliable SoVI mapping, and social vulnerability at the local level is recommended.

References

1. Gaire, S., Delgado, R.C., González, P.A.: Disaster risk profile and existing legal framework of Nepal: floods and landslides. Risk Manage. Healthcare Pol. **8**, 139 (2015)
2. MoHA (Ministry of Home Affairs) and DPNet (National Disaster Preparedness Network—Nepal) (2009) National Disaster Report 2009: The Hazards and Vulnerability. Kathmandu: Ministry of Home Affairs (MoHA) and National Disaster Preparedness Network— Nepal (DPNet)
3. Koirala, P.K.: Country profile: Nepal. Disaster Management Instution and System in Nepal (2014)
4. Gautam, D.: Assessment of social vulnerability to natural hazards in Nepal. Nat. Hazard. **17**, 2313–2320 (2017)
5. Dewan, T.H.: Societal impacts and vulnerability to floods in Bangladesh and nepal. Weather Clim Extremes **7**, 36–42 (2015)
6. Mosquera-Machado, S., Dilley, M.: A comparison of selected global disaster risk assessment results. Nat. Hazards **48**(3), 439–456 (2009)
7. United Nations Development Program/Bureau of Crisis Prevention and Recovery (UNDP/BCPR): Reducing disaster risk, A challenge for development, New York (2004)
8. Noji, E.K.: Public health in the aftermath of disasters. BMJ **330**(7504), 1379–1381 (2005)
9. Watson, J.T., Gayer, M., Connolly, M.A.: Epidemics after natural disasters. Emerg. Infect. Dis. **13**(1), 1e5
10. Aryal, K.R.: The history of disaster incidents and impacts in Nepal 1900–2005. Int. J. Disaster Risk Sci. **3**(3), 147–154 (2012)
11. Cutter, S.L., Boruff, B.J., Shirley, W.L.: Social vulnerability to environmental hazards. Soc. Sci. Q. **84**(3), 242 (2003)
12. Frigerio, I., Ventura, S., Strigaro, D., Mattavelli, M., De Amicis, M., Mugnano, S., Boffi, M.: A GIS-based approach to identify the spatial variability of social vulnerability to seismic hazard in Italy. Appl. Geogr. **74**, 12–22 (2016)
13. Flanagan, B.E., Gregory, E.W., Hallisey, E.J., Heitgerd, J.L., Lewis, B.: A social vulnerability index for disaster management. J. Homeland Secur. Emerg. Manag. **8**(1) (2011)
14. Yoon, D.K.: Assessment of social vulnerability to natural disasters: a comparative study. Nat. Hazards **63**(2), 823–843 (2012)
15. Chakraborty, J., Montz, B.E., Tobin, G.A.: Population evacuation: assessing spatial variability in geophysical risk and social vulnerability to natural hazards. Nat. Hazards Rev. **6**(1), 23–33 (2005)
16. Wu, S.Y., Yarnal, B., Fisher, A.: Vulnerability of coastal communities to sea-level rise: a case study of Cape May County, New Jersey, USA. Clim. Res. **22**, 255–270 (2002)
17. Aksha, S.K., Juran, L., Resler, L.M., Zhang, Y.: An analysis of social vulnerability to natural hazards in Nepal using a modified social vulnerability index. Int. J. Disaster Risk Sci. **10**(1), 103–116 (2019)
18. Cutter, S.L., Mitchell, J.T., Scott, M.S.: Revealing the vulnerability of people and places: a case study of Georgetown County, South Carolina. Ann. Assoc. Am. Geogr. **90**(4), 713–737 (2000)

19. Martin, S.A.: A framework to understand the relationship between social factors that reduce resilience in cities: application to the City of Boston. Int. J. Disaster Risk Reduct. **12**, 53–80 (2015)
20. Chaulagain, H., Gautam, D., Rodrigues, H.: Revisiting major historical earthquakes in Nepal: Overview of 1833, 1934, 1980, 1988, 2011, and 2015 seismic events. In: Impacts and Insights of the Gorkha Earthquake, pp. 1–17 (2018)
21. NPC. Nepal Earthquake 2015. Post Disaster Needs Assessment. Sector Reports. Kathmandu: National Planning Commission, Government of Nepal (2015)
22. Chen, W., Cutter, S.L., Emrich, C.T., Shi, P.: Measuring social vulnerability to natural hazards in the Yangtze River Delta region, China. Int. J. Disaster Risk Sci. **4**(4), 169–181 (2013)
23. Dwyer, A., Zoppou, C., Nielsen, O., Day, S., Roberts, S.: Quantifying social vulnerability: a methodology for identifying those at risk to natural hazards (2004)
24. Philip, D., Rayhan, M.: Frustration-aggression hypothesis: examination and reformulation. Psychol. Bull. **106**(1), 59 (1989)
25. Philip, D., Rayhan, M.I.: Vulnerability and Poverty: what are the causes and how are they related. ZEF Bonn, center for Development Research, University of Bonn (2004)
26. Berkowitz, L.: Frustration-aggression hypothesis: examination and reformulation. Psychol. Bull. **106**(1), 59 (1989)

Sustainable Urban Drainage System to Avoid Flooding of Rain Origin and Improving Green Areas, Lima, Peru

López Amaro, José Luis, Villavicencio Cuya, Raquel Lorena, Silva Dávila, and Marisa Rosana

1 Introduction

"Lima is now the fifth-largest city in Latin America, home to almost one-third of Peru's population, and—with an annual rainfall of just 25 mm—the world's second-biggest desert city after Cairo. Since 1950, the mass migration of rural people from the country's highlands has increased its population nine times over. In the past 30 years, the urbanized area has expanded by more than 200 km^2" [1].

For this study, the area around Mateo Pumacahua Av. has been selected. It is in the district of Villa El Salvador in Lima, on the Pacific coast of Peru, whose location is shown in Fig. 1.

Villa El Salvador is located at an altitude of approximately 177 meters above sea level, which is characterized primarily by a sandy desert relief and an arid subtropical climate [2]. In 2018, the authorities carried out paving works to improve pedestrian and vehicular access in this area. Later, in August 2018, several news agencies reported that the main streets of both districts Villa María del Triunfo and Villa El Salvador (see Fig. 1) were flooded due to the continuous drizzles, provoking accidents, and affecting many people. An event is shown in Fig. 2.

In 2017, no such damage was observed in the area, even though from January to March, there were a series of extreme rainfall events on the coast of Peru, which caused serious floods with hundreds of human victims and billions of dollars in economic losses [3].

L. Amaro · J. Luis (✉) · V. Cuya · R. Lorena · S. Dávila · M. Rosana
Peruvian University of Applied Sciences (UPC), Lima, Peru
e-mail: U201524428@upc.edu.pe

R. Lorena
e-mail: u201525303@upc.edu.pe

M. Rosana
e-mail: pccimasi@upc.edu.pe

I. Pal and S. Kolathayar (eds.), *Sustainable Cities and Resilience*, Lecture Notes in Civil Engineering 183, https://doi.org/10.1007/978-981-16-5543-2_10

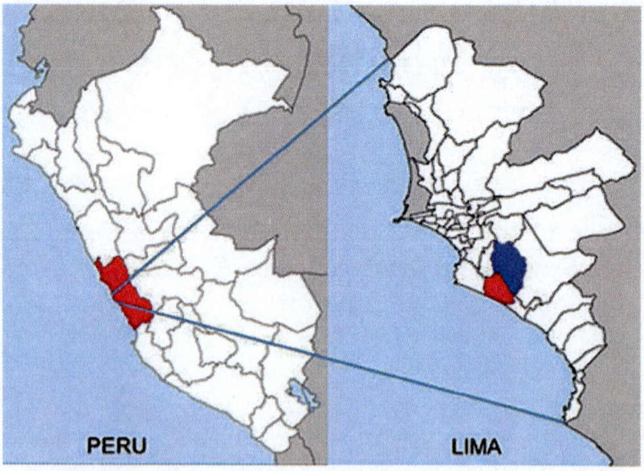

Fig. 1 Left: Map of Peru showing Lima, the capital city. Right: Location of the study site: district of Villa El Salvador (Red). Also, it is shown the neighborhood district Villa María del Triunfo (Blue)

Fig. 2 Mateo Pumacahua Av. was flooded by heavy rains that fall in the districts of Villa María del Triunfo and Villa El Salvador. September 2018

The main objective of this study is to provide an engineering solution to the problem with a sustainable urban drainage system (SUDS). It consists of a set of elements of the drainage network that will allow the collection, transport, decontamination, retention, infiltration, and rainwater evacuation sustainably. Besides this, the collected water will distribute to increase the green spaces of the district.

In the Villa El Salvador Urban Development Plan to 2021, it was reported that the district has 2 m^2 of green area per inhabitant and according to the recommendation by the World Health Organization results in a deficit of 6 m^2 per inhabitant. It also indicates that there are public areas in avenues and parks that can be used for the progressive implementation of green areas. To this goal, it is planned to promote the reuse of properly treated wastewater in existing treatment plants. Also, it is required cooperation with neighboring districts [4].

Successful storm water management requires a transition from conventional to sustainable storm water management over time. The changes must be developed in a socio-institutional framework based on three pillars: cultural-cognitive, normative, and regulatory changes. The elements within the three pillars can work simultaneously and synergistically to achieve widespread change in which the urban and regional planning system can play a crucial role [5].

Also, it is very important to include in the study all the constraints for the implementation of SDUS. Istianto et al. [6] reported their research to solve flood problems in the northern part of Jakarta, Indonesia. One of the main constraints to increasing the volume of storage for controlling floods is the availability of land. It is because most of the areas have been occupied for residential, commercial, and industrial purposes. Meanwhile, the use of the public space available as a park or garden in these areas is not suitable. However, they concluded that it is possible to solve recurring floods using a sustainable urban development system (SUDs) application of a temporary retention basin in the urban area known as an urban polder. This requires a combination of the appropriate design of polder water management components (levees, open water storage, pumping capacity, and outlet, including the adequate selection of both their size and location). Likewise, it is necessary to ensure correct operation: adequate initial condition and permanent maintenance of the drainage systems. Also, the final solution must consider the economic and environmental analysis of the area [6].

2 Determination of Rainfall Excess

2.1 Watershed Characteristics

The watershed of study belongs to Virgen de Lourdes creek. It is in the Villa María del Triunfo district, Lima-Peru, and belongs to the Pacific hydrographic basin. It is showed in Fig. 3 including the study zone, in green, that has a slope of 4%.

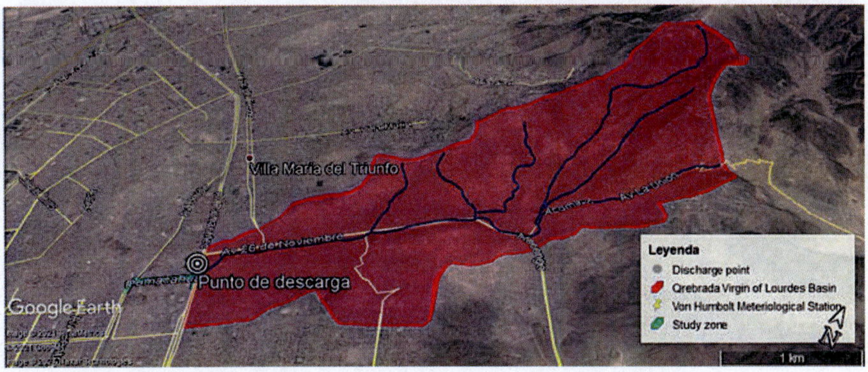

Fig. 3 Watershed limits of Virgen de Lourdes Lima-Peru creek and location of the study zone

The area of the basin is 9.56 km^2, the main channel length is 7.86 km. Its altitude ranges from 160 to 950 meters above sea level with an average bed slope of 10.0%. Their surface is composed of 19% of sand, 23% of gravel, and 58% rock. The entire area is urbanized being 19% for commercial business use and 81% for the residential area.

2.2 Analysis of Maximum Precipitation

There is no rainfall observation station in the study watershed, but it is considered that its precipitation regime is explained by existing measurement data in Von Humboldt meteorological station which is operated by the National Hydrological and Meteorological Service, SENAMHI [7]. This station is located approximately 9 km to the north, in 247 meters above sea level. It has registered daily precipitation with a recording period from 1964 to 2018.

A seasonal regime is observed in which the maximum values of rain appear in January, February, and March. On August 8, 2018, accumulated rainfall depth was 0.2 mm in the early morning but caused flooding in the study area. The observed values of maximum precipitation in 24 h per year have been used for the calculations. The average value of 2.4 mm with a standard deviation of 1.8 mm has been calculated. It was observed that the maximum precipitations of the measurement data did not occur in the years in which the ENSO phenomenon cause serious damages on the Peruvian coast.

For our study, it was considered a useful life of 25 years and an admissible risk of failure of 0.64, obtaining a return time design of 25 years. The Gumbel frequency curve is used for calculating the corresponding maximum precipitation in 24 h obtaining 6.0 mm.

2.3 Design Peak Discharge Estimation

There is no either hydrometric observation station, so the catchment modeling technique was used for estimating 25-year peak discharge.

It was performed precipitation-discharge modeling using a hydrological modeling system, HEC-HMS [8, 9], for the study basin, considering the recommended methodology in Annex N° 01—Hydrology of Standard OS.060 Urban Drainage [10]. A design histogram was carried out using the alternate block technique.

The Soil Conservation Service-curve number method was used for estimating losses. A curve number equal to 91 was identified using soil characteristics. For transforming excess precipitation into a direct runoff hydrograph, the Soil Conservation Service unit hydrograph method was utilized with a concentration–time of 125 min. The computed peak discharge shows a good match with the computed flow using a regional study based on observed flow in several hydrometric stations.

Therefore, the calculated design discharge for a 25-year return period is equal to 0.3 m^3/s with a runoff volume of 6 700 m^3.

3 Determination of the Water Requirement for the Green Area

The study section has a length of 502 m and an available area of 23,217 m^2 was obtained to implement the green spaces considering the land located in the avenues and parks. It is shown in Fig. 4, and the direction of the pluvial flow is also indicated. The area was determined using the plan of the project "Improvement of the Vehicular and Pedestrian Walkability of the Mateo Pumacahua Avenue section between Av. Separadora Industrial and Av. Pachacútec, District of Villa El Salvador—Lima" [11].

In the district of Villa El Salvador, there is a diversity of ornamental plants with low water consumption, of which Aptenia Cordifolia (known as "señoritas") and Ficus, as shown in Fig. 5, were chosen to use in the selected areas. They are very easy to grow and highly resistant to drought and grow quickly in humid soil. These ornamental plants have also been planted on the green coast with good results.

The measurement monthly climatic data in Von Humboldt meteorological station of rainfall, maximum temperatures, minimum temperatures, relative humidity, and

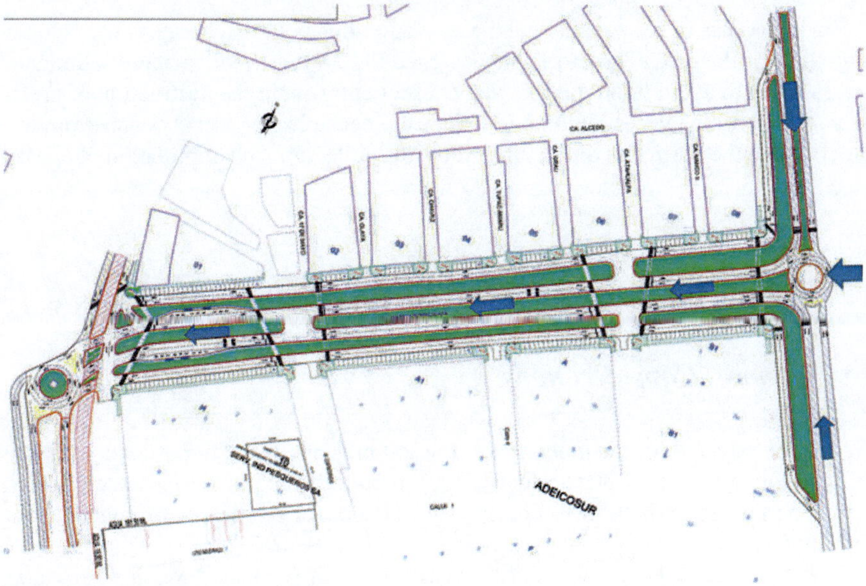

Fig. 4 Green areas and flow direction

Fig. 5 Aptenia Cordifolia (left) and Ficus (right)

wind speed was used for calculation of reference evapotranspiration (ETo) using the CROPWAT (water requirement of the crop) software [12].

Using the crop coefficient of each species, it was calculated the monthly real evapotranspiration by multiplication. Besides, it is necessary to multiply by the efficiency of irrigation.

In Lima, there are long periods with almost zero rainfall; however, there are some periods in which heavy rainfall occurs. Therefore, the water requirement was calculated if the effective precipitation was zero obtaining the maximum water requirement. In humid periods, the rains are enough to satisfy this requirement. Continuous maintenance is necessary by those in charge of the municipality to ensure that the green areas do not disappear due to lack of water.

But in the case of heavy rains and the corresponding peak discharge, water volume is too big for the area. Therefore, the designed SUDS must reduce and/or eliminate excess water to avoid flood damage and consequently avoid the destruction of green areas. A detention tank is required to reduce the peak discharge that comes from the upper part of the basin and ditches that allow the collection and elimination of excess water.

4 Urban Systems Alternatives Sustainable Drainage

4.1 General Considerations

In the zone selected for the green space, the soil presents granular particles with low to light contents of non-plastic fines with a predominance of sandy material. The corresponding slope is 4.58%. The selected factors for the design of channels are shown in Table 1.

It was considered four alternatives of SDUS: rain gardens, French filter water drains, green gutter and vegetation, and detention tanks. Their characteristics are explained below.

Table 1 Selected factors for the design of channels

Factor	Value
25-year peak discharge	300 l/s
Side slope	2:1 (Sandy soil: only channels of a trapezoidal section)
Manning's roughness coefficient	0.02–0.05 (vegetation cover). It was adopted 0.02
Maximum velocity	1.45 m/s
Minimum velocity	0.8 m/s

4.2 Bioretention Areas/Rain Gardens

After the channel is excavated, the area will be divided into three parts; the center will be filled with farmland and the sides with crushed stone which will allow the treatment and runoff infiltration. Subsequently, it will proceed with the planting of Ficus every 6 m and 4 units of Aptenia Cordifolia per m^2, as shown in Fig. 6.

The main advantages of rain gardens are it improves urban landscaping, and its maintenance cost can be incorporated into the general management of urban landscapes; removes urban pollutants efficiently; reduces the runoff coefficient and the volumes of water generated; makes it easier to detect and eliminate elements that hinder its operation.

The disadvantages are that they are not suitable in steep areas, and there is a risk of blockage in the connection with the outlet collector are. As far as its maintenance is expensive, the elimination must be permanent of the residues that obstruct the circulation of the water, periodically prune grass and repair eroded or damaged areas [13].

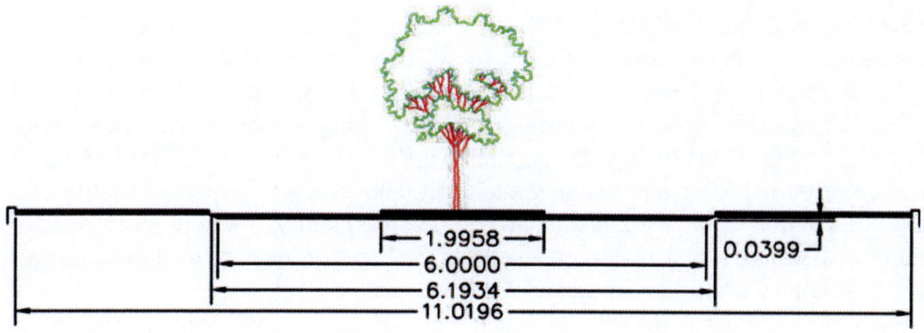

Fig. 6 SUDS rain gardens

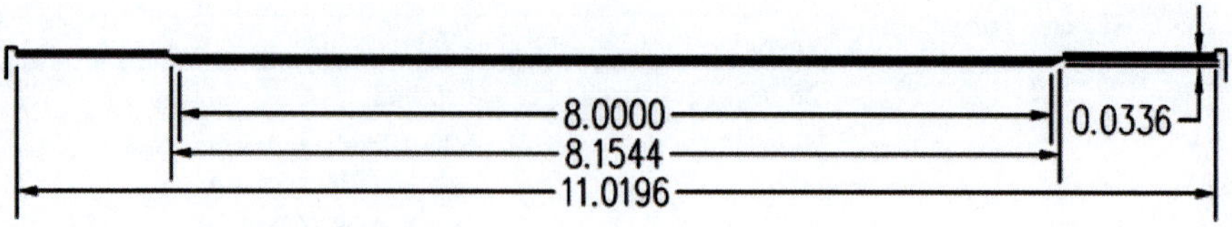

Fig. 7 French SUDS filter drains

4.3 *Filtering Water Drains*

After the channel is excavated, it will be filled with draining granular material that will allow to capture and filter the runoff from contiguous impervious surfaces, promoting the infiltration of the runoff volumes, see Fig. 7.

They are suitable for treating runoff from large adjacent impervious surfaces, helping to reduce runoff with infiltration, facilitating the evaporation of water. They are of easy construction and low cost. A pre-treatment of the water is very effective. They are easily integrated into the landscape and can have a great aesthetic value are their main advantages. The disadvantages are they are not suitable in steep areas; they are not suitable in places where the incoming water has a high pollution load and an aquifer is nearby, and they do not significantly attenuate runoff water volume or peak flow. On the other hand, its maintenance leads to inspect and clean them periodically; they need to care for the present vegetation and repair the eroded and/or damaged areas [13].

4.4 *Green Gutters and Vegetation*

Runoff water flows slowly through them, allowing the entrained particles and pollutants to precipitate and settle, later it will infiltrate into the soil and be captured to satisfy the water requirements of the plants, but for this, the soil must be well-drained.

Since the slope of the study terrain is high, a channel section with a very wide base was proposed to obtain velocities that were not erosive, to which the vegetation also contributes. After the planning, stakeout, and excavation of the canal, excessive compaction of the soil must be avoided. Subsequently, it will proceed with the planting of trees every 6 m along the berm in the border crown, as well as 4 units per m^2 of Aptenia Cordifolia throughout the berm area, see Fig. 8.

Advantages: Easy to incorporate into the landscape; good removal of urban pollutants; they reduce the runoff coefficient and the volumes of water generated; they have low cost and are easy to detect and eliminate those elements that hinder their operation. Disadvantages: they are not suitable in steep areas; the option of putting trees to treat them as garden areas is very limited or not convenient, and there is a risk of blockage in the connection with the outlet collector. Maintenance: elimination of waste and any element that obstructs the circulation of water; periodically prune the

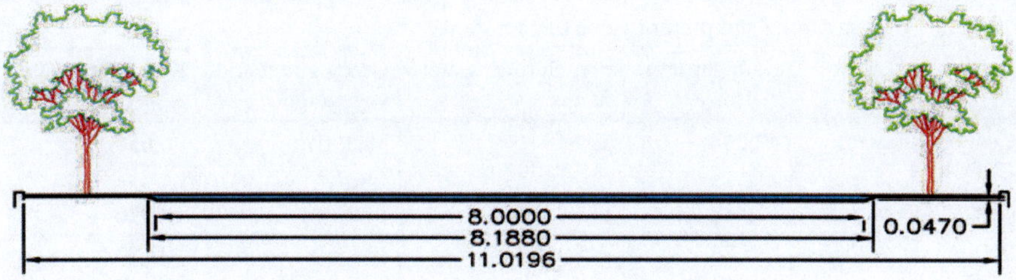

Fig. 8 SUDS green gutter and vegetation design

grass and remove debris; cleaning sewer inlets of debris and sediments and repairing eroded or damaged areas [13].

4.5 Detention Tank

A water reservoir will be proposed to temporarily store the volumes of runoff generated upstream, promoting sedimentation, and thus reducing pollution. The deposit will be located on Av. Pachacútec with a volume of 300 m^3, whose area is 1200 m^2 and a height of 0.25 m. The bed will be covered with the Aptenia Cordifolia, see Fig. 9.

Due to space and budget constraints, the tank has been dimensioned only for 300 m^3, while the design runoff volume is 6700 m^3. Therefore, it will not completely reduce the flooding, but it will reduce the peak of the avenue and increase the time to the peak, which will allow other non-structural prevention measures that have been proposed in the Villa El Salvador Disaster Prevention Plan [2].

Main advantages: It reduces the volume of rainwater generated by runoff, allowing them to be stored and reused for irrigation, the installation of the system is cheap. Disadvantages: If it is necessary to install a treatment system, it becomes quite expensive; sometimes a pumping system is required and is not aesthetically attractive. For their maintenance, they must be inspected and cleaned periodically [13].

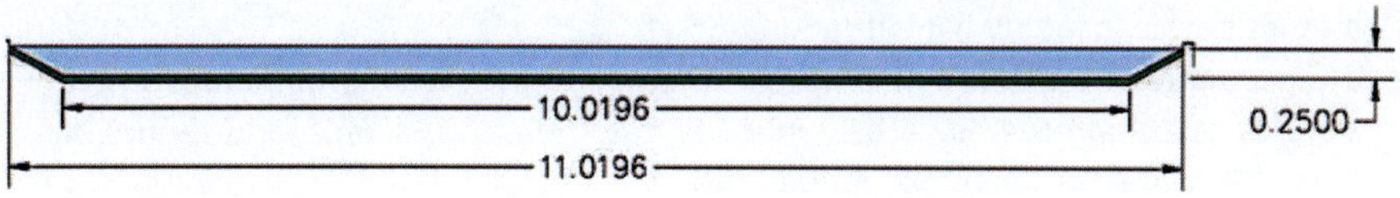

Fig. 9 Design of SUDS detention tanks

Table 2 Costs summary and present value of cost

Item	Rain gardens	French filter water drains	Green gutter and vegetation	Detention tanks
Cost of installation, USD	42,664	52,660	38,390	36,650
Cost of O&M, USD	11,880	11,825	11,119	10,976
Present value costs, USD	135,838	145,408	125,598	122,738

5 Economic and Environmental Evaluation of the Alternatives

5.1 Economic Evaluation

For the economic evaluation of the project, the economic present value of cost indicator was calculated. For this, a 25-year analysis period, the installation cost in year zero, and the annual operation and maintenance cost were considered. Results are shown in Table 2.

The detention tank is a complementary alternative to the others that are exclusive between them. Because the best alternative according to the economic criteria is the one with the lowest present value of cost; among this collection and driving systems, the green gutter and vegetation is selected.

5.2 Environmental Evaluation

All the SUDS alternatives allow two main benefits to the environment: control of the quantity and quality of the water coming from the runoff and improving the landscaping. In addition to [13]:

- Reduction of pollution that reaches the receiving environment.
- Less interference in the natural regimes of the receiving water masses.
- Reduction of the "heat island" effect in cities, counteracting the increase in temperature caused by asphalt and concrete surfaces.
- The recharge of aquifers using these techniques can solve environmental problems such as marine intrusion, subsidence, degradation of wetlands, and reduction of base flows of river channels, among others.

5.3 Selected Alternative

According to the above, for the study site, the detention tanks and green gutter and vegetation, which are complementary, were selected. Figure 10 shows the transversal section of the road with the green gutters (yard), and Fig. 11 shows the plan view.

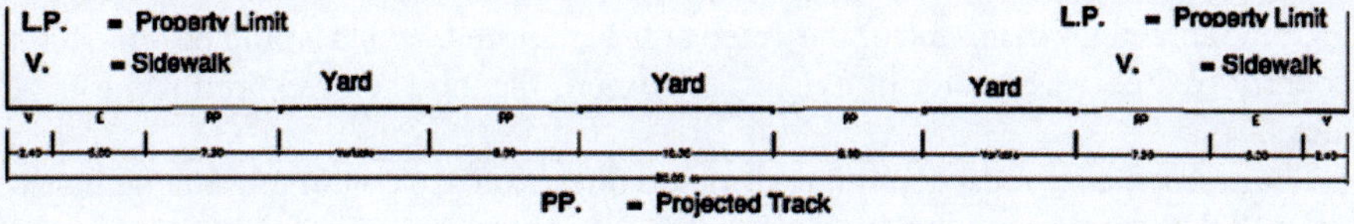

Fig. 10 Road section

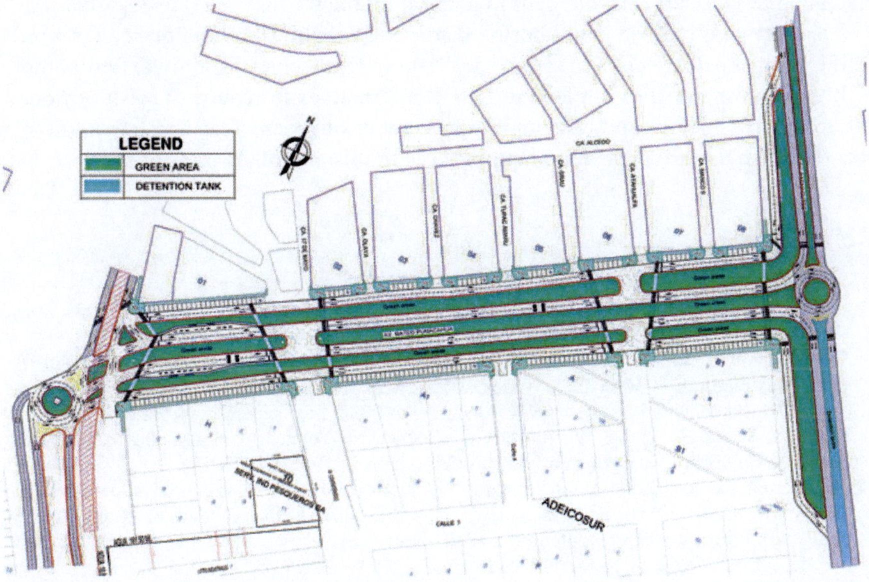

Fig. 11 Design of SUDS selected alternative

As it was mentioned for the maintenance of the green areas, additional irrigation is required. For this, the municipality's logistical management must be compromised to use the treated water in the existing sewage treatment plant for irrigation using tank trucks.

Besides, to achieve good results, it is necessary to complement these measures with the following: proper maintenance of the SDUS components, rigorous control of irrigation and maintenance of green areas, adequate institutional management for the fulfillment of the objectives, and compliance with Villa El Salvador plans of both Urban Development and Disaster Management.

6 Conclusions and Recommendations

The green ditches with detention tank where the adopted solution will allow it to channel the volume of runoff water, and the use for irrigation of the green areas.

Complementary irrigation of the green area is required, which would be done with properly treated wastewater in existing treatment plants when required using tank trucks.

The municipality must commit to the good operation and continuous maintenance of the SDUS system and logistics to irrigate the green area.

The entire drainage system, conventional and SDUS, must be part of the long-term planning of the Villa El Salvador district in which both the Urban Development Plan and the Disaster Management Plan are integrated.

Also, cooperation with neighboring districts is required. Among these, it is worth highlighting the district of Villa María del Triunfo from which the pluvial flow comes.

It is recommended to implement SUDS alternatives to reduce the risk of floods due to the urbanization process and improvement of green areas and to avoid their occurrence in the urban development process in this and other areas of the city.

References

1. FAO Homepage. http://www.fao.org/ag/agp/greenercities/en/GGCLAC/lima.html. Last accessed 14 March 2021
2. Municipalidad Distrital de Villa el Salvador: Plan de prevención y reducción de riesgos de desastres distrital de Villa El Salvador 2019–2022". Peru (2019). CENEPRED Homepage https://sigrid.cenepred.gob.pe/sigridv3/documento/8033. Last accessed 14 March 2021
3. Son, R., Wang, S., Tseng, W., Barreto, C., Becker, E., Yoon, J.: Climate diagnostics of the extreme floods in Peru during early 2017. Clim. Dyn. **54**, 935–945 (2020)
4. Municipalidad de Lima. Plan integral de Desarrollo de Villa el Salvador al 2021. Instituto Metropolitano de planificación. Peru (2006). Municipalidad de Lima Homepage http://imp.gob.pe/wp-content/uploads/2020/09/villa_el_salvador_plan_integral_de_desarrollo.pdf, last accessed 14 March 2021
5. Goulden, S., Portman, M., Carmon, N., Alon-Mozes, T.: From conventional drainage to sustainable stormwater management: beyond the technical challenges. J. Environ. Manage. **219**, 37–45 (2018)
6. Istianto, H., Suryadi, F., Hamim S.: Potentials and constraints of urban polder development in Jakarta, Indonesia. Case study: Rawa Badak polder. In: Proceedings of the 37th IAHR World Congress, Malasia (2017)
7. SENAMHI Homepage: Datos hidrometeorologicos https://www.senamhi.gob.pe/?&p=estaciones. Last accessed 14 March 2021
8. USACE-HEC Hydrologic modeling system HEC-HMS technical reference manual: US Army Corps of Engineers, Hydrologic Engineering Centre. HEC), Davis, USA (2000)
9. USACE-HEC Hydrologic modeling system HEC-HMS v3.2 user's manual. US Army Corps of Engineers, Hydrologic Engineering Center (HEC), Davis, USA (2008)
10. Ministerio Vivienda, Construcción y Saneamiento. Norma OS.060 Drenaje Pluvial Urbano. Ministerio Vivienda, Construcción y Saneamiento. Peru (2006)
11. Municipalidad de Villa El Salvador. "Mejoramiento de la Transitabilidad Vehicular y Peatonal de la Av. Mateo Pumacahua tramo Comprendido Entre La Av. Separadora Industrial y la Av. Pachacutec, Distrito De Villa El Salvador - Lima - Lima". Peru (2017)
12. FAO Homepage. CROPWAT. http://www.fao.org/land-water/databases-and-software/cropwat/es/
13. Sud Homepage. http://sudsostenible.com/beneficios-de-los-suds/, last accessed 14 March 2021

Innovative Construction Interventions

Replacement of River Sand with Coal Bottom Ash as Fine Aggregate in Cement Mortar

Sharmili Wasnik, G. S. Pavan, and Susant Padhi

Notations

A_c	Cross-sectional area of cube
E	Modulus of elasticity of mortar
E_{ms}	Average secant modulus of elasticity of mortar
E_s	Secant modulus of elasticity of mortar
f_c	Compressive strength of cube at 28 days
f_{mc}	Average compressive strength of cube
P	Maximum load at failure
ε	Strain in mortar
ε'_p	Peak strain corresponding to peak stress at 30%
σ_{mp}	Average peak compressive stress of mortar
σ	Compressive stress of mortar
σ'_p	Peak stress at 30

1 Introduction

India is one of the top five largest coal producing countries in the world. In India, coal is the prime source of electricity generation and accounted for nearly 71% of the total electricity produced. Estimation of coal resources in India has reached to 319.02 billion tonnes in 2018 from 298.14 billion tonnes in 2013. Nearly 200 kgs of coal is

S. Wasnik · S. Padhi
Department of Civil Engineering, Shiv Nadar University, Chennai, India

G. S. Pavan (✉)
Department of Civil Engineering, National Institute of Technology Karnataka, Surathkal, India
e-mail: pavan.gs@nitk.edu.in

I. Pal and S. Kolathayar (eds.), *Sustainable Cities and Resilience*, Lecture Notes in Civil Engineering 183, https://doi.org/10.1007/978-981-16-5543-2_11

needed to produce 1 tonne of cement and 300–400 kgs of coal is required to produce 1 cubic metre of concrete. By virtue of thermal power stations, mankind is making the best utilization of coal while assuring that electricity reaches to every household, industry, commercial buildings and every other sector. However, thermal power plants are significantly causing severe impact on health and environment. Thermal power plants, in the course of electricity generation, emit various harmful gases such as carbon dioxide, sulphur dioxide, nitrogen dioxide and particulate matters (PM), which are leading to the increase in global warming. The burning of coal emits hazardous air pollutants that can spread for hundreds of kilometres. Exposure to these pollutants can damage people's cardiovascular, respiratory and nervous systems, increasing the risk of lung cancer, stroke, heart disease, chronic respiratory diseases and lethal respiratory infections.

Moreover, the residues if left untreated are dumped in the landfills which can cause serious hazards as the residues, namely coal bottom ash (CBA) and fly ash, are composed of many oxides and other components. Coal bottom ash is granular and porous particles which are greyish in colour. It has a very rough surface, interlocking characteristics and the particle size ranges from fine gravel to fine sand. The chemical composition of CBA corresponds to fly ash but typically contains greater quantities of carbon. Fly ash is soft fine-powdered material and has pozzolanic characteristics. It is a fine-grained material having mostly spherical, ranging between 10 and 100 micron, and glassy particles. Generally Indian coal contains about 30–45% ash. Considering a thermal power plant of 2000 MW capacity, the daily ash production will be about 13,200 tonnes out of which 2200 tonnes will be CBA and remaining 11,000 tonnes will be fly ash. The amount of coal ash produced per day is estimated to be 2.6 million tonnes out of which 0.43 million tonnes is the estimated production of CBA. Environmentally safe disposal of large quantities of coal ash is not only tedious and expensive, but also possesses its own set of challenges. Disposal of ash results in increased land usage, capital loss to power plants, health hazards, ecological imbalances and related environmental problems. Therefore research studies on utilization of industrial byproducts like flyash, mine tailings, copper slag, coal bottom ash as construction materials is necessary and studies have been carried out in literature towards this direction [1–8]. Due to these real challenges and to address the problems of pollution caused by the same, government agencies are mandating the usage of fly ash-based building products such as cement or concrete, fly ash bricks, blocks and tiles, in all construction projects, in road or flyover embankment construction and in reclamation of low-lying areas within 100 km of radius of a thermal power plant. From 55.79% in 2011, the utilization has been increased up to 78.19% till the first half of 2020. This indicates that there has been a significant reduction in the improper disposal of fly ash.

There is, however, fewer data available on the utilization of coal bottom ash. The present study is based on the incorporation of CBA in masonry mortar. Because of the improper treatment leading to detrimental consequences, disposal of coal bottom ash has become a primary concern in India. In addition, sand is a limited resource and is depleting day by day. Mining of aggregates has surged beyond the purview of the given guidelines and thus threatening the environment. The excessive demand has led to the scarcity of sand. Hence, a substitute for sand has become thoroughly indispensable.

The aim of this study is to assess the suitability of CBA as a fine aggregate in masonry mortar, either with complete replacement or partial replacement of river sand. The main objective of this study is the evaluation of mechanical properties such as compressive strength and modulus of elasticity of mortar composed of CBA as a fine aggregate.

2 Experimental Programme

Mortar adopted for masonry construction comprises of binder, fine aggregate and water. Ordinary portland cement of grade 43 was utilized as a binder for the preparation of masonry mortar. The ratio of cement to fine aggregate adopted for masonry mortar in this study is 1:6 by weight. River sand, which is generally utilized as a fine aggregate in cement mortar mix, is now replaced with coal bottom ash in percentages of 25, 50, 75 and 100. Series of experiments are conducted to investigate the properties of cement mortar with partial replacement of sand with coal bottom ash as a fine aggregate. Coal bottom ash procured from National Thermal Power Corporation Limited (NTPC) in Dadri and locally available river sand are employed.

All the combinations of cement mortar mixes adopted in this study are listed in Table 1. Mix CBA0 represents the control mix, where the abbreviation CBA denotes coal bottom ash and the following number indicates the percentage of coal bottom ash in fine aggregate. Similarly, the numbers 25, 50, 75 and 100 represent the percentage content of coal bottom ash in fine aggregate of mixes CBA25, CBA50, CBA75 and CBA100, respectively.

Particle size distribution (PSD) of fine aggregates containing varying proportion of sand and coal bottom ash was carried out through sieve analysis conforming to Indian Standard specifications (I.S. 2116–1980 and I.S. 2386 (Part 1)—1963). The fineness modulus (FM) of river sand and coal bottom ash is 2.09 and 1.2, respectively. Figure 1 shows the PSD curves obtained for combination of sand and coal bottom ash. Water/cement (W/C) ratio was determined by the mortar flow table test as per the guidelines laid by BS: 4551—1980 and IS: 5512—1969. Figure 2 shows the

Table 1 Mix proportion of mortar

Mix design	Cement	Fine aggregate	Sand (%)	Coal bottom ash
CBA0	1	6	100	0
CBA25	1	6	75	25
CBA50	1	6	50	50
CBA75	1	6	25	75
CBA100	1	6	0	100

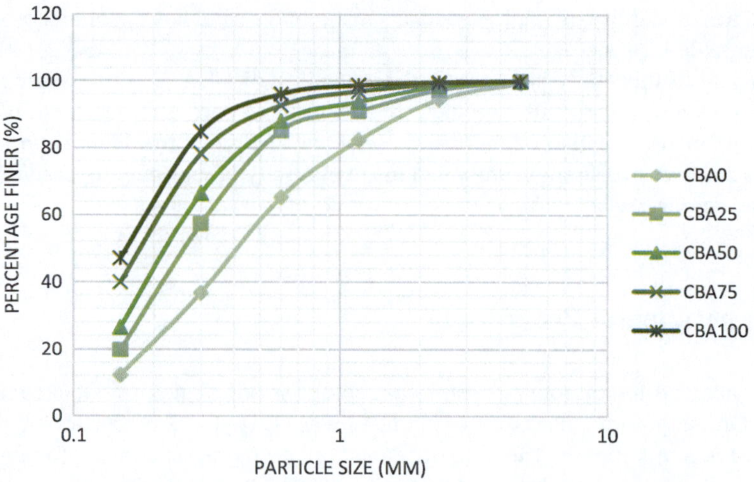

Fig. 1 Particle size distribution of fine aggregate used in different mortar mixes

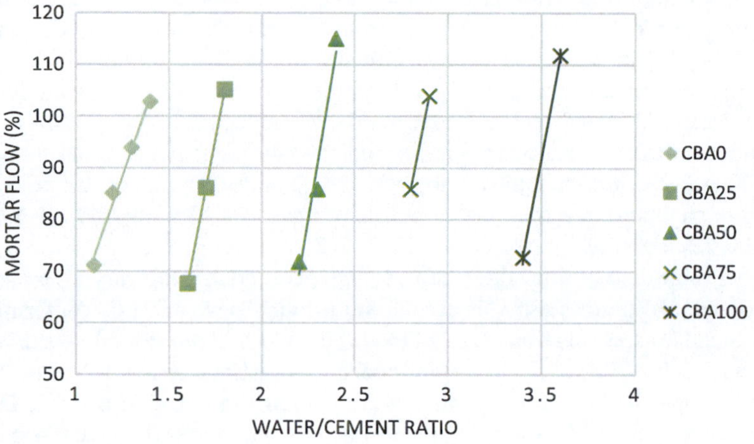

Fig. 2 Mortar flow versus water/cement ratio for different mortar mix

plot of percentage mortar flow versus water/cement ratio. The water/cement ratio corresponding to a flow of 85% is determined and is listed in Table 3.

2.1 Casting of Specimens

A total of 15 cubes, each of size 70.5 mm, was cast (three per mix) to determine the compressive strength of the cement mortar. A total of twenty cylinders, each of size 150 mm × 300 mm, was cast (four per mix) to determine the modulus of elasticity of cement mortar. The inner surface of cubical and cylindrical moulds was lubricated properly and filled with mortar mix in three layers followed by tamping 25 times using a tamping rod. All the fifteen cubes and twenty cylinders were demoulded after 24 h of casting and stored for curing in a water tank for a period of 28 days. Specimens casted for all the experiments are denoted as follows:

Specimens (cubes/cylinders) of mix CBA0 are represented as A_1, A_2, A_3, etc.
Specimens (cubes/cylinders) of mix CBA25 are represented as B_1, B_2, B_3, etc.
Specimens (cubes/cylinders) of mix CBA50 are represented as C_1, C_2, i_3, etc.
Specimens (cubes/cylinders) of mix CBA45 are represented as D_1, D_2, D_3, etc.
Specimens (cubes/cylinders) of mix CBA100 are represented as E_1, E_2, E_3, etc.

2.2 Testing of Specimens

After 28 days of curing, the cubes were subjected to compression under the CTM at a constant loading rate of 0.01 mm/s, maximum load at which the cube fails and cracks were recorded. Figure 3 shows the testing of cubes under compression testing machine (CTM).

The cylinders were tested under the universal testing machine (UTM) of capacity 1500 KN. The cylinder was placed in the centre and fixed properly between the two plattens of the UTM. Compressive load was applied at a rate of 3 microns/second. Two dial gauges were attached to the cylinder with the help of L angles to determine deformation. Distance between the two dial was considered as gauge length and used

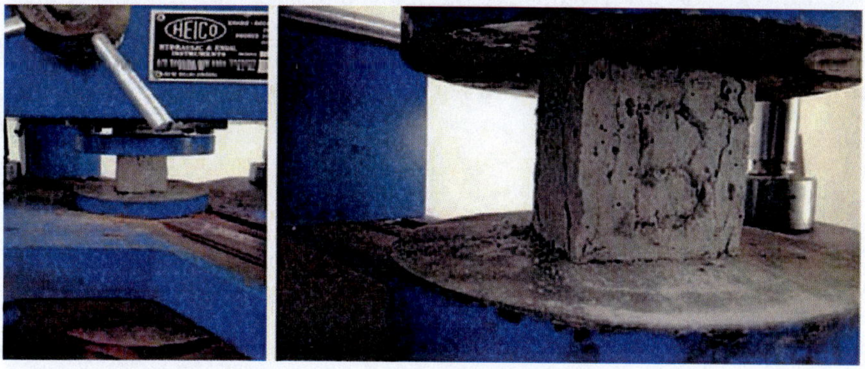

Fig. 3 Compressive strength of cubes at 28 days

Fig. 4 Set-up of cylinder specimen A under UTM

to determine axial strain. The deformation was recorded with every 4 KN of load until it reaches the peak stress and plummets till the breaking point. The set-up of cylinder in the UTM is depicted in Fig. 4.

3 Results and Discussions

This section discusses the results obtained from the different experiments conducted on masonry mortar. Figure 1 depicts the variation in the distribution curves of different combinations of the fine aggregate considered in this study. Mix CBA100 exhibits the size of particles finer than the control mix. Besides, the curves of other three mixes lie between the mix CBA100 and CBA0. Thus, from these curves, it can be interpreted that as the content of coal bottom ash increases in a mix, the particles become finer. Table 2 shows the fineness modulus of all the five combinations. The

Table 2 Fineness modulus of different fine aggregate combinations

Mix type	Fineness modulus
CBA0	2.09
CBA25	2.49
CBA50	2.26
CBA75	1.93
CBA100	1.2

Table 3 Details of water/cement ratio for different mortar mixes

Mix type	Mortar flow (%)	Water/cement ratio
CBA0	85.16	1.22
CBA25	86.16	1.69
CBA50	85.91	2.26
CBA75	85.91	2.8
CBA100	84.48	3.46

flow characteristics of mortar mix were determined by mortar flow table test. It was observed that the requirement of water kept increasing with the addition of coal bottom ash. W/C ratio for mix CBA100 and CBA0 was 3.46 and 1.22 for a similar flow value of 85%, whereas the w/c ratio of other mixes falls in between. Table 3 shows the details of water/cement ratio. A similar experiment was conducted on different grades of sand in cement mortar and cement—lime mortar [9]. In their study, it is explained that mortar with finer-sized particles requires more water to attain similar flow values when compared to medium-sized particles in cement or cement—lime mortar. Since the *w/c* ratio is increasing linearly with the percentage of coal bottom ash, it can be said that as the particle size becomes finer the water demand surges. Bottom ash particles possess a tendency of holding water within as its surface texture is rough and porous which affects the mortar flow [10]. The plot of mortar flow against water/cement ratio is depicted in Fig. 2.

3.1 Compressive Strength of Mortar

Results of mortar cubes subjected to compression are shown in Table 4. A continuous decline in compressive strength of mortar (f_c) was observed. The average compressive strength (f_{mc}) of each cubes containing coal bottom ash was lower than the control mix (CBA0). Cubes cast out of mix CBA100 were the most fragile as it did not gain sufficient strength within 24 h of casting, therefore crumbled during the process of demoulding. Moreover, those which sustained further disintegrated in water in the process of curing. This explains that mix CBA100 requires longer duration than 24 h to attain sufficient strength. The f_{mc} of mix CBA75 was observed to be the lowest. The average strength of mix CBA25 was lower than the control mix, however higher among those containing some percentage of coal bottom ash. The reduction in strength can be attributed to the higher w/c ratio due to the fineness of particles which results in a reduction in dry density and thus lower compressive strength [9]. Reference [11] considers the loss in strength due to increase in water demand and poor interlocking between the aggregate particles when replaced with coal bottom ash. Table 4 shows f_c and f_{mc} at 28 days. The compressive strength is calculated as per IS: 2250 -1981 and is given by (Fig. 5)

Table 4 Compressive strength of cement mortar cubes

Specimen	Surface area (mm^2)	Load (KN)	Compressive strength (MPa)	Mean compressive strength (MPa)
A_1	4970.25	44.8	9.01	9.54
A_2	4970.25	45.6	9.17	
A_3	4970.25	51.9	10.44	
B_1	4970.25	37.7	7.58	7.10
B_2	4970.25	35.6	7.16	
B_3	4970.25	33.5	6.74	
B_4	4970.25	34.3	6.9	
C_1	4970.25	34	6.84	5.38
C_2	4970.25	21.3	4.28	
C_3	4970.25	25	5.029	
D_1	4970.25	19.6	3.94	3.90
D_2	4970.25	17.7	3.561	
D_3	4970.25	20.9	4.2	
E_1	4970.25	26.4	5.31	5.31

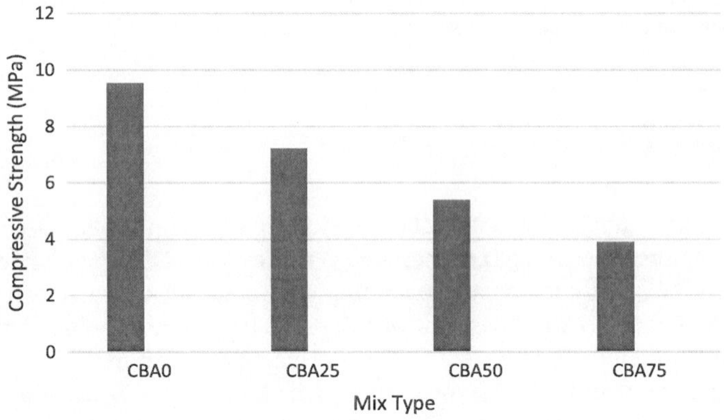

Fig. 5 Graph showing compressive strength of cubes at 28 days

$$f_c = \frac{P}{A_c} \tag{1}$$

where

f_c compressive strength of cube at 28 days in N/mm^2,
P maximum load at failure in N,
A_c cross-sectional area of cube

Table 5 Elastic properties of cement mortar cylinders

Specimen	Peak stress	30% peak stress	Secant modulus
CBA0	8.092	2.428	2686.675
CBA25	6.304	1.891	618.718
CBA50	2.150	0.645	1087.202
CBA75	1.471	0.441	445.848

3.2 Modulus of Elasticity of Masonry Mortar

The elastic properties of mortar combinations are shown in Table 5. The average peak stress (σ_{mp}) of control mix was found to be 8.092 MPa at a peak strain of 0.00185. The average secant modulus (E_{ms}) of the control mix and mix CBA75 was 2686.675 MPa and 445.848, respectively. The modulus of elasticity (E) of the control mix measured at 28 days was the highest and mix CBA75 exhibited the lowest. The modulus of elasticity of mix CBA50 was slightly higher than mix CBA25. The durability and stiffness of mix CBA0 were found to be quite durable as compared to the other mortar mixes which have low resistance to deformation. [11] reported that the huge variation is accredited to the fineness of the aggregate. As the content of coal bottom ash in mortar increases, the modulus of elasticity decreases. [10] reported in their study that mortar having fine bottom ash aggregates exhibits lower modulus of elasticity as compared to normal fine aggregates, i.e. sand. Higher porosity and higher *w*/*c* ratio resulted in lower modulus of elasticity and strength. However, partial replacement of coal bottom ash, i.e. mix CBA50, provides slightly higher durability in comparison with mix CBA25 and mix CBA75. Figure 6 shows the stress–strain curve of masonry mortar. The modulus of elasticity is calculated using the Hooke's law given below

$$E = \frac{\sigma}{\varepsilon} \tag{2}$$

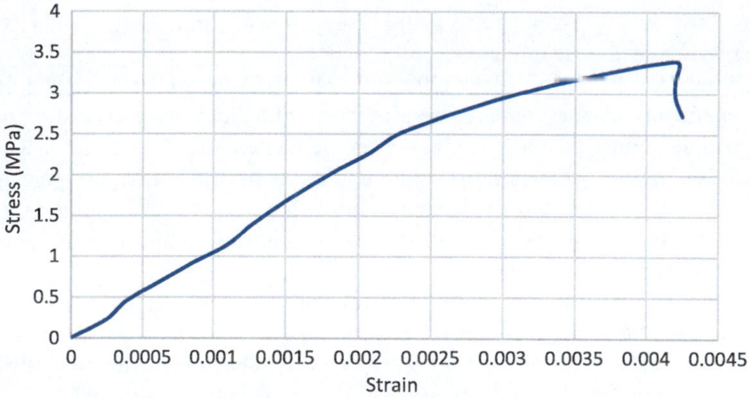

Fig. 6 Stress–strain curve for CBA50

where

E modulus of elasticity of mortar in N/mm^2,
σ stress in N/mm^2 and
ε strain. The secant modulus of elasticity at 30% peak stress is expressed as follows:

$$E_s = \frac{\sigma_p'}{\varepsilon_p'} \quad (3)$$

where

E_s secant modulus of elasticity,
σ_p' peak stress at 30% and
ε_p' peak strain corresponding to peak stress at 30%.

4 Conclusion

The fineness modulus of three samples of river sand was 2.09, 2.11 and 2.19 that falls in the range of 2.2–2.6 exhibiting fine sand, which is reliable to be used in the masonry works. From the PSD curves of all combinations, it can be understood that mix CBA25, CBA50, CBA75 and CBA100 are finer in comparison with the control mix (CBA0). The fineness modulus begins to plunge with the increase in the coal bottom ash. Table 2 shows the fineness modulus obtained for all the mortar mixes. However, mix CBA25 and CBA50, having fineness modulus 2.49 and 2.6, respectively, fit in the range of 2.2–2.6 whereas mix CBA75 and CBA100, containing higher percentages of coal bottom ash, exhibit a fineness modulus of 1.93 and 1.2, respectively. This indicates that mix CBA25 and CBA50 are the most suitable to be used as a substitute for river sand as a fine aggregate while mix CBA75 and CBA100 are finer which seem less reliable.

The water/cement ratio of mix CBA100 and CBA0 was evaluated as 3.46 and 1.22, respectively. Higher water/cement ratio could be attributed to be the fineness of the particles. Finer the particles, higher is the water/cement ratio. Table 3 shows the reduction in water/cement ratio with the increase in coal bottom ash percentage. The particles of coal bottom ash have a rough and porous texture due to which the particles absorb and hold water within and release it during compaction which makes it difficult to work with.

The compressive strength (f_c) of the control mix is higher than all the other combinations. The average compressive strength (f_{mc}) of mix CBA75 and control mix was found to be 3.90 and 9.54 MPa, respectively. The lower strength is attributed to the excess amount of water absorbed by the coal bottom ash particles. Higher porosity and water/cement ratio lead to a lower compressive strength.

The modulus of elasticity (E) of mix CBA0, CBA25, CBA50 and CBA75 measured at 28 days was 1519.20, 722.210, 791.583 and 338.560 MPa, respectively. This explains that the durability of control mix is higher in comparison with other mortar mix. Mix CBA75 fails to offer higher resistance, and its lower durability is attributed to the presence of higher percentages of coal bottom ash due to its low fineness modulus. Mix CBA50 offers the best durability among the other mortar combination containing coal bottom ash.

From the results obtained, it is understood that the performance of mortar mix CBA75 and CBA100 was not remarkable. Mostly due to the higher content of coal bottom ash, they could not offer higher strength, stiffness and durability. Mortar mix CBA25 and CBA50 seem to render lower strength than the control mix; however, they provide higher strength than mix CBA75 and CBA100 in every aspect. Strength and deformation of CBA25 and CBA50 are sufficient for them to be used as mix proportion for masonry mortar.

References

1. Ghafoori, N., Bucholc, J.: Investigation of lignite-based bottom ash for structural concrete. J. Mater. Civil Eng. **8**(3), 128–137 (1996)
2. Humam, T., Siddique, R.: Properties of mortar incorporating iron slag. Leonardo J. Sci. **1**(23), 53–60 (2013)
3. Kim, H.K.: Utilization of sieved and ground coal bottom ash powders as a coarse binder in high-strength mortar to improve workability. Constr. Build. Mater. **91**, 57–64 (2015)
4. Lal, D., Chatterjee, A., Dwivedi, A.: Investigation of properties of cement mortar incorporating pond ash–an environmental sustainable material. Constr. Build. Mater. **209**, 20–31 (2019)
5. Naganathan, S., Mohamed, A.Y.O., Mustapha, K.N.: Performance of bricks made using fly ash and bottom ash. Constr. Build. Mater. **96**, 576–580 (2015)
6. Shakir, A.A., Naganathan, S., Mustapha, K.N.: Properties of bricks made using fly ash, quarry dust and billet scale. Constr. Build. Mater. **41**, 131–138 (2013)
7. Singh, G., Siddique, R.: Effect of iron slag as partial replacement of fine aggregates on the durability characteristics of self-compacting concrete. Constr. Build. Mater. **128**, 88–95 (2016)
8. Singh, M., Siddique, R.: Effect of coal bottom ash as partial replacement of sand on properties of concrete. Resour. Conserv. Recycling **72**, 20–32 (2013)
9. Reddy, B.V., Gupta, A.: Influence of sand grading on the characteristics of mortars and soil-cement block masonry. Constr. Build. Mater. **22**(8), 1614–1623 (2008)
10. Kim, H.K., Jeon, J.H., Lee, H.K.: Flow, water absorption, and mechanical characteristics of normal-and high-strength mortar incorporating fine bottom ash aggregates. Constr. Build. Mater. **26**(1), 249–256 (2012)
11. Siddique, R., Kunal.: Design and development of self-compacting concrete made with coal bottom ash. J. Sustain. Cement-Based Mater. **4**(3–4), 225–237 (2015)
12. Al-Jabri, K.S., Al-Saidy, A.H. and Taha, R.: Effect of copper slag as a fine aggregate on the properties of cement mortars and concrete. Constr. Build. Mater. **25**(2), 933–938 (2011)

Strength and Durability Properties of Alkali-Activated Fly Ash Earth Bricks

G. S. Vasavi, R. Mourougane, and G. S. Pavan

1 Introduction

Bricks, generally employed in masonry wall construction, constitute a major portion of a structure. Different types of bricks such as conventional burnt clay bricks, concrete blocks, hollow bricks, sun dried bricks, fly ash bricks, porotherm bricks, and sand-lime bricks are used in construction purposes. But, burnt clay bricks constitute a significant percentage of use among all the bricks available. These bricks are being used from many generations. But, the production of these bricks consumes a large amount of energy as the bricks are burnt at high temperatures in a kiln.

In today's era, we are constantly searching for more sustainable and energy-efficient alternatives to prevent the exploitation of natural resources. Soil-stabilized blocks are a good option to replace the burnt clay bricks and are currently being used in many parts of the world successfully. These blocks are manufactured by compacting the processed soil (mixture of soil stabilizer—generally cement and water) manually or automated using a brick-making machine [1]. The embodied energy of these bricks, though is lower when compared to that of burnt clay bricks, the use of cement again adds to the carbon footprint and increases the embodied energy [2]. To address this problem, alkali-activated bricks look like a promising alternative. Instead of cement in soil-stabilized blocks, alkali-activated binder can be used. Generally, pozzolanic materials such as fly ash, metakaolin, and rice husk ash along with the alkali activator are used as binder. The alkali activation of these pozzolanic materials is also known as geopolymerization.

In the current study, fly ash is adopted as the pozzolanic material. Fly ash is the by-product of coal-based thermal power plant. Indian coal contains ash content in

G. S. Vasavi · R. Mourougane
Ramaiah Institute of Technology, Bengaluru, India

G. S. Pavan (✉)
National Institute of Technology Karnataka, Surathkal, India
e-mail: pavan.gs@nitk.edu.in

I. Pal and S. Kolathayar (eds.), *Sustainable Cities and Resilience*, Lecture Notes in Civil Engineering 183, https://doi.org/10.1007/978-981-16-5543-2_12

the range of 30–45% [3]. Hence, enormous quantities of fly ash is being produced in India every year. In the year 2018–19, more than 200 million tonnes of fly ash was generated. The percentage of utilization of fly ash amounts to 77.59% [3]. Though significant amount of fly ash is being utilized, 100% utilization is not happening, and heaps of fly ash are already present in fly ash ponds near thermal power plants. Therefore, fly ash can be used to produce alkali-activated bricks.

Soil-stabilized bricks show satisfactory strength, durability, and thermal conductivity characteristics and can serve as a potential alternative to conventional burnt clay bricks [4]. It also is environmentally friendly and sustainable alternative. Soil with non-expansive clay mineral up to 16% can be considered for manufacturing the soil-stabilized bricks. Sandy soils serves the purpose better than clayey soils [5]. Alkali activation of fly ash involves dissolution of Si and Al ions from fly ash, condensation of precursor ions into monomers followed by polymerization of monomers, and hardening of gel [6].

This study focuses on producing alkali-activated fly ash earth bricks using soil, fly ash, M-sand, alkali activator, and water and assessing its strength and durability characteristics which are investigated by conducting relevant tests on the bricks.

2 Experimental Programme

The experimental programme involved procurement of materials, characterization of soil by performing various tests on soil, deciding on mix proportion, casting of specimens, and curing and testing of bricks by conducting relevant tests. Two variants of bricks were produced, one set with the use of M-sand and another without M-sand. Bricks with M-sand are designated as SFM bricks, and bricks without M-sand are designated as SF bricks.

2.1 Characterization of Soil

The soil was procured from a site at Konankunte, Bengaluru, for the current investigation. It is reddish-brown in colour and feels sandy. Figure 1 shows the picture of soil considered for investigation. Various tests were conducted on soil to estimate the different parameters.

The soil contained 57% sand, 27% silt, and 14% clay. The liquid limit was found to be 42%, and plastic limit was found to be 22%. The plasticity index was estimated to be 20%. The specific gravity of soil was estimated to be 2.64. Figure 2 shows the grain size distribution curve or particle size distribution curve obtained by performing both sieve analysis and hydrometer analysis.

The different characteristics of soil are tabulated in Table 1 given below. According to IS 1498–1970 [12] soil classification, the soil was classified to be clayey sand which can also be abbreviated as SC.

Fig. 1 Soil

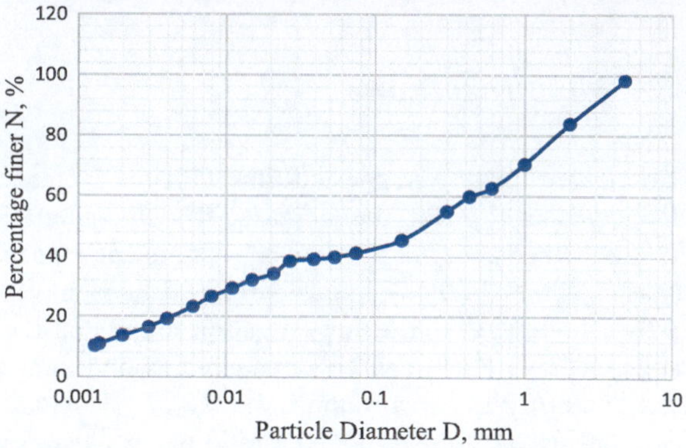

Fig. 2 Particle size distribution curve

Table 1 Soil properties

Parameter	Result
Soil composition	
Gravel (>4.75 mm)	1.61%
Sand (>75μ and <4.75 mm)	57.08%
Silt (>2 μ and <75 μ)	27.54%
Clay (<2 μ)	13.77%
Atterberg limits	
Liquid limit	41.80%
Plastic limit	21.60%
Plasticity index	20.20%
Natural water content	8.60%
Specific gravity of soil	2.64
Specific gravity of soil passing 75 μ	2.5

Fig. 3 Compaction curve

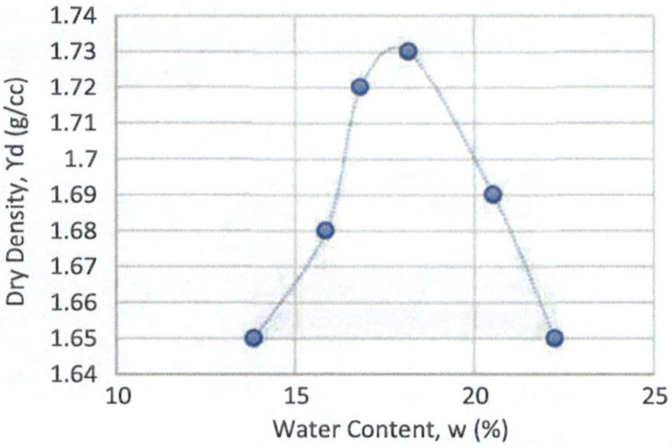

2.2 *Mix Proportion*

Earlier investigation suggested that with the increase in fly ash content, strength also increases [7]. But, the increased fly ash content demands more alkali activator solution which does not prove to be economical. Hence, to make it more economical, fly ash/soil ratio of 0.4 was considered which means 40% of soil mass was considered as fly ash content. Earlier investigation found that the Na2SiO3/NaOH ratio in the range of 0.5–1 [7–9] can be considered as optimum value, and the optimum ratio of activator/fly ash was found to be 0.6 [7, 8]. Alkali activator/fly ash ratio of 0.6 was considered. Na_2SiO_3/NaOH ratio of 1 was considered. This means that equal weights of sodium hydroxide and sodium silicate were considered for the preparation of alkali activator solution.

For bricks with M-sand, 40% of the cumulative mass of soil and M-sand was considered as fly ash amount. Soil and M-sand were added in 1:1 ratio for producing bricks with M-sand. The (alkali activator)/fly ash ratio of 0.6 was followed. The optimum moisture content for the designed mix was determined by conducting the standard compaction test, and it was found to be 18%. This included the water in alkali activator, water content present in soil, and additional water added to the mix. The compaction curve can be seen in Fig. 3. The additional water of 6% of (soil + fly ash) was considered for bricks without M-sand, and 6% of (soil + m-sand + fly ash) was adopted for bricks with M-sand content.

2.3 *Casting Procedure*

The manufacture of bricks involved soil, fly ash, alkali activator, M-sand, and water. We have seen the characterization of soil in Sect. 2.1. Class F fly ash is used in the manufacture of bricks as it contains high-reactive alumina and silica required for alkali activation. Alkali activator is a combination of sodium hydroxide and sodium silicate solution. Sodium hydroxide was used in the pellets form. In the current investigation, 10 M NaOH solution was used. Sodium hydroxide solution was prepared 24 h prior to testing. Laboratory-grade sodium silicate was used in the preparation of alkali activator. Prior to casting, both sodium silicate and sodium hydroxide solution were mixed which is known as alkali activator. Two variants of bricks were produced, one set with the use of M-sand (SFM) and another without M-sand (SF). Density of the wet mix of the brick was maintained at 2000 kg/m^3.

Compacted alkali-activated fly ash earth bricks of size 114 mm × 102 mm × 230 mm (4.5″ × 4″ × 9″) were prepared using a manually operated brick-making machine. The brick-making machine was of the fixed stroke kind and thus produced bricks of equal thickness. The manufacture of bricks can be condensed into the following steps:

1. Soil was first sieved through a 5 mm sieve to avoid gravelly stones and other large particles.
2. Soil, fly ash, and M-sand were first dry mixed thoroughly.
3. Required amounts of sodium hydroxide, sodium silicate, and excess water were taken in a bucket and mixed thoroughly to produce a homogeneous alkali activator solution.
4. The solution was added to the mixture of soil and fly ash or soil, fly ash, and M-sand for bricks with the use of M-sand. With the help of spades, the mixture was thoroughly mixed, and some lumps were unavoidable due to the presence of viscous sodium silicate.
5. Required amount of the damp mixture was added into the brick mould of the machine, and the lid was closed. The compaction was carried out by the movement of a piston attached to the toggle lever.
6. The lid was opened, and the brick was ejected out and stacked in a ventilated place. The bricks were allowed to dry in open atmosphere until the day of testing. Figures 4, 5, 6, 7, 8, 9, 10, and 11 represent the stages involved in casting of bricks.

2.4 *Tests Conducted*

Various tests were conducted on these bricks to assess its strength and water absorption characteristics. Dry and wet compressive strength tests, flexure test, and split tensile test were conducted to determine its strength characteristics, and complete saturation test was conducted to determine the water absorption characteristic which is also a measure of durability property. For dry compressive strength test, five bricks in both the variants were considered. For remaining tests, three bricks for each variant were considered.

Fig. 4 Sieving of soil

Fig. 5 Soil and fly ash mixture

3 Testing Procedures

This section details the procedures adopted in performing the tests.

3.1 Dry Compressive Strength

The bricks were first oven-dried at 60 °C for 24 h. The frogs of the brick were filled with Plaster of Paris to get a levelled surface. Later, the bricks were placed in

Fig. 6 Soil, M-sand, and Fly ash mixture

Fig. 7 Addition of alkali activator

Fig. 8 Filling of brick mould

Fig. 9 Brick ejection

Fig. 10 SF bricks

Fig. 11 SFM bricks

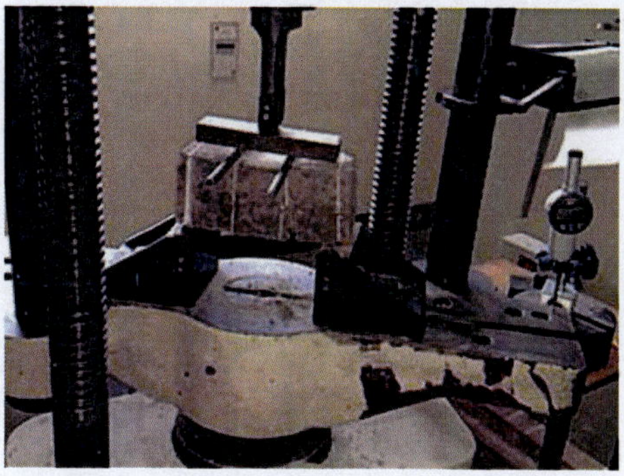

Fig. 12 Failure of brick in flexure test

compression-testing machine in between two plywood sheets. The ultimate crushing load was noted, and dry compressive strength was calculated.

3.2 *Wet Compressive Strength*

This test was performed according to [10]. The bricks were first immersed in water for 24 h. After removing the specimen from water bath, excess moisture was wiped out, and the frogs were filled with Plaster of Paris. Later, the bricks were tested in the compression-testing machine. The ultimate load was noted, and the wet compressive strength was determined.

3.3 *Flexure Test*

Flexural strength was assessed by adopting four-point bending test. Ambient cured bricks were used for this test. The load was applied through the two steel bearings on the upper bedding surface of the brick. The load at failure was used to calculate the modulus of rupture or flexural strength of the brick. Figure 12 shows the failure pattern of the brick.

3.4 *Split Tensile Test*

This test was conducted according to [11]. Ambient cured bricks were used for the test. The frogs were first filled with Plaster of Paris to achieve a levelled surface. Steel loading pieces of 6 mm width and 4 mm depth were placed on the central lines of two bedding faces, and this assembly was placed in the compression-testing machine. The load at failure was used for calculating the splitting tensile strength of

Fig. 13 Split tensile test setup

the bricks. Figure 13 shows the setup of the test, and a vertical failure at the central span of the brick was observed which splits the brick into two equal halves.

3.5 Complete Saturation Test

Ambient cured bricks were oven-dried at 100 °C for a day. Later, it was weighed and immersed in water for 24 h. It was taken out and weighed again. Later, the water absorption of the bricks was determined by utilizing the difference in weights of the two.

Table 2 Characteristics of SFM and SF bricks

Characteristic	Bricks with M-sand (SFMB)	Bricks without M-sand (SFB)
Dry compressive strength (MPa)	8.84	8.41
Wet compressive strength (MPa)	6.27	5.69
Dry density (g/cc)	1.83	1.83
Split tensile strength (MPa)	0.55	0.49
Water absorption (%)	10.91	10.5
Flexural strength (MPa)	0.93	0.91

4 Results and Discussions

Table 2 below represents the different characteristics of the bricks with and without M-sand. The values are obtained by performing the respective tests according to standard specifications. The dry compressive strength varies from 8 to 10 MPa. The average dry compressive strength of the SF bricks was found to be 8.41 N/mm^2. The average dry compressive strength of the SFM bricks was found to be 8.84 N/mm^2.The wet compressive strength is around 70% of the dry compressive strength. The average wet compressive strength of the SF bricks was found to be 5.69 N/mm^2. The average wet compressive strength of the SFM bricks was found to be 6.27 N/mm^2. The splitting tensile strength of the bricks is in the range of 0.47–0.55 MPa. The flexural strength of these bricks ranges from 0.85 to 1.01 MPa. The average dry density of the bricks was found to be 1.83 g/cc.

Compressive strength of these bricks varied from 8 to 10 MPa in dry condition, which is similar to the compressive strength of the first class burnt clay bricks which vary from 7 to 10.5 MPa. The water absorption of the bricks is less than 15%. Hence, these bricks are promising alternatives to common burnt clay bricks. Alkali-activated fly ash earth bricks also are more sustainable and have lower carbon footprint, and results in saving of transportation costs as the bricks are produced at site.

It was observed that the performance of SFM bricks with M-sand as fine aggregate is slightly higher strength than that of bricks without M-sand, though the difference is marginal. SFM bricks also possessed perfect edges in comparison to SF bricks. The process of casting and compaction of the wet mix was also an easier process, when M-sand was used. Hence, M-sand can be adopted along with soil during the production of fly ash bricks. Efflorescence was observed on the surface of the both the types of bricks, being less in SFM bricks. This is an issue which has to be addressed.

5 Concluding Remarks

Considering strength, durability, and economic reasons, the following ratios were decided:

1. Fly ash/soil ratio is 0.4
2. Alkali activator/fly ash ratio is 0.6
3. Sodium silicate/sodium hydroxide is 1
4. Ambient curing for 28 days.

Soil and fly ash alkali-activated bricks can be used as an alternative to traditional burnt clay bricks as the strength parameters of these bricks are in the same range or even slightly higher than that of burnt clay bricks. However, further research needs to be carried out in absorption characteristics, shear strength, masonry characteristics, and durability characteristics. Further study is required to find the optimum soil: M-sand ratio and also methods to minimize the efflorescence. Alkali-activated fly ash earth bricks are also sustainable as there is no usage of cement, bricks are produced at site using the locally available soil, and no burning of bricks in a kiln is required.

References

1. Reddy, B.V.V.: Pressed soil-cement block: an alternative building material for masonry. Sustain. Constr. **1**, 425–433 (1994)
2. Venkatarama Reddy, B.V., Prasanna Kumar, P.: Embodied energy in cement stabilised rammed earth walls. Energy Build. **42**, 380–385 (2010). https://doi.org/10.1016/j.enbuild.2009.10.005.
3. I. Central Electricity Authority, Report on Flyash Generation at Coal/Ligninte Based Thermal Power Stations and its Utilization in the country for the year 2018–19, New Delhi, 2020
4. Riza, F.V., Rahman, I.A., Zaidi, A.M.A.: Preliminary study of compressed stabilized earth brick (CSEB). Aust. J. Basic Appl. Sci. **5**, 6–12 (2011)
5. Reddy, B.V.V., Lal, R., Rao, K.S.N.: Optimum soil grading for the soil-cement blocks. J. Mater. Civ. Eng. **19**, 139–148 (2007). https://doi.org/10.1061/(ASCE)0899-1561(2007)19:2(139)
6. Mustafa Al Bakri, A.M., Kamarudin, H., Bnhussain, M., Nizar, I.K., Mastura, W.I.W.: Mechanism and chemical reaction of fly ash geopolymer cement- a review. J. Chem. Inf. Model. **53,** 1689–1699 (2013). https://doi.org/10.1017/CBO9781107415324.004
7. Leong, H.Y., Ong, D.E.L., Sanjayan, J.G., Nazari, A.: Strength development of soil-fly ash geopolymer: assessment of soil, fly ash, alkali activators, and water. J. Mater. Civ. Eng. **30** (2018). https://doi.org/10.1061/(ASCE)MT.1943-5533.0002363
8. Sukmak, P., Horpibulsuk, S., Shen, S.L.: Strength development in clay-fly ash geopolymer. Constr. Build. Mater. **40**, 566–574 (2013). https://doi.org/10.1016/j.conbuildmat.2012.11.015
9. Morsy, M.S., Alsayed, S.H., Al-Salloum, Y., Almusallam, T.: Effect of sodium silicate to sodium hydroxide ratios on strength and microstructure of fly ash geopolymer binder. Arab. J. Sci. Eng. **39**, 4333–4339 (2014). https://doi.org/10.1007/s13369-014-1093-8
10. Bureau of Indian Standards, IS 3495 Parts 1–4, 1992 : Methods of Tests of Burnt Clay building brick
11. Bureau of Indian Standards, IS:5816–1999, 2004: Splitting tensile strength of concrete, Bur. Indian Stand. Dehli. (2004)
12. Bureau of Indian Standards, IS:1498–1970: Classification and identification of soils for general engineering purposes

Experimental Optimization of GGBS Fly Ash-Based Geopolymer Concrete Paver Blocks

Debjit Mitra Roy, Satadru Das Adhikary, and Piyali Sengupta

1 Introduction

Geopolymer concrete (GPC), which is termed as an "inorganic polymer," is known as a "green" composite of having immense possibility in the construction field with the help of sustainable byproducts [1]. The production of GPC is performed with natural resources of geological origin activated by industrial by-products such as fly ash and GGBS to provide a high silica and aluminum source with calcium. The production of GPC utilizing the recycled industrial by-products such as fly ash and GGBS declines the manufacturing of ordinary portland cement (OPC) [2–4]. In the development of pavement roads, paver blocks are significantly applicable. The precast paver blocks are primarily considered in the surface of the pavement and roadside footpath [5, 6]. In comparison with the conventional concrete, GPC paver blocks are more reliable in medium-traffic areas due to long-durability and abrasion-free surface. So, thorough experimental investigation is required before designing the cost-effective cementless GPC mix to ensure the strength requirement of GPC paver blocks. With the enhancement of strength of cementless GPC mix, the paver blocks don't go through crack development quickly, maintaining curing free and economical production. Existing literature [7–10] accounted for that GPC paver blocks achieved better resistance to chemical attacks, higher compressive and flexural strength, in comparison with OPC blocks of the similar mix proportioning.

Therefore, in the present study, an extensive experimental investigation is carried out using various mix proportions of GPC to identify the design mix considering the compression testing results and other key parameters of GPC with low-cost technology [11]. The mechanical property of GPC paver blocks was investigated with different loading levels for GGBS-FA 50 and 60% and AAS/B 0.45 and 0.55. In this study, 15 geopolymer paver block samples are cast and cured at an ambient temperature of 27 °C ± 2 °C before evaluating the compressive strength of the

D. M. Roy (✉) · S. Das Adhikary · P. Sengupta
Indian Institute of Technology (ISM), Dhanbad, Jharkhand, India

I. Pal and S. Kolathayar (eds.), *Sustainable Cities and Resilience*, Lecture Notes in Civil Engineering 183, https://doi.org/10.1007/978-981-16-5543-2_13

samples in the universal testing machine. It is observed that 50% GGBS and 0.425% AAS/B produce optimum design mix in terms of compressive strength by considering economical aspects.

2 Materials

2.1 *Coal-Fired Fly Ash*

Fly ashes originating from flue gases through electrostatic precipitator in dry form, consist of fine particles predominantly spherical in shape, mostly glassy and possessing suitable pozzolanic property. It is known as silica fly ash or class F fly ash if the calcium oxide (CaO) has remained under 10%. Fly ash (class F) is usually produced by burning anthracite or bituminous coal, which possesses pozzolanic properties. When the reactive calcium oxide (CaO) is greater than 10%, it is considered calcareous fly ash or class C fly ash [12]. This fly ash usually manufactured from burning lignite or sub-bituminous coal has both pozzolanic and hydraulic properties. Depending on the amount of unburned carbon, the color of fly ash can range from grey to black. In the current study, Class F fly ash (low-calcium) from Maithan Power Limited, Jharkhand, was utilized.

2.2 *Ground Granulated Blast Furnace Slag (GGBS) [13]*

Ground granulated blast furnace slag (GGBS) is originated as a by-product from the iron and steel industry where raw materials (i.e., iron ore, coke and limestone) are burnt in the furnace at a temperature of about 1500 °C. It leads to the production of molten iron and molten slag after melting. As molten slag can be lighter than molten iron, it goes up by floating above molten iron. The granulation is then accomplished by rapidly cooling the molten slag using high-pressure water or steam. The rapid cooling of molten slag ensures production of smaller glassy granular particles that typically have a 95% calcium-alumino-silica content. The obtained granulated slag is then dried and ground into a fine powder with the required fineness. The off-white or near-white colored GGBS has cementitious properties (Fig. 1).

As stated in the literature [1], fly ash and GGBS are activated by using an alkali-activated solution (AAS) to produce GPC. The GGBS-Fly ash having a high alumina-silica source is considered in the present study as per ASTM C618. The chemical component (mass % as oxide) of the fly ash are 0.5–1% CaO, 50–55% SiO_2, 30–35% Al_2O_3, 5–6% Fe_2O_3, 0.5% MgO, 0.5% Na_2O and a loss on ignition of 0.40. The major chemical oxide component by weight of the GGBS included 35–40% CaO, 30–35% SiO_2, 15–20% Al_2O_3, 2.5–5% Fe_2O_3, 5–10% MgO, 0.3% Na_2O and a loss

(a) (b)

Fig. 1 **a** Fly ash (FA); **b** GGBS

on ignition of 0.30. The activator was an alkali solution made from a combination of NaOH and Na_2SiO_3.

As per IS 3812 (Part 1): 2013 [12], FA consists of both silica and alumina. According to IS 12089: 1987 [13], GGBS contains silica, alumina and calcium. Both the FA and GGBS are considered in a combined form of silicate, alumina and calcium materials for the production of GGBS Fly ash-based geopolymer concrete (GPC). Chemically and geologically, aluminum oxide (Al_2O_3), silicon dioxide (SiO_2) and calcium oxide (CaO) are naturally derived from FA and GGBS and activated by an alkaline activator solution (AAS). The chemical reaction is generally expressed as $xAl_2O_3.ySiO_2.\ zH_2O$.

2.3 *Alkali Activator Solution (AAS)*

The alkaline activator solution [14] is the combined form of sodium hydroxide (SH or NaOH) solution and sodium silicate (SS or Na_2SiO_3) solution.

Generally, the sodium hydroxides (SH) are readily obtained in solid shape as in pellets or flakes. The purchasing amount of the SH is primarily dependent on its purity. Therefore, SH (of purity 94–96%) is recommended for purchase at the lowest cost. By mixing SH with pure water, a homogeneous sodium hydroxide solution with preferred molarity (M) is prepared. The SH solution is used to activate the sodium silicate. In the present study, the SH pellets were used are shown in Fig. 2a [15].

Sodium silicate (SS) utilized as a liquid bonding parameter by the detergent manufacturer or textile developer is readily obtained as a liquid-gel form. SS is considered for the development of GPC.

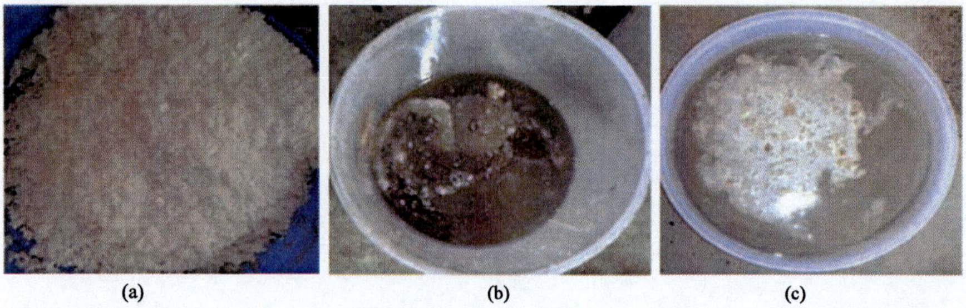

Fig. 2 **a** Sodium hydroxide (SH) pallets. **b** Sodium silicate (SS) liquid. **c** Alkali-activated solution (AAS)

For the current study, sodium hydroxide (NaOH) having a specific gravity of 1.6 g/mL (for pellets) and sodium silicate (Na_2SiO_3) with a concentration of 1.4 g/mL (for solution) were used. The alkaline activator solution (AAS) was carried out to allow adequate time to complete the polymerization reaction. 7 Molar (M) concentrations of sodium hydroxide solutions were used to prepare with portable water by mixing the required weight of pallets of 95–98% purity. The oxide proportions of Na_2SiO_3 solution was 30.0% SiO_2, 11.6% Na_2O and 58.4% water. The weight ratio of SiO_2 to Na_2O of Na_2SiO_3 solution was 2.6. The final prepared AAS is presented below in Fig. 2c.

2.4 Aggregate

2.4.1 Fine Aggregate

In this present study, the fine aggregates are collected from Mayurakshi river sand. The fine aggregate (sand) belonging to Zone II was regarded as per Table 3.3 of IS: 2386–1968 Part III. The fine aggregates as per Tables 4 and 5 of IS 383—2016, passing through IS Sieve 4.75 mm, obtained from Jharkhand are used for this study [17] (Table 1).

2.4.2 Coarse Aggregate (CA)

Readily available crushed stones of coarse aggregate (CA) conforming to nominal size 12.5 mm as per IS 383—2016 [18] are considered for the present study. Physical components of CA of 12 and 6 mm size as per Tables 2 and 3 of IS 383—2016 are mentioned below.

Table 1 Physical components of CA of 12 and 6 mm size and FA of 4.75 mm

Sl. No	Descriptions	Values obtained			IS 383 recommended
		CA-12 mm	CA-6 mm	Fine Agg. – 4.75 mm	
1	Specific gravity	2.72	2.84	2.54	2.5–3
2	Water absorption, %	2.4	3.23	5.23	0.1–2
3	Loose bulk density, kg/m^3	1340.5	1460.5	1590	–
4	Compacted bulk density, %	1570.5	1640.5	1650	–
5	Impact value, %	9.7	–		<30
6	Crushing strength value, %	7.1	–		<30
7	Abrasion value, %	24.7	–		<35

Table 2 Combined gradation of CA of 12-mm and 6 mm size

Sl. No	Sieve size (mm)	Passing Percentage %		Passing Percentage %			Recommendation for passing % of nominal size (12 mm
		Fraction (1) 12 mm	Fraction (2) 6 mm	Fraction (1) 85%	Fraction (2) 15%	Combined 100%	
1	40	100	100	85	15	100	100
2	20	100	100	85	15	100	100
3	12.5	95.5	100	81.17	15	96.17	90–100
4	10	79.7	100	67.74	15	82.74	40–85
5	4.75	0.45	63.2	0.38	9.48	9.86	0–10

Table 3 Mix proportioning of GPC

Descriptions	Cement	Fine Aggregate	Coarse Aggregate	Water	Admixture
Mix proportion by weight	1.0	1.826	2.53	0.425	NA
Material quantities in kg/m^3	438	800	1107	186	NA

3 Experimental Programme

3.1 Mix Proportion

In this series, GGBS and FA were equally proportioned of the total binder and AAS/B was considered as 0.425. and SS/SH was kept as 2 with 7 M of SH. Based on unit

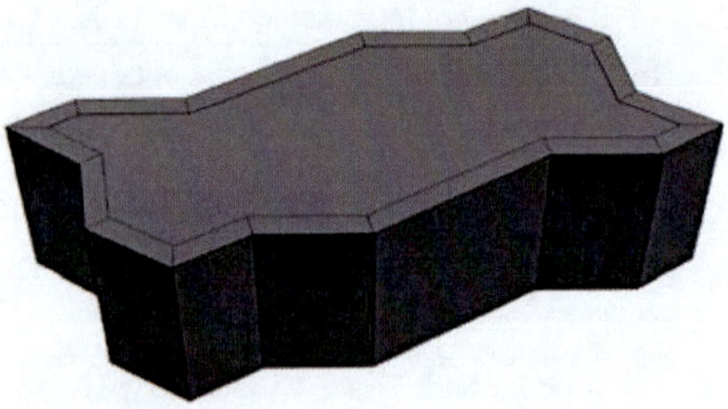

Fig. 3 Tested Sample mold size with a surface area of 0.322578 square meters with a thickness of 80 mm

volume (1 m^3), the ingredients of the proposed GPC were proportioned and the entire binder quantity (GGBS + FA) was kept identical as 438 kg/m^3. The amount of GGBS and FA particles was calculated by considering the mix-design method as per IS 10262: 2019 [19]. The total quantity of aggregates was determined from the remaining amount after deducting the binder content. Portable water was considered for the mixing while extra water or super plasticizer was not considered in this present study.

3.2 Testing Program

The methods of preparing the proposed GPC are as follows. At first, the FA, GGBS, fine and coarse aggregates were placed in a pan-mixer and dry-mixed for 2 min to make it a homogeneous mixture. Then, the already-prepared AAS was added to the solid mixture and mixed for another 2–3 min. Fresh GPC was then immediately placed in paver block molds of surface area of 0.322578 square meters per single piece and thickness of 80 mm. The paver block mold is shown in Fig. 3.

According to IS-516: 1959 [20], all specimens were cured at a temperature of 27 °C ± 2 °C and a relative humidity level of 60% ± 5% in a concrete laboratory. After one day, the casted samples were opened and kept aside at the laboratory under the same environmental condition up to the testing date and compressive strength of the samples was determined at 28 days from the day of casting by using a Universal testing machine of 3000 kN capacity with a constant loading rate of 4500 N/s as per the IS: 516–1959 [20].

4 Results and Discussions

This study was carried out to define the optimum mix proportion of GPC and its corresponding compressive strength, which was experimented for different loading levels based on weight. Based on the performance, GGBS and fly ash with 50–50% showed to the GPC mix sample of paver block to analyze the optimum composition of SS/SH and AAS/Binder to develop the best quality product.

Fig. 4 Testing of paver block under the universal testing machine at the laboratory

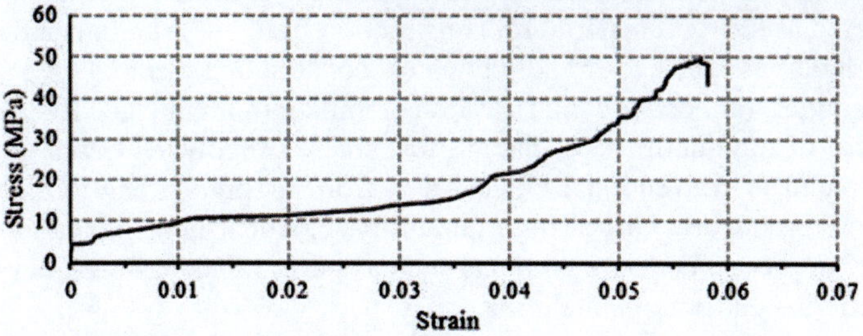

Fig. 5 Stress versus strain curve of GPC

The compressive strength tests of GPC and OPC paver blocks were performed as per IS 3495(Part I):1992 (Figs. 4 and 5).

From the compression test, it was observed that the deformation at the load application stage is equal to the deformation after a specific increment of load depending on the shape of the paver block. At the applied load of 110 kN, it was noticed that the resistance to strain deformation was independent of the tested paver block's shape. It was further noticed that all the blocks exhibited similar deformation behavior up to 1.5 mm deflection. Since Coro-lock blocks presented the highest load-carrying capability with a magnitude of more than 1600 kN, the Coro-lock shape was found the most suitable for paver blocks (Table 4).

A cost comparison was made between $(OPC_{CC})_{\text{Paver block}}$ and $(GPC)_{\text{Paver block}}$ of M-40 grade concrete on the basis of compressive strength and production cost [10]

Table 4 Cost comparison of GPC with conventional concrete

Class designation [M-40]	Strength of paver block w.r.t production cost	
	Strength (compressive) (MPa)	Production cost (Rupees)
$(OPC_{CC})_{Paver\ block}$	40	3861
$(GPC_{CC})_{Paver\ block}$	49	3550

5 Conclusions

The GPC paver blocks were casted with the percentages of fly ash and GGBS of (50 and 60%, respectively, and SS/SH as 2. From the experimental investigation, the optimum design mix was developed considering GGBS and fly ash of 50–50%, 0.425% AAL/Binder ratios and NaOH molarity of 7 M. Based on the design mix-proportions, it was concluded that the paver blocks with 50% GGBS in ambient temperature provide optimum performance considering the strength and minimum deformation after load application. From the compression testing results, GPC can be considered as the right material for construction from both strength and deformation considerations. The eco-friendly low-cost construction material uses minimum natural resources (almost less than 25%) and emits minimum CO_2. Because of the nominal use of minimum use of natural resources, geopolymer-based composites were compared to conventional OPC. Apart from its high strength, this material requires low production cost at high durability. Practical implementation of GPC technology in precast concrete products such as paver block has been developed in Indian scenarios under ambient curing.

Acknowledgements The authors express sincere thanks to Indian Institute of Technology (ISM), Dhanbad and Bengal Institute of Technology and Management (BITM), Santiniketan, Bolpur for providing the necessary support and facilities during the research work.

References

1. Juenger, M.C.G., Winnefeld, F., Provis, J.L., Ideker, J.H.: Advances in alternative cementitious binders. Cem. Concr. Res. **41**, 1232–1243 (2011). https://doi.org/10.1016/j.cemconres.2010.11.012
2. Saha, S., Rajasekaran, C.: Enhancement of the properties of fly ash-based geopolymer paste by incorporating ground granulated blast furnace slag. Constr. Build. Mater. **146**, 615–620 (2017). https://doi.org/10.1016/j.conbuildmat.2017.04.139
3. Deb, P.S., Nath, P., Sarker, P.K.: The effects of ground granulated blast-furnace slag blending with fly ash and activator content on the workability and strength properties of geopolymer concrete cured at ambient temperature. Mater. Des. **62**, 32–39 (2014). https://doi.org/10.1016/j.matdes.2014.05.001
4. Nath, P., Sarker, P.K.: Flexural strength and elastic modulus of ambient-cured blended low-calcium fly ash geopolymer concrete. Constr. Build. Mater. **130**, 22–31 (2017). https://doi.org/10.1016/j.conbuildmat.2016.11.034

5. Mccann, D.M., Forde, M.C.: Review of NDT methods in the assessment of concrete and masonry structures. NDT E Int. **34**, 71–84 (2001). https://doi.org/10.1016/S0963-8695(00)00032-3
6. Ling, Y., Wang, K., Fu, C.: Shrinkage behavior of fly ash based geopolymer pastes with and without shrinkage reducing admixture. Cement Concr. Compos. (2019). https://doi.org/10.1016/j.cemconcomp.2019.02.007
7. Mehta, A., Siddique, R., Ozbakkaloglu, T., Uddin, F., Shaikh, A., Belarbi, R.: Fly ash and ground granulated blast furnace slag-based alkali-activated concrete: mechanical, transport and microstructural properties. Constr. Build. Mater. **257**, 119548 (2020). https://doi.org/10.1016/j.conbuildmat.2020.119548
8. Basil, M.M., Abraham, R.: Study on geopolymer concrete used for paver blocks. Int. J. Innov. Res. Adv. Eng. **3**(9), 62–66. https://doi.org/10.6084/m9.figshare.4052382; Aaron, D.V., D'Souza, D.N., Kaliveer, N., Satish, K.T., Amar, S.M.: Geopolymer paver blocks. In: Proceedings of International Conference on Advances in Civil Engineering, pp. 173–178 (2012)
9. Banupriya, C., John, S., Suresh, R., Divya, E., Vinitha, D.: Experimental investigations on geopolymer bricks/paver blocks. Indian J. Sci. Technol. **9**(16), 1–5 (2016). https://doi.org/10.17485/ijst/2016/v9i16/92209
10. Muthukumar, M., Mareeswaran, L., Subburaman, S., Jeyaprakash, B.: Evaluation on compressive strength of paver block using geopolymer. Int. J. Recent Trends Eng. Res. **5**(3), 12–20 (2019). https://doi.org/10.23883/IJRTER.2019.5018.GFC14
11. Khan, H.A., Nanda, R.P.: Out-of-plane bending of masonry wallette strengthened with geosynthetic. Constr. Build. Mater. **231**, 117198 (2020). https://doi.org/10.1016/j.conbuildmat.2019.117198
12. IS:3812 (Part-1): Specification for Pulverized fuel ash, For Use as Pozzolana in Cement, Cement Mortar and Concrete (2013)
13. IS: 12089: Specification for granulated slag for the manufacturer of portland slag cement (1987)
14. Soutsos, M., Boyle, A.P., Vinai, R., Hadjierakleous, A., Barnett, S.J.: Factors influencing the compressive strength of fly ash based geopolymers. Constr. Build. Mater. **110**, 355–368 (2016). https://doi.org/10.1016/j.conbuildmat.2015.11.045
15. Ismail, I., Bernal, S.A., Provis, J.L., San Nicolas, R., Hamdan, S., Van Deventer, J.S.J.: Modification of phase evolution in alkali-activated blast furnace slag by the incorporation of fly ash, Cem. Concr. Compos. **45**, 125–135 (2014). https://doi.org/10.1016/j.cemconcomp.2013.09.006
16. Prusty, J.K., Pradhan, B.: Multi-response optimization using Taguchi-Grey relational analysis for composition of fly ash-ground granulated blast furnace slag based geopolymer concrete. Constr. Build. Mater. **241**, 118049 (2020). https://doi.org/10.1016/j.conbuildmat.2020.118049
17. IS 1199: Indian Standard Methods of sampling and analysis of concrete (2004)
18. IS: 383: Coarse and fine aggregate for concrete—specification (2016)
19. IS 10262: Concrete Mix Proportioning—Guidelines (2019)
20. IS 516: Method of Tests for Strength of Concrete (1959)

An Experimental Study of Creep Behavior for Disturbance of Unsaturated Expansive Clay Soil

Nariman Hisham Halalo

1 Introduction

Soil swelling issues are considered an important issue from an economic point of view [1], and in terms of the wide spread of this soil in various parts of the world and the specificity of this soil when exposed to moisture as it is distinguished from other soils in that its size increases when it absorbs water and shrinks when it is exposed to heat [2].

The behavior of puffy soils is a reflection of the capillary property of the soil under the influence of periodic pathways of wetting and drying due to natural environmental fluctuations [3]. This behavior is related to the disturbance of the sample when it is transported to the laboratory, which is the result of changes in the stress position of the soil, changes in the moisture content and the ratio of the voids, chemical changes. The change of the stress position in the soil is related to the water saturation state of the soil, which is either completely saturated soil or partially saturated soil [4].

In many references, the basic concepts of secondary consolidation were described, However, no attempt was made to calculate the amount of secondary consolidation [1, 2, 5–8] Although the secondary consolidation rate is assumed to be constant over time, field and laboratory measurements indicate that the secondary consolidation curves may have different shapes depending on soil characteristics and test conditions [7–11]. The primary variable of quantitative analysis of secondary consolidation is the secondary consolidation index that can be expressed in terms of the percentage of blanks or the amount of volume change [6–10].

N. H. Halalo (✉)
Department of Geotechnical Engineering, Civil Engineering Faculty, Damascus University, Damascus, Syria

I. Pal and S. Kolathayar (eds.), *Sustainable Cities and Resilience*, Lecture Notes in Civil Engineering 183, https://doi.org/10.1007/978-981-16-5543-2_14

Fig. 1 Disturbed samples from the field

Fig. 2 Remolded samples in laboratory (Proctor sample, Oedometer sample)

2 The Importance and Objectives of Research

The importance of research and its objectives of this research are to provide a clear view of the impact of sample disturbance on the variables of the primary and secondary consolidation indicators of partially saturated metastatic soil for soil samples that are disturbed and compared to samples that have been reconstructed in the laboratory according to the regular Proctor experience [9].

Two types of samples were used [12]:

1. Disturbed samples from field No. /4/.
2. Remolded samples in laboratory No. /4/ (Figs. 1 and 2).

3 Test Material

Soil samples were brought from Damascus countryside and Sweida Governorate in Syria, samples were extracted by a simple drilling mechanism at a depth (60 cm) below the dust layer, samples were brought in sealed plastic bags and two samples were examined for each site and the samples were labeled as Table 1 [4].

Table 1 Soil sample sites

Location code	Location number	Soil color
A-1	Um Rwaq village—Sweida—Syria	Dark Brown
A-2	Um Rwaq village—Sweida- Syria	Dark Brown
B-1	Deir al-Hajar village—Ghazlania—Damascus countryside- Syria	Light Brown
B-2	Deir al-Hajar village—Ghazlania—Damascus countryside- Syria	Light Brown

4 Test Equipment and Test Procedure

The method in this research, in the first phase, is based on all tests with the basic physical, classified, mathematical, mechanical, and chemical properties of the soil used in this research, taking into account the requirements of ASTM [4], and in the second stage the relationship between the amount of sample reduction (urinary distortions) was represented. With time logarithm, according to the time logarithm method, the secondary consolidation index values are recursive from the primary consolidation index values; the calculation of the amount of total consolidation settlement and the calculation of the time required for the start of the secondary consolidation settlement in the Oedometer test [13] for laboratory soil samples are compared with sample problems.

Tests were conducted within the Oedometer test on a sample (consolidation–drained) [14]. Dimensions of the inner diameter of the Oedometer ring (5.0 cm), ring height (1.99 cm).

- The sample is tested in a confined-sided manner, without lateral expansion and coaxial with increased loads.
- A constant load of (24) hours has been applied and readings have been taken after each of the applied loads have been applied, and at each load value, we take readings that reflect the amount of decrease in the sample height (urinary distortions) with the time taken at each reading of the sample height.

- **Sample formation Method**

The sample is formed according to American standard (ASTM D-2435) [13] which requires a path from the soil on sieve number (#4) and dried in the oven at a temperature (105–110 m), and the sample similar to the original sample is formed by pinning it into the mold (Oedometer ring) within three layers where each layer has 25 strokes, so we get the sample that is similar to the sample formed according to the regular Proctor. The sample formed by the American standard (ASTM D-2435) obtained by the regular projector (optimal humidity and colonic dry weight), where the first and the last 3 cm was first scraped is from a sample of a projector template so that the middle portion remains and from which the Oedometer can be sampled.

- American standard (ASTM D-2435) fully sets the composition of the samples so that all of the samples that are damaged and formed have the same optimal humidity and volumetric weight, so they can be taken as a group and a single study and effect spirit as one on all samples.

The degree of water saturation of the soil samples was determined from the direct shear experiment by knowing the specific weight, the optimum humidity and the dry volumetric weight according to the Proctor experiment of the soil sample from the relation:

$$S_r = \frac{w * G_s}{e} = \frac{w * G_s}{\frac{G_s * \gamma_w}{\gamma_e} - 1}$$

After all, water saturation (S_r) (%), relative specific weight G_s, dry volumetric weight of soil (γ_d) g/cm^3, the volumetric weight of a water unit, (γ_w) g/cm, optimum humidity (proctor moisture) (W),: porosity coefficient (e).

Study of the relationship between the first consolidation index (C_c) and the degree of water saturation of the soil (S_r) (Table 2).

- Study the relationship between the first consolidation index (C_c) and the secondary consolidation index (C_α).

Table 2 The results of the water saturation values of the soil samples within Oedometer experience

Location code		Disturbed sample	Remolded sample	Unit
A-1	The primary degree of saturation	90.3	72.7	%
	The final degree of saturation	102.5	77.6	%
A-2	The primary degree of saturation	84.1	72.2	%
	The final degree of saturation	91.8	77.8	%
B-1	The primary degree of saturation	61.9	46.1	%
	The final degree of saturation	67.9	47.2	%
B-2	The primary degree of saturation	66.7	47.4	%
	The final degree of saturation	75.0	48.8	%

5 Results and Discussion

5.1 The Results of Physical and Mechanical Properties

The size of particles for each of the studied soils was determined by the sieve granular analysis method for particles with diameters greater than (0.075 mm) according to (ASTM-D422) and by sedimentation method (hydrometer) for particles with diameters less than (0.075 mm) according to (ASTM-D1140). The results were as shown in Table 3 (Fig. 3).

The soil was classified for the studied samples according to the US Standard Soil Classification (USC) using Casagrande's plasticity chart [11] as follows:

1 High plasticity (CH) soil (Umm Rawq village), Site (A).
2 Low plasticity (ML) soils (Deir Al Hajar village), Site (B) (Fig. 4).

Table 3 The physical and mechanical properties of expansive soil samples

ASTM	Sample (A)	Sample (B)	Units
Natural moisture content (W)	37.1	23.57	%
Specific gravity (G_s)	2.68	2.71	g/cm^3
Unit weight (γ)	1.62	1.57	g/cm^3
Dry unit weight (γ_d)	1.18	1.27	g/cm^3
Void ratio (*e*)	1.266	1.133	–
Porosity (*n*)	0.559	0.531	–
Degree of Saturation (S_r)	78.45	56.38	%
Liquid Limit (LL)	65.92	42.73	%
Plastic Limit (PL)	33.33	36.76	%
Shrinkage Limit (SL)	21.54	33.3	%
Liquid Index (LI)	0.12	0.48	–
Plastic Index (PI)	32.59	5.97	%
Shrinkage Index (SI)	11.79	3.46	%
Consistency Index(CI)	0.88	0.52	–
Activity (PI/C)	0.437	0.176	–
Percent finer passing No.200 sieve	97.54	78.18	%
Clay content (*C*)	74.56	33.88	%
Silt content (*M*)	25.44	47.12	%
Sand content (*S*)	2.46	19	%

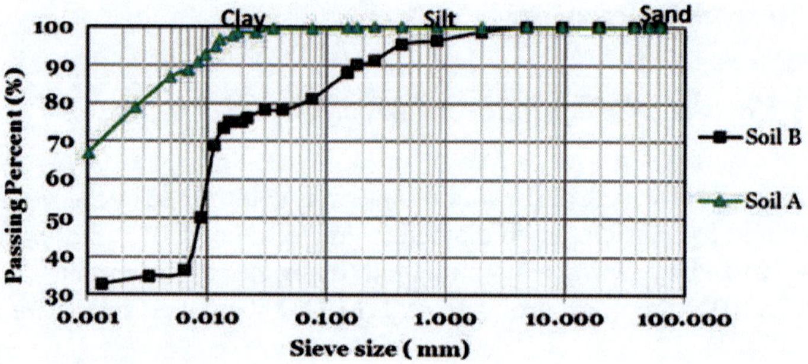

Fig. 3 Hydrometer grain analysis curve

Fig. 4 Soil classification chart by code chart for Casagrande [11]

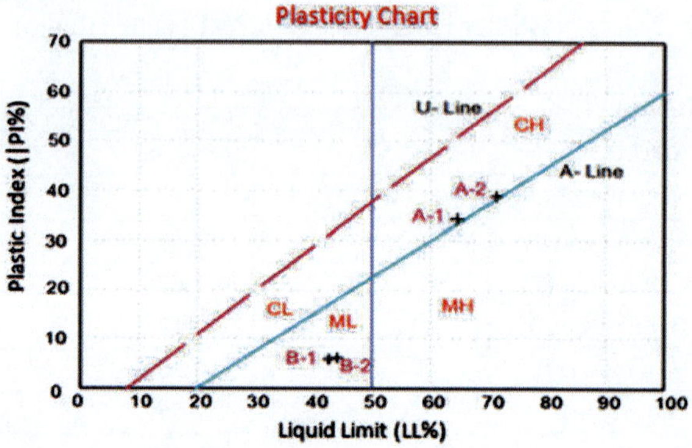

5.2 *The Systemic Proctor Experiment Results*

The purpose of the Proctor experiment is to determine the maximum dry density and ideal humidity of the studied samples in order to form samples according to the American standard (ASTM D-2435) (Fig. 5).

Fig. 5 (A-B) systematic proctor soil test curves

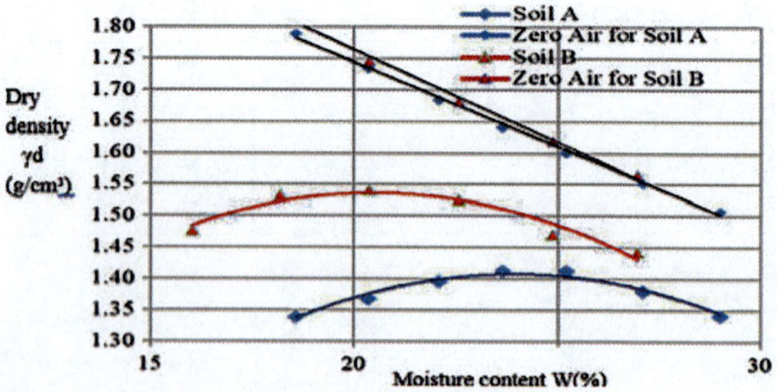

5.3 Chemical Test Results

Table 4 shows the results of the soil chemical properties used in research by identifying the organic content and determining the pH of the soil types used in this research.

Figures 6 and 7 represent the mineralogical analysis of samples using X-ray diffraction, which was performed in the Public Institution of Geology [15].

The most abundant clay mineral for both sites is the mineral quartz, which has a weak relationship with water as the presence of quartz in high proportions in the two types of soil explains the low plasticity values, especially since quartz is known to be an anti-sintering of soil particles and thus gives low plasticity [21] as in the site's soil. (B). (A) the site soil samples contain agaric bulging minerals such as montmorillonite and kaolinite but in low proportions (Table 5).

Table 4 The results of the chemical properties of the soil used in the research

ASTM	Site (A)	Site (B)	Units	observations
Organic content	6.8	4.6	%	Greater than (2%) and soil is rich in organic matter
PH value	5.5	6	–	The soil is acidic

Fig. 6 Seismic analysis of soil sample /A/

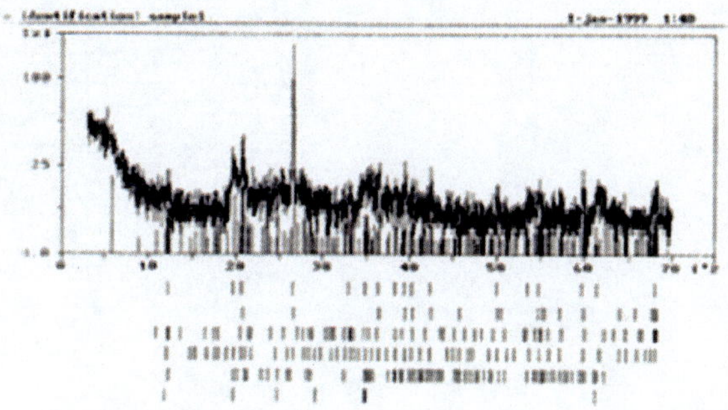

Fig. 7 Seismic analysis of soil sample /B/

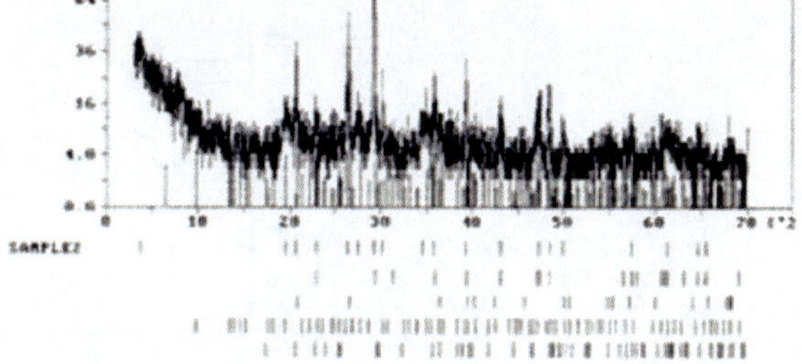

Table 5 Expansive soil mineral composition

Mineral composition	
Soil (A)	Quartz—Phillipsite—Gismodine—Kaolinite-Montmorillonite
Soil (B)	Calcite—Quartz—Mordenite—Forsterite

5.4 The Results of the Oedometer Tests

The figures show the deformation patterns recorded for the studied samples with the logarithm of time according to the method (Casagrande and Fadum, 1940) drawn with semi-logarithmic coordinates.

- **High Plasticity Soil (A-1)** (Figs. 8 and 9)
- **High Plasticity Soil (A-2)** (Figs. 10 and 11)
- **Low Plasticity Soil (B-1)** (Figs. 12 and 13)

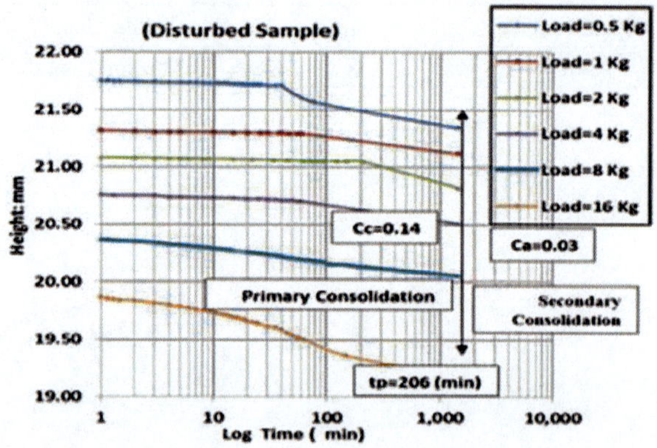

Fig. 8 Oedometer Test: The deformations recorded with the logarithm of time—disturbed sample (A-1)

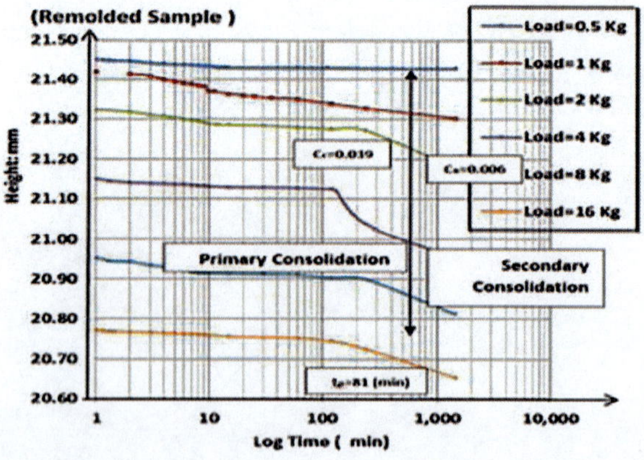

Fig. 9 Oedometer Test: The deformations recorded with the logarithm of time—remolded sample (A-1)

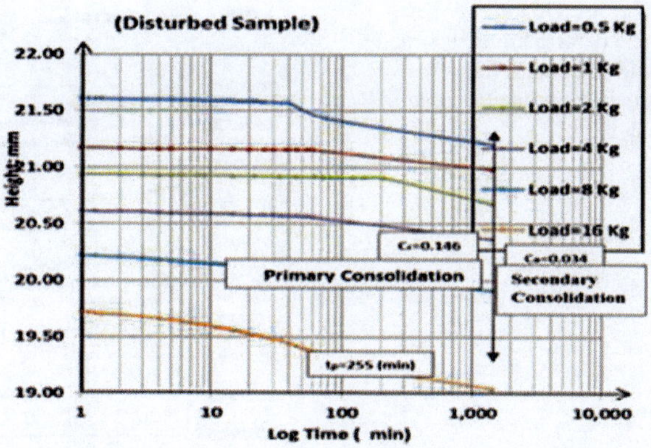

Fig. 10 Oedometer Test: The deformations recorded with the logarithm of time—disturbed sample (A-2)

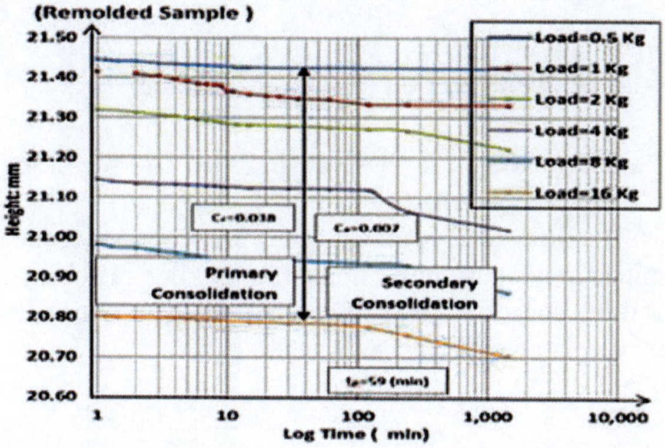

Fig. 11 Oedometer Test: The deformations recorded with the logarithm of time—remolded sample (A-2)

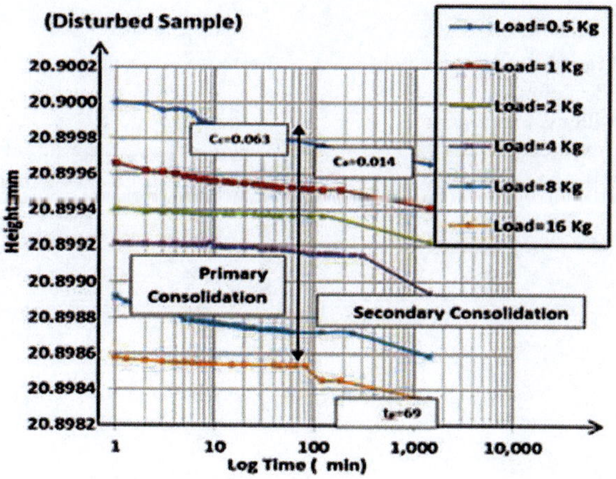

Fig. 12 Oedometer Test: The deformations recorded with the logarithm of time—disturbed sample (B-1)

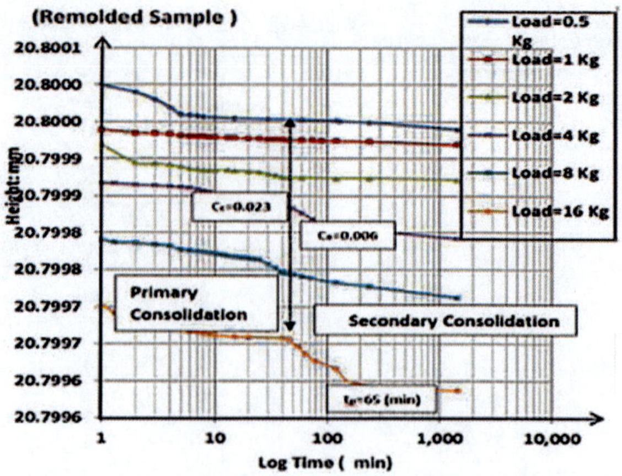

Fig. 13 Oedometer Test: The deformations recorded with the logarithm of time—remolded sample (B-1)

- **Low Plasticity Soil (B-2)** (Figs. 14 and 15).

Calculate the time needed to begin the secondary Consolidation Settlement (Creep settlement) (t_p)

The time value needed to start the secondary consolidation settlement was determined in a logarithm of time after the time factor values of the consolidation degree and the consolidation coefficient were determined [9]. The time of the relationship is calculated (Tables 6 and 7):

$$t_p = \frac{H_t^2 * T_v}{c_v}$$

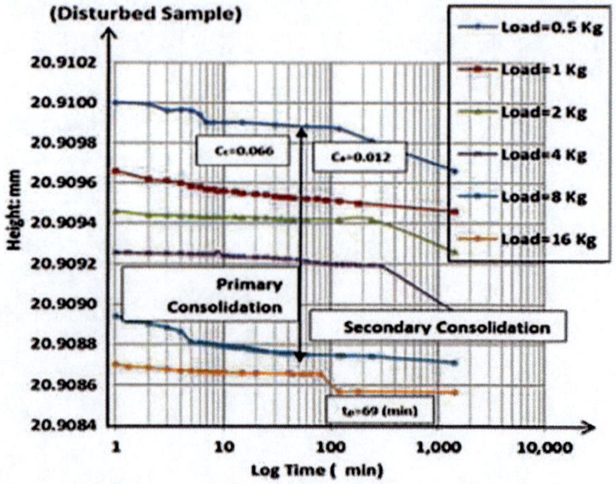

Fig. 14 Oedometer Test: The deformations recorded with the logarithm of time—disturbed sample (B-2)

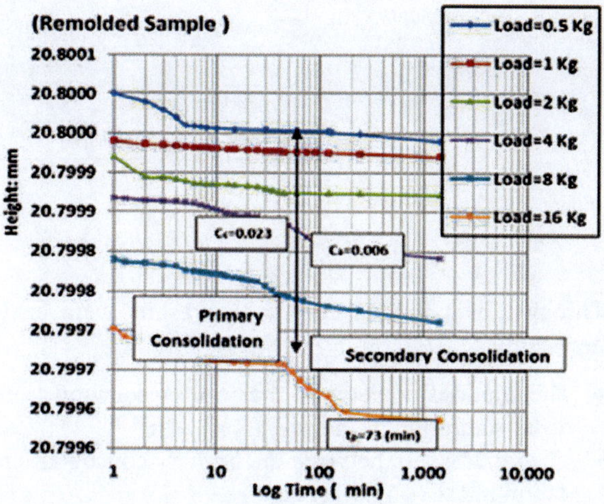

Fig. 15 Oedometer Test: The deformations recorded with the logarithm of time—remolded sample (B-2)

Table 6 The results of soil types used in the research to calculate creep time (t_p)

ASTM	Site (A)	Site (B)	Units	observations
Organic content	6.8	4.6	%	Greater than (2%), and soil is rich in organic matter
PH	5.5	6	–	The soil is acidic

Table 7 The results of soil types used in the research to calculate Creep time (t_p)

Location	Disturbed sample (t_p) (mm)	Remolded sample (t_p) (mm)
A-1	206	81
A-2	255	59
B-1	69	65
B-2	69	73

According to the tests carried out in this research, the time required for the end of the first consolidation settlement varies according to the plasticity of the soil. For low plasticity soil samples, whether the sample is damaged or problematic, the time required for the end of the first compressive settlement appears after $t_p = 69$(min).

This is quite different for high plasticity soil samples since the time required for the end of the first compressive subsidence appears after $t_p = 230$(min) (for disturbed—high plasticity samples) compared to the time required for the end of the first compressive subsidence (for remolded samples—high plasticity) $t_p = 70$(min), as the following table illustrates (Table 8).

Table 8 The secondary consolidation settlement (creep settlement) (t_p)

Sample type	High plasticity soil	Low plasticity soil
	t_p = (min)	t_p = (min)
Disturbed Sample	230	69
Remolded Sample	70	69

6 Results

Through the previous curves, the results were analyzed to find the correlation between:

- The relationship between the primary consolidation index (C_c) and the degree of water saturation of the soil (S_r).
- The relationship between the primary consolidation (C_c) and secondary consolidation index (C_α) (Fig. 16).

The results show an increase in soil Consolidation (C_c) with an increase in the degree of soil saturation through the following relationships:

High Plasticity Soils
(For Disturbed samples) $C_c = 0.015 * S_r - 1.308$ where $R^2 = 0.525$ (coefficient of correlation between variables).

(For Remolded samples) $C_c = 0.004 * S_r - 0.314$ where $R^2 = 0.824$ (coefficient of correlation between variables).

Low Plasticity Soils
(For Disturbed samples) $C_c = 0.012 * S_r - 0.742$ where $R^2 = 0.423$ (coefficient of correlation between variables).

(For Remolded samples) $C_c = 0.021 * S_r - 1.005$ where $R^2 = 0.825$ (coefficient of correlation between variables).

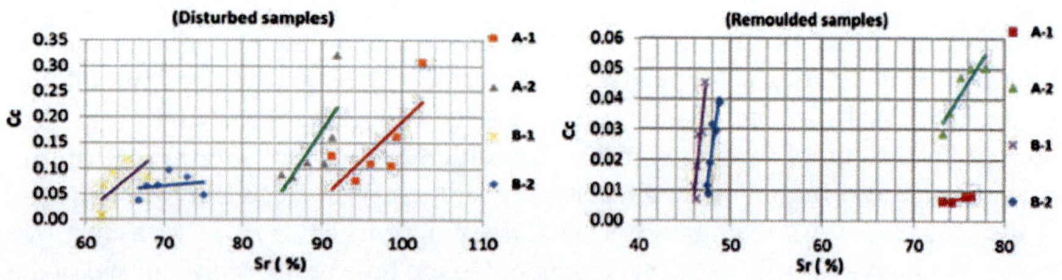

Fig. 16 The relationship between the primary consolidation index (C_c) and the degree of saturation (S_r)

The Relationship Between the Primary Consolidation (C_c) and Secondary (C_α) Consolidation Index:

The relationship between the primary (C_c) and secondary Consolidation index (C_α) is shown in the following figures: (25), (26), (27), (28), (29), (30), (31), (32).

Through these the values, a mathematical relationship is derived to calculate the value of the primary and secondary Consolidation index.

High Plasticity Soils

(For Disturbed samples) $C_\alpha = 0.308 * C_c - 0.011$ where $R^2 = 0.993$ (coefficient of correlation between variables).

(For Remolded samples) $C_\alpha = 0.332 * C_c - 0.006$ where $R^2 = 0.747$ (coefficient of correlation between variables).

Low Plasticity Soils

(For Disturbed samples) $C_\alpha = 0.218 * C_c - 0.005$ where $R^2 = 0.891$ (coefficient of correlation between variables).

(For Remolded samples) $C_\alpha = 0.560 * C_c - 0.007$ where $R^2 = 0.766$ (coefficient of correlation between variables) (Fig. 17).

The data values for both types of samples were analyzed to estimate the sample quality using several criteria $\left(\frac{\Delta e}{e_0}, \frac{C_c}{1+e_0}, \text{OCR}, \varepsilon_v\right)$. The sample quality index (SQD) defined by Terzaghi et al. [3] in addition to Lunne et al.(1997) [5].

Tables 9, 10 and 11 show the amount of decline occurring for each sample and the classification of sample quality.

After all, the primary consolidation settlement S_c.

The secondary consolidation settlement: S_a.

The total consolidation settlement $S_T = S_c + S_a$.

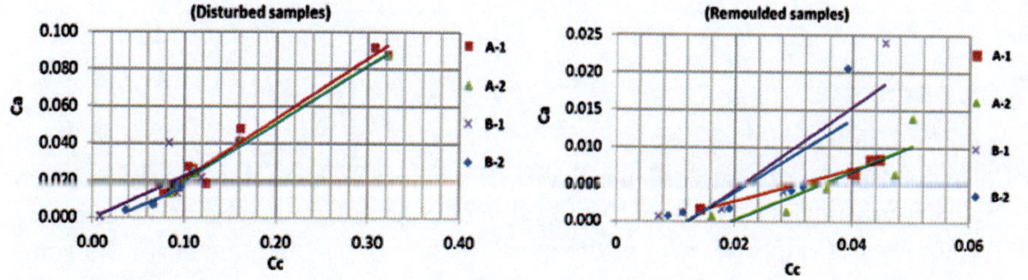

Fig. 17 The relationship between the primary consolidation index (C_c) and secondary (C_α) consolidation index

Table 9 The results of the total consolidation Settlement (S_T)

Location code	(S_c) (mm) Disturbed sample	(S_a)(mm) Disturbed sample	(S_c)(mm) Remolded sample	(S_a)(mm) Remolded sample	(S_T)(mm) Remolded sample	(S_T)(mm) Remolded sample
A-1	4.017	2.756	2.462	1.933	6.773	4.395
A-2	4.698	2.57	3.271	0.267	7.268	3.538
B-1	5.549	1.152	1.172	0.398	6.701	1.570
B-2	6.423	1.275	1.13	0.406	7.698	1.536

Table 10 The Sample Quality Index (SQD) for disturbed samples

Location code	OCR	ε_v (%)	$\frac{\Delta e}{e_0}$	SQD
	Disturbed sample			
A-1	2.03	31.041	0.253	Very poor sample
A-2	1.89	33.308	0.230	Very poor sample
B-1	1.44	30.710	0.112	Poor sample
B-2	1.18	35.280	0.109	Poor sample

Table 11 The Sample Quality Index (SQD) for remolded samples

Location code	OCR	ε_v(%)	$\frac{\Delta e}{e_0}$	SQD
	Remolded sample			
A-1	1.82	20.140	0.080	Good sample
A-2	1.21	16.217	0.078	Good sample
B-1	1.45	7.190	0.037	Very Good sample
B-2	1.57	7.040	0.038	Very Good sample

7 Conclusions

- The disturbance of samples leads to different values in the design and the effect of this disturbance is by increasing the value of consolidation characteristics represented by the values of each of the consolidation index, the amount of the volumetric distortion, the amount of the primary and secondary consolidation settlement for disturbed samples compared to the decrease in their value for laboratory remolded samples.
- The degree of sample disturbance can be estimated from the volumetric distortion ratio (ε_v), as with increasing turbulence in the sample, the amount of volumetric distortion increases.
- The effect of disturbance to the sample increases with a decrease in the plasticity index. The effect on soils with low plasticity (PI = 6%) is more pronounced than on soils of high plasticity (PI = 32%).

- There is a linear relationship between the primary consolidation index (C_c) and the degree of soil saturation, as with increasing the degree of soil saturation (S_r), the consolidation index increases.
- There is a linear relationship between the primary consolidation index (C_c) and the secondary consolidation index (C_α). The mathematical relationships have been proposed. These the relationships help in knowing the value of the secondary consolidation index for the soil of the studied area from the primary consolidation index by using the Oedometer test.
- The index of secondary consolidation (C_α) is constant during the second consolidation index (C_α) for all load levels and does not depend on stress. The slope of this line is defined by time and according to the type of expansive soil.

References

1. Fredlund, D.G., Rahardjo, H.: Soil Mechanics for Unsaturated Soils. Wiley, New York (1993)
2. Das, B.M.: Principles of Geotechnical Engineering, 4th edn. PWS Publishing Company, Boston, MA, USA (1998)
3. Terzaghi, K., Peck, R.B., Mesri, G.: Soil Mechanics in Engineering Practice. Wiley (1996)
4. ASTM—American Society for Testing and Materials
5. Escario, S., Saez, J.: The shear strength of partly saturated soils. Geotechnique **36**(4), 453–456 (1986)
6. Gan, J.K.M., Fredlund, D.G., Rahardjo, H.: Determination of the shear strength parameters of an unsaturated soil using the direct shear test. Can. Geotech. J. **25**(8), 500–510 (1988)
7. H. A. Amundsen1, V. Thakur and A. Emdal -Sample disturbances in block samples on low plastic soft clays. Norwegian University of Science and Technology (2016)
8. John, D.N., Debora J.M.: Expansive Soils Problems and Practice in Foundation and Pavement Engineering. Wiley, p 255 (1992)
9. Lambe, T.W., Whitman, R.V.: Soil Mechanics. Wiley, New York, USA (1969)
10. Michael Long, BE, MEngSc, Ph.D., CEng, MICE, MIEI- Sampling disturbance effects on medium plasticity clay/silt. J. Geotech. Geoenviron. Eng. ASCE **159**(2), 99–111 (2004)
11. Tanaka, H.: Sample quality of cohesive soils. Lessons from three sites, Ariake, Bothkennar and Drammen. Soils Found. **20**(4), 57–74 (2000)
12. Christopher, A.B., Tuncer, B.E., Craig, H.B., David, M.M.: Geological and physical factors affecting the friction angle of compacted sands. J. Geotech. Geoenviron. Eng. **134**(10), 1476 1489 (2008)
13. ASTM -D2435. Standard Test Methods for One-Dimensional Consolidation Properties of Soils Using Incremental Loading. American Society for Testing and Materials (1989)
14. Budhu, M.: Soil Mechanics and Foundations. Wiley, New York, NY, USA (2000)
15. Caruso, A., Tarantino, A.: A shear box for testing unsaturated soils from medium to high degree of saturation. Geotechnique **54**(4), 281–284 (2004)
16. Chaminda Pathma Kumara Gallage and Taro Uchimura- Direct shear testing on unsaturated silty soils to investigate the effects of drying and wetting on shear strength parameters at low suction. Soils and foundations. J. Geotech. Geoenviron. Eng. **50**(1), 161–172 (2010)
17. Clayton, C.R.I., Matthews, M.C., Simons, N.E.: Sampling and sample disturbance. Inst. Civ. Eng. Gotech. Eng. (1999)
18. Tanaka, H., Sharma, P., Tsuchida, T., Tanaka, M.: Comparative study on sample quality using several types of samplers. Soils Found. **36**(1), 57–68 (1996)

Practical Design of Stone Column in Predicting Settlement Performance

Kok Shien Ng and Yee Ming Chew

1 Introduction

Stone column (also termed as vibro replacement, vibro displacement or granular pile) has been generally recognized as a useful technique to improve the weak ground. This technique requires large size columns of granular material to be inserted into the ground by means of special vibrators or by augered-casing system to form a stiffer composite structure with the surrounding soils. The beneficial effects of stone column in treating the soft soils include the decrease in total settlement together with rapid consolidation time, increase in bearing capacity and shear resistance as well as the reduction of liquefaction risk. The success of implementing the stone column in the construction of road embankment, airfield, bridge abutments and lightweight structures over the past 40 years has increased its reputation as one of the most versatile and cost-effective ground improvement techniques.

End bearing columns are mostly used in the design, but occasionally floating stone columns are adopted [1]. Figure 1 shows the foundation supported by the end bearing columns and the floating columns. Long-term settlement is observed for foundation supported by the floating columns due to the untreated zone below the column toe. Therefore, the design of floating type of stone column must be carefully carried out. Stone columns are arranged in either small group or large group. Small column group are used to support smaller foundation with smaller loadings while large column group are used to support widespread loading such as embankment and mat foundation. Due to different geometries, the failure modes of stone columns are different as they exist complex load sharing and strain compatibility between foundation, stone columns and the surrounding soils [2]. Therefore, the designs of small and large group of stone columns have to be treated separately.

K. S. Ng (✉) · Y. M. Chew
Cawangan Pulau Pinang, Universiti Teknologi MARA, Kampus Permatang Pauh, Penang, Malaysia
e-mail: ngkokshien@uitm.edu.my

I. Pal and S. Kolathayar (eds.), *Sustainable Cities and Resilience*, Lecture Notes in Civil Engineering 183, https://doi.org/10.1007/978-981-16-5543-2_15

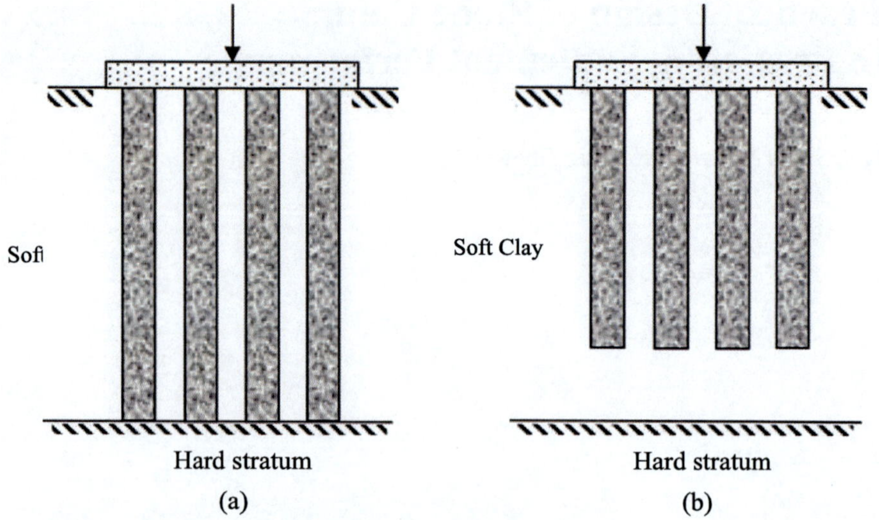

Fig. 1 **a** End bearing columns and **b** Floating columns

This paper presents simple yet practical design methods for both the end bearing and floating stone columns for large and small column group. In stone column design, the settlement and the bearing capacity are two main design criteria to be fulfilled. The settlement criterion is normally of bigger concern compared to the bearing capacity criterion. Hence, the methods introduced here focus mainly on the estimation of settlements of stone column improved ground. They are based on previous works of author [3–5]. The validity of the methods has been proven using different approaches such as field results, analytical solutions and laboratory works. This paper shows the comparison of those methods using two design examples with the objective to ease the use of the methods among the practitioners.

2 Settlement Prediction

Methods to predict settlement performance of stone column improved ground can be divided into two categories: large column group and small column group. It is sometimes difficult to decide which categories are the problem lies in. Thus, it is suggested that when the foundation size or breadth (B) of the foundation is 1 time or more than the thickness of the soft soil layer (H), it should be categorized as large column group. Otherwise, it falls under the small column group category. For example, a 5 m wide foundation supported by stone columns (regardless of end bearing or floating type) installed in a 10 m thick soft soil is categorized as small column group. On the other hand, a 20 m wide embankment on 10 m thick soft soil is considered as large column group.

2.1 Large Column Group

In this category, three methods are introduced to predict the consolidation settlement of stone column improved ground under widespread loading. Stone columns under widespread loading are normally regarded as infinite column grid. All three methods are derived from the numerical results of unit cell model. Unit cell model is commonly used to analyze the performance of the central columns as the representative of the large column group performance [6]. The central columns experience minor bulging along the depth and some punching if the columns are floating. The methods to be discussed here consider cases where groundwater table is on the ground surface to represent a more critical condition.

Method A

The numerical results of unit cell model for floating stone columns of varied length ratio, β (defined as the ratio of column length, L over soft soil thickness, H) was produced by Ng and Tan [3]. End bearing columns ($\beta = 1.0$) exhibited the best performance in controlling the settlement and consolidation time while short floating columns resulted in larger induced settlement and slower consolidation rate. The untreated zone significantly influences the settlement performance of the improved ground. Beside the column length, other factors such as area replacement ratio (α, defined as area of column over the influential area), column's friction angle (ϕ'), loading intensity (q), post-installation earth pressure (K) were also found to be influencing. With that, the authors produced design equations Eqs. (1) and (2) to predict the settlement improvement factor (n) of stone columns under widespread uniform loading. Settlement improvement factor (n) is the ratio of the settlement without improvement over the settlement with improvement.

$$n = n_0\left[1 - \left(C_\alpha + C_\phi + C_q + C_K\right)\right] \quad (1)$$

$$n_0 = 9.43\alpha^2 + 1.49\alpha + 1.06 \quad (2)$$

where n_0 is the basic improvement factor derived from the end bearing column results with parameters $\alpha = 0.11$, $\phi' = 40°$, $q = 100$ kPa and $K = 0.7$. Parameters C_α, C_ϕ, C_q, and C_K are the correction factors for area replacement ratio, column's friction angle, loading intensity and post-installation earth pressure can be obtained from Fig. 2. The range of parameters in the design charts covers most of the practical values used in the design of stone columns.

Method B

Simplified homogenization method was developed by Ng and Tan [4] to predict the settlement of stone column improved ground. In this method, the improved ground is treated as a composite ground with equivalent stiffness E_{eq} obtained from Eqs. (3) and (4).

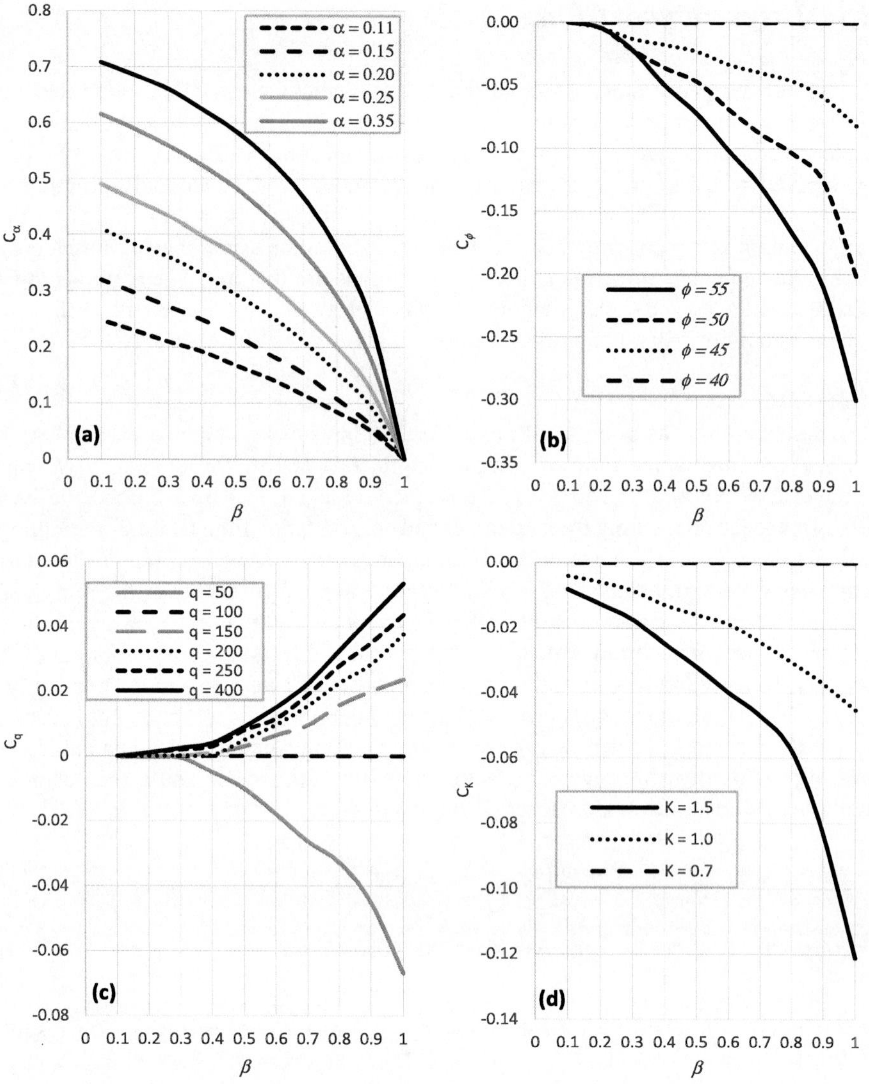

Fig. 2 Correction factors for C_α, C_ϕ, C_q, and C_K

$$E_{eq} = \frac{E_{comp}}{N_{corr}} \tag{3}$$

$$E_{comp} = \alpha\, E_c + (1 - \alpha) E_s \tag{4}$$

where E_c = stiffness of column, E_s = stiffness of the surrounding soil, E_{comp} = composite stiffness and N_{corr} = yielding correction factor. The composite stiffness if

Table 1 Yielding correction factor, N_{corr}

N_{corr}	$\phi' = 40^o$	$\phi' = 45^o$	$\phi' = 50^o$
$q = 25$ kPa	1.30	1.25	1.20
$q = 100$ kPa	1.60	1.45	1.25
$q = 400$ kPa	1.75	1.60	1.40

used for stone column design is found to underestimate the actual settlement of stone columns improved ground. This is because the stone column is not a fully elastic material but with significant yielding occurring along the column length. As the induced load increases, subsequent plastic straining also occurs in the surrounding soil. Therefore, N_{corr} is introduced to account for this yielding effect, as tabulated in Table 1. The values in this table are conservatively adopted from the results in Ng and Tan [4] by assuming same yielding intensity happens in the end bearing columns and the floating columns. Many studies have shown that the stiffness of column should be limited to 10 to 20 times the stiffness of surrounding soil. In this method, the column stiffness is set to be 10 times the soil stiffness. Therefore the Eq. (3) can be rewritten as

$$E_{comp} = (1 + 9\alpha)E_s \tag{5}$$

Subsequently, the settlement, S either for the improved ground or the unimproved ground can be easily obtained from one-dimensional analytical calculation where $S = qH/E_{oed}$ where H is the thickness of soft soil, while E_{oed} is the oedometer modulus or constraint modulus. The relationship between Young's modulus (E) and oedometer modulus is given in Eq. (6).

$$E_{oed} = \frac{(1 - \nu)E}{(1 - 2\nu)(1 + \nu)} \tag{6}$$

In this method, the equivalent permeability of the improved ground, k_{eq} can be obtained from Eq. (7).

$$k_{eq} = \frac{K_{composite} \cdot k_s}{1 \times 10^{-4}} \quad \text{(m/day)} \tag{7}$$

where k_{eq} is equivalent permeability, while $K_{composite}$ is the composite permeability obtained from Fig. 3. The value of $K_{composite}$ depends on the permeability ratio (k_c/k_s; k_c = permeability of stone column, k_s = permeability of surrounding soil) and the area replacement ratio (α).

Method C

The third method was presented by Ng [5] to predict the settlement of floating columns given in terms of settlement ratio (S_f/S_{uc}, where S_f is the settlement of

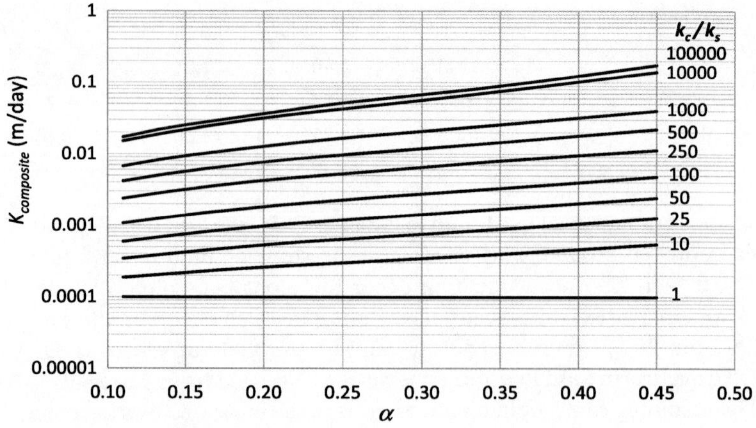

Fig. 3. Composite permeability, $K_{\text{composite}}$

floating column group and S_{uc} is the settlement of end bearing columns. The settlement ratio can be predicted using Eq. (8) by combining the effect of area replacement ratio (α), column's friction angle (ϕ) and length ratio (β). The settlement of floating column can then be calculated when the settlement of end bearing column is known which it can be predicted from the above Method A or Method B.

$$\frac{S_{\text{f}}}{S_{\text{uc}}} = 1 + \left[7.9\alpha^{1.4} + 0.029(\varphi - 40°)\right](1 - \beta) \tag{8}$$

2.2 *Design Example 1*

A 20 m × 20 m reinforced concrete water tank structure is founded on a raft foundation above an 8.0 m thick soft soil. The coefficient of volume compressibility (m_v) of soil is 0.00025 m^2/kN and the unit weight γ_b of soil is 16 kN/m^3. The stone columns of 1.0 m diameter are installed in square grid with spacing of 2.0 m (equivalent to α = 0.2). The friction angle and unit weight of column material is 45° and 20 kN/m^3 respectively. The total applied loading is 150 kPa. Compute the settlement of the improved ground for (a) end bearing columns, (b) floating columns with 6 m length.

Solution (a)

Settlement of unimproved ground, S_o:

$$E_{\text{oed}} = 1/m_v = 1/0.00025 = 4000\,\text{kN/m}^2$$

$$S_o = qH/E_{oed} = 150(8)/4000 = 0.3\,\text{m}$$

Settlement of improved ground for end bearing columns, S_{uc}:

Method A:

From Fig. 2: $C_\alpha = 0\ (\beta = 1)$, $C_\phi = -0.08, C_q = 0.024$, $C_K = -0.045$ (assume K= 1)

$n_O = 9.43\alpha^2 + 1.49\,\alpha\emptyset + 1.06 = 9.43(0.2)^2 + 1.49(0.2) + 1.06 = 1.735.$

$n = n = n_o\,[1 - (C_\alpha + C_\phi + C_q + C_K)] = 1.735[1-(0-0.08 + 0.024-0.045)] = 1.91.$

$S_{uc} = S_o/n = 0.3/1.91 = 0.157$ m.

Method B:

From Table 1 with interpolation: $N_{corr} = 1.475$.

Equations (3) and (4) can be written in the form of oedometer modulus instead of secant modulus.

$$E_{comp} = (1 + 9\alpha)E_s = (1 + 9 \cdot 0.2)4000 = 11200\,\text{kN/m}^2$$
$$E_{eq} = E_{comp}\,/N_{corr} = 11200/1.475 = 7593\,\text{kN/m}^2$$
$$S_{uc} = qH/E_{eq} = 150(8)/7593 = 0.158\,\text{m}$$

Method A and Method B yield the same results for this case. Comparison was made with Priebe [7] and Aboshi et al. [8] where the predicted settlement were 0.12 m and 0.189 m, respectively. Aboshi's method requires the input of stress concentration ratio (n_s) which typical range from 2 to 5 and in this case $n_s = 4$ is used. The results of above methods lie between these solutions.

Additionally, if time for 90% consolidation is to be predicted, one may use Method B to estimate the t_{90}. In this case, say $k_s = 0.0002$ m/day and $k_c = 0.05$ m/ day, the permeability ratio would be $k_c/k_s = 250$. From Fig. 3, the $K_{comp} = 0.0043$ m/day is obtained and then from Eq. (7) the equivalent permeability would be.

$$k_{eq} = \frac{K_{composite} \cdot k_s}{1 \times 10^{-4}} = \frac{0.0043 \cdot 0.0002}{0.0001} = 0.00086\ \text{m/day}.$$

Subsequently,

$$c_v = \frac{E_{eq,oed} \cdot k_{eq}}{\gamma_w} = c_v = \frac{7593 \cdot 0.0086}{10} = 6.53\,\text{m}^2/\text{day}$$

If, the drainage path is $d = H$ (1-way drainage); for 90% consolidation, t_{90} is then calculated as.

$t_{90} = 0.848\,d^2/c_v = 0.848(8)^2/6.53 = 8.31$ days

Hence, ninety percent of consolidation settlement is completed in about 8 days. This shows the effectiveness of stone columns in speeding up the consolidation rate.

Using Han and Ye [9] analytical method, the time required for the above example would be around 7 days which is slightly faster than the current method. This is because their method does not consider the yielding effect which slow down the consolidation rate. On the other hand, use of simplified solution by CUR191 for equivalent permeability of prefabricated vertical drain (PCD) design yielded $t_{90} =$ 4.7 days. This substantial difference in result is probably due to the assumption of infinite drainage capacity in the PVD.

Solution (b)

Settlement of improved ground for floating columns of 6 m length, S_f:

Method A:

$\beta = L/H = 6/8 = 0.75$; $n_o = 1.735$ and $S = 0.3$ m from (a) above.
From Figs. 3, 4, 5 and 6: $C\alpha = 0.18$, $C_\phi = -0.04$, $C_q = 0.012$, $C_K = -0.052$ (assume $K = 1$)

$$n = n_o\left[1 - \left(C_\alpha + C_\phi + C_q + C_K\right)\right]$$
$$= 1.735[1 - (0 - 0.04 + 0.012 - 0.052)] = 1.56$$

$$S_f = S/n = 0.3/1.56 = 0.192\,\text{m}$$

Method B:

Divide the calculation into two zone: treated (H_1) and untreated (H_2).
Settlement of untreated zone, $S_o = qH_1/E_{oed} = 150(2)/4000 = 0.075\,\text{m}$.
Settlement of untreated zone, $S_t = qH_2/E_{eq} = 150(6)/7593 = 0.119\,\text{m}$

$$S_f = 0.075 + 0.119 = 0.194\,\text{m}$$

Method C:

$$\frac{S_f}{S_{uc}} = 1 + \left[7.9\alpha^{1.4} + 0.029(\varphi - 40°)\right](1 - \beta)$$
$$= 1 + \left[7.9 \times 0.2^{1.4} + 0.029(45 - 40)\right](1 - 0.75) = 1.244\,\text{m}$$

Taking $S_{uc} = 0.157$ m from Method A for end bearing columns.

$$S_f = 1.244(0.157) = 0.195\,\text{m}$$

All three methods produced similar settlement results for floating stone columns. This is not surprising since all these methods are developed from the same numerical

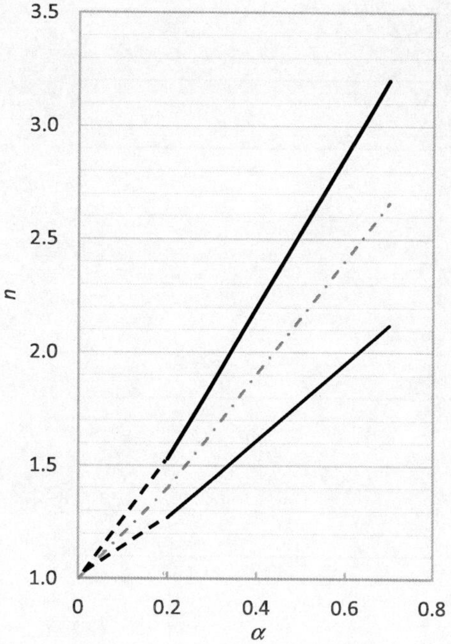

Fig. 4 *n* for floating columns with L_{opt}

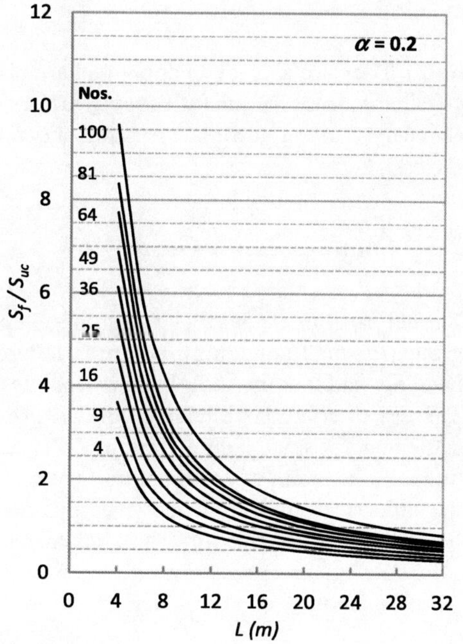

Fig. 5. Settlement ratio of $\alpha = 0.2$

Fig. 6 Settlement ratio of $\alpha = 0.4$

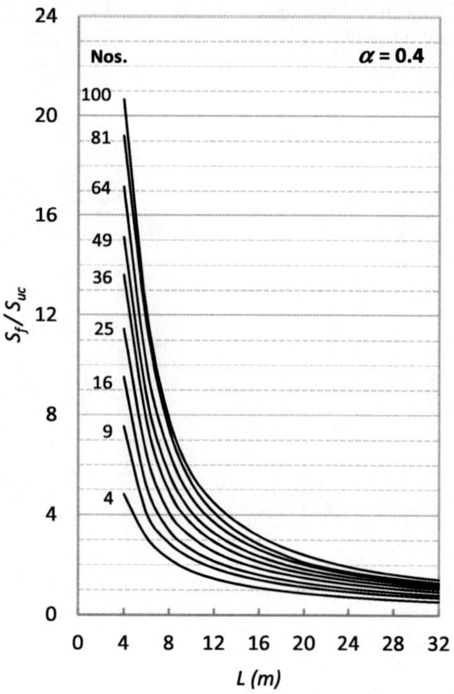

result. They are all easy to apply and are practical with given design charts besides considering most of the influencing factors. Method B has an additional advantage as it can be used to estimate the consolidation time of stone columns.

2.3 Small Column Group

The deformation modes for small stone column group are different from large column group. In small column group, three failure modes are involved namely shearing, bending, and bulging. If column is floating, additional punching mode is observed. The use of solutions for large column group would lead to the overprediction of settlement for small column group [2]. Three methods are introduced here to predict drained settlement of small column group. The proposed methods are derived from the numerical analysis where the soil models are characterized by linear-elastic perfectly plastic behavior. The studied range for the loading intensity is from 0 to 150 kPa, post-installation earth pressure $K = 0.7$ and friction angle of column, $\phi' = 40°$.

Method 1
Tan et al. [2] conducted numerical analyses on small column group ranging from 4 to 100 numbers per group subjected to uniform loading on circular foundation. The drained performance of the floating stone columns was studied. Effect of the area replacement ratio (α, defined as the ratio of column area to the foundation area, ranging from 0.2 to 0.7), the loading intensity (q = 50–150 kPa) and the column's friction angles ($\phi' = 40°$ to 55°) on the settlement improvement factor (n) were investigated. The results are shown in Fig. 4 for columns achieving the critical (optimum) length. Column with length longer than optimum confers no advantage. The optimum length (L_{opt}) can be taken as

$$L_{opt} = (2\alpha + 0.8)D \tag{9}$$

where D is the diameter of the foundation (for square foundation, replaced D with $1.13B$, where B is breadth of the footing). The results are bounded in the shaded area and the average line (shown as dotted line) can be interpreted as

$$n = 2.4\alpha + 1 \tag{10}$$

The above solution can also be applied to the end bearing columns provided that the length of the stone columns has exceeded the optimum length.

Method 2
This method is related to the Method C above where the drained settlement of the floating columns (S_f) in the small group can be predicted using the settlement of end bearing column from the large group (S_{uc}). The relationship is shown in Figs. 5, 6 and 7 in terms of settlement ratio S_f/S_{uc} for area replacement ratio of 0.2, 0.4 and 0.6 respectively. The details of the findings can be referred to Ng [5]. Once the settlement ratio is found, the S_f can be calculated after obtaining S_{uc} from either Method A or Method B prescribed above for the large column group.

2.4 *Design Example 2*

A group of 16 stone columns with area replacement ratio, α of 0.3, loading intensity, q of 100 kPa, and length, L of 12 m is to be installed in a thick soft ground (H = 15 m) with oedometer modulus of E_{oed} of 5000 kPa. The size of the foundation is 6.5 m × 6.5 m. Predict the settlement of this floating stone column group, S_f.

Solution
Method 1

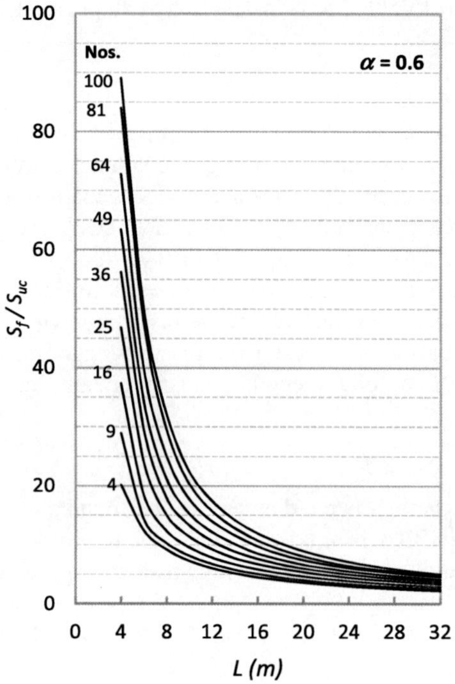

Fig. 7 Settlement ratio of $\alpha = 0.6$

$$L_{\text{opt}} = (2\alpha + 0.8)(1.13B) = (2 \times 0.3 + 0.8)(1.13B) = 10.28\,\text{m} < L\ldots\text{OK!}$$

Settlement of unimproved ground, S:

Numerical analysis was conducted to obtain the settlement of the shallow foundation for the unimproved ground. The drained settlement obtained was 0.28 m. For hand calculation, Boussineq's method can be used. The calculation is shown here:

Divide the soil layers into three layers (5 m each). Obtain the stress increase ($\Delta\sigma$) at mid-depth of each layer (z).

$$\Delta\sigma_1 = 81.5\,\text{kPa},\ \Delta\sigma_2 = 27.3\,\text{kPa},\ \Delta\sigma_3 = 11.60\,\text{kPa}$$

$$\begin{aligned} S = S_1 + S_2 + S_3 &= \frac{\Delta\sigma_1 \cdot H_1}{E_{\text{oed}}} + \frac{\Delta\sigma_2 \cdot H_2}{E_{\text{oed}}} + \frac{\Delta\sigma_3 \cdot H_3}{E_{\text{oed}}} \\ &= \frac{(81.5 + 27.3 + 11.6)(5)}{5000} = 0.12\,\text{m} \end{aligned}$$

which is much smaller than predicted by numerical analysis, i.e., $S = 0.28$ m. This is probably because the above equation does not consider the yielding effect and lateral deformation. The plastic straining in soil is more significant when the soil is very soft as in this case. Furthermore, the use of secant Young' modulus E instead of E_{oed} maybe more appropriate to account for lateral strain, resulting in $S = 0.16$ m when

E is taken as 3704 kN/m^2 (taking $\nu' = 0.3$). In the subsequent calculation for stone column improved ground, the numerical result was used.

Settlement of improved ground of floating columns, S_f:

$$n = 2.4\alpha + 1 = 2.4(0.3) + 1 = 1.72$$
$$S_f = S/n = 0.28/1.72 = 0.16\,\text{m}$$

Method 2:

The settlement of unimproved ground under *oedometric* condition, S

$$S = qL/E_{oed}\ 100/5000 \times 12 = 0.24\,\text{m}$$

The settlement of improved ground with end bearing columns, S_{uc}

$$n_0 = 9.43(0.3)^2 + 1.49(0.3) + 1.06 = 2.356$$
$$S_{uc} = S/n_0 = 0.24/2.356 = 0.102\,\text{m}$$

Settlement of floating stone column, S_f in small group:
From Fig. 9 (for $\alpha = 0.2$) and 10 (for $\alpha = 0.4$), the obtained settlement ratio are $S_f/S_{uc} = 1.2$ and $S_f/S_{uc} = 2.1$ respectively.
For $\alpha = 0.3$, take average of the two: $S_f/S_{uc} = 1.65$

$$S_f = S_f/S_{uc} \times S_{uc} = 1.65 \times 0.102 = 0.168\,\text{m}$$

In summary, the settlement predicted by Method 1 is in close agreement with Method 2. The limitation of Method 1 is that it requires the settlement result of the unimprovement ground which conventional analytical solution may produce unconservative results especially when the loading is high where the significant yielding in the soil has taken place. On the other hand, Method 2 does not require the prediction of settlement for unimprovement ground thus making the solution faster to obtain.

3 Conclusion

In this paper, three methods have been introduced for large stone columns group and two methods for small stone columns group. The methods are derived from numerical analyses where the effect of influencing factors on the settlement performance are taken into accounts. All the methods are simple and yet practical to use. Design examples are given for comparisons between each method. Engineering discretion should be exercised for the usage of the methods for cases out of the studied range.

References

1. Ong, D.E.L., Sim, Y.S., Leung, C.F.: Performance of field and numerical back-analysis of floating stone columns in soft clay considering the influence of dilatancy. Int. J. Geomech. **18**(10), 04018135 (2018)
2. Tan, S.A., Ng, K.S, Sun, J.: Column group analyses for stone column reinforced foundation. In: From Soil Behavior Fundamentals to Innovations in Geotechnical Engineering. GSP 233, pp. 597–608. ASCE, Reston (2014)
3. Ng, K.S., Tan, S.A.: Design and analyses of floating stone columns. Soils Found. **54**(3), 478–487 (2014)
4. Ng, K.S., Tan, S.A.: Simplified homogenization method in stone column design. Soils Found. **55**(1), 154–165 (2015)
5. Ng, K.S.: Settlement ratio of floating stone columns for small and large loaded areas. J. Geo Eng. **12**(2), 89–96 (2017)
6. Balaam, N.P., Booker, J.R.: Analysis of rigid rafts supported by granular piles. Int. J. Numer. Anal. Meth. Geomech. **5**(4), 379–403 (1981)
7. Priebe, H.J.: The design of vibro replacement. Ground Eng. **28**, 31–37 (1995)
8. Aboshi, H., Ichimoto, E., Enoki, M., Harada, K.: The "Compozer"-a method to improve characteristics of soft clays by inclusion of large diameter sand columns. In: Proceedings of International Conference on Soil Reinforcement, pp. 211–216 (1979)
9. Han, J., Ye, S.L.: Simplified method for consolidation rate of stone column reinforced foundations. J. Geotech. Geoenviron. Eng. **127**(7), 597–603 (2001)

Using Prestressed Reinforced Concrete Piles as Basement Walls for High-Rise Buildings

Thi My Dung Do and Thanh Quang Khai Lam

1 Introduction

Vietnam's major cities, such as Hanoi City, Ho Chi Minh City. The land area is rapidly exhausted, land prices are rising, population is rising, demand for people's housing is increasing, so the solution to building high-rise apartment buildings, shopping centers, leasing office buildings, these buildings are now built to have basements, using the basement space as a garage and technical rooms are the solutions that invest. The complicating problem is that deep excavation hole construction solutions in narrow ground can change the state of stress deformation in the environment surface the excavated area and likely change the level of groundwater. This leads to ground displacement, which can cause subsidence, if we do not have effective design and construction solutions, damage to neighboring buildings.

In the study [1], the authors presented possible causes of potential problems, risks commonly encountered in construction and at the same time offered common treatment solutions to be overcome in the construction of barrette walls in Vinh Long City, Vinh Long Province and analyze and suggest solutions for the construction of barrette walls suitable for the Vinh Long City area, then construct the general method of constructing the poured barrette wall in place and assemble the basement for high rise buildings in Vinh Long City for construction.

Research [2–5] analyzed the effectiveness of the use of prestressed reinforced concrete piles in the overall protection of the soil as an embankment structure, providing three directions. Using prestressed reinforced concrete piles, build projects for the embankment: do not strengthen, strengthen the wall with anchors, and strengthen the foundation by pre-compression. The results of the study show that the use of prestressed reinforced concrete piles with the correct length can bring

T. M. D. Do · T. Q. K. Lam (✉)
Mien Tay Construction University, 20B Pho Co Dieu street, Vinh Long, Vietnam

I. Pal and S. Kolathayar (eds.), *Sustainable Cities and Resilience*, Lecture Notes in Civil Engineering 183, https://doi.org/10.1007/978-981-16-5543-2_16

great economic efficiency, reducing the cost by almost 60% compared to the traditional method. The authors analyzed the advantages, disadvantages, and the construction method of prestressed reinforced concrete piles as basement walls for high-rise buildings on the basis of the analyzed studies and this paper.

2 Materials and Methods

Many cases are now built with basements or semi-basements to provide more space for use as parking spaces, storage, etc., when constructing buildings (high-rise and low-rise). It involves the depth of the basement and the depth of the foundation, so that the level of excavation is relatively deep and wide (digging the entire surface area of the work); it is therefore easy to cause neighboring works to subside.

However, the loads are complex due to neighboring structures that place pressure on the excavation. Since then, for example, multi-story buildings, single-story buildings such as brick construction, fake casting house, shallow foundation solution, foundation on natural foundation completely depend on the size and structure of neighboring buildings, etc. Therefore, both the designer and the construction unit need to provide specific measures from excavation work to build the basement and to withstand the building loads using the diaphragm wall system to the bracing system. Adjacent caused while digging. It should be a specialist agency to devise this measure and should be checked previous to construction. It is important that the investor and the construction unit first conduct a structural survey to see how specific the structure of adjacent buildings (a unit with a design function) is. Since then, his method of construction has been reasonable.

There are many different solutions for constructing basements at present, but the common ones are the following measures: open method of excavation, wall method in soil, topdown method, etc. (Fig. 1). Depending on the actual situation, each method has advantages and disadvantages, depending on the required options for the construction unit. The authors analyze the benefits and disadvantages of each method in this paper and then suggest effective construction methods [6–12].

3 Results and Discussion

3.1 Construction of the Basement of High-Rise Buildings According to the Open Method of Excavation

The open method of excavation is divided into two methods:

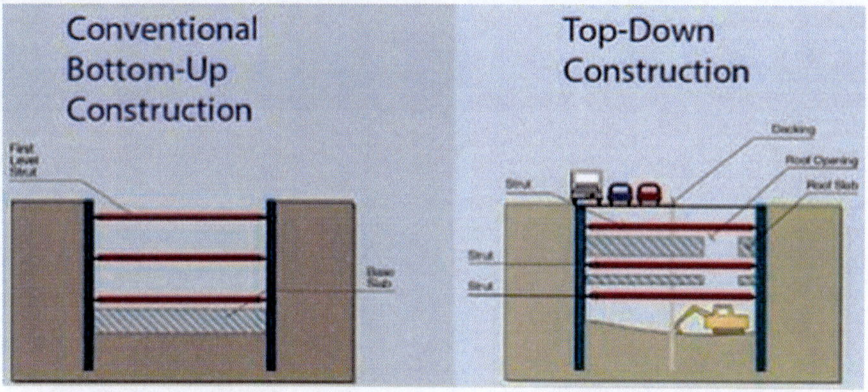

a) Conventional bottom-up construction b) Top-down construction

c) Open method of excavation

Fig. 1 Some different solutions for constructing basements

3.1.1 Digging Without Wall Support for the Holes

This method is to dig with a certain excavation coefficient (this coefficient depends on each type of soil) to the design elevation, measures such as spraying concrete, concrete coating, etc., may be used for the hole wall. This method is mostly applied to large-area projects; the deep groundwater level, this water has little influence on earthwork construction; the geology of engineering at a good construction site; the depth of digging is not great (1–2 basements).

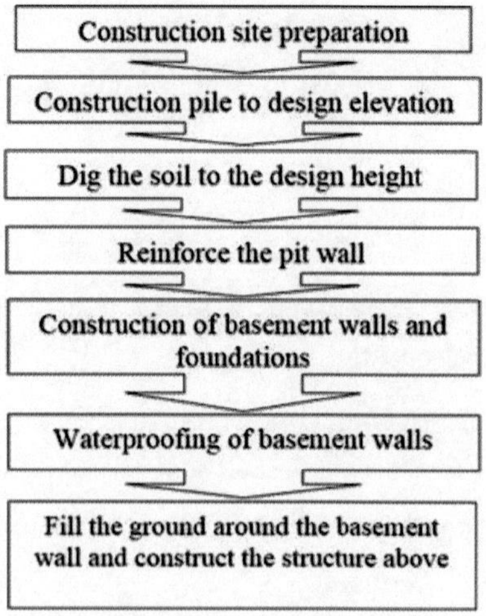

Fig. 2 Excavation process does not have a support system for the pit wall

The construction process of this method is (Fig. 2).

There are the following advantages to the excavation process without a pit wall support system:

- It is very convenient for the construction process due to the construction of an open excavation for the whole project.
- Do not use measures to protect the wall of an excavation hole, thus reducing the cost of construction.
- In the case of good soil, the depth of the underground water is deep and the soil can first be excavated to the base of the scaffold before the piles are constructed. Since it is not trapped at the top of the pile, the excavation work is made simpler (can be completely mechanized). It should be noted, however, that the groundwater level must be higher than the scaffold's foundation. In addition, when pressing the pile, there is no need to use negative piles.

Disadvantages

- This method is only applicable in places with a large work plan, no neighboring works (because the excavation with an excavation coefficient).
- Applicable only for strong geological works, deep underground level of water.
- The depth of the basement is not high
- As mentioned above the excavation method without the pit wall support system has advantages and disadvantages, it is mostly constructed in crowded construction conditions in big cities; nowadays, so the construction cannot follow this method.

3.1.2 Open Excavation with Wall Support System

Open excavation with a wall support system is a technique in which people build a wall system with earth cement piles, walls with reinforced concrete piles, walls with Larsen piles, etc. After completion of the pile and wall construction, dig to an acceptable depth (this depth will be measured in advance), then install the diaphragm wall support system and continue digging and re-installing the system. After that the basement wall is usually about 1 m away from the diaphragm wall, creating the foundations, basement floors, basement walls, columns, and the next floor. When the construction of the basement wall is finished, it is also very simple for persons to sweep and paste additional waterproofing layers on the inside and outside of the wall, expansion joints, stop circuits, settlement slots, etc.

In general, with this method of construction, we can more carefully check the quality of concrete and waterproof for the basement structures since most basement walls and floors are covered in a "raincoat" layer. But the Building Unit rarely follows this technique nowadays, since it is very costly to make the support structure for the diaphragm wall, it is a difficult problem to determine the safe design of the diaphragm wall. On the other hand, the construction of deep pit excavation is also very easy to cause problems such as subsidence of neighboring buildings and collapse of the pit wall with weak geological conditions, high groundwater level, and construction in narrow conditions. However, some companies also follow this method due to the ease of construction.

3.2 Construction of the Basement of a High-Rise Building According to the Method of Wall in Soil

Construction of the basement of a high-rise building according to the wall-in-soil technique means that the construction unit must first carry out the construction of the basement wall (barrette wall construction) before the construction of the excavation, then dig soil inside the wall to the basement bottom. Therefore the effect of soil pressure on this wall is very high. In order to strengthen the wall, the following solutions are often used by people: the use of a steel support system. This construction method is simple, but when the hole is dug deeper, it requires a lot of support, but it can be recovered and reused for another building after using it (Fig. 3).

The disadvantage of this method is that since the project has a wide diameter, the support system takes up a lot of space in the excavated hole, this support system becomes difficult, affecting the construction process greatly. It is necessary and valuable to predict the displacement of the diaphragm wall in each construction process, to ensure that the construction unit has chosen the correct construction plan, as well as to proactively avoid and cope with the complex changes that take place during the construction of deep foundation holes. In addition, it is also very complicated to calculate the stability for this support system; nowadays, there is also software to

Fig. 3 Method of wall in soil

support the measurement of displacement and the design of supporting structures, construction measures for works like Plaxis with contact with the surface. However, the results are also influenced by some input factors when simulating Plaxis, such as

- It requires really reliable geological data, and laboratory tests give high-reliability results (simulating almost natural lying soil sample) such as triaxial compression test with UU diagrams, CU and CD = > high cost.
- The construction time is quick or slow with regard to the immediate or long-term soil behavior, but shear parameters can be used with and without drainage.
- Groundwater level: The building phase will change the groundwater level considering the underground water pump, leading to a change in wall tension.
- To suit the real construction conditions, it is important to simulate the construction phases.
- The geology can be uneven for large sites, leading to simulations that offer unrepresentative results in any one section.
- The measurement model in Plaxis must be chosen to fit the actual working conditions of the ground.

3.3 Construction of the Basement of a High-Rise Building According to the Topdown Method

Topdown construction technology (from top to bottom) is the construction technology that varies from conventional construction technology from bottom to top for the included in of the work following the topdown method. Construction Technology Topdown can also construct the basements below the building's complete foundation and the building's foundation and build some of the building's upper floors. There are currently many companies that can perform very well in Construction Topdown, including a few traditional companies such as Bachy Solatance, Delta, and Longgiang. Building the basement of a high-rise building with topdown building technology has advantages such as

- Topdown construction technology does not require a wide excavation area of the foundation or reduces the cost of building temporary soil digging walls or the cost of creating temporary support framework structures for later retention of walls as basement walls.
- Topdown building technology is done quickly because the foundation and basement construction are both carried out on the upper floors at the same time, thus saving construction time, leading to faster progress in construction.
- During excavation and basement construction, there is no need for an auxiliary support system to support the basement wall, thereby minimizing costs for this auxiliary support system, thus saving construction costs.
- Topdown construction technology is a technology that uses a basement beam system to build a wall system when excavating soil in combination with basement walls, thereby completely resolving the stability of retaining walls when digging. As this wall and floor, girder system is very firmly constructed, and this is the basement floor of the building's wall and girder system.
- The scaffolding system is not available in the basement girder system construction, because the basement beam system is built on the ground.
- Topdown construction technology when first constructing the ground-floor girder system and then constructing the basement works, thus reducing the effect (if any) of weather factors.

This Topdown-based construction method is now very widely used in the construction of basements, because of the very high hardness of the floor, the displacement of the diaphragm wall when digging the soil is small and easy to control. Using the retaining walls as a wall later for the basement. Before pouring concrete, care must be taken to ensure a very good quality of concrete and especially waterproof gaskets at positions between barrettes, expansion joints, wall-to-floor link, between beams and columns, etc. Before pouring concrete, it must be properly installed and accepted. It must be re-installed if there is any deviation.

Concrete is formed by the process of surging mortar, and bentonite is the wall-keeping solution. Disadvantages of this method: When constructing the basement column system is complicated, it is difficult to link the beams with the columns and walls in the basement. Since digging in enclosed space, through pits, excavation is difficult, so it is very cramped and difficult to mechanize. Construction conditions in closed tunnels have a major impact on workers' health and productivity, so artificial ventilation and lighting systems are necessary. Due to the design of the mortar process, the quality of concrete is difficult to control, so the quality depends a lot on the quality of the system to create the walls, the quality of concrete input, the managers' qualifications, the employees' hands, etc. Some projects have to adopt the reverse waterproofing process, resulting in the low and costly waterproofing effect, the quality of the concrete directly affects the waterproofing. Sometimes, they have to do it again and again.

From the above analysis, the authors realize that it is especially important to choose a basement construction solution to be safe for neighboring works, easy construction, and save investment costs. Prestressed prefabricated reinforced concrete piles are first made by P.S. Mitsubishi (Japan) more than 50 years ago. Since 1999, prestressed precast reinforced concrete piles have been applied in Vietnam at Phu My thermal work in the main and branch canal system, carrying water to cool the gas turbines from the Thi Vai River. Presently, prestressed, precast reinforced concrete piles are also used in the Kien Giang Province to protect the Can Tho River banks, Tra Noc River, and reclamation works.

With different shapes such as flat and waveform, piles are designed. There are connected joints at the boundary of each pile so that they are connected together to form a retaining wall when lowering them; based on this feature, many units have researched and proposed plans to use prestressed concrete piles to create diaphragm wall structure combined with topdown construction to build basements for high-rise buildings with basements. This technique has been used in a number of high-efficiency projects, such as

- A mix of on-site poured concrete and semi-assembled concrete is being constructed at the Mo Market Trade Center project using the new technology of prestressed reinforced concrete piles for the construction of three basements. The reinforced concrete wall has a thickness of 550 mm. During pile pressing, the average depth must be force-resistant, noise-proof, and vibration-free. Vinaconex Xuan Mai chose a prefabricated pile solution without digging soil, not mixing on-site concrete, then pressing 349 bored piles with a hydraulic press with a capacity of 600–800 tons, 784 piles to a depth of 20 m.
- CT2 Building project, Ngo Thi Nham-Ha Dong, the building was constructed on a land area of about 10,000m^2, built with 2 basements, 25 floors, Vinaconex Concrete investor and Xuan Mai Construction Joint Stock Company, the project uses a total of 511 piles equal to 257 m, the pile is 15 m long, the construction time is 8 m long/day.
- HEMISCO Building project, Sa La - Ha Dong, the building was built on a land area of nearly 2,810m^2, designed with 2 basements, 29 floors, investor: Song Da 101 Joint Venture and Electrical Machinery Construction Ha Tay, the project uses a total of 376 piles, equivalent to 189 m, long piles of 13 m, etc.

The construction process for the basement wall consists of three phases:

Fig. 4 Construction of pile pressing

Phase 1: Preparation work
Phase 2: Construction of piles
Phase 3: Construction of the basement and the above structure.

Phase1: Work on preparation

1. Locate the hole for the drill.
2. Drilling to create holes.
3. Locate the pile.
4. Move the machine into the place of pressing.

Phase 2: Pile presses construction (Fig. 4)

5. To load the pile into the clamp, use a crane or an equivalent.
6. Until pressing the pile to the design depth, modify the vertical position and position of the pile.

Notes: Check the pile back 3 cm from the guide wire of the pile wall during the pressing process, always machine-check the vertical of the pile. Construction measures for consultancy work and approval by investors need to be prepared by contractors.

7. The second pile compressor connects to the first pile, pushes to the design height of the first pile, and then stops pressing.
8. To continue pressing in this position, pull the second section of the pile out and move the machine to the new center.

Fig. 5 Pile segments 2,3, etc., is repeated

9. With pile segments 2,3,4,5, etc., the pressing process is repeated (Fig. 5)

Phase 3: Constructing the basement and the structure above

10. After pressing the prestressed concrete piles to the designed height for the entire circumference of the building, we continue to construct the top-wall girder framework, and the location of the displacement monitoring points must be marked at the top of the pile wall during the construction process (Fig. 6).
11. Digging soil to the bottom of the first basement, pay attention to arrange temporary water collection holes during construction (Fig. 7).
12. Construction of beams and floors for basement 1 to support piles (Fig. 8).
13. Dig soil to the bottom of the foundation height (for a house with 2 cellars, if there are 3 basements, this step is to enter the basement height 2). Notice that during the construction process, it is important to arrange trenches and temporary water collection holes. (Fig. 9).
14. Construction of foundations, foundation beams (the lower piles are constructed first) (Fig. 10).
15. Construction of basement columns 2, beams, and basement floors 1 (Fig. 11).
16. Construction of 150-thick waterproof concrete around the basement (Fig. 12).
17. Columns, beams, floors 1, and floors above are constructed.

The following advantages are the use of prestressed concrete piles for the diaphragm wall structure:

Fig. 6 Construct the top-wall girder

Fig. 7 Dig soil to the height of the first basement

- High-precision prestressed concrete piles are made in the factory, using high-grade concrete combined with a high-strength cable to save materials in order to control the quality of the concrete from the factory.
- Construction of specialized equipment, particularly coastal piles, to ensure vertical, tight, low noise, and eco-friendly activity.
- Short construction time, 8 to 15 m long/day can be conducted, thus shortening construction progress of underground work.

Fig. 8 Construction of beams, floors for basement 1

Fig. 9 Dig soil to the bottom of the foundation height

- When digging, the use of pre-tensioned concrete piles combined with semi-topdown construction would save all the costs.
- Waterproof treatment: Treatment with the waterproofing concrete layer of the inner surface of the basement wall improves the looks compared to other choices.
- Quick execution, saving the cost of investment in construction work.

Fig. 10 Construction of foundations, foundation beams

Fig. 11 Construction of basement columns 2, beams, and basement floors 1

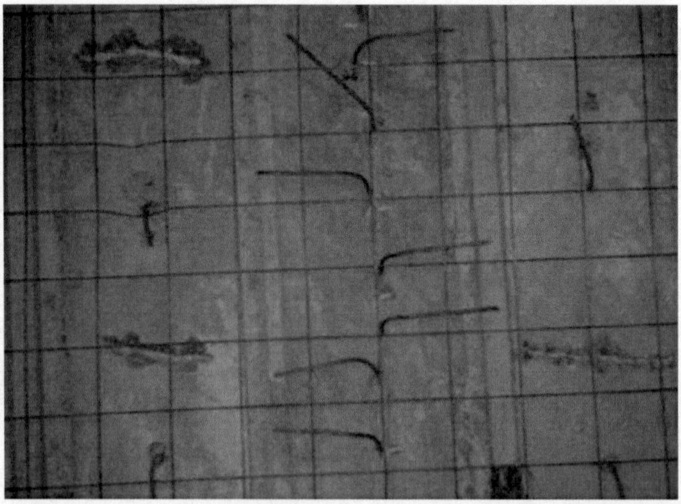

Fig. 12 Install steel on the inside of the piles wall and pour a layer of concrete 15 cm waterproofing

- When combining pile pressing with lead drilling, construction has no effect on neighboring works.

There are also some disadvantages, in addition to the above advantages on prestressed precast reinforced concrete piles, such as the pile length is reduced so that it cannot be used for houses with many basements (>3 basements), equipment construction must be modern because high precision is needed for construction. Before pressing the pile, drilling must be completed first, so the progress of construction is relatively slow. Difficult to construct according to curves with small radius, the depth of the pile is limited by complicated connection details.

4 Conclusion

1. The use of prestressed reinforced concrete piles saves and lowers the project's cost, bringing the construction to high quality. Besides that, shorten the time, improve the construction and ensure the work's aesthetics, engineering, and quality. In the construction revolution, this is the optimal solution.
2. Applying prestressed reinforced concrete pile technology to the basement wall design for high-rise buildings lets design engineers have more choices in the design calculation and comparison process, allowing investors to choose the right use and financing plan to help improve the project's investment effectiveness. However, in order to increase the length of the piles as well as study to improve suitable construction technology, it is important to have in-depth studies.

References

1. Huynh, H.C.: Construction process of building Barrette wall for high-rise buildings in Vinh Long city. Master thesis. Hanoi Architectural University (2016)
2. Truong, T.N.: Evaluation of the applicability of prestressed reinforced concrete piles for river embankment structures in Can Tho area. Master thesis, National University of Civil Engineering (2014)
3. Do, T.L.: Research on the use of pre-stressed reinforced concrete piles (Mekong Delta basin). Ho Chi Minh City University of Technology (2008)
4. Le, X.K.: Research on the application of prestressed reinforced concrete piles to protect river banks in urban areas. J. Sci. Technol. Water Resour. Environ. **43,** (2013)
5. Nguyen, B.V.: Prestressed reinforced concrete piles, applied in embankments with soft ground. J Constr Sci Technol 1 (2015)
6. Nguyen, B.K.: Construction of urban underground construction according to open excavation method. Construction Publishing House (2012)
7. Nguyen, V.Q.: Guielines for the design and construction of Barrettes diaphgram wall and ground anchors. Construction Publishing House (2010)
8. Prestressed Concrete Pile, Technical Instructions, Beton 6 Joint Stock Company (2014)
9. Pre-stressed Concrete Pile, Technical Instructions, Vina - PSMC Precast Concrete Company Limited
10. Tran, H.Q., Lam, T.Q.K., Do, T.M.D.: Model of prefabricated concrete frame in the condition of southern Vietnam. In: E3S Web of Conferences, vol. 135, p. 03043, (2019). https://doi.org/10.1051/e3sconf/201913503043
11. Do, T.M.D., Lam, T.Q.K.: Quality of construction works at the design phase. Lecture Notes Civil Eng **70**, 15–24 (2020). https://doi.org/10.1007/978-3-030-42351-3_2
12. Do, T.M.D., Nguyen, T.C., Lam, T.Q.K.: Investigating the effectiveness of insulation for walls of buildings in Vietnamese climatic condition. IOP Conf. Series: Mater. Sci. Eng. **869**, 032008, 2020. https://doi.org/10.1088/1757-899X/869/3/032008

Analytical Study on Single Span Reinforced Concrete Beam with Continuous Spiral Reinforcement Under Pure Torsion

M. Prakash, V. Deepakraj, N. Parthasarathi, and S. Srinivasa Senthil

1 Introduction

In reinforced concrete (RC) beams, crucial factors such as flexural, shear, and torsional resistance play a vital role. Torsional moments are developed in RC members so that the torsional capacity of the members needs to be enhanced to avoid structural damage, deterioration due to amplified loading scenarios. In reinforced concrete structural systems, torsion has been considered the least important, but they are required to study for modern structural configurations. The brittle nature of concrete under the development of shear stress due to torsion is an essential aspect of seismic analysis and design. The rebar arrangement of a beam with the continuous spiral reinforcement is represented in Fig. 1. The general design methods are complex to ignore the torsional moments produced due to the monolithic construction. However, the theory of elasticity could precisely estimate the torsional strength of the sections developed from homogeneous materials.

Although it is very complicated to evaluate the heterogeneous reinforced concrete's torsional strength, the problem becomes more intense as these members hardly undergo pure torsion. Instead, they put through bending, shear, and torsion. Torsion is always an unavoidable and essential structural aspect. This research concentrates on a single span beam's torsional behavior with inclined shear reinforcement with a different configuration of angle to understand the beam's failure mechanism under pure torsion. The utilization of continuous spiral reinforcement will result in better ductile performance than the conventional one.

M. Prakash (✉) · V. Deepakraj · N. Parthasarathi · S. S. Senthil
Department of Civil Engineering, College of Engineering and Technology, SRM Institute of Science and Technology, SRM Nagar, Kattankulathur, Tamil Nadu 603 203, India

I. Pal and S. Kolathayar (eds.), *Sustainable Cities and Resilience*, Lecture Notes in Civil Engineering 183, https://doi.org/10.1007/978-981-16-5543-2_17

Fig. 1 Beam with continuous spiral reinforcement

2 Previous Works in Torsional Beam

The use of optimum angle of reinforcement is better resistant against torsion [1]. Abdeldayem Hadhooda et al. observed the torsion in concrete beams reinforced with GFRP spirals [2]. Jeyashree T.M. and M Nethaji experimented on the flexural behavior of RC beams with special forms of shear reinforcement [3]. Ammar N. Hanoon and Haider A. Abdulhameed evaluated electricity absorption in CFRP-strengthened spans strengthened concrete beams under pure torsion [4]. Tuan-Anh Nguyen et al. performed a more desirable finite detail version for reinforced concrete beneath torsion with steady fabric parameters. [5]. Mohammed Sirage et al. observed the vital impact of concrete cowl on bolstered concrete beams underneath pure torsion [6]. Nasim Shatarat et al. studied the shear capability of reinforced concrete beams with two distinctive tendencies [7]. Mohammad Amin Azimi et al. investigated the continuous square spiral transverse reinforcements underneath seismic overall performance of RC beam–column connections for low ductility instances [8]. Piero Colajanni et al. investigated the shear capacity of concrete beams in two different inclinations stirrups [9].

3 Analytical Program

3.1 Design Criteria

For the analysis of structural members, a finite element software ABAQUS (FEA) has been chosen. In the preprocessing work, the general properties of elements, i.e., concrete grade M30 grade of concrete and Fe415 grade of reinforcements, were used along with those other nonlinear properties of concrete. Then the element was modeled, and configuration reinforcement arrangement was placed into the element, and the interaction was provided to acts as a homogenous element, and the element was finely meshed to obtain the accurate results.

3.2 *Description of Specimens*

For the analysis, four reinforced concrete beams with M30 grade of concrete were used for this investigation. The cross sections of the beams were 230 × 300 mm at a span of 2000 mm. Figure 2 shows the different configurations of rebar arrangement. Beam (B1) was designed with 2 nos—12 mm Ø bars at top and bottom, 8 mm Ø @150 mm c/c stirrup (conventional type 90°) are provided, beam (B2) was designed with 2 nos—12 mm Ø bars at top and bottom, 8 mm Ø @150 stirrup (inclined type 80°) are provided, beam (B3) was designed with 2 nos—12 mm Ø bars at top and bottom, 8 mm Ø @150 mm stirrup (inclined type 75°) are provided, beam (B4) was designed with 2 nos—12 mm Ø bars at top and bottom, 8 mm Ø @150 mm stirrup (Inclined type 45°) are provided, these details are presented in Table 1.

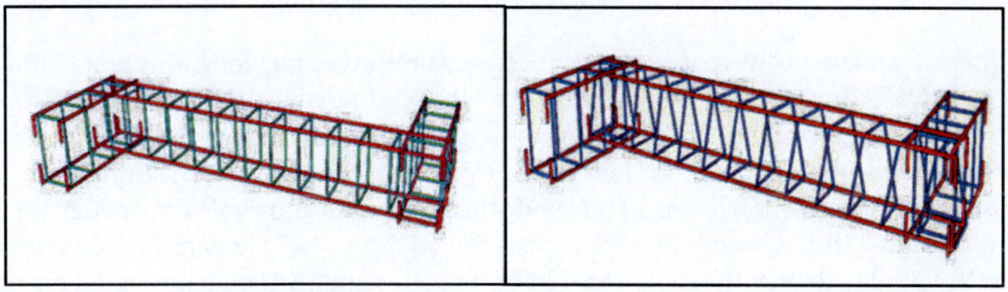

a) B1 - Conventional type Stirrups (90°)

b) B2 - Inclined type Stirrups (80°)

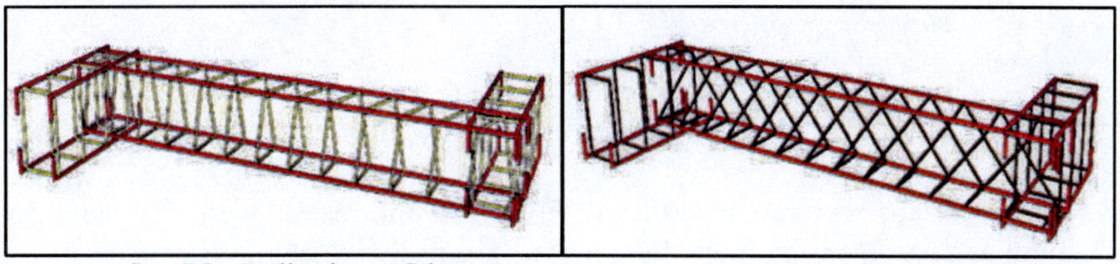

d) B3 - Inclined type Stirrups (75°)

c) B4 - Inclined type Stirrups (45°)

Fig. 2 Rebar arrangement

Table 1 Reinforcement details of the members

Beam ID	Cross section	Longitudinal Reinforcement	Transverse Reinforcement	Type	Concrete and Steel Grade
B1	230 mm × 300 mm	2–12 mm Ø at top 2–12 mm Ø at Bottom	8 mm Ø at 150 mm c/c	Conventional (90°)	M30 & Fe 415
B2				Inclined (80°)	
B3				Inclined (75°)	
B4				Inclined (45°)	

Table 2 Reinforcement details of the members

Beam ID	Initial Crack load (kN)	Ultimate Crack load (kN)	Maximum Crack Width (mm)	Initial Cracking Torque (kNm)	Ultimate Cracking Torque (kNm)
B1	9.5	12	0.24	16.91	26.70
B2	9.5		0.24	16.91	
B3	11.5		0.19	22.40	
B4	11.5		0.19	22.40	

4 Results and Discussion

4.1 Load Versus Crack

The crack pattern on the specimens was obtained under the crack/crushing plot option in ABAQUS. Initial crack can be defined as the load at which the extreme tension fiber reaches the modulus of rupture. As the applied torque increases, the spiral cracks developed approximately at 45° and spread over the test region. The maximum crack observed at specimen (B1) is 0.24 mm under the maximum allowable limit of 0.3 mm. From Table 2, it is observed that the specimen (B1) and (B2) undergo initial crack due to initial cracking torque of 16.91 kNm and (B3) and (B4) undergo initial crack due to initial cracking torque of 22.40 kNm. The beam (B1, B2, B3, and B4) observed the same ultimate cracking torque was obtained as 26.70 kNm.

4.2 Load Versus Deflection

All four specimens were analyzed using FEM, and the results were compared for maximum deflection and ultimate load-carrying capacity. Specimens were analyzed for pure torsion loading condition. Fig. 3 shows the span vs. deflection curve of beams. From Fig. 3, it is observed that specimen (B3 and B4), i.e., specimen with (45° and 75°) inclined continuous stirrups has an increase in load-carrying capacity and a significantly less deflection of 0.0011 mm when compared to the specimen with conventional type stirrup. Under flexural loading conditions, the maximum allowable deflection 6.6 mm for the span of 2 m. Here, due to pure torsion condition, the maximum deflection in specimens is 0.0013 mm. Therefore, all the specimens are safe under deflection.

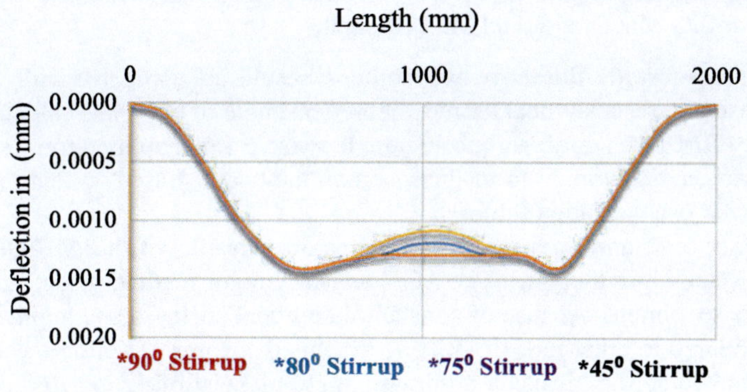

Fig. 3 Span versus central deflection in 'mm'

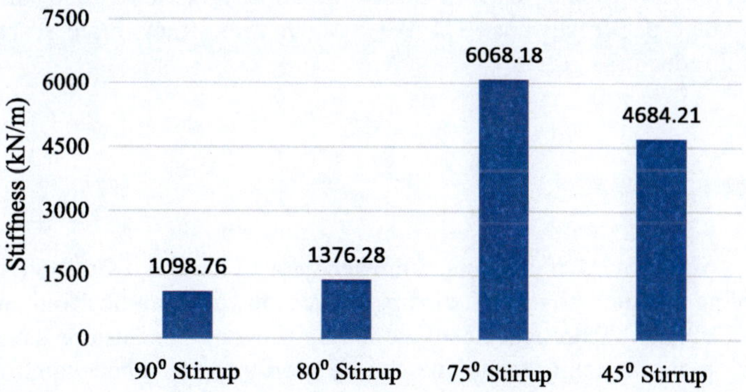

Fig. 4 Comparative stiffness between different configuration of stirrups

4.3 Stiffness

The torsional stiffness of the torsional beam is calculated based on the torsional equation; Figure 4 shows the comparative torsional stiffness between different configurations of stirrups for the ultimate torque. When compared to conventional type stirrup specimen (B1), specimen (B3), i.e., 75° inclined stirrups has 81% more stiffness.

4.4 Torque Versus Angle of Twist

Many theories are proposed to find the torsional strength of RC rectangular beam here following methods are used for comparison of analytical torsion with theoretical calculations.

- Based on ACI 318–99

- Based on J.G MacGregor and M.G Ghonelm

In this present study, the above two methods' results are compared with analytical results. Based on the analytical torque, theoretical angle of twist was calculated. The ABAQUS FEM gives direct torque and angle twist. So, the applied torque depends on the iteration and amount to load applied on the beam. Based on that theoretical angle of twist is calculated (Table 3).

Figures 5, 6, 7, and 8 represent the comparative results of the ABAQUS FEA, ACI, and MacGregor methods. The ACI and MacGregor method results are similar for all four specimens because of constant theoretical stiffness and applied torque. When compared to other models (B3), 75° inclined stirrups performed well against the torsional force; i.e., it has a minimum angle of twist of 4.4×10^{-3} radians/m for the maximum applied torque of 26.70 kN.m. Other specimens (B1, B2, and B4) carry the angle of twist of $(24.3, 19.4, \text{and } 5.8) \times 10^{-3}$ radians/m. Comparatively, specimen (B3) resists nearly 81% of twist angle for the applied torque compared to a conventional type 90° stirrups (B1). We know that if stiffness increases, the angle of twist will reduce.

5 Conclusion

The study's main aim is to understand the behavior of torsion, twist, and deflection under varying inclined angles of stirrups. Based on the analytical research using ABAQUS FEA and theoretical calculation, the following conclusions have arrived. So finite element modeling technique can be used to better understand of torsion failure mechanism.

1. The elements (B3 and B4) having inclined stirrups have significantly less deflection (0.0011 mm) than the allowable deflection (6.6 mm).
2. The element (B3), which has 75° inclined stirrups, has a minimum angle of twist of 4.4×10^{-3} radians /m other specimens (B1, B2, and B4) carries the angle of twist of $(24.3, 19.4, \text{and } 5.8) \times 10^{-3}$ radians/m. Thus, comparatively, specimen (B3) resists nearly 81% of Twist angle for the applied torque compared to a conventional type 90° stirrups (B1).
3. The element (B3), which has 75° inclined stirrups, has maximum stiffness of 6068.18 kN/m. Other specimens (B1, B2, and B4) have the stiffness of (1098.76, 1376.28, 4684.21) x kN/m. Comparatively, specimen (B3) resists nearly 81% of twist angle for the applied torque as compared to a specimen with conventional type 90° stirrups (B1).

Table 3 Analytical and theoretical results from each beam under a different angle of stirrups

Torque Tu in kN.m	The angle of twist Θ in (radians/m) × 10^{-3}			The angle of twist Θ in (radians/m) × 10^{-3}			The angle of twist Θ in radians/m) × 10^{-3}			The angle of twist Θ in radians/m) × 10^{-3}		
	90°			80°			75°			45°		
	FEM	ACI	MAG	FEM	ACI	MAG	FEM	ACI	MAG	FEM	ACI	MAG
26.70	24.3	16.5	23.1	19.4	16.5	23.1	4.4	16.5	23.1	5.8	16.5	23.1

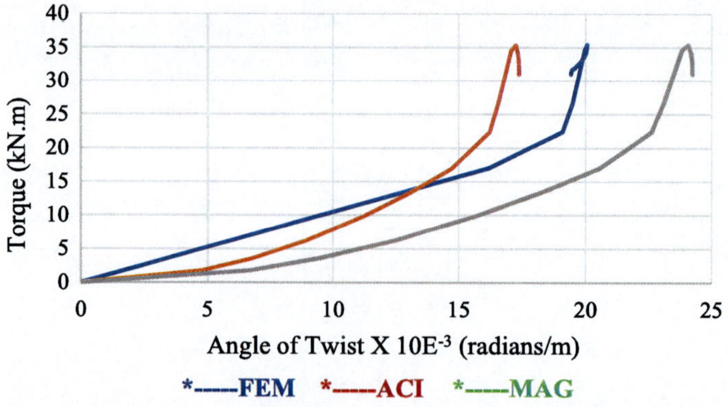

Fig. 5 Comparative results of conventional 90° stirrups

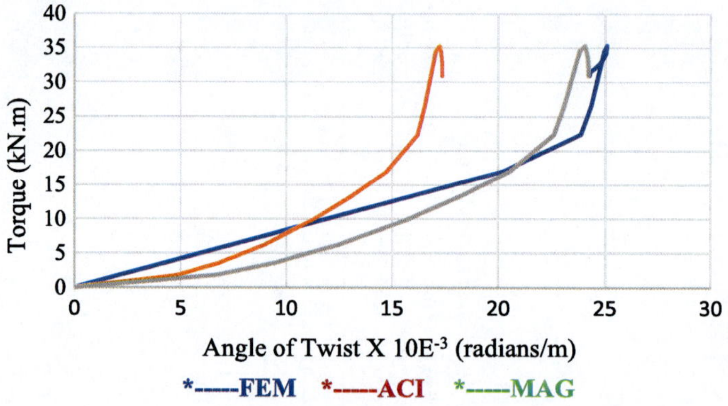

Fig. 6 Comparative results of inclined 80° stirrups

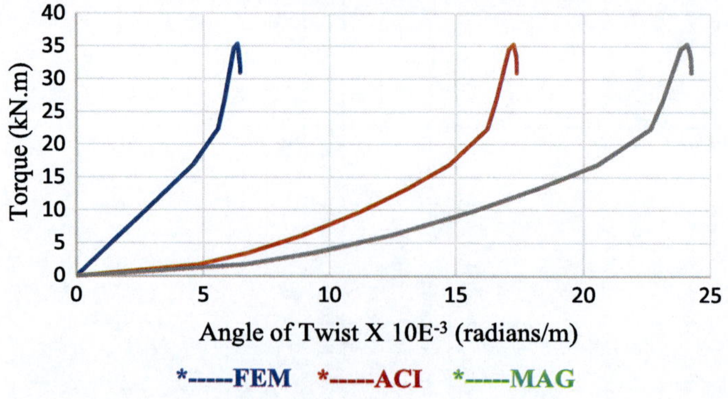

Fig. 7 Comparative results of inclined 75° stirrups

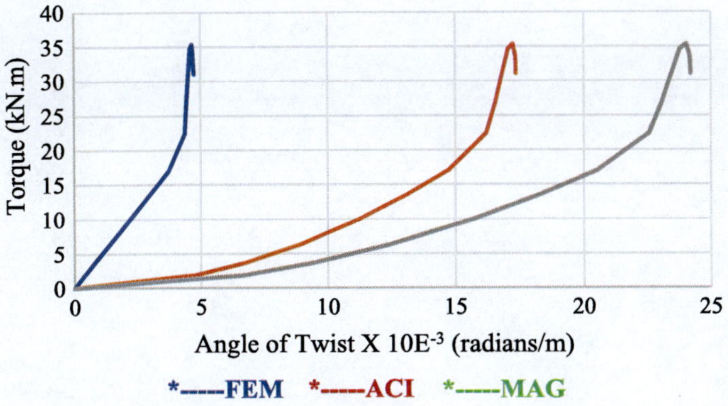

Fig. 8 Comparative results of inclined 45° stirrups

References

1. Had hood, A., Gouda, M.G., Agamy, M.H., Mohamed, H.M., Sherif, A.: Torsion in concrete beams reinforced with GFRP spirals. Eng. Struct. **206**, 110174 (2020). ISSN 0141-0296https://doi.org/10.1016/j.engstruct.2020.110174
2. Jeyashree, T. M., Nethaji, M.: Experimental investigation on flexural behavior of reinforced concrete curved beams with different types of shear reinforcement. Int. J. Emerging Technol. **11**(3), 615–618 (2020)
3. Ammar, H., Ali, A., Haider, A., Saad, K.: Energy absorption evaluation of CFRP-strengthened two- spans reinforced concrete beams under pure torsion. Civil Eng. J. **5**, 2007–2018 (2019). https://doi.org/10.28991/cej-2019-03091389
4. Nguyen, T.-A., Nguyen, Q.-H., Somja, H.: An enhanced finite element model for reinforced concrete members under torsion with consistent material parameters. Finite Elements Anal. Des. **167**, 103323 (2019). ISSN 0168–874X, https://doi.org/10.1016/j.finel.2019.103323
5. Ibrahim, M.S., Gebreyouhannes, E., Muhdin, A., Gebre, A.: Effect of concrete cover on the pure torsional behavior of reinforced concrete beams. Eng. Struct. **216**(2020), 110790 (2019). ISSN 0141-0296,https://doi.org/10.1016/j.engstruct.2020.110790
6. Shatarat, N., Katkhuda H., Abdel-Jaber, M., Alqam, M.: Experimental investigation of reinforced concrete beams with spiral reinforcement in shear. Constr. Build. Mater. **125**, 585–594 (2016). ISSN 0950–0618, https://doi.org/10.1016/j.conbuildmat.2016.08.070
7. Azimi, M., Adnan, A.B., Sam, A.R.B.M., Mohammadamin, R.H.A., Azlan, A., Tahir, M., Abdul Rahman, M.S., Muiz, S.A.R.S.: Seismic performance of ductility classes medium RC beam-column connections with continuous rectangular spiral transverse reinforcements. Latin Am. J. Solids Struct. **12**.https://doi.org/10.1590/1679-78251387
8. Colajanni, P., La Mendola, L., Mancini, G., Recupero, A., Spinella, N.: Shear capacity in concrete beams reinforced by stirrups with two different inclinations. Eng. Struct. **81**, 444–453 (2014). ISSN 0141–0296. https://doi.org/10.1016/j.engstruct.2014.10.011
9. Gunasekaran, K., Ramasubramani, R., Annadurai, R., Prakash Chandar, S.: Study on reinforced lightweight coconut shell concrete beam behavior under torsion. Mater. Des. **57**, 374–382 (2014). ISSN 0261–3069, https://doi.org/10.1016/j.matdes.2013.12.058

Investigation of Wind Loads on Setback Building Using Computational Fluid Dynamics

Vigneshwaran Rajendran and S. Prabavathy

1 Introduction

For the past couple of decades, there is tremendous growth in the construction of tall building around the world. According to the Council on Tall building and Urban Habitant (CTBUH), an organization used for collecting and storing tall building database, reported that there are about 5387 buildings of height greater than 300+m. For the design of such tall building, wind loads are the major design parameter to be considered. Since most of the wind engineering code provision lags in pressure coefficient (C_p) for irregular plan shape building, it is important in the current scenario to investigate the wind loads for buildings with shapes either numerically or experimentally. Many comparative studies have been done by the previous researcher using wind tunnel testing and computational fluid dynamics (CFD) in order to investigate the accuracy of the results between experimental and numerical analysis. Researcher [1] investigated the dynamic behaviour of the façade under wind flow. The façade design in the building is decided based on the minimum wind pressure acting on the buildings. Indeed, many researchers worked on the buildings with different shapes, on this sphere author [2] investigated the mean (C_p) on 'E' plan shape building using CFD and wind tunnel testing. The C_p on each face of the are investigated and the error analysis (R^2, SSE, RMSE) is done between wind tunnel and CFD, the obtained results are within the permissible limits. Similarly, researchers [3] investigated the wind on 'Z' plan shape building. The simulation is carried using a CFD package available in ANSYS software. The simulations are carried for the wind angles 0° to 150° at an interval of 30°. The flow separation behind the bluff bodies is well captured using CFD for 'Z' plan shape building. The author [4] made a critical review on the investigation of the wind-induced loads and response of the tall building. The widely used methods are numerical simulations, wind tunnel

V. Rajendran (✉) · S. Prabavathy
Department of Civil Engineering, Mepco Schlenk Engineering College, Sivakasi, Tamil Nadu, India

I. Pal and S. Kolathayar (eds.), *Sustainable Cities and Resilience*, Lecture Notes in Civil Engineering 183, https://doi.org/10.1007/978-981-16-5543-2_18

testing and field investigation. Moreover, author [5] has investigated the wind flow around the building with corner modification such as recessed and chamfered model building, and the obtained results are compared with CAARC building models. The investigation is done using CFD and the turbulence model shear stress transport (SST) is used. The corner modification in building models tends to reduce the formation of the recirculated regions behind the building.

From the above study, this paper presents with the novelty in investigating the wind flow around the setback buildings using CFD. The simulation is carried for 0° wind angles and the pressures on each faces of the buildings are obtained.

1.1 Details of the Model

For the present study, the building models are scaled in the ratio 1:300. Further, the building setback is done at two different levels in building height. The total height of the building is 300 mm, the first setback is provided at a height of about 150 mm from the base, and the second setback is provided at a height of about 225 mm from the base. Figure 1a shows the 3D view of the setback building using solid work software. Figure 1b shows the plan view of the setback building. Figure 1c shows the elevation view of the setback building.

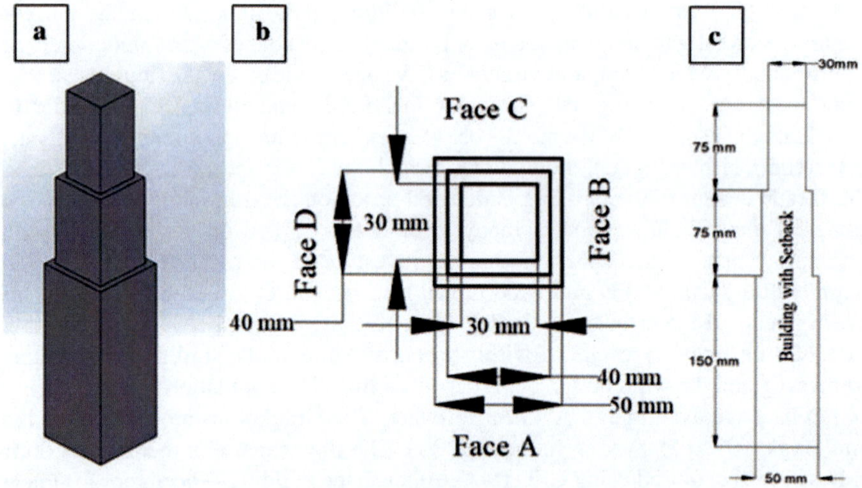

Fig. 1 Typical perspective view of setback building **a** 3D view, **b** plan view, **c** elevation view

2 Numerical Simulation

In order to investigate the pressure coefficient (C_p) on the setback building using CFD, turbulence model k-ε is used. Many past literatures [6] investigate that the turbulence model k-ε is very much useful for external flow simulation. The governing equation involved is momentum equation, energy equation and continuity equation.

2.1 Computational Domain

In order to study the wind flow behaviour, the building model is placed inside the computational domain. The computational domain is built on size 5H in the upstream and side faces of the computational domain and 15H in the downstream of the computational domain. Where H represents the height of the computation domain. As per the guidelines provided by the past researchers [7, 8], the provided size should be sufficient enough to develop the flow inside the domain and to study the physics of the wind flow behind the buildings. Figure 2 shows the details of the computational domain.

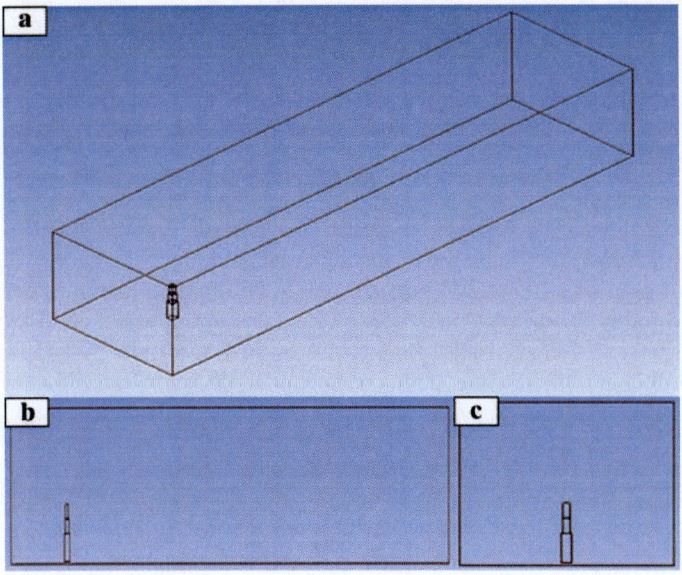

Fig. 2 Computational domain **a** 3D view, **b** side view, **c** front view

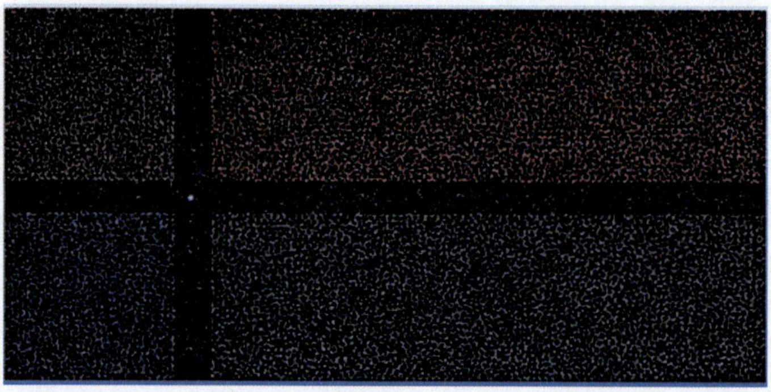

Fig. 3 Meshing of the computational domain

2.2 Meshing

The meshing of the building models is done in the forms of grid and very fine mesh is adopted in the wall of the buildings model in order to capture the pressure contour being developed on the surface of the building. Grid independency study is done with three category such as fine, medium and course. Further, fine mesh is used for the capturing the fluid flow and pressure developed on each faces of the building. Figure 3 shows the mesh pattern using computational domain.

2.3 Validation

The numerical study is validated based on the pressure coefficient (C_p) obtained from the square building with the code values available in IS 875(Part 3) 2015. The building of plan size 100 × 100 mm and height 500 mm and the turbulence model k-ε is used for external flow simulation. Table 1 shows the variation of the pressure coefficient (C_p) between Ansys Fluent 19.2 and IS 875 (part 3) 2015. A similar validation study is done by the previous researcher [3] for 'Z' plan shape building.

Table 1 Pressure coefficient (C_p) between numerical and codal provisions

Wind loading code	Face A (C_p)	Face B(C_p)	Face C(C_p)	Face D(C_p)
ANSYS Fluent 19.2	0.90	−0.20	−0.86	−0.85
IS:875(Part 3) 2015	0.80	−0.25	−0.80	−0.80

3 Result and Discussion

The investigation of the wind load on square building with setback is done using computational fluid dynamics (CFD). The simulation is carried for the 0° wind angle at a wind velocity of 10 m/s, and the pressure coefficient (C_p) is calculated using the Eq. (1).

$$C_{p,\text{mean}} = \frac{p1 - p_0}{\frac{1}{2}pU_2^H} \tag{1}$$

where p is the point pressure taken from the building surface, p_o is the pressure at reference height, ρ is the density of the atmospheric air as 1.225 kg/m^3. U_H is the mean velocity at the height H. Figure 4 shows the direction of the wind towards setback building.

3.1 *Pressure Contour*

The outcome of the wind flow around the building results in the formation of the pressure on each faces of the building. The pressure contour on each face of the building depends upon the shape, height and configuration of the building. For the present study, the pressure contour on each face of the building surface is observed. In order to study the wind pressure distribution on setback buildings, the building is divided into three segments from the base, from 0 to 15 cm face A1, 15 to 22.5 cm face A2, 22.5 to 30 cm face A3. Similarly for the sides faces B1, B2, B3; D1, D2,

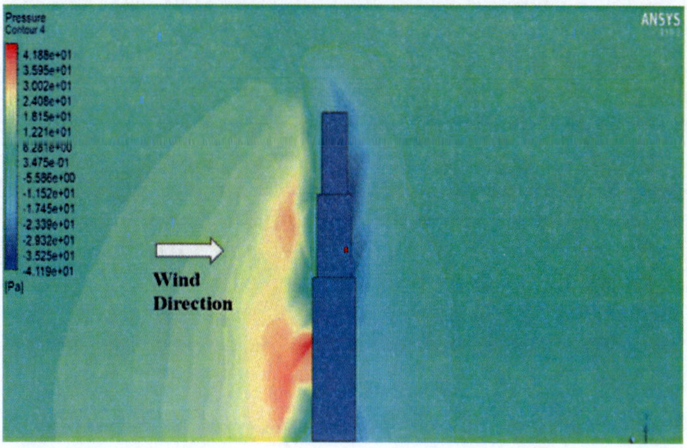

Fig. 4 Wind direction toward the building

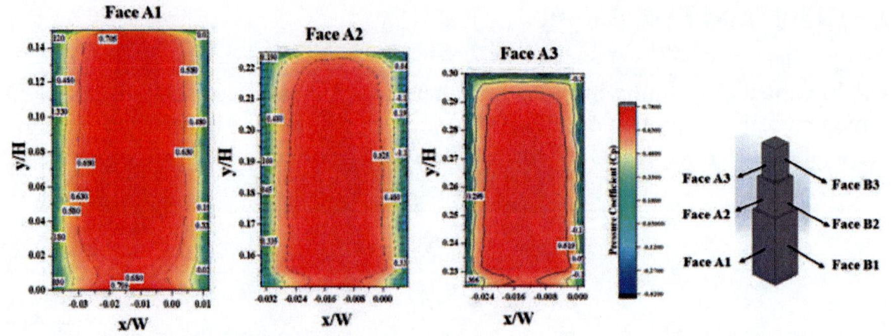

Fig. 5 Formation of the pressure contour for the faces A1, A2, A3

D3 and in leeward faces C1, C2, C3. When the wind hits the building face A, the wind diverges and the side face B and D and the leeward face D experiences negative pressure.

3.1.1 Pressure Contour Distribution for the Face A

Figure 5 shows the formation of the pressure contour on face A of the setback building. Since face A is subjected to windward direction, it experiences positive pressure. On the face A1 as shown in the Fig. 5, experiences a very high positive pressure on the central core and the pressure gradually decreases along the side edges in face A1. Further, the pressure contour obtained in the face A2 is similar to the face A1 and the pressure decreases along the side edges. Whereas for the face A3, the pressure contour decreases along the top and the side edges of this face A3. As noticed by the previous researcher [8], the positive pressure decreases along the top and side edges due to the diversion of the wind.

3.1.2 Pressure Contour Distribution for the Face B

Figure 6 shows the formation of the pressure contour on face B of the setback building. Since face B lies on the side faces (i.e. leeward), it experiences negative pressure coefficient on all the faces B1, B2 and B3. On the face B1, the pressure increases along the edges and further low pressure is observed in the face B2 (which is based on the blue patch indication in the face B2) as shown in the Fig. 6. Moreover, for the face B3, very low pressure is observed, and the pressure increases along the edges.

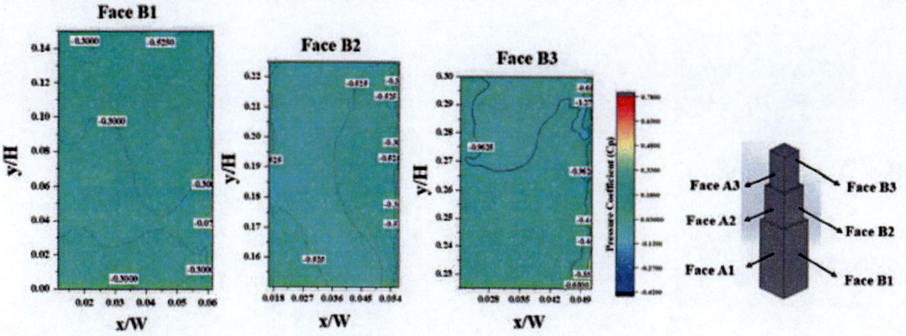

Fig. 6 Formation of the pressure contour for the faces B1, B2, B3

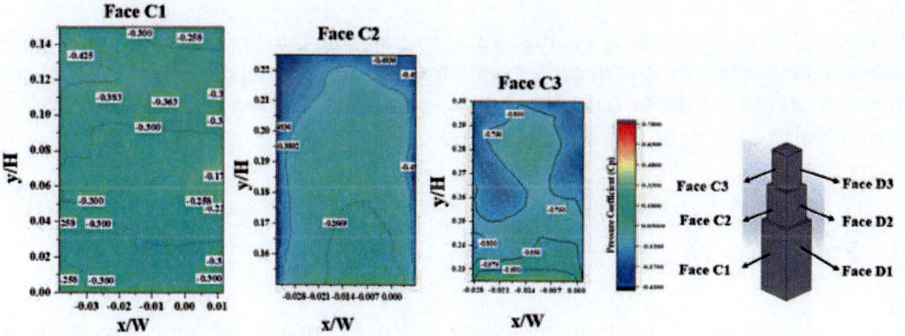

Fig. 7 Formation of the pressure contour for the faces C1, C2, C3

3.1.3 Pressure Contour Distribution for the Face C

Figure 7 shows the formation of the pressure contour on the face C of the setback building. Face C lies completely on the leeward side. When the wind falls in the face A, the wind diverges and leads to the formation of low pressure on the face C1, C2 and C3. The faces C2 and C3 observe a very low pressure and lead to the formation of wake regions. The formation of the negative pressure for setback building is less when compared to the regular square and rectangular building.

3.1.4 Pressure Contour Distribution for the Face D

Figure 8 shows the formation of the pressure contour on the face D of the setback building. Since face D lies in sideward direction, it experiences negative pressure. In the face D1, blue colour contour patches are formed in the upper position which indicate very low pressure, and the lower half of the face D1 green contour are

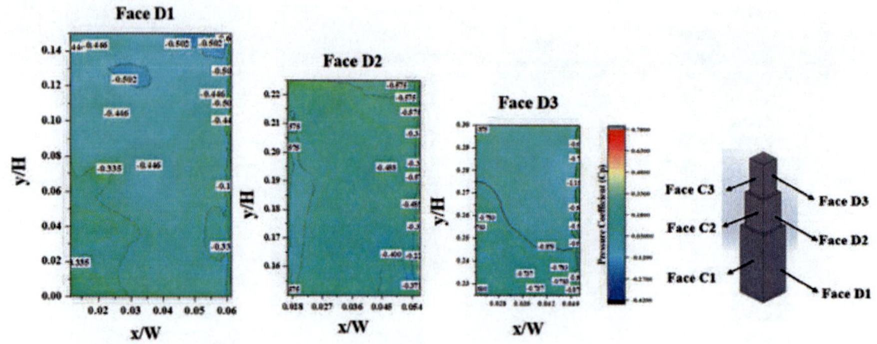

Fig. 8 Formation of the pressure contour for the faces D1, D2, D3

observed due to stagnation on the wind. Further, in the face D2, the wind gets stagnated experiences low negative pressure, and the pressure increases along the edges. Moreover, in the face D3, low pressure is observed in the face and the pressure increase along the right edges of the face D3.

4 Flow Pattern

The wind pressure acting on the building surface depends on the wind flow pattern. When the wind hits the building surface, it deviates in multiple directions. Figure 8 shows the wind flow pattern for the setback building. The number 1 to 5 in Fig. 8 represents the physics of the wind flow. Moreover, the number 1 in the figure represents the incoming flow of the wind towards the building. Number 2 represents the free streamline flow, i.e. Movement of the wind without touching the building, usually occurs at the top of the building. Number 3 represents the stagnant flow. Number 4 represents horseshoe vortices and number 5 represent the expansion of the wake (Fig. 9).

5 Conclusion

From the above study, the following conclusion are made.

- The pressure on each face of the building is obtained with the help of CFD, the maximum pressure is observed in windward face, i.e. face A, and the negative pressure is observed in the leeward face, i.e. face D, and side faces, i.e. face B and C.
- The pressure on the building surface is very much essential for the design of cladding on the building surface; pressure at each point on the building faces

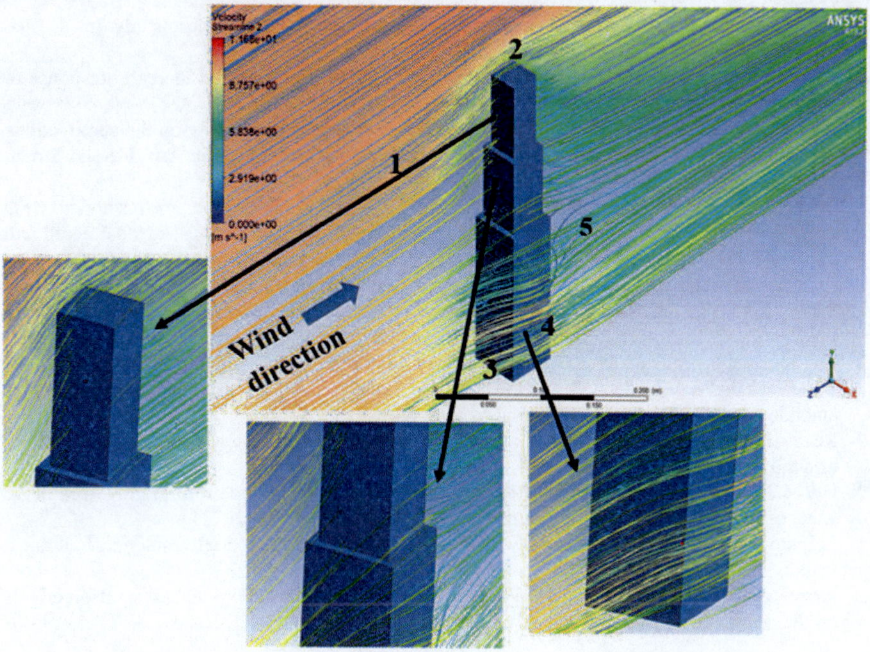

Fig. 9 Wind flow pattern around the setback building

are needed. And it is important to capture the pressure at the corners of the building surface since there are variation in the pressure along the edges and the core regions. From the observation of the present study, the negative pressure intensifies at the edges and moderate at the centre for the windward direction.

- Further, the pressure positive pressure increases along the edges and moderate at the centre for the leeward direction.
- The turbulence model k-ε is used for the external flow simulation, and it is good to capture the flow streamlines and its behaviour when it hits the setback building.

Acknowledgements The author would like to thank Institute of Engineers India (IEI India) Project ID: DR2020004 for the finical support and Computational facilities provided by Mepco Schlenk Engineering College, Sivakasi, India.

References

1. Ding, F., Kareem, A.: Tall buildings with dynamic facade under winds. Engineering **6**(12), 1443–1453(2020)
2. Bhattacharyya, B., Dalui, S.K.: Investigation of mean wind pressure on 'E' plan shaped tall building. Wind Struct. **26**(2), 99–114(2018)

3. Paul, R., Dalui, S.K.: Wind effects on 'Z' plan shaped tall building: a case study. Int. J. Adv. Struct. Eng. 319–335(2016)
4. Hou, F., Jafari, M.: Investigation approaches to quantify wind-induced load and response of tall buildings: a review. Sustain. Cities Soc. **62**, 2210–6707 (2020)
5. Alminhana, G.W., Braun, A.L., Loredo-souza, A.M.: A numerical study on the aerodynamic performance of building cross sections using corner modification. Latin Am. J. Solid Struct. **15**(7), 1–18(2018)
6. Thordal, M.S., Bennetsen, J.C., Capra, S., Kragh, A.K., Koss, H.H.: Towards a standard CFD setup for wind load assessment of high-rise buildings: part 2—blind test of chamfered and rounded corner high-rise buildings. J. Wind Eng. Ind. Aerodyn. **205**, 0167–6105 (2020)
7. Franke, J., Hirsch, C., Jensen, A.C., Krus, H.W., Schatzmann, M., Westbury, P.S., Miles, S.D., Wisse, J.A., Wright, N.G.: Recommendations on the use of CFD in wind engineering. In: Presented at COSTA Action C14, Impact of wind and strom on City life and Build Environment, Von Karman Institute for Fluid Dynamics, Rhode-Saint-Genese, Belgium, May pp. 5–7, (2004)
8. Suresh Kumar, K.: Wind loading on tall buildings: review of Indian standards and recommended amendments. J. Wind Eng. Ind. Aerodyn. **204**, (2020)
9. Revuz, J., Hargreaves, D.M., Owen, J.S.: On the domain size for the steady-state CFD modelling of a tall building. Wind Struct. **15**(4), 313–329 (2012)
10. Ge, M., Gayme, D.F., Meneveau, C.: Large-eddy simulation of wind turbines immersed in the wake of a cube-shaped building. Renew. Energy **163**, 1063–1077 (2021)
11. Leighton, S., Cochran.: Ten questions concerning wind effects on supertall residential buildings. Build. Environ **169**, (2020)
12. Avini, R., Kumar, P., Hughes, S.J.: Wind loading on high-rise buildings and the comfort effects on the occupants. Sustain. Urban Areas **45**, 378–394 (2019)

A Review on Role of Pavement Materials on Urban Heat Island Effects

Mahesh Mungule and Kannan K. R. Iyer

1 Introduction

The impact and scale of modifications in urban climate contributing to urban heat island (UHI) effect have been studied by several researchers [7, 11, 12, 19, 25]. These studies identified difference in the reflectivity of rural and urban surfaces as a dominant parameter driving the modification in urban climate. The reflectivity of materials measured in terms of surface albedo suggests low reflectivity (dark colour) materials such as asphalt absorb and store higher amount of energy than natural or high reflectivity (light-coloured) materials [21, 29, 32]. Albedo, the most critical parameter influencing ground surface temperature ranges between 0.05–0.4 for bare soil and 0.15–0.45 for sand. In comparison, urban surfaces have albedo values of 0.05–0.10 for asphalt and 0.25–0.4 for concrete [15, 5]. An analysis of surface temperature over 11-year period of observation for four different surface types, namely clayey soil, sand, grass and asphalt, exhibited the highest temperature for asphalt and the lowest temperature for grass [5]. This observation is consistent with other studies that reported asphalt has the highest contribution to increasing urban temperature followed by concrete, natural ground and vegetation [13].

Studies evaluating the impact of albedo values on air temperature concluded that decrease in albedo value by 0.01 can increase air temperature by 0.2–0.25 °C [32]. The increase in air temperature increases the electricity consumption, with every 0.6^0C rise in temperature increases the electricity consumption by 2% [30]. Further, correlation of albedo values with CO_2 emission suggested that 0.01 increase in albedo value per square metre area can decrease CO_2 emission by 2.55 kg [1].

M. Mungule · K. K. R. Iyer (✉)
Department of Civil Engineering, Institute of Infrastructure, Technology, Research and Management, Ahmedabad 380026, India
e-mail: kannaniyer@iitram.ac.in

M. Mungule
e-mail: maheshmungule@iitram.ac.in

I. Pal and S. Kolathayar (eds.), *Sustainable Cities and Resilience*, Lecture Notes in Civil Engineering 183, https://doi.org/10.1007/978-981-16-5543-2_19

Studies conducted to evaluate the land use pattern for four different cities in USA reported that impervious surfaces representing pavement and roofs account for 56% of the landscape [2, 26]. Since the impervious surfaces are due to engineered materials like asphalt and concrete, it is pertinent to review the thermal performance of these materials and corroborate attempts at improving the performance, thus contributing to reduction in UHI and enhancement in sustainability of cities. The large part of impervious surfaces observed in above studies were pavement surfaces that include roads (15%), parking lots (12%), sidewalks (5%) and driveways (3%). Also, pavement surfaces can be assumed to have same elevation, thereby partly simplifying the comparison and assessment on performance of different materials. The present study summarizes the performance of pavement materials and their effects on UHI, and recent developments towards enhancing sustainability.

2 Pavement Materials

Pavement temperature is important as it influences the distress in pavement materials and can cause adverse impacts to commuters and environment. For rigid pavement (viz., concrete pavement), high temperature contributes to warping and curling, whereas for flexible pavement (viz., asphalt pavement), it increases the risk of rutting. The albedo values for asphalt are reportedly in the range from 0.05–0.10, whereas for concrete, the range is 0.25–0.4 [15, 5]. With ageing, the albedo values of asphalt increase to 0.15–0.20 and for concrete decrease to 0.25–0.30. For a freshly paved pavement, difference in albedo values of asphalt and concrete translates to 3-10^{0}C higher temperature for asphalt [34]. Despite the advantage of lower surface temperature, share of concrete in comparison with asphalt as pavement material is relatively low. In USA, only 10% of the total paved surfaces is made up of concrete [16].

In addition to higher surface temperature, temperature gradient is higher for asphalt pavement than concrete pavement. A typical depth-wise temperature distribution recorded for asphalt pavement over 24-h period in an earlier study indicated the temperature gradient upto 21 cm depth below asphalt [5]. Higher temperature gradient within the pavement material influences pavement stability and can lead to degradation. Asphalt being the popular pavement material, and subjected to higher surface temperature and temperature gradient, has received considerable attention from researchers. The essential focus of researchers has been to monitor surface temperature, quantify its effect on UHI and develop models for life cycle assessment of pavement. The summary of such studies with significant outcomes is outlined in Table.1 Accurate prediction of surface temperature and temperature gradient is important to understand durability and help in the selection of pavement material [6, 14].

In comparison with asphalt, concrete pavement is subjected to lower surface temperature and lower gradient. As a result, it experiences lower temperature stress thereby reducing maintenance cost and increasing the life cycle of pavement.

Table 1 Summary of studies on thermal behaviour of pavement surfaces

References	Materials	Remarks
[10]	Asphalt, granite slab, terracotta bricks, interlocking blocks	• Amongst all pavement materials considered, asphalt pavement exhibits considerably higher temperature over other pavement materials • The effect of pavement surface is also noted to influence air temperature upto 2 m height from the surface • Optical properties, namely albedo and emissivity, have been considered responsible for influencing daytime and night-time maximum temperature
[31]	Asphalt-black, Asphalt light coloured, grass	
[3]	Asphalt, concrete and bare soil	
[20]	Asphalt and concrete	
[17]	Concrete	
[22]	Asphalt	
[33]	Asphalt and concrete	
[8]	Asphalt	
[35]	Asphalt soil and grass	

However, very few attempts have been made to analyse concrete pavement. A theoretical study quantifying the effect of increase in albedo suggested that increasing albedo value by 0.25 for 1250 km^2 would contribute to significant energy savings along with smog and delay reduction [5].

3 Parameters Influencing Pavement Performance

Given the importance of surface temperature, several attempts have been made to develop prediction models. Two broad categories of models are employed, namely analytical and statistical models [18, 23, 27]. Analytical models are based on heat transfer theories and correlate thermal properties of pavement with those of incident radiation. Statistical models, on the other hand, rely on the relationship between measured pavement surface temperature and climate data but ignores thermal properties of pavement.

Heat transfer-based predictive model proposed by Gui et al. [8] can be used to assess the influence of thermal and optical properties of pavement material on influencing the surface temperature. In order to rank the effect of each parameter, influence of the studied parameter on surface temperature has been considered while all other parameters are kept constant. The effect of each material and parameter is then discussed by studying the normalized value of the parameter with maximum and minimum surface temperature. In order to obtain normalized parameter, the maximum value of optical and thermal properties considered are albedo: 0.5, emissivity: 1, thermal conductivity: 2.59 $Wm^{-1}\ °C^{-1}$, heat capacity: 2.80 $MJm^{-3}\ °C^{-1}$. Figure 1 shows the plot between maximum/minimum temperature with normalized values of albedo, conductivity, heat capacity and emissivity, based on data presented by Gui et al. [8]. From the figure, it has been observed that the most significant parameter influencing maximum surface temperature is albedo followed by thermal

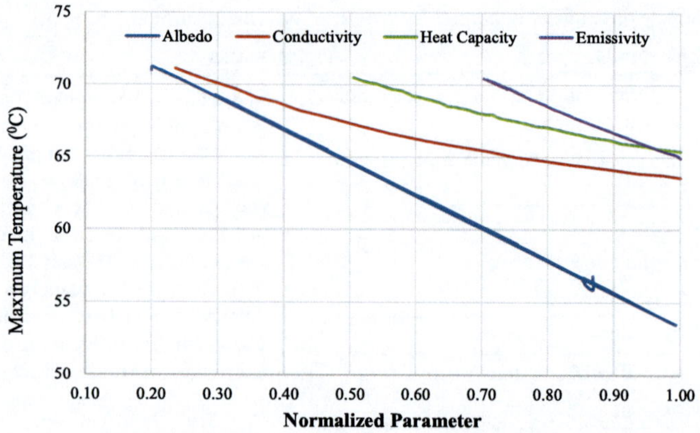

(a) Effect on maximum surface temperature

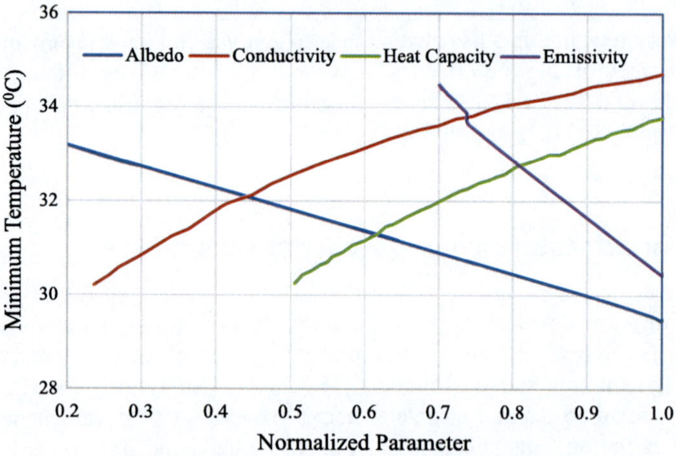

(b) Effect on minimum surface temperature

Fig. 1 Effect of thermal and optical properties on pavement surface temperature

conductivity, heat capacity and emissivity. In contrast, for minimum temperature, the significant parameters have been reported as emissivity and albedo. Further, thermal properties, namely thermal conductivity and heat capacity, are inversely proportional to minimum surface temperature. In order to compare the performance of common pavement materials, namely asphalt and concrete, the thermal and optical properties of the representative composition of the materials that are generally considered are shown in Table 2.

Table 2 Thermal and optical properties of asphalt and concrete surfaces

Properties	Hot mix asphalt	Cement concrete
Density (Kg/m^3)	2250	2400
Heat capacity (MJm^{-3} °C^{-1})	2.06	2.11
Conductivity (Wm^{-1} °C^{-1})	1.21	1.7
Albedo	0.05–0.10*	0.25–0.4*
Emissivity	0.85	0.9

*Ageing may increase albedo values of asphalt while it may reduce the albedo values of concrete

Considering the properties in Table 2, the lower surface temperature for concrete pavement as compared to asphalt as reported in several experimental studies is justified. The higher values of albedo not only benefit reduction in daytime temperature but also contribute to reduction in night-time temperature [1, 13]. Attempts at improving thermal conductivity and heat capacity may benefit with reduction in daytime temperature, but the increase in night-time temperature will offset part of the gain. Thus, most significant parameter that will control thermal performance is identified as albedo value of pavement material [2, 34, 16].

In addition to material properties, the pavement surface temperature is also influenced by incident solar radiation. Influence of solar radiation on maximum and minimum surface temperature of pavement is observed to be proportional and linear [23]. The effect of input radiation on maximum temperature is more significant as compared to minimum temperature. The incident radiation cannot be directly controlled and so attempts at improving the thermal performance of pavement have largely focused on improving the material properties. In material properties, albedo value of pavement material has the highest influence on controlling maximum surface temperature and thus received maximum attention from researchers. All attempts at modifying pavement surface characteristics have been with a view to enhance albedo value. Few attempts have also been made towards improving other parameters, namely heat capacity and thermal conductivity.

3.1 Development of Modified Pavement Materials

The effect of pavement surface temperature on life cycle and enhancement of UHI resulted in developments of modified surfaces for asphalt as well as concrete pavements. The surface modification for pavement was done through the application of reflective coatings and engineering the material composition to achieve higher albedo values [21, 29, 9, 28]. The primary approach applied largely for asphalt pavement was to enhance solar reflectance using white coating on the surface. The results from these studies suggest beneficial role of reflective coatings in reducing surface temperature. However, limitations in terms of higher initial cost, construction delay, reduced skid and abrasion resistance often contribute to reduced lifespan. Further, the properties of reflective coatings degrade rapidly with ageing and weathering effects [15, 28]. The glare from reflective coated surface and vehicle headlight is higher and may cause visibility issues for persons with eye cataract.

Another approach at enhancing surface reflectivity was through engineering the composition of pavement materials to achieve higher albedo values. Albedo values of asphalt and concrete pavement modified using reflective coatings and by engineering the material composition as reported in the literature are summarized in Table 3 [21, 29, 24, 17, 35, 9, 28]. It can be clearly seen that the albedo values can be altered with modification of the surface or by altering the composition. In case of asphalt, application of reflective coating is more widespread than altering the composition. The percentage benefit obtained in terms of increased albedo value is higher with reflective coating than with new compositions. However, the shortened lifespan of reflective coatings is still a challenge, and thus, these engineered compositions appear as a more viable solution. Attempts at improving the albedo values of concrete pavement are at nascent stage with more focus on developing suitable composition than altering its surface. Limited attempts corroborated in the literature and also shown in Table 3 highlight the potential gain in albedo values that can be achieved if the composition is effectively engineered.

Table 3 Albedo values of modified asphalt and concrete pavement

Parent Material	Modified material	Mean Albedo
Asphalt and Hot Mix Asphalt (HMA)	Asphalt Rubber Thin	0.11
	Asphalt Rubber Thick	0.13
	Asphalt Rubber Thin with White Paint	0.13
	HMA Thick	0.17
	HMA Thin	0.2
	HMA Thick with White Paint	0.25
	HMA Thin with White Paint	0.25
	Asphalt Rubber Thick with White Paint	0.26
Plain Cement Concrete (PCC)	Crumb Rubber Concrete	0.32
	12' Concrete	0.36
	Fly Ash Concrete	0.25
	Slag Concrete	0.55
	Fly Ash and Slag blended concrete	0.4

As evident in the albedo values of normal and modified concrete, the mineral admixtures can be effectively used to modify the reflectivity values. The choice and replacement percentage of mineral admixtures will dictate the net gain or loss in albedo values. As evident for slag concrete in Table 3, for right combination of mineral admixtures, higher values of albedo can be achieved. Further, the albedo values of modified concrete are substantially higher than that of modified asphalt suggesting considerably lower temperatures can be achieved by using modified concrete pavement. In addition to lower surface temperatures, modified concrete have been shown to enhance concrete strength and durability, thereby contributing to increased lifespan and reduced emissions. Apart from the benefits highlighted, modified concrete is both economical and less energy-intensive.

3.2 Concluding Remarks

The role of pavement materials on influencing UHI has been recognized, and attempts have been made to study, quantify and improve the thermal performance of pavements. The present work summarizes the developments in thermal assessment of pavement and correlates the experimental observation on surface temperature with the material and surface properties. The correlation justifies the focus of most researchers on improving the thermal performance by enhancing the albedo value of the material. The literature studies reported suggest considerable attention being paid to improve thermal performance of asphalt pavement. The values of albedo for plain concrete are often higher than asphalt as well as the modified asphalt pavement. Despite the obvious advantage with concrete, little attention has been paid to improve albedo values of concrete. Limited experimental investigations on modified concrete pavement highlight the potential gain that can be achieved. The utilization of admixtures for enhancing the albedo value of concrete pavement can be seen as an effective way to reduce the UHI effects, especially in urban areas. Studies on enhancing the albedo values of concrete pavement would play an important role towards making cities more sustainable.

References

1. Akbari, H., Rosenfeld, A., Taha, H.: Recent development in heat island research: technical and policy. In: Proceedings of the Workshop on Heat Islands, Feb 23–24, 1989, Berkeley, California. pp. 14–30, (1989)
2. Akbari, H., Matthews, H.D., Seto, D., The long-term effect of increasing the albedo of urban areas. Environ. Res. Lett. 7 (2), 024004, (2012).
3. Asaeda, T., Ca, V.T., Wake, A.: Heat storage of pavement and its effect on the lower atmosphere. Atmos. Environ. **30**(3), 413–427 (1996)
4. Cambridge Systematics, Inc. Cool Pavement Report: EPA Cool Pavements Study—Task 5 (.pdf). Draft Report. Environmental Protection Agency, Washington, DC, (2005)

5. Cermak, V., Bodri, L., Kresl, M., Dedecek, P., Safanda, J.: Eleven years of ground-air temperature tracking over different land cover types. Int. J. Climatol. **37**(2), 1084–1099. https://doi.org/10.1002/joc.4764
6. Diefenderfer, B.K., Al-Qadi, I.L., Diefenderfer, S.D.: Model to predict pavement temperature profile: development and validation. J. Transp. Eng. **132**(2), 162–167 (2006)
7. Grimmond, C.S.B., Okeand, T.R., Cleugh, H.A.: The role of rural in comparisons of observed suburban rural flux differences. Exchange processes at the land surface for a range of space and time scales. In: Proceedings of the Yokohama Symposium, July (1993)
8. Gui, J., Phelan, P., Kaloush, K., Golden, J.: Impact of pavement thermophysical properties on surface temperatures. J. Mater. Civ. Eng. **19**(8), 683–690 (2007)
9. Guntor, N.A.A., Md Din, M.F., Ponraj, M., Iwao, K.: Thermal performance of developed coating material as cool pavement material for tropical regions". J. Mater. Civ. Eng. **26**(4), 755–760 (2014)
10. van der Hage, J.C.H.: Interpretation of field measurements made with a portable albedometer. J. Atmos. Oceanic Tech. **9**(4), 420–425 (1992)
11. Heisler, G.M., Grant, R.H., Grimmond, S., Souch, C.: Urban forests–cooling our communities? In: Inside urban ecosystems. In: Proceedings of the 7th National Urban Forest Conference, 31.-34, New York, Sept 12–16, (1996)
12. Huang, J., Akbari, H., Taha, H., Rosenfeld, A.: The potential of vegetation in reducing summer cooling loads in residential buildings. J. Clim. Apl. Meteorol. **26**, 1103–1106 (1987)
13. Ikechukwu, E.E.: The effects of road and other pavement materials on urban heat island (a case study of Port Harcourt city). J. Environ. Prot. **6**, 328–340 (2015)
14. Jansson, C., Almkvist, E., Jansson, P.-E.: Heat balance of an asphalt surface: observations and physically-based simulations. Meteorol. Appl. **13**, 203–212 (2006)
15. Li, H., Harvey, J., Kendall, A.: Field measurement of albedo for different land cover materials and effects on thermal performance. Build. Environ. **59**, 536–546 (2012)
16. Manzouri, T.P.: Pavement temperature effects on overall urban heat island, Masters of Science thesis, Arizona State University, USA, (2013)
17. Marceau, M.L., VanGeem, M.G.: Solar reflectance of concretes for LEED Sustainable Sites Credit: Heat Island Effect. PCA R&D Serial No. 2982. Portland Cement Association, Skokie, I, (2007)
18. National Cooperative Highway Research Program NCHRP. "Guide for mechanistic-empirical design of new and rehabilitated pavement structures." Final Rep. No. 1–37A, Part 2: Design Inputs, Chapter 3: Environmental Effects, Transportation Research Board, National Research Council, Washington, D.C, (2004)
19. Oke T.R.: Boundary layer climates. William Clowes and Sons Limited, London, 372 pp, (1978)
20. Pomerantz, M., Akbari, H., Chen, A., Taha, H., Rosenfeld, A.H.: Pavement materials for heat island mitigation. Report LBL-38074. Lawrence Berkeley National Laboratory, University of California, Berkeley, CA, (1997)
21. Pomerantz M., Pon B., Akbari H. and Chang S.C.: The effect of pavements' temperatures on air temperatures in large cities (LBNL–43442). Lawrence Berkeley National Laboratory, Berkeley, CA, (2000)
22. Puttonen, E., Suomalainen, J., Hakala, T., Peltoniemi, J.: Measurement of reflectance properties of asphalt surfaces and their usability as reference targets for aerial photos. IEEE Trans. Geosci. Remote Sens. **47**(7), 2330–2339 (2009)
23. Qin, Y.: Pavement surface maximum temperature increases linearly with solar absorption and reciprocal thermal inertial. Int. J. Heat Mass Transf. **97**, 391–399 (2016)
24. Reza, F., Boriboonsomsin, K.: Pavements made of concrete with high solar reflectance. In: Eco-Efficient Materials for Mitigating Building Cooling Needs, (2015)
25. Robaa, S.M.: Urban–suburban/rural differences over Greater Cairo. Egypt. Atmósfera **16**, 157–171 (2003)
26. Rose, L.S., Akbari, H., Taha, H.: Characterizing the Fabric of the Urban environment: a case study of Greater Houston. Lawrence Berkeley National Laboratory, Texas. Berkeley, CA, (2003)

27. Schindler, A.K., Ruiz, J.M., Rasmussen, R.O., Chang, G.K., Wathne, L.G.: Concrete pavement temperature prediction and case studies with the FHWA HIPERPAV models. Cem. Concr. Compos. **265**, 463–471 (2004)
28. Sen, S., King, D., Roesler, J.: Structural and environmental benefits of concrete inlays for pavement preservation. Airfield Highw Pavements (2015)
29. Synnefa, A., Santamouris, M., Livada, I.: A study of the thermal performance of reflective coatings for the urban environment. Sol. Energy **80**, 968–981, (2006)
30. TERI, Final report on urban planning characteristics to mitigate climate change in context of urban heat island effect. The Energy and Resources Institute (2017)
31. Taha, H., Sailor, D., Akbari, H.: High-Albedo materials for reducing building cooling energy use. Energy (1992)
32. Taha, H.: Modeling the impacts of large–scale albedo changes on ozone air quality in the South Coast Air Basin. Atmos. Environ. **31**, 1667–1676, (1997)
33. Takebayashi, H., Moriyama, M.: Study on surface heat budget of various pavements for urban heat island mitigation. Adv. Mater. Sci. Eng. 2012, (2012)
34. USDOT. Highway Statistics: US Department of Transportation, p. 2013. DC, Washington (2012)
35. Yilmaz, H., Toy, S., Irmak,M A., Yilmaz,S., Bulut, Y.: Determination of temperature differences between asphalt concrete, soil and grass surfaces of the city of Erzurum, Turkey. Atmósfera **21**(2), México Abr. (2008)

Studies on Strip Footings Resting on Lateritic Slopes

A. Anjali, J. Jayamohan, and S. S. Rageena

1 Introduction

Several studies have been conducted to analyze bearing capacity, settlement, and failure of footing resting on horizontal surfaces by various researchers [8, 9, 11–13]. Comparing to those studies, research related to foundations resting on or near the slopes is limited. The terrains in many places in the state of Kerala are sloping in nature, and hence, foundations of structures are located on or near the slopes. Building foundations, bridge pier which is situated in slopes are such examples. Bearing capacity is the major factor for any type of foundations. In this regard, the assessments of bearing capacity for foundations which are situated on slope surface or crest of slope become complicated.

Shields et al. [10] reported that the ultimate bearing capacity increases by 2 to 3 times when the embedment depth of footing was doubled. The field investigation of [6] on the behavior of inclined footings at a natural slope reported that compared to horizontal ones, inclined footing with anchors is having more bearing capacity.

The effect of depth which is replaced with sand layer and offset distance between footing and slope crest under sinusoidal loading determined by El Sawwaf [7]. Settlement reduction with increasing bearing capacity was found out. Numerical analyses on ısolated shallow footings resting near the crest of slope were conducted using 3D finite element analysis by Acharyya and Dey [1, 2], and various parameters were studied. They found out that beyond the critical ratio of 4 for square and 6 for strip, and there behavior was approaching similar to footing placed at horizontal

A. Anjali (✉) · S. S. Rageena
Department of Civil Engineering, Marian Engineering College, Thiruvananthapuram, Kerala, India

J. Jayamohan
Department of Civil Engineering, LBS Institute of Technology for Women, Poojappura, Thiruvananthapuram, Kerala, India
e-mail: jayamohanj@lbsitw.ac.in

I. Pal and S. Kolathayar (eds.), *Sustainable Cities and Resilience*, Lecture Notes in Civil Engineering 183, https://doi.org/10.1007/978-981-16-5543-2_20

ground. Also used soft computing and optimization techniques instead of standard techniques. Interference of strip footings located on slope crest was studied by Acharyya and Dey [3–5] and reported that, after a critical spacing ratio of 8, the interference of footing disappears.

Most of the researchers have investigated the behavior of footings resting on the crest of the slope. In urban areas having sloping terrain, the footings may be placed at various levels along a slope. In this paper, the load–settlement behavior of footings resting at various levels along a slope is investigated by carrying out a series of laboratory-scale load tests on model footings. The influences of different parameters like eccentricity of footing from the edge and slope angle are investigated. Finite element analysis is carried out using the FE program PLAXIS 2D *connect edition V20,* and the results are compared with those obtained from the laboratory model tests.

2 Laboratory Model Tests

The experimental program reported in this paper involves a series of laboratory-scale load tests on model footings resting at various levels along the slope surface was carried out at Geotechnical Research Lab of LBS Institute of Technology for Women, Thiruvananthapuram.

2.1 Lateritic Soil

The material used for slope preparation is locally available lateritic soil, and the properties of soil are given in Table 1.

Table 1 Properties of soil used in model tests

S. No.	Properties	Values
1	Specific Gravity	2.6
2	Uniformity Coefficient (Cu)	6.67
3	Coefficient of Curvature (Cc)	1.2
4	Optimum Moisture Content (%)	15.5
5	Dry Unit Weight (kN/m^3)	18.84
6	Angle of Shearing Resistance, ϕ (°)	32
7	Cohesion, c (kN/m^2)	5
8	IS Designation	SW

2.2 *Test Setup*

The load tests are conducted in a combined test bed and loading frame assembly. The test beds are prepared in a tank which is designed keeping in mind the size of the model and the zone of influence. The dimensions of the test tank are 1000 mm length × 750 mm width × 750 mm depth, which has 23-cm-thick brick masonry walls on four sides. The loading tests are carried out in a loading frame fabricated with ISMB 300. The loads are applied using hand-operated mechanical jacks of capacity of 50 kN. The applied loads are measured using proving rings of capacities 50 kN. The settlements of the two model strip footings are measured using two dial gauges each of 0.01 mm sensitivity kept diametrically opposite to each other. Drawing of the test setup is shown in Fig. 1 and photograph in Fig. 2.

Fig. 1 Drawing of test setup

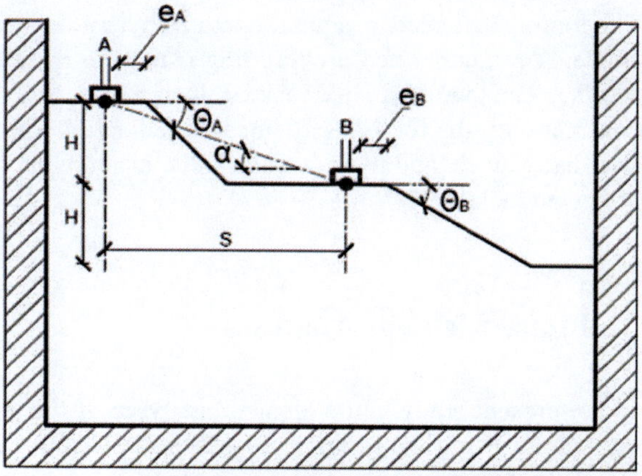

Fig. 2 Photograph of test setup

2.3 Preparation of Slope

At first, the lateritic soil is filled in the test tank to the required level of slopes with compaction done in layers of 10 cm thickness. The water content of the soil is maintained constant throughout the test procedure. To achieve the desired density of the soil, the layered filling technique is used. The lateritic soil is compacted by ramming.

2.4 Testing Procedure

After preparing the slope, the slope surface is leveled, and the footing is placed at two levels of slope exactly at the center of the loading jack to avoid eccentric loading. The footing is loaded by a hand-operated hydraulic jack supported against a reaction frame. A pre-calibrated proving ring is used to measure the load transferred to the footing. The load is applied in small increments. Each load increment is maintained constant until the footing settlement is stabilized. The settlement is measured using two dial gauges, and their average value is adopted. The details of testing program are given in Table 2.

3 Finite Element Analyses

In the present study, finite element analyses of the strip footing resting slopes are carried out using the program PLAXIS 2D *Connect Edition V20* which is a finite element software package. Mohr–Coulomb model with drained condition is used in this study to simulate the soil behavior as it is the simplest model which is based on the basic soil parameters that can be obtained from direct shear tests: cohesion intercept and internal friction angle. Since strip footing is used in this study, a plain strain model is adopted for the analyses. Here the footings are modeled with plate elements.

Table 2 Experimental program

Type	Series	e_a (cm)	e_b (cm)	θ_a (°)	θ_b (°)
Lateritic soil	A	5	0,2.5,5	30	30
	B	0,2.5,5	5	30	30
	C	5	5	30	30, 45, 60
	D	5	5	30, 45, 60	30

In the present study, instead of modeling the footing, the indentation caused by the footing or settlement of the footing is simulated using nonzero prescribed displacements. The initial geostatic stress states for the analyses are fixed according to the gravity loading. The soil is modeled using 15 noded triangular elements. A medium mesh size is adopted for the soil. The outer boundaries of the mesh are drawn to the same dimensions as that of the tank used for model tests. The boundary conditions are made in such a way that the displacement of the bottom boundary is restricted in all directions, while at the vertical sides displacement is restricted only in the horizontal direction. To simulate the interaction between the footings and surrounding soil, interface elements are provided between the footings and surrounding soil.

4 Results and Discussions

4.1 *Effect of Distance of Footing from Edge of Slope (Eccentricity)*

Vertical stress versus normalized settlement curves of footings resting on various levels of slopes with varying eccentricity values are shown in Figs. 3 and 4. The

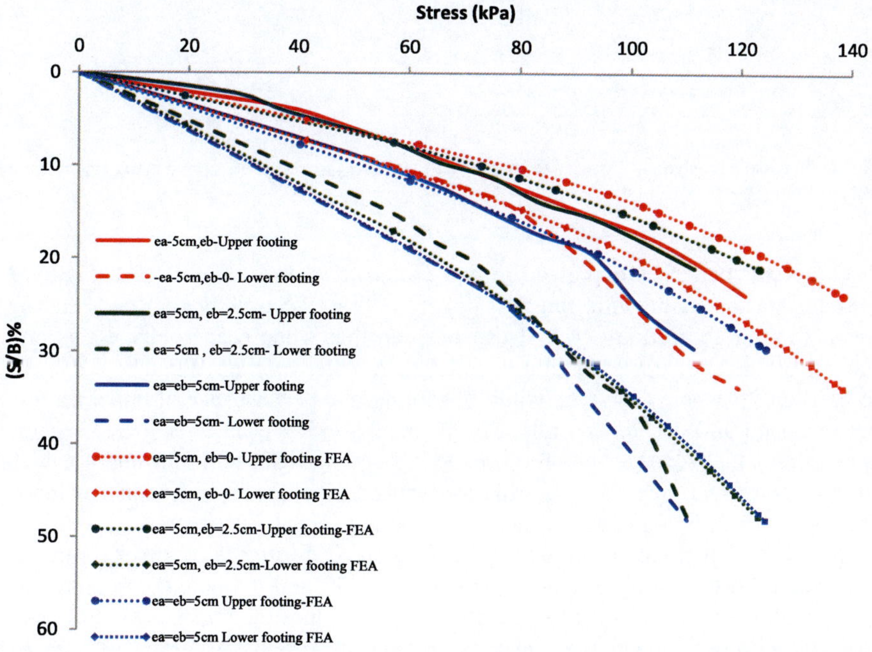

Fig. 3 Vertical stress versus normalized settlement curves e_a = constant, e_b= varying ($\theta_a = \theta_b = 30°$)

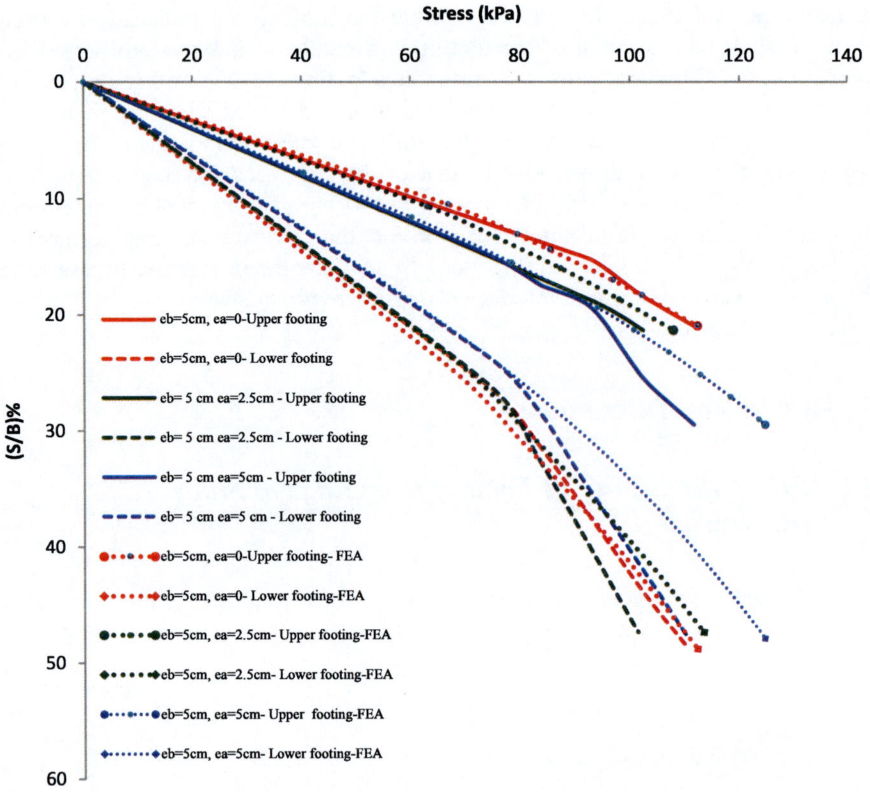

Fig. 4 Vertical stress versus normalized settlement curves e_b = constant, e_a= varying ($\theta_a = \theta_b = 30°$)

footing settlement S is expressed in non-dimensional form as S/B (%). Figure 3 presents the behavior with constant upper footing eccentricity (e_a) and varying lower footing eccentricity (e_b). It can be seen that when eccentricity e_b is zero, the performance of upper footing is better than lower footing. When e_b increased to 0.5B and B where (B) is the width of footing, the performance of upper footing gets reduced and due to the influence of upper footing even after increasing the eccentricity, the performance of lower footing getting reduced. From Fig. 4, which represents vertical stress versus normalized settlement of footing with constant lower footing eccentricity (e_b) and varying upper footing eccentricity, (e_a).

It is seen that when eccentricity e_a is zero the performance of upper footing is better than lower footing. When e_a increased to 0.5B and B where (B) is the width of footing, the performance of upper footing gets reduced, but due to the influence of upper footing even with the constant lower eccentricity, the performance of lower footing getting increased. There is a reasonably good agreement of FEA results with experimental results.

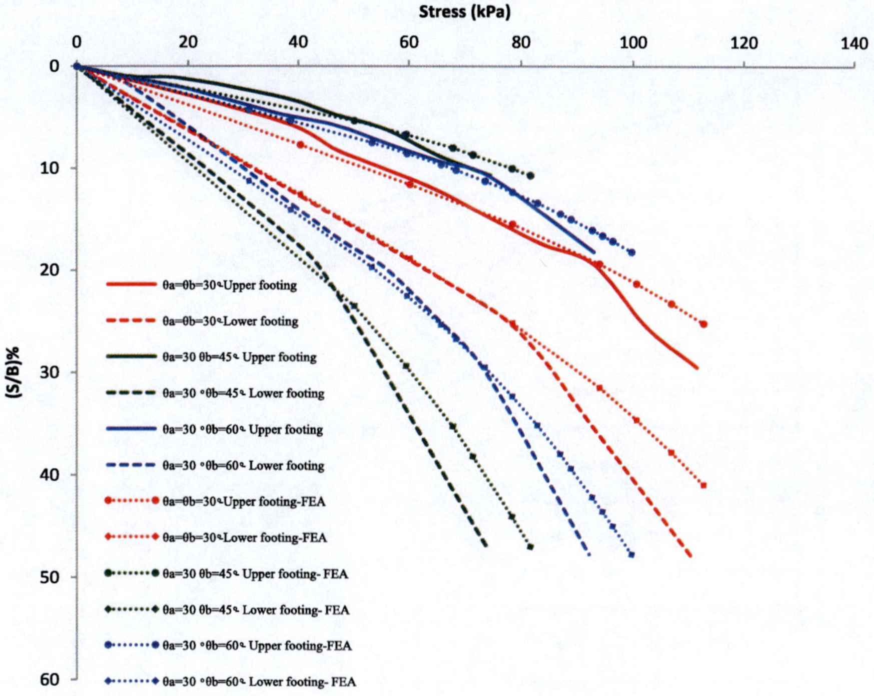

Fig. 5 Vertical stress versus normalized settlement curves $\theta_a = 30°$ and $\theta_b =$ varying ($e_a = e_b =$ 5 cm)

4.2 Effect of Slope Angle

Vertical stress versus normalized settlement curves of footings resting on various levels of slopes with varying slope angle values are shown in Figs. 5 and 6.

From Fig. 5 which presents the vertical stress versus normalized settlement of footing with constant slope angle for top level of slope, θ_a and varying slope angle for bottom level of slope, θ_b. It can be seen that when top and bottom slope angles were same (30°), the performance of upper footing is better than lower footing. The reduction in lower footing performance is due to the influence of upper footing. When the lower slope angle changes to 45° the upper footing performance increased and lower footing gets reduced. But when the angle increased more than 45° the performances got reverse action, i.e., upper footing performance reduced and lower gets increased. From Fig. 6, which represents the vertical stress versus normalized settlement of footing with constant slope angle for bottom level of slope, θ_b and varying slope angle for top level of slope, θ_a. It can be seen that when top and bottom slope angles were same (30°), the performance of upper footing is better than lower footing. The reduction in lower footing performance is due to the influence of upper footing. When the upper slope angle changes to 45° the upper footing performance

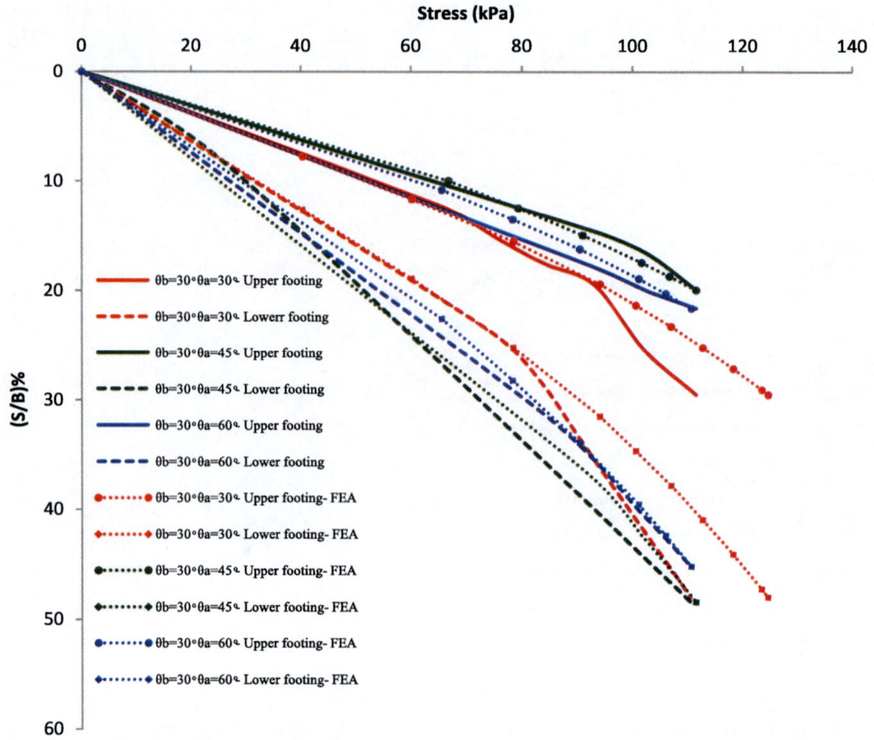

Fig. 6 Vertical stress versus normalized settlement curves $\theta_b = 30°$ and θ_a = varying ($e_a = e_b =$ 5 cm)

increased and lower footing gets reduced. But when the angle increased more than 45°, the performances got reverse action, i.e., upper footing performance reduced and lower gets increased. There is a reasonably good agreement of FEA results with experimental results for both cases. In all the cases, lower footing performance is lower than upper footing; it is mainly due to the influence of upper footing on the lower footing, and further, it causes differential settlement for footings.

It is observed from the results that the load settlement behavior of strip footings resting on lateritic slopes is influenced by eccentricity of footing from edge of slope and slope angle. The failure loads of the upper footing and lower footing will be nearly equal when either both upper and lower eccentricities are equal or both the upper and lower slope angles are equal. So in such cases both the footing will fail at same time, and a spike is observed in corresponding load settlement graphs.

4.3 Contact Pressure Distributions Beneath Footing

4.3.1 Effect of Eccentricity

Figure 7 represents the normal stress distribution beneath the footing with constant upper footing eccentricity (e_a) and varying lower footing eccentricity (e_b). The normal stress is found to be minimum at the center of footing and maximum at the ends of footing. It can be seen that, in lateritic soil, increasing the lower footing eccentricity has effect on normal stress distribution. Normal stress is found to be decreasing for upper footing up to a value of lower footing eccentricity 0.5B where B is the width of footing, and decreasing for lower footing with the increasing of lower footing eccentricity (e_b).

Figure 8 represents the normal stress distribution beneath the footing with constant lower footing eccentricity (e_b) and varying upper footing eccentricity (e_a). The normal stress is found to be minimum at the center of footing and maximum at the ends of footing. It can be seen that, in lateritic soil, increasing the upper footing eccentricity has effect on normal stress distribution. Normal stress is found to be increasing for lower footing and upper footing with the increasing of upper footing eccentricity (e_a).

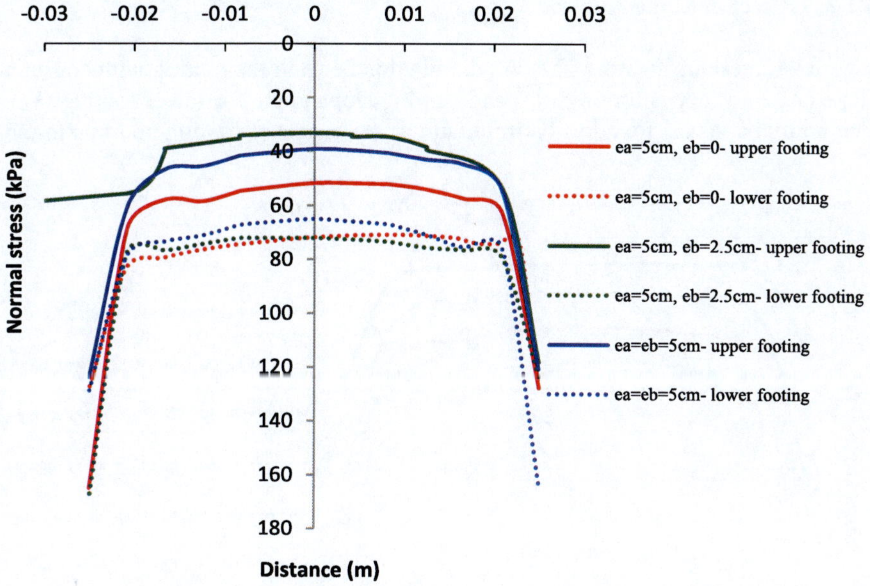

Fig. 7 Normal stress distribution beneath footings: e_a = constant, e_b= varying ($\theta_a = \theta_b = 30°$)

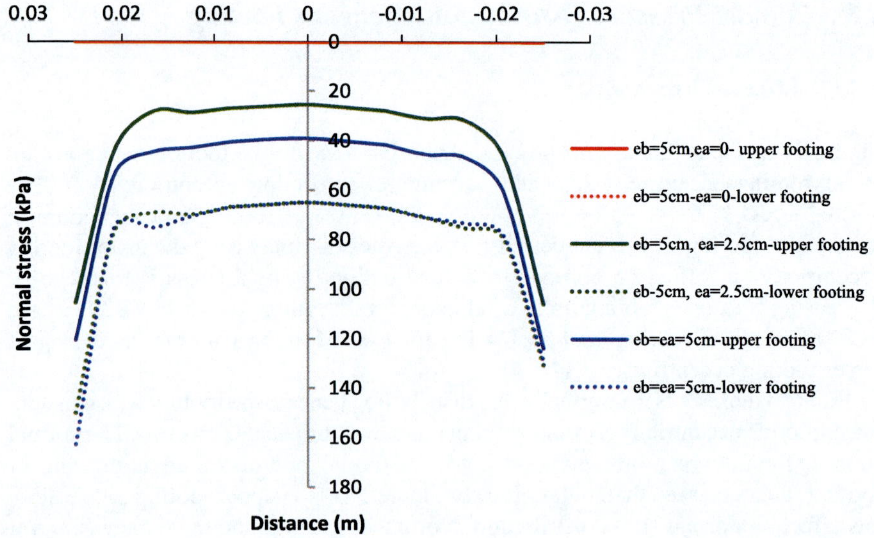

Fig. 8 Normal stress distribution beneath footings: e_b = constant, e_a= varying ($\theta_a = \theta_b = 30°$)

4.3.2 Effect of Slope Angle

Figure 9 represents the normal stress distribution beneath the footing with constant slope angle for upper footing (θ_a) and varying slope angle for lower footing (θ_b). The normal stress is found to be minimum at the center of footing and maximum

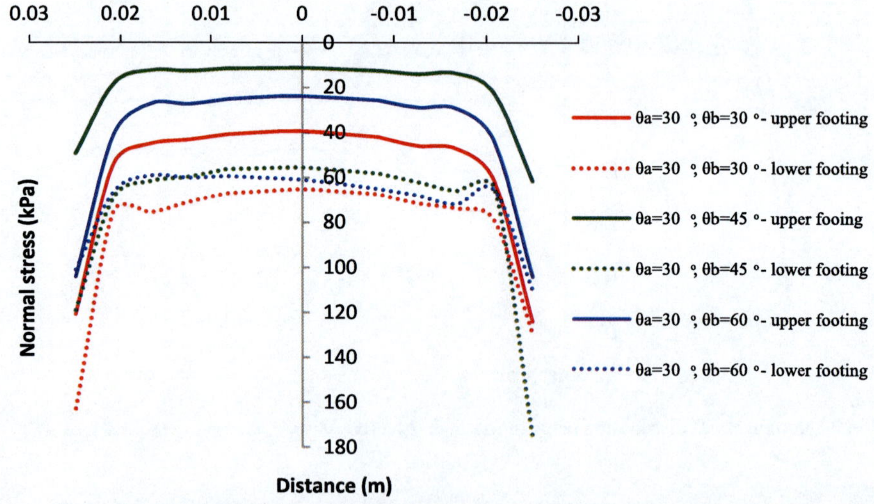

Fig. 9 Normal stress distribution beneath footings: θ_a = constant θ_b = varying ($e_a = e_b = 5$ cm)

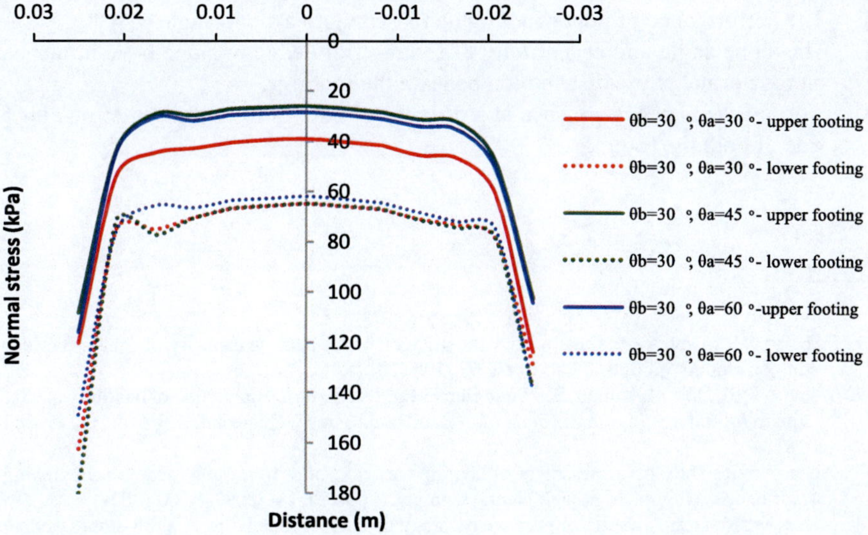

Fig. 10 Normal stress distribution beneath footings: $\theta_b =$ constant $\theta_a =$ varying ($e_a = e_b = 5$ cm)

at the ends of footing. In lateritic soil, increasing the slope angle for lower footing has effect on normal stress distribution. Normal stress is found to be decreasing for lower and upper footing up to a value of $\theta_b = 45\ ^\circ$.

Figure 10 represents the normal stress distribution beneath the footing with constant slope angle for lower footing (θ_b) and varying slope angle for upper footing (θ_a). The normal stress is found to be minimum at the center of footing and maximum at the ends of footing. In lateritic soil, increasing the slope angle for upper footing has effect on normal stress distribution. Normal stress is found to be decreasing for upper footing up to a value of $\theta_a = 45\ ^\circ$ and decreasing the normal stress for lower footing with the increase of θ_a.

In all the cases, normal stress peaks are seen at a distance of 0.02 m from the center on either side of both the footings.

5 Conclusions

Based on the results obtained from experimental and numerical studies, the following conclusions can be made on the load–settlement behavior of strip footings resting at various levels on lateritic slopes.

1. Eccentricity and slope angle are major factors which affect the load settlement behavior of footing resting on slopes.
2. The upper footing adversely affects the load settlement behavior of lower footing.

3. The performance of upper footing increased up to a slope angle of 45°.
4. The slope angle and eccentricity of footing from edge of slope have influence on the normal stress distribution beneath the footings.
5. Normal stress peaks are seen at a distance of 0.02 m from the center on either side of both the footings.

References

1. Acharyya, R., Dey, A.: Finite element investigation of the bearing capacity of square footings resting on sloping ground. INAE Let **2**, 97–105 (2017)
2. Acharyya, R., Dey, A., Kumar, B.: Finite element and ANN-based prediction of bearing capacity of square footing resting on the crest of c-φ soil slope. Int. J. Geotechn. Eng. **14**(2), 176–187 (2018)
3. Acharyya, R., Dey, A.: Assessment of bearing capacity for strip footing located near sloping surface considering ANN-model. Neural Comput. Appl. **31**(11), 8087–8100 (2018)
4. Acharyya, R., Dey, A.: Suitability of the typology of shallow foundations on hill-slopes. Indian Geotechn. J. **49**(6), 635–649 (2019)
5. Acharyya, R., Dey, A.: Assessment of bearing capacity and failure mechanism of single and interfering strip footings on sloping ground. Int. J. Geotechn. Eng. 1–12 (2018). https://doi.org/10.1080/19386362.2018.1540099
6. Clark, J.I., Mckeown, S., Crawford, C.B.: Field measurements of the behaviour of inclined footings on a natural slope. Can. Geotech. J. **25**, 662–674 (1988)
7. El Sawwaf, M.A., Nazir, A.K.: Cyclic settlement behavior of strip footings resting on reinforced layered sand slope. J. Adv. Res. **3**(4), 315–324 (2012)
8. Hansen, B.J.: A revised and extended formula for bearing capacity. Danish Geotechnical Institute, Bulletin No. 28, Copenhagen (1970)
9. Meyerhof, G.G.: The ultimate bearing capacity of foundations. Geotechnique **2**, 301–332 (1951)
10. Shields, D.H., Scott, J.D., Bauer, G.E., Deschenes, J.H., Barsvary, A.K.: Bearing capacity of foundation near slopes. In: Proceedings of the 10th International Conference on Soil Mechanics and Foundation Engineering. Japanese Society of Soil Mechanics and Foundation Engineering, vol. 2, pp. 715–720. Tokyo, Japan
11. Skempton, A.W.: The bearing capacity of clay, Building Research Congress, England (1951)
12. Terzaghi, K.: Theoretical soil mechanics. Wiley, New York, USA (1943)
13. Vesic, A.S.: Analysis of ultimate loads of shallow foundation. J. Soil Mech. Foundation Division. ASCE **99**(SM1), 45–73 (1973)

Waste Management and Disaster Risk Reduction

Effective Utilization of Waste Plastic Bottle as Geocell in Road Pavement: A Numerical Study

Rohan Deshmukh, Gopal Patil, Urvi Bhatt, Ritik Shingote, and Tejas Patil

1 Introduction

The biggest problem faced by the human race in the present era is dealing with the waste materials being generated at an alarming rate. Due to its production at an alarming rate, it has become a headache for mankind. The major waste material being used is waste plastic bottles. Their disposal is such a huge mess that it has become very difficult to find a solution. Different techniques are being used for finding a solution. The present study reviews the best solution available for this problem. At the same time, there is another issue of low bearing capacity of soft soil and loose sand. There is a need for ground improvement to make structures safe resting on such soil. A new sustainable method to improve the quality of soil can be developed with the help of waste plastic bottles. Nowadays there is a shortage of land for the disposal of waste. Through the use of waste products, this problem can be countered effectively. This will reduce land pollution which is the effect of excessive dumping of waste. The sustainable methods may give rise to small-scale industries which may process the waste material for use. This will boost the economy as well as employ a large number of people. The use of the waste material to make stabilized soil can be helpful to reduce the harm to the environment, provide employment as well as provide an opportunity for small as well as large scale industries and boost the economy of the country. Because of this several past researchers have attempted to use these waste plastic bottles as geocell reinforcement for ground improvement.

R. Deshmukh (✉) · G. Patil · U. Bhatt · R. Shingote · T. Patil
Department of Civil Engineering, Terna Engineering College, Nerul, Navi Mumbai 400706, India

I. Pal and S. Kolathayar (eds.), *Sustainable Cities and Resilience*, Lecture Notes in Civil Engineering 183, https://doi.org/10.1007/978-981-16-5543-2_21

2 Literature Review

2.1 Experimental Studies on Waste Plastic Bottle Geocell (WPBG)

Ramesh et al. [1] performed an experimental study on geocell made up of PET (polyethylene terephthalate) and on soil subgrade of fiber-reinforced under static and repetitive load. Improvement was observed in CBR, Resilient Modulus (M_R), and UCS values when soil is reinforced with basalt fiber. When a PET bottle of 80 mm diameter is provided at 1/3H from the top increment is observed in CBR and M_R. Adhana and Mandal [2] carried out an experimental study on fly ash steep slopes on the soft foundations with and without reinforcement to check the stability of the steep slope. Results indicated that both geocell strips and geogrid strips made of waste plastic bottle have a better load-carrying capacity than the unreinforced soil. Thakare and Sonule [3] analyzed and carried on a detailed experimental study on all the possible uses of plastic waste bottles for the improvement of soil bearing capacity. The study observed that the Bearing Capacity Ratio (BCR) was increased for the spacing of *B*/8 between the layers. The maximum benefits were obtained at the space of *B*/8. The study concluded that the optimum results were obtained by providing an *L*/*B* ratio of 2 with 3 reinforcing layers and with the spacing of *B*/8. Lal and Mandal [4] carried out an experimental investigation and suggested that the maximum load for failure was more for the geocell mattress compared with the strip geocell when used as reinforcement.

2.2 Experimental and Numerical Studies on WPBG

Nadaf Maheboobsab and Mandal [5] have conducted numerical analyses using finite-element analysis (FEA) PLAXIS 3D. Results from both experimental and numerical studies confirmed that the cellular reinforcement made with a waste plastic bottle allows much better load distribution over a larger area. Lal and Mandal [4] studied the "performance of cellular reinforcement (geocell) in fly ash under the static triaxial test. With an increase in the height of cellular reinforcement for both the diameters of reinforcement the Peak deviator stress and shear strength parameters are found to increase. Similar results are also obtained with Finite element simulation carried out by using PLAXIS 2D." Dutta and Mandal [6] "carried out Numerical analyses of cellular mattress-reinforced fly ash beds overlying soft clay. Results indicated an improvement in footing capacity by approximately 1.4 times larger over fly ash bed with the presence of a single geotextile separator."

3 Numerical Analysis

3.1 Optimization of WPBG in PLAXIS 2D

Model of standard triaxial loading frame of diameter 75 mm and height 150 mm is created in Plaxis 2D (Figs. 1 and 2) wherein clayey soil sample material properties are incorporated. The numerical simulation is carried out for 50 mm diameter of geocell and varying height of geocell as 10, 20, 30, and 40 mm. Axis symmetric model was considered with Hardening soil model and 15 node triangular elements. Material possessed undrained behavior. The reduction factor *R*-inter is 0.87 considered for reinforced soil. The axial stiffness of geocell is taken as 1000 kN/m. In the model, horizontal and vertical fixities are incorporated to visualize the failure pattern. 150 kPa of cell pressure is applied. The graph is plotted between deviator stress (kPa) and axial strain for different geocell heights. One layer of reinforcement was kept at the center of the model and two-layer of reinforcement were placed at the 1/3rd from top and bottom. Unit weight of 12.2 kN/m^3, elasticity modulus (E_{50}) of 2000 kPa were considered for the soil in the analysis.

With an increase in the height of reinforcement geocell, the deviator stress value and respective strain value were seen increasing (Fig. 3). The optimal depth at which the deviator stress value came maximum is when the two-layer reinforcement was placed at one-third height from top and bottom (Fig. 4). It is been analyzed that the optimized diameter and height of geocell is 50 mm and 40 mm respectively which is used in PLAXIS 3D for static cyclic loading analysis.

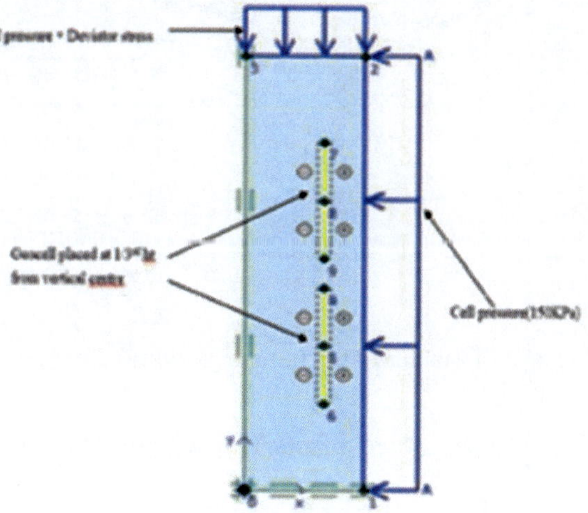

Fig. 1 Triaxial test in PLAXIS 2D

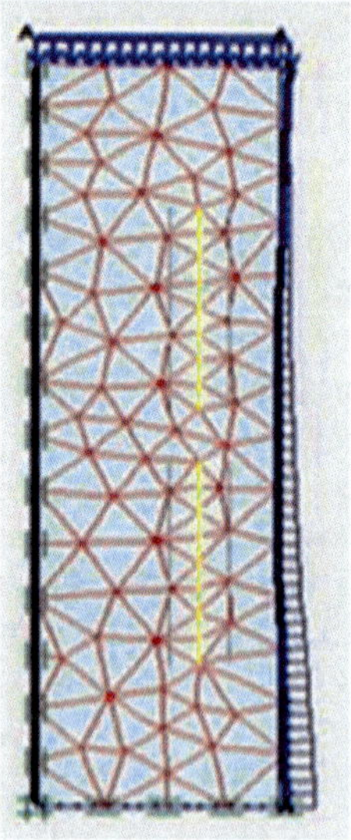

Fig. 2 Deformed mesh

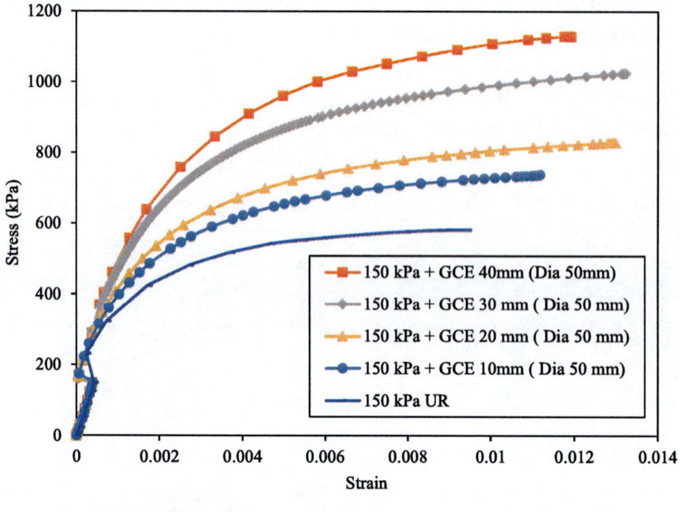

Fig. 3 Result of triaxial test when single WPBG (50 mm Dia.) placed at the center

3.2 *Static Cyclic Loading Analysis in PLAXIS 3D*

A hardening soil model of size 6 m × 6 m is created in PLAXIS 3D (Fig. 5). Through this analysis Deformation, Horizontal Tensile strain, Vertical Compressive strain, Traffic Benefit ratio (TBR), and Rutting Reduction ratio (RRR) is simulated. An analysis is done for subgrade CBR of 3% and the performance of WPBG is

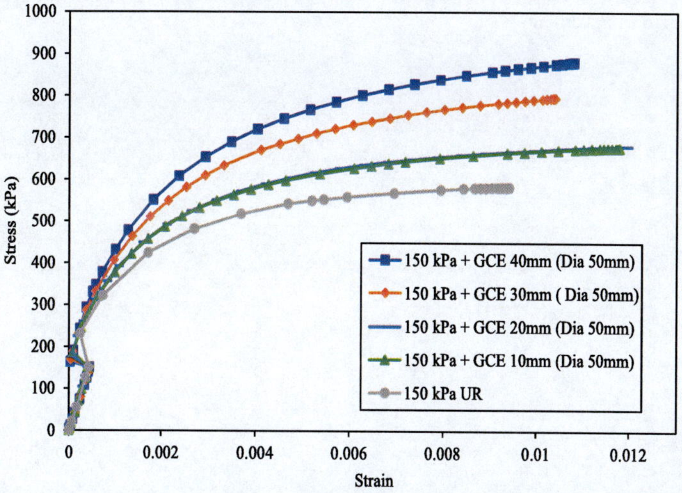

Fig. 4 Result of triaxial test when double WPBG (50 mm Dia.) placed at the 1/3rd from top and bottom

Fig. 5 3D Pavement model

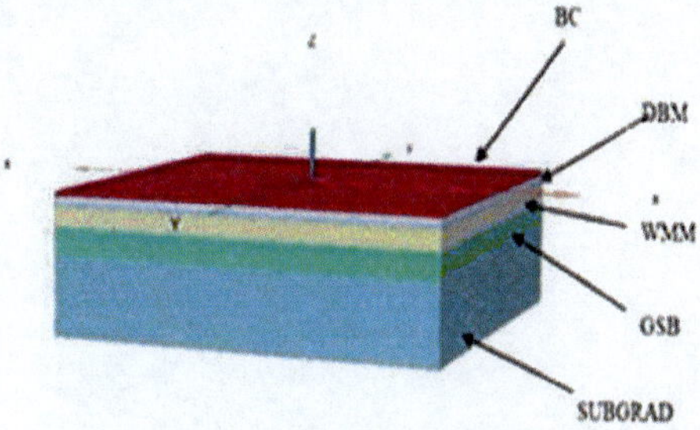

checked against an unreinforced model. Also, a comparison is done between WPBG, conventional geocell of diameter 150 and 350 mm, and geogrids of 20 and 40 kN/m tensile strength (Figs. 6, 7, and 8). Static cyclic loading with intensity 565 kPa of a total of 100 cycles is applied on the top of the pavement model to simulate actual vehicular traffic. The deformed pavement model for geogrid, WPBG, and geocell is shown in Fig. 9a–c, respectively.

The angle of internal friction (Ø) for subgrade, GSB, and WMM are considered as 30°, 43° and 48°. Material properties for all the pavement layers are given in Table 1. Reinforcement properties are given in Table 2. The result of numerical analysis is plotted in terms of the graph between the number of loading cycles and deformation (Fig. 10), number of loading cycles and tensile strain (Fig. 11), number of loading cycles and compressive strain (Fig. 12), rut depth against TBR (Fig. 13) and number of loading cycles and RRR (Fig. 14) for subgrade CBR of 3%.

Fig. 6 Geogrid in Pavement

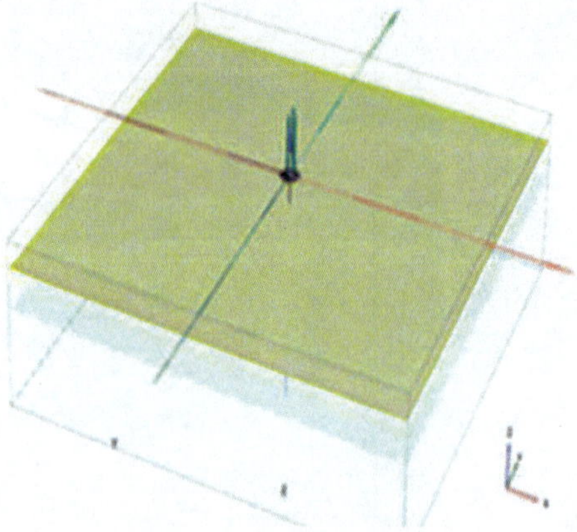

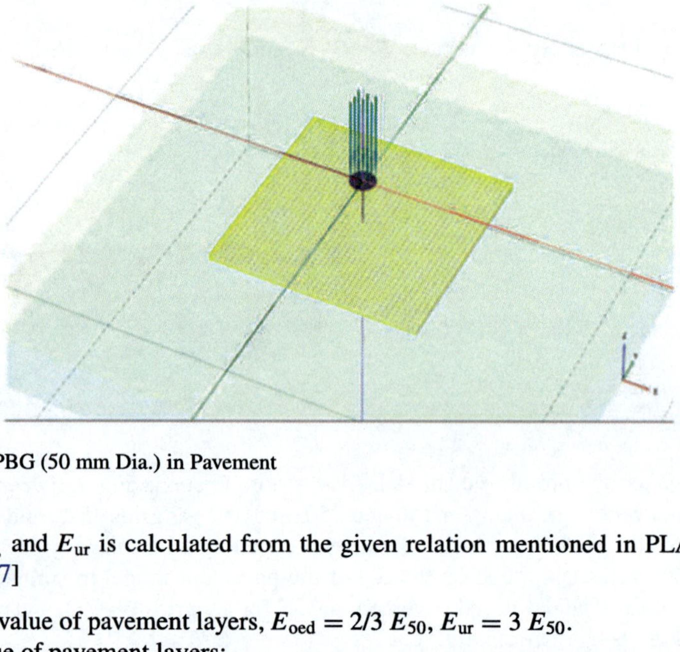

Fig. 7 WPBG (50 mm Dia.) in Pavement

E_{50}, E_{oed}, and E_{ur} is calculated from the given relation mentioned in PLAXIS 3D Manual [7]

$E_{50} = E$ value of pavement layers, $E_{oed} = 2/3\ E_{50}$, $E_{ur} = 3\ E_{50}$.

E value of pavement layers:

$E_{subgrade} = 10 \times$ Subgrade CBR, $E_{GSB} = 0.2 \times (\text{GSB thickness})^{0.45} \times E_{subgrade}$, $E_{granular} = 0.2 \times (\text{GSB} + \text{WMM thickness})^{0.45} \times E_{subgrade}$. E_{BC} and E_{DBM} value is taken from IRC-37 2012 [8] considering reference temperature as 35 °C and VG30 Bitumen grade.

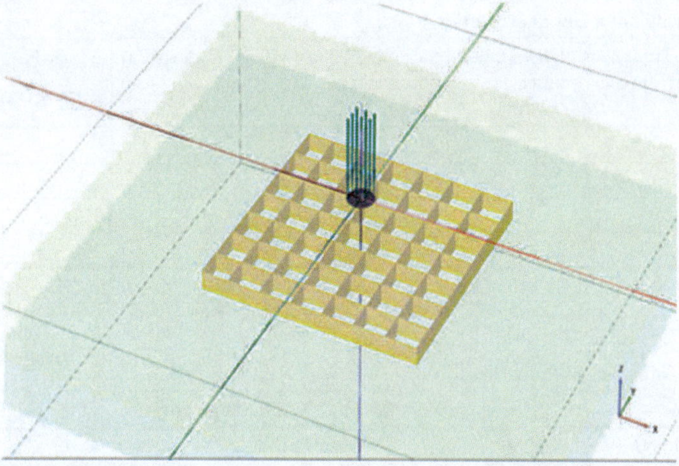

Fig. 8 Geocell in pavement

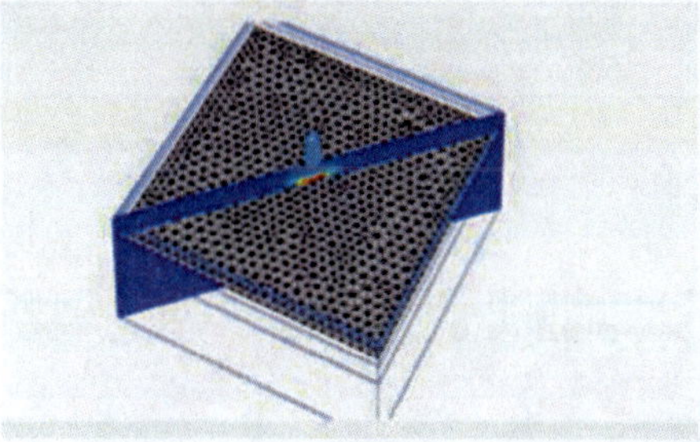

a) b)

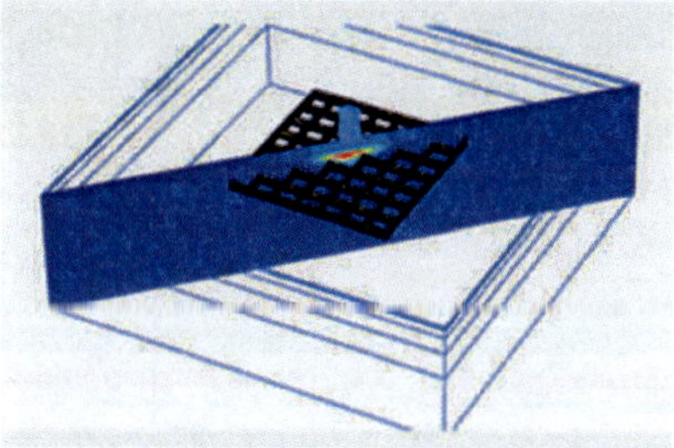

c)

Fig. 9 Deformed pavement model for **a** geogrid, **b** WPBG, and **c** geocell

Table 1 Pavement material properties

Material	Thickness (mm)	Unsaturated unit wt (kN/m^3)	Saturated unit wt. (kN/m^3)	E_{50} (MPa)	E_{oed} (MPa)	E_{ur} (MPa)	E (MPa)
BC	40	23.3	23.3	–	–	–	1700
DBM	135	22.6	22.6	–	–	–	1700
WMM	250	21.8	23	109	72.66	327	–
GSB	380	21.49	23	86	57.33	258	–
Subgrade	1000	17.90	21.2	30	20	90	–

Table 2 Reinforcement Properties

Reinforcement	Axial Stiffness (kN/m)	Cell diameter (mm)	Cell height (mm)
Geogrid 20 kN/m	350	–	–
Geogrid 40 kN/m	700	–	–
Geocell (Dia. 150 mm)	1000	150	150
Geocell (Dia. 350 mm)	1000	350	150
WPBG	50	60	50

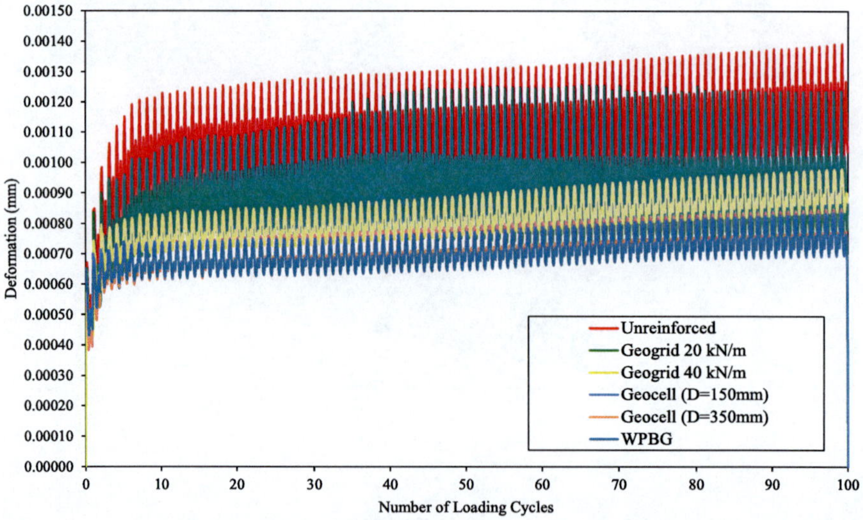

Fig. 10 Deformation against number of loading cycles

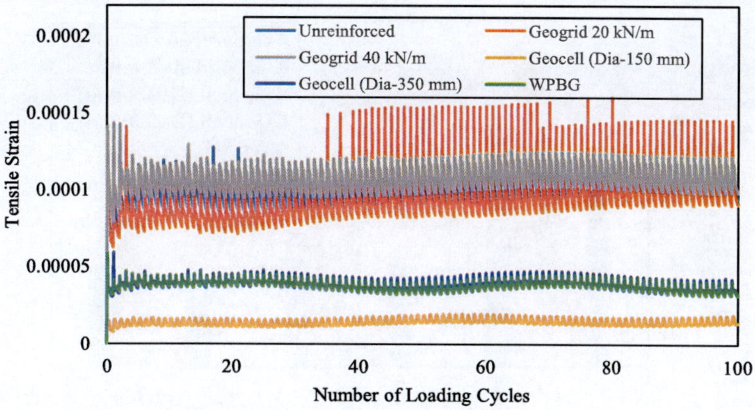

Fig. 11 Horizontal tensile strain against number of loading cycles

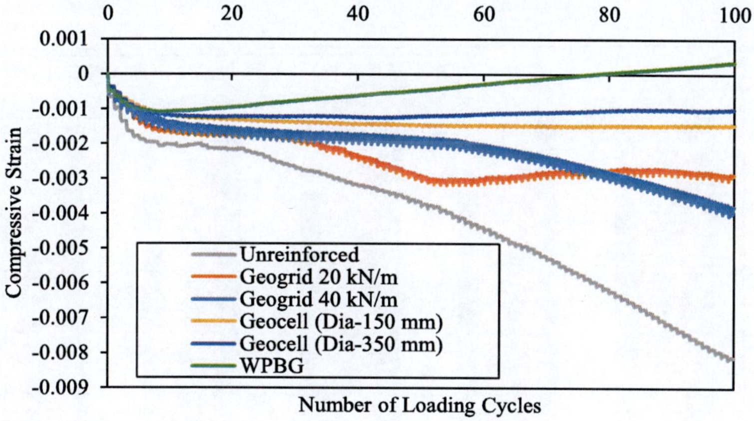

Fig. 12 Vertical compressive strain against number of loading cycles

4 Results and Discussion

From Fig. 10 it is observed that with comparison to all the types of reinforcement, conventional geocell is performing well against deformation under static cyclic loading. WPBG also performing well against deformation compared to geogrid reinforcements. Similarly, conventional geocell also performing well against fatigue and rutting failure by reducing horizontal tensile and vertical compressive strain, respectively in the pavement. WPBG shows the least vertical compressive strain at the top of the subgrade amongst all types of reinforcement.

When WPBG is encased in the model there is a decrement in deformation by 37% as compared to the unreinforced model. Also, WPBG is seen effective than the

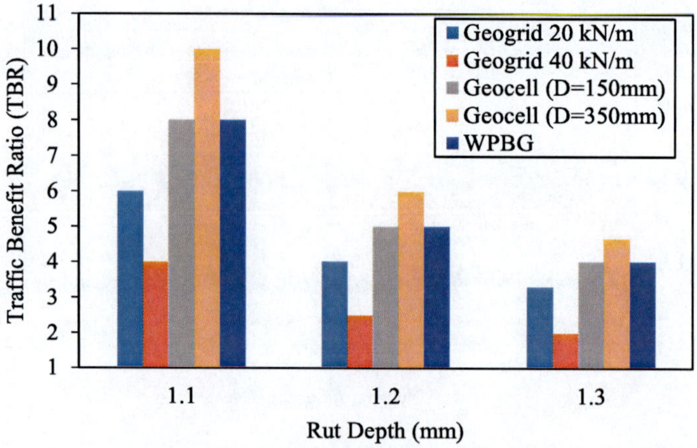

Fig. 13 Traffic benefit ratio (TBR) against rut depth (mm)

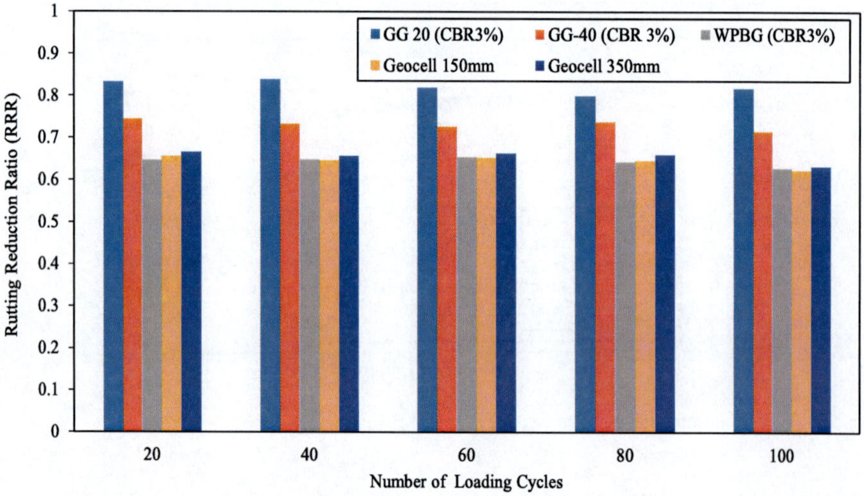

Fig. 14 Rutting reduction ratio (RRR) against number of loading cycles

geogrids of both tensile strengths of 20 and 40 kN/m and Conventional geocell of size 350 mm.

4.1 *Traffic Benefit Ratio (TBR) and Rutting Reduction Ratio (RRR)*

The benefits of WPBG are also confirming with two new parameters namely, traffic benefit ratio (TBR) and rutting reduction ratio (RRR). TBR and RRR values for all the reinforcement are calculated based on Eqs. (1) and (2), respectively.

$$\text{TBR} = \frac{\text{Number of Loading Cycles for Reinforced Section}_{\text{at perticular rut depth}}}{\text{Number of Loading Cycles for Unreinforced Section}_{\text{at same rut depth}}} \tag{1}$$

From Eqs. (1) and (2) it is confirmed that the higher the TBR value higher the benefits of reinforcement because more no. of cycles are needed for a reinforced section than the unreinforced section for a particular deformation and lower the RRR value higher the benefits of reinforcement because less deformation is needed for a reinforced section than the unreinforced section for a particular loading cycle.

From Fig. 13 it is again confirmed that the conventional geocell is better reinforcement material for pavement due to its high TBR value. WPBG also shows a good TBR value compared with geogrid reinforcement. Both conventional geocell and WPBG shows lower RRR value compared with geogrid reinforcement (Fig. 14).

$$\text{RRR} = \frac{\text{Rut Depth for Reinforced Section}_{\text{at perticular loading cycle}}}{\text{Rut Depth for Unreinforced Section}_{\text{at same loading cycle}}} \tag{2}$$

5 Conclusions

Conclusions of finite element analysis for WPBG are summarized below,

- Using Finite element analysis in PLAXIS 2D, it is seen that when the height of the reinforcement increased the value of deviator stress and respective strain value increased. The maximum value of deviator stress was obtained at the optimal depth when the reinforcement was placed at one-third height from top and bottom. From a 2D analysis, the optimum height and diameter of the geocell are 60 and 50 mm.
- In static cyclic loading analysis, WPBG shows less deformation than geogrids and geocell of diameter 350 mm. Also, WPBG shows 37% less deformation than the unreinforced pavement model.
- WPBG shows the least vertical compressive strain at the top of the subgrade amongst all types of reinforcement. Similarly, WPBG shows less horizontal tensile strain than the geogrid reinforcement.
- TBR value of WPBG is 2 times than geogrid of 40 kN/m tensile strength and 1.25 times than geogrid of 20 kN/m tensile strength.

- Both conventional geocell and WPBG shows lower RRR value compared with geogrid reinforcement.

References

1. Ramesh, A., Nageshwar Rao, C., Kumar, M.: Experimental study on geocell and of fibre reinforced soil sub-grade under static and repetitive load. In: Sundaram, R., Shahu, J., Havanagi, V. (eds) Geotechnics for Transportation Infrastructure. Lecture Notes in Civil Engineering, vol. 29. Springer, Singapore (2019)
2. Adhana, S., Mandal, J.N.: Reinforced fly ash slope using different geosynthetics. In: Proceedings of Indian Geotechnical Conference, pp. 15–17, Kochi (2011) (Paper No. J 194)
3. Thakare, S.W., Sonule, S.K.: Performance of plastic bottle reinforced soil. Int. J. Eng. Innov. Res. **2**(3), 207–210 (2013)
4. Lal, B.R.R., Mandal, J N.: Study of cellular reinforced fly ash under triaxial loading conditions. Int. J. Geotech. Eng. **7**(1), 91–104 (2013)
5. Nadaf, M.B., Mandal, J.N.: Numerical simulation of loading strip footing resting on cellular mattress and strips: reinforced fly ash slopes. Int. J. Geosyn. Ground Eng. **3**, 26, (2017)
6. Dutta, S., Mandal, J.N.: Model studies on geocell reinforced fly ash bed overlying soft clay. J. Mater. Civil Eng. ASCE **28**(2) (2016)
7. PLAXIS 3D: Material Models Manual (2013)
8. IRC 37: Guidelines for the design of flexible pavements. In: Indian Road Congress (IRC) (2012)

Impacts of Temple Waste on the Environment and Its Mitigation

S. S. Jahagirdar, V. K. Patki, G. J. Kilkarni, and S. B. More

1 Introduction

We give generously to our gods and that is evident from our temple offerings. It is estimated that some 800 million tons of flowers, including roses and yellow marigolds, are offered across the temples mosques, and gurudwaras in the country. Along with flowers vermilion packets, plastic incense packets, and bangles made of synthetic material are also generated. However, when these generous offerings turn into colossal waste, it creates a tricky problem that is detrimental to our environment.

The City of Solapur, located in the southern part of Maharashtra, touching the borders of the States Karnataka and Telangana is an abode to the Lord Siddheshwara. The annual pilgrimage is held in January which attracts lakhs of devotees. The number swings from hundreds to thousands during the rest of the year. Devotees present offerings to the gods which is nothing but solid waste after it has served its purpose. Moreover, the waste generated daily at the temples and periodic religious gatherings intensifies the complexity of the waste making it harder for the local bodies to handle and dispose of it. The quantities of the waste and its intricacy increase during the peak festive seasons and pilgrim gatherings. Modernization and economic development have enhanced the living standard of people through urbanization which led to an increase in the quantity and complexity of the waste generated in the cities [6]. In India, the management of urban solid waste (which includes wastes generated from temples) is considered an increasing and serious environmental problem that requires immediate resolution [5]. This investigation has been done to minimize the implication on the MSW management due to temple wastes. Improper disposal of such waste pollutes all the vital components of the

S. S. Jahagirdar (✉)
Professor in Civil Engineering and Dean R and D, NKOCET, Solapur, MH, India
e-mail: shrikantjahagirdar@orchidengg.ac.in

V. K. Patki · G. J. Kilkarni · S. B. More
Professor in Civil Engineering, NKOCET, Solapur, MH, India

I. Pal and S. Kolathayar (eds.), *Sustainable Cities and Resilience*, Lecture Notes in Civil Engineering 183, https://doi.org/10.1007/978-981-16-5543-2_22

living environment, and this problem is more acute in developing nations like India than developed nations due to the adaption of technologies in waste management in the latter ones [4]. Mostly, pollution is caused due to littering, which is the improper and careless way of disposing of small amounts of wastes [2]. Littering is a special type of waste that largely occurs in the premises of temples, the places where the devotes temporarily reside during the pilgrimage as well as during any other public gatherings. The daily solid waste management regime carried by the local bodies is burdened due to this special kind of solid and liquid waste effluents from the temples which contain organic as well as inorganic refuse such as floral offerings (Nirmalya), coconut shells, fruits, polythene bags, plastic, Abhishekam, etc. An attempt has been made to dispose or reuse such type of temple wastes at the source itself and reduce the burden on the MSW management which is carried by the Solapur Municipal Corporation.

Figure 1 shows elements of temple waste commonly observed in temples. Wastes arising from temples can be categorized as liquid, solid, and gaseous (air emissions).

Few other factors are affecting these waste discharges, namely habits of devotees, vendors outside the temple and characterization of products, priests and their habits, etc. Therefore, an attempt has been made to collect, characterize, and analyze two types of waste discharges, i.e., solid waste and liquid waste. The objective of this

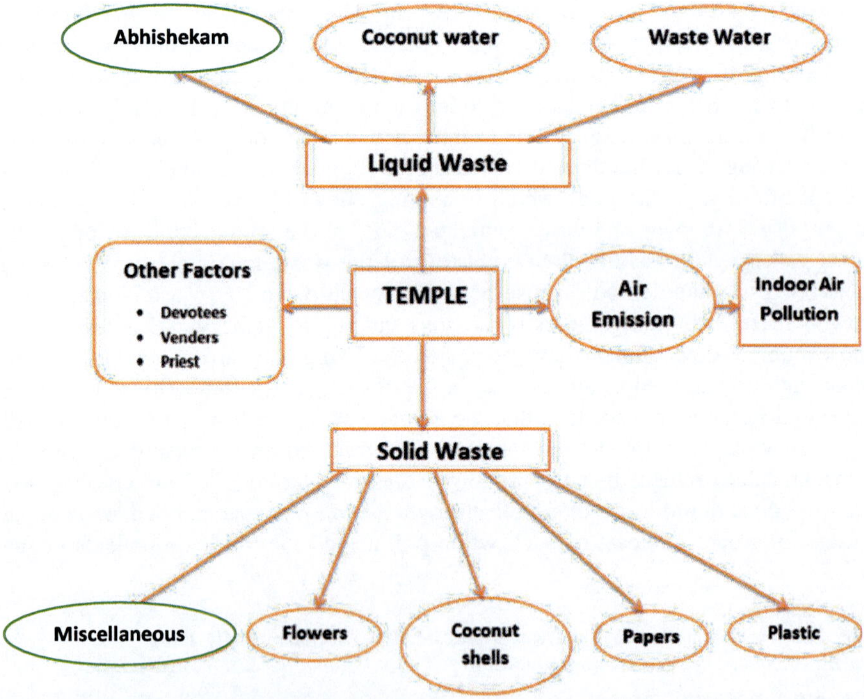

Fig. 1 Elements of temple waste management

study was to collect, segregate, and analyze solid waste samples and to collect and analyze liquid waste samples for different water quality parameters.

2 Materials and Methods

For this study, three major temples were chosen from Solapur City, Maharashtra, India for the sampling process from where the waste samples had been collected. The following Fig. 2 shows the location of the three sampling sites; Lord Siddheshwara Temple (LST), Rupa Bhavani Temple (RBT), and Mashrum Ganpati Temple (MGT).

Solid and liquid wastes samples waste were collected from the above temples on:

- Peak day (PD)—a day having the highest rush
- Average day (AD)—a day/days having the moderate rush
- Lean day (LD)—a day/days having less crowd.

Collected samples were analyzed for pH, chloride content, alkalinity, hardness, turbidity, conductivity, and dissolved oxygen tests as per standard procedures [1]. On the other hand, solid waste samples were collected and segregated in the laboratory,

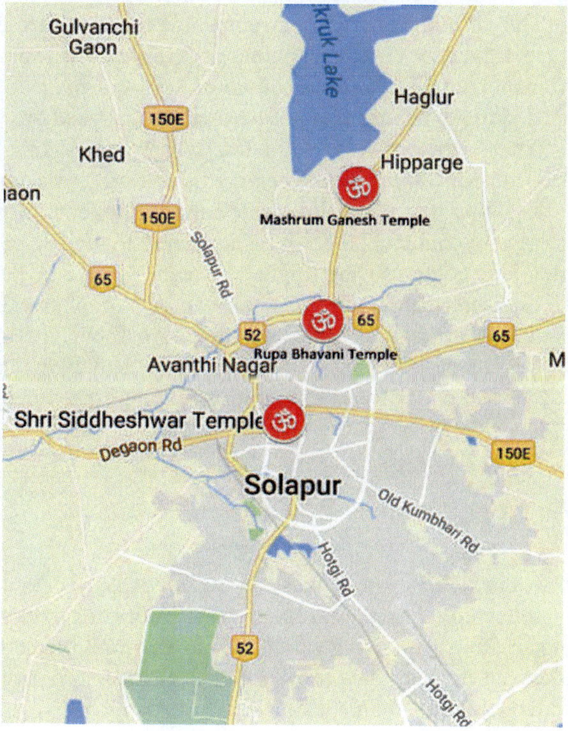

Fig. 2 Maps of Solapur City indicating sampling locations

and its physical and chemical properties such as density, moisture content, volatile content, carbon content, and nitrogen content were determined.

3 Results and Discussions

Tables 1, 2, 3, 4, 5 and 6 show results of comparative analysis of fresh water and liquid waste generated during pooja rituals from three temples.

pH of freshwater before Abhishekam is in the alkaline range. However, after Abhishekam pH reduces because of the addition of milk and curd to water and comes down to near neutral range. Average values are within BIS 10500 limits. The change in pH can be attributed to ingredients used during Abhishekam such as milk, curd, Chandana, Haldi, and Kumkum.

The alkalinity of water before Abhishekam is near to BIS 10500 standards but after Abhishekam alkalinity of water increases because of the use of milk and curd during the *t* process of Abhishekam.

The hardness of water after Abhishekam increases.

The turbidity of Abhishekam water increases many folds. An increase in turbidity can be attributed to the use of various materials used during Abhishekam such as milk, curd, Chandana, Haldi, and Kumkum.

The addition of milk and curd along with water during Abhishekam depletes oxygen present in water as time progresses. All samples showed DO is absent in Abhishekam water after a couple of hours. It indicates that biological treatment is needed to Abhishekam water before it is disposed of.

After segregation of solid waste samples organic and inorganic contents was determined. Table 6 shows the organic and inorganic contents of solid waste samples.

Results show that solid waste samples collected from all three temples are rich in organic content. This can be attributed to the use of flowers, tulsi, coconuts, etc. Inorganic contents were might be because of plastic bags, ash, etc.

Characterization of solid waste samples for physical and chemical properties was done, and average results are presented in Table 7.

Results indicate that C/N ratio is more than 20 indicating the flower waste can be composted easily by using traditional composting methods.

4 Conclusion

Instead of letting tons of waste going down the river or any other water body, it should be collected at one place, sent to waste processing units where it undergoes the process of shredding, following which the bio-enzymes break down complex waste particles to create organic manure. Floral waste has a sufficient C/N ratio which is required for composting. Abhishekam water needs biological treatment before it is disposed into the environment. A separate study is required to study air emissions from holy

Table 1 pH of Abhishekam water

Temple	No. of samples								BIS-10500 standard [3]
	Water before Abhishekam				Wastewater after Abhishekam				
	1	2	3	Average	1	2	3	Average	
LST	10.3	8.7	8.2	9.0	9.0	5.5	7.2	7.2	6.5 to 8.5
MGT	11.3	9.1	10.5	10.3	5.9	8.4	6.1	6.8	
RBT	10.5	9.8	10.2	10.1	8.3	6.1	6.4	6.9	

Table 2 Alkalinity of Abhishekam water

Temple	No. of samples								BIS-10500 standard
	Water before Abhishekam			Average	Waste water after Abhishekam			Average	
	1	2	3		1	2	3		
LST	288	252	292	277.3	428	340	360	442.6	200–600
MGT	200	280	240	240	290	320	480	363.3	
RBT	230	258	235	241	370	310	345	341.6	

Table 3 Hardness of Abhishekam water

Temple	No. of samples								BIS-10500 standard
	Water before Abhishekam			Average	Waste water after Abhishekam			Average	
	1	2	3		1	2	3		
LST	336	408	436	393.3	488	424	428	446.6	200–600
MGT	328	104	336	256	348	180	360	369.3	
RBT	280	250	230	253	285	260	250	265	

Table 4 Turbidity of Abhishekam water

Temple	No. of samples								BIS-10500 standard
	Water before Abhishekam				Waste water after Abhishekam				
	1	2	3	Average	1	2	3	Average	
LST	2.8	2.6	2.4	2.6	1151	954	1287	1130.6	1–5 NTU
MGT	1.2	2.5	1.4	1.7	960	823	850	877.6	
RBT	1	0	1.2	0.7	1000	1020	1070	1030	

Table 5 DO of Abhishekam water

Temple	No. of samples								BIS standard
	Water before Abhishekam			Average	Waste water after Abhishekam			Average	
	1	2	3		1	2	3		
LST	5.6	4.3	7.6	5.86	0	0	0	0	Min. 4
MGT	5.8	6.4	6.5	6.23	0	0	0	0	
RBT	6.1	6	6.8	6.3	0	0	0	0	

Table 6 Percentage of organic and inorganic content of solid waste

No.	Temple	Total weight of sample (Kg)	Organic content (%)	Inorganic content (%)
1	LST	3	98.8	1.2
		3	99.33	0.67
		3	96.97	3.03
2	RBT	3	98	2
		3	98	2
		3	93.47	6.52
3	MGT	3	98.14	1.86
		5	99.02	0.98
		3	99.3	0.70

Table 7 Some typical characteristic of temple waste

S. No.	Parameter	Flower waste
1	Density	36.4 kg/m^3
2	Moisture content	85.42%
3	pH	5.5
4	Carbon content	45.1%
5	Nitrogen content	2.2%
6	C:N ratio	>20

places. If liquid waste, solid waste, and air emissions from temples are taken care of then all temples might become sustainable and smart temples.

References

1. APHA Standard Methods for the examination of water and wastewater. American Public Health Association (2012)
2. Al-Khatib, I.A., Arafat, H.A., Daoud, R., Shwahneh, H.: Enhanced solid waste management by understanding the effects of gender, income, marital status, and religious convictions on attitudes and practices related to street littering in Nablus—Palestinian territory. Waste Manag. **29**(1), 449–455 (2009)
3. BIS 10500: Drinking water—specifications. Bureau of Indian Standards, New Delhi (2012)
4. Kaushik, S., Joshi, B.D.: A comparative study of solid waste generation at Mansa Devi and Chandi Devi Temples in the Shiwalik Foothills, during the Kumbh Mela 2010. In: Report and Opinion, vol. 4(8), pp. 39–42 (2011)
5. Sushil, B., Joshi, B.D.: An assessment of solid waste generation during the Religious occasions at Haridwar city. IJENAS **3**(4), 1–7 (2014)
6. Verma, A., Bhonde, P.B.K.: Optimization of municipal solid waste management of Indore city using GIS. Int. J. Emerg. Technol. **5**(1), 194–200 (2014)

Age and Household Solid Waste Arising in Suburban Malaysia: A Statistical Approach

Dani Irwan, Noor Ezlin Ahmad Basri, Kohei Watanabe, and Ali Najah Ahmed

1 Introduction

A society that is characterized as ageing is a slow and looming disaster to modern centres of population as it signifies a decline in commercial activity, a shrinking labour force and also increased healthcare costs. An increase in the average age of a society signifies not only a slowdown in the working-age population, but also a curtailment of economic output. Compounded with the main cost of an ageing society, namely elevated healthcare expenditures, all of the aforementioned are major obstacles for any city to be truly termed as sustainable.

Additionally, the world's population is ever-expanding. In 2011, the world's population hit the 7 billion mark and in 2013, it stood at 7.3 billion [1]. It is forecasted that by the year 2100, Earth will have 10 billion inhabitants [2]. With this tremendous population growth, there has been and will be a commensurate increase in global municipal solid waste (MSW) generation [3], resulting in copious quantities of solid waste that is in need of suitable means of handling, processing and ultimately final disposal.

The issues brought on by the absence of such appropriate solid waste management (SWM) measures are multitudinous. Amongst others, these adverse effects

D. Irwan (✉)
Civil Engineering Department, Kulliyyah of Engineering, International Islamic University of Malaysia, 10, 50728 Kuala Lumpur, Malaysia
e-mail: dani@iium.edu.my

N. E. A. Basri
Civil and Structural Engineering Department, Faculty of Engineering and Built Environment, Universiti Kebangsaan, Bangi, Selangor, Malaysia

K. Watanabe
Department of Sociology, Taekyo University, Hachioji, Tokyo, Japan

A. N. Ahmed
Institute of Energy Infrastructure, Universiti Tenaga Nasional, Kajang, Selangor, Malaysia

I. Pal and S. Kolathayar (eds.), *Sustainable Cities and Resilience*, Lecture Notes in Civil Engineering 183, https://doi.org/10.1007/978-981-16-5543-2_23

include environmental pollution [4], degradation of public sanitation [5], compromised quality of life [6] and also a largely irreversible depletion of scarce natural resources and land area [7].

According to [8, 9], these aforementioned issues are especially prevalent in developing countries, whereby constraints such as inadequate funding, lack of proper facilities and technical expertise coupled with common public apathy leads to an aggravation of the aforementioned issues [10]. Evidently, these problems need to be solved in an efficient and integrated manner not only for the purpose of diminishing MSW amounts and hence pollution prevention, but also to avoid depleting already scarce natural resources and to maintain as high a quality of life as humanly possible, both of which are goals that are strongly relatable amidst today's current climate of heightened environmental awareness.

Generally, in Malaysia, urban areas have higher rates of MSW generation as compared to their rural counterparts; furthermore, MSW generation in Peninsular Malaysia is higher than in Sabah and Sarawak [11]. There are a variety of reasons that this could be attributed to, although many of them are beyond the extent of this research and will be precluded from further discussion. Nevertheless, it is important to note that 75% of the Malaysian population dwells in cities and nearby suburban residential areas. Thus, there is great impetus to focus on the household solid waste (HSW) component of Malaysian MSW, in particular those originating from suburban areas.

The bulk of MSW being produced now consists of household waste. In an extant study by [12], it was stated that 60–90% of MSW is made up of refuse from households. Another study by [11] reported that a total of 33,130 tonnes of MSW is generated daily in the whole of Malaysia, 21,627 tonnes (65.3%) of which is comprised of HSW. Therefore, sufficient priority should be given to the household waste stream when planning for and managing waste in a holistic manner. Even so, to date there has been little discussion made pertaining to household waste specifically. Discourse has instead been focused on MSW which includes industrial, commercial and institutional (ICI) sources. Clearly, predicting and managing refuse from households is a central element in the field of solid waste management, one that holds a profound interest to the study authors.

Previous studies have suggested that demographic and socioeconomic factors do affect waste generation rates, chief of which being affluence. According to [13], the waste output of a country depends on the mean living standards of its people. A research conducted by [14] investigated the effects a household's economy on its production of refuse. The effect of income on MSW generation rates for various economic levels on a national scale was researched by [15]. More recently, [16] studied the composition of wastes produced by households based on their mean annual earnings.

However, income level is not the sole factor thought to have an impact on waste output. A preceding study by [17] looked into the relationship between electrical power consumption and residential solid waste arising while [18] investigated how per capita HSW is related to education and household size, amongst other demographic factors. In another study, [19] explored the connection between education level and employment status to solid waste generation.

More pertinently, [20] posited that age also has a bearing on the quantity of waste a person produces. In a study by [21], it was reported that households with retirement-age elderly residents produced more HSW per capita when compared to households without. Still, the findings of a study conducted by [22] found no rigid correlation between age and per capita HSW arising.

The question of whether age is in any way related to HSW arising has always been of singular interest to the study authors. Thus, in this research, the scope of work is focused on utilizing age as a demographic variable in an effort to achieve the main objective; that is to determine if age is significantly related to daily HSW generation per capita in selected suburban Malaysian households through quantitative testing based on established statistical methods.

2 Methods

This study is an exploratory undertaking that makes use of primary data acquired directly from field work involving a combination of methods, specifically door-to-door waste weighing and face-to-face interviews with the household residents or respondents based on set questionnaire survey forms.

This study was conducted in Bandar Baru Bangi, Putrajaya and Kajang, three neighbouring suburban townships in the state of Selangor that are approximately 30 kms away from the city centre of the capital, Kuala Lumpur. Ostensibly, the areas' local councils and their appointed waste management concessionaires provide kerbside solid waste collection from households purportedly 3 times a week, either on Mondays, Wednesdays and Fridays or on Tuesdays, Thursdays and Fridays.

The local climate throughout the study areas is tropical, characterized by high humidity and precipitation levels throughout the year. Rainfall is brought on by frequent rainstorms and two monsoon seasons, one running from March to April and the other from October till November, with the penultimate month of the year also being its wettest one. The study areas are also subject to a dry season, which typically occurs from May until September. June is theoretically the driest month of the year; nevertheless, it still receives around 10 rain days.

Table 1 shows the selected study areas, the housing types present and the number of houses of each type on which this study was performed. In all, 423 households were selected and sampled from during the course of this study. To best capture varying socioeconomic levels, the selected study areas were deliberately chosen so that they consisted of housing elements which are discrete and discontinuous from each other. The individual households selected for this study were picked mainly due to ease of logistics and the need to collude with the timing of waste collection vehicles; this is to enable the study team to work as efficiently as possible within certain constraints that were imposed with regards to available manpower, transportation and time. The

Table 1 Types and number of sampled households in selected study areas

Area	Housing type	Sampled N	N post-sporadic exclusion (APP)
Bandar Baru Bangi	Bungalows	47	29
	Terraced	138	84
Putrajaya	Bungalows	25	13
	Semi detached	51	30
	Terraced	66	26
Kajang	Terraced	60	19
	Flats	36	18
Total		423	219

selected households are presumed to be analogous to and representative of the entire neighbourhood and other similar housing projects in other localities in the country.

The first stage of the study is the waste sampling phase. Primary waste generation data from the households under study was obtained by direct door-to-door weighing of HSW for 2 weeks and done every day consecutively from Mondays to Saturdays. The resultant is 12 days' worth of HSW sampling data which constitutes one sampling phase conducted concurrently in all study areas. Overall, a total of 4 sampling phases were conducted in pre-determined months over a span of 13 months.

The justification for a two-week waste sampling period is provided in Table 1 of MS2505:2012 by [23], which stated that the recommended minimum waste sampling period for an academic study pertaining to solid waste is one week. Furthermore, it is important to note that the study team carefully precluded carrying out the waste sampling work in months with long school holidays and festive celebrations such as Chinese New Year and Eid so as to avoid introducing any outliers into the data.

The door-to-door HSW weighing activities were conducted by 3 fieldwork teams with a team having 2 to 3 personnel each. Every team was also equipped with standardized digital electronic weighing scales with a maximum capacity of 40 kg, a resolution of 10 g and a readability of 10 g. The bin sampling method as applied by [24] was utilized for this study due to its accuracy in identifying the exact source of the refuse sampled and also because it ensures that waste is not amalgamated during sampling.

Standard operating procedures during the waste sampling phase is for the fieldwork team members to identify, withdraw and then uncover the waste bin of the house currently being sampled, extricate all the waste that has been discharged within and then weigh the same using the provided digital scale. Waste that are loose or those that has been scattered inside the waste bin were repackaged in new litter bags and then weighed whenever possible. Materials that were sorted into distinct categories, meaning those that were obviously sorted with a purpose to facilitate recycling activities, were weighed and recorded separately from commingled waste. The standard operating procedure outlined above deviates slightly for flats. This is because the building utilizes a built-in chute system which conveys the waste deposited into the chutes on the upper floors directly to the ground-level communal waste receptacle, from which the fieldwork team members would extract the accumulated waste from before proceeding to repackage them if necessary and weighing the same.

With regards to demographic data of the sample, a face-to-face questionnaire survey was performed on all of the sampled households immediately prior to the third sampling phase. The face-to-face method was employed because it has been shown to be the most reliable questionnaire survey approach, especially in the collection of socioeconomic figures [25]. The survey ran for a period of 16 days and involved 11 trained student enumerators who worked under supervision for 277 man-hours to complete this phase of the study. Together, they managed to survey 69.03% of the sampled households successfully.

The questionnaire survey form used by the enumerators consists of 24 questions pertaining to the various demographic, socioeconomic status and waste management habits of the household under study. Amongst the more pertinent questions in the questionnaire survey is regarding the number of persons living in the house, which is crucial for calculating per capita HSW generation. Another salient question asks the respondent to state the approximate combined monthly income of their entire household.

Every effort was made to sample all the houses that were selected. For example, call cards were left in the mailboxes of houses that were vacant and those that seemed to be devoid of occupants even after repeated visits by the enumerators. Call cards were also given to uncooperative or aggressive respondents. These call cards implored the reader to visit the URL of a website which in turn leads to a link that enables them to download a soft copy of the survey questionnaire form. The respondent can then fill out the questionnaire form at their convenience after which they were instructed to submit the filled-out form to the study team via email.

The demographic data acquired from the questionnaire survey exercise is then studied in parallel with the waste arising figures from the 4 waste sampling phases to determine the daily per capita HSW arising and age characteristic of the sampled households. Analyses were carried out via established parametric statistical testing methods such as independent samples T-test and Pearson's product-moment bivariate correlational assessment primarily using IBM SPSS Statistics version 21 and also Microsoft Office Excel.

3 Results and Discussion

The initial step undertaken prior to statistical analyses is data refinement, whereby outliers and unreliable data points were omitted from the dataset in order to preserve accuracy and avoid erratic results in the analysis. Houses that were vacant or that discharged waste too sporadically such as guesthouses and houses that were only occupied on the weekends were precluded from the final data analysis. A house is designated as being sporadic if it has more than 4 zero readings in one sampling phase.

After all sporadic and questionable premises were excluded, together with households for which there were no questionnaire survey data, the number of houses that remained for analysis was presented earlier in Table 1. This set of houses is identified as the All Periods Present (APP) dataset. The APP dataset signifies houses that consistently produced good reliable data and discharged waste regularly throughout the whole 13 months sampling period. Overall, a total of 4623 distinct data points or measurements from 219 households present in the APP dataset were analysed in this study. Table 1 of Malaysian Standard MS2505:2012 by [23] states that the number of samples required for the purpose of an academic study on HSW is 50 premises minimum, which this study's 219 premises far exceed.

The dependent variable (DV) in this study is 'Avg', a continuous variable and a direct representation of daily per capita HSW generation of a household (also denoted as a 'case' in SPSS) averaged over the 4 waste sampling phases. This DV is in units of kilograms per capita per day (kg/cap/day). There are two independent variables (IV) used in the analysis, namely 'Weighted Avg Age' and 'Code Avg Age', both of which rely heavily on demographic data acquired from the questionnaire survey.

To better represent the age variable of the sample in the SPSS data editor and analysis, the weighted average age of each household was calculated by applying the ensuing method. In the questionnaire survey form, the question regarding age of the household's residents were given as six age brackets with predefined set age ranges (newborn–3, 3–18, 19–29, 30–44, 45–58 and above 58-years old). Respondents answer by indicating the number of the residents that belong in each applicable age bracket. Then, the number of people in every age bracket was multiplied by the mid-point or the median of the said bracket's age range. For example, for the 19–29-years old age bracket, the median value is 24-years old. If there are two people in the household that belong to this age bracket, then the result of the multiplication is equal to 48. Note that the upper range of the 'above 58-years old' age bracket was set at 73-years old, which is the Malaysian life expectancy at birth [26]. The products of these multiplications for every age bracket were subsequently summed up and finally divided by the total number of residents in that particular household to obtain the weighted average age of the same and therefore the IV 'Weighted Avg Age'.

The other IV employed in the analysis, 'Code Avg Age' is a categorical IV whose synthesis is based on the aforementioned weighted average age variable. It separates the sample into younger and older households based on the Malaysian population's median age of 27.1 years [27] as a line of demarcation. Households with a weighted average age of 27.1 years and below were designated as 'younger' households, whilst those that had a weighted average age exceeding 27.1 years were denoted as 'older' households. The presence and absence of elderly people aged 58-years old and above in each sampled household were also recorded and derived into a separate categorical IV titled 'Elder'. Similarly, the presence or absence of toddlers were codified in another categorical IV titled 'Toddler'. Here, toddlers are defined as children who are under 3 years of age.

To determine if younger households discharge HSW at a similar rate to older households, an independent samples T-test was performed. Table 2 shows the results of the independent samples T-test between the DV 'Avg' and IV 'Code Avg Age'. 95

Table 2 Independent T-test result of 'Avg'-'Code Avg Age'

'Code Avg Age' (IV)	n	Mean of 'Avg' DV (kg/cap/day)	Std deviation	Sig	Eta^2
Younger	95	0.33	0.21	0.52	0.0019
Older	124	0.35	0.2		

households of the sample (43.4%) were designated as younger households, whilst 124 (56.6%) were denoted as older households.

It could be seen that the older households group had a higher mean value along the DV 'Avg' of 0.35 ± 0.2 kg/cap/day. In comparison with the group consisting of younger households had a DV mean value of 0.33 ± 0.21 kg/cap/day. This essentially means that older households discharged more HSW per capita daily as compared to their younger counterparts. However, this discrepancy was found to be statistically insignificant because the p-value is in excess of 0.05. This means that at the 95% confidence interval, no statistically significant difference between the groups is perceptible. Therefore, the null hypothesis is accepted and favoured over the alternative hypothesis.

The eta square value for this T-test pairing was calculated to be 0.0019, which according to [28] signifies that the effect size of a household's age on their HSW generation is very small. When expressed as a percentage, only 0.19% of the variance in the DV 'Avg' is explained by the IV 'Code Avg Age'.

In an effort to investigate whether households that had elderly residents generated more or less waste than households without, another independent samples T-test was carried out. The result of the T-test between 'Avg' and the IV 'Elder' is presented in Table 3. It was observed that the group of households with elderly residents present had a higher rate of daily HSW generation per capita (0.39 ± 0.23 kg/cap/day) when compared to households in which elderlies are absent from (0.33 ± 0.2 kg/cap/day).

Nonetheless, similar to the findings of the previous T-test, the difference between the two groups was deemed to be insignificant from a statistical standpoint due to the p-value surpassing 0.05. Likewise, the null hypothesis was accepted and favoured over the alternative hypothesis in this T-test pairing. With regard to effect size, the eta squared value returned was 0.01. This denotes a small effect size [28] and that 1% of the variance in the DV 'Avg' can be explained by the IV 'Elder'.

Conversely, the question of whether households in which toddlers are present as residents produced more solid waste per capita than their contemporaries was also explored. Table 4 shows the output from the independent T-test between the DV 'Avg' and the IV 'Toddler'.

Table 3 Independent T-test result of 'Avg'-'Elder'

'Elder' (IV)	n	Mean of 'Avg' DV (kg/cap/day)	Std. deviation	Sig	Eta^2
Absent	180	0.33	0.2	0.1	0.01
Present	39	0.39	0.23		

Table 4 Independent T-test result of 'Avg'-'Toddler'

'Toddler' (IV)	*n*	Mean of 'Avg' DV (kg/cap/day)	Std. deviation	Sig	Eta^2
Absent	176	0.34	0.2	0.96	0.0014
Present	43	0.34	0.22		

It can be seen from Table 4 that both groups of households that are being tested, namely households with toddlers as residents and those without, have similar values for per capita daily waste arising and that is 0.34 ± 0.2 kg/cap/day. As to be expected from the comparable mean values along the DV, this independent samples T-test also returned an insignificant finding with a *p*-value of 0.96.

In terms of effect size, the eta squared for this T-test was calculated to be 0.0014. Based on [28], this signifies a very small effect size. Consequently, the toddler variable is only able to explain 0.14% of the variance in the DV.

In view of the above test results, it would only be natural to wonder about households that have neither toddlers nor elders as occupants and also those that have both. To address this line of inquiry, the one-way analysis of variance (ANOVA) is another statistical test that was conducted on the study sample. Through utilization of the one-way ANOVA, comparisons along the DV could be made between 3 or more groups within the sample. In this instance, the sample was split into 4 groups: one group of households having only elders (but no toddlers) as part of their occupants, one group of households with only toddlers (but no elders), another group in which both toddlers and elders are present, whilst the last group consists of households that have neither amongst their residents. Table 5 presents the results of this one-way ANOVA.

Scrutiny of Table 5 shows that neither elders nor toddlers are present as occupants in the majority of households in the sample. This is followed by households that count either toddlers or elders as part of their occupants, both of which are more or less equally divided in terms of numbers. Households that have both toddlers and elders are in the absolute minority, numbering only 3. In terms of per capita daily waste generation, the group of households with both toddlers and elderly occupants have a slightly lower mean value along the DV 'Avg' (0.3 ± 0.44 kg/cap/day) as compared to the group of households that have neither (0.33 ± 0.19 kg/cap/day). Figure 1 is the means plot of this ANOVA test, which is a graphical representation of the abovementioned findings. It is of probable note that the HSW generation rates

Table 5 Output of one-way ANOVA

ANOVA grouping	*n*	Mean of 'Avg' DV (kg/cap/day)	Std. deviation	Sig	Eta^2
Have elders	36	0.39	0.24	0.3	0.0169
Have toddlers	40	0.34	0.23		
Have both	3	0.3	0.44		
Have neither	140	0.33	0.19		

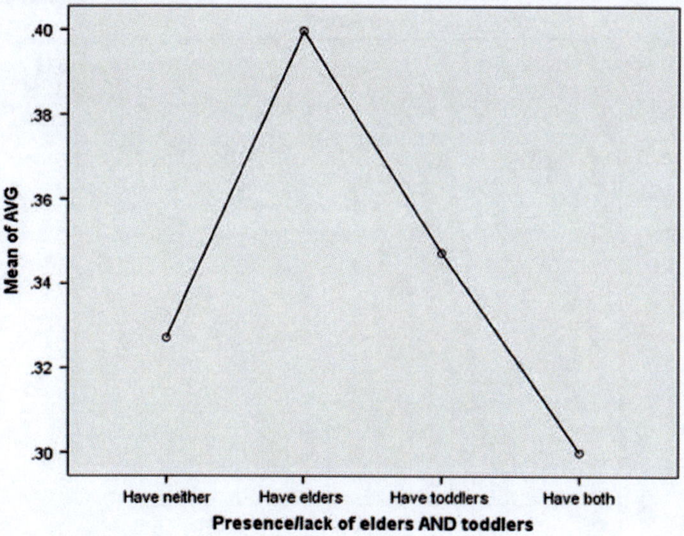

Fig. 1 Means plot of the one-way ANOVA

of the group of households with both toddlers and the elderly and those without both are lower than the other two groups, namely households that have just toddlers but no elderly and vice versa.

Still, the significance value of this ANOVA test as shown in Table 5 is in excess of 0.05, therefore denoting statistical insignificance. This indicates that no statistically significant difference exists amongst the mean scores of the four household groups being compared. The resultant eta squared value calculated for this ANOVA test signifies a small effect size according to the aforementioned guideline by [28].

With respect to the correlation between HSW arising and age, the salient variable should be first considered. Figure 2 is the histogram for the continuous scale IV 'Weighted Avg Age', in which it can be seen that the main cluster of the calculated weighted average age values is concentrated around the 30 years mark. This corresponds with the mean value of the weighted average age of the sampled households which is 30.13 years with a standard deviation of ±9.68 years. Hence, it would seem that the study sample has a mean weighted average age that is generally slightly above the reported median age of the Malaysian population, which is 27.1 years.

Figure 3 illustrates the scatterplot of the bivariate Pearson's correlational analysis between the DV 'Avg' and the IV 'Weighted Avg Age'. Visual inspection yields the fact that the data points in the scatterplot are quite dispersed and does not conform

Table 6 'Avg'-'Weighted Avg Age' correlation analysis results

n	*r*	*p*	Coeff. of determination	% of shared variance
219	0.11	0.1	0.01	1

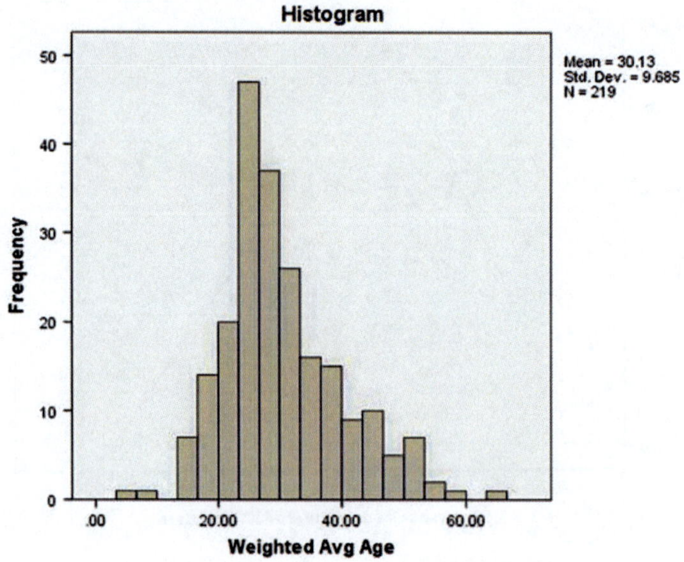

Fig. 2 Histogram of the IV 'Weighted Avg Age'

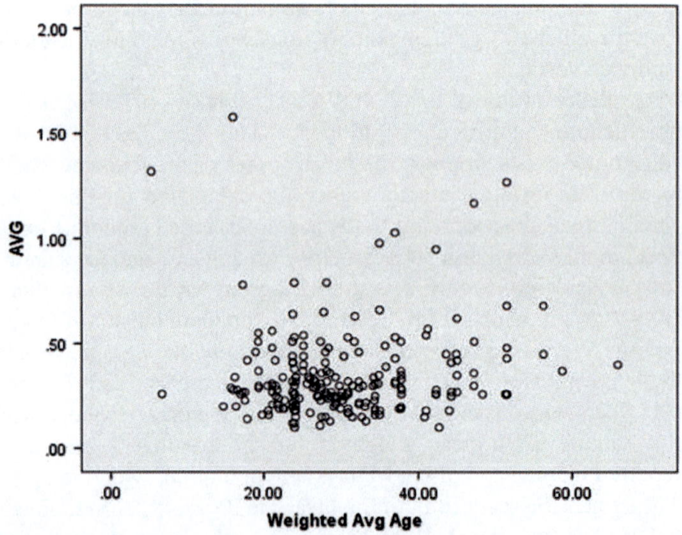

Fig. 3 Correlation scatterplot of 'Avg'-'Weighted Avg Age'

to a linear line. Most data points are clustered around the 30-years old mark and the lower range of the DV 'Avg'. Further numerical scrutiny is warranted and is duly provided by examining Table 6.

Table 6 presents the results of the bivariate Pearson's correlational analysis between the DV 'Avg' and the IV 'Weighted Avg Age'. The Pearson's Correlation Coefficient value, *r*, of 0.11 indicates that a positive correlation exists between the said variables. However, the significance value returned for this variable pairing is 0.1, denoting insignificance. Hence, the null hypothesis stating that no statistically significant relationship between the two variables is perceptible is hereby accepted and favoured.

Considering the abovementioned *r* value and based on guidelines laid out by [28] suggests that the relationship between per capita HSW generation and the weighted average age of a household has a small effect size. By squaring the *r* value, we arrive at the coefficient of determination that is calculated to be 0.01. Converting it to percentage form gives 1%. This means that the DV 'Avg' and the IV 'Weighted Average Age' share only 1% of their variance. In other words, there is not much overlap between these two variables [29]. Alternatively, it can be said that a household's weighted average age helps explain 1% of the variance in its per capita HSW generation rate.

The preceding statistical testing yields several important outcomes. More specifically, it was revealed throughout the course of this research that older households and households with elderly residents both discharged HSW at slightly elevated rates as compared to their respective counterparts. Identical HSW arising rates were observed in households with and without occupants belonging to the toddler age group. In addition, it was also determined that households with both the elderly and toddlers as part of their residents discharged HSW at a very similar rate to those households with neither. Lastly, there is evidence to suggest a positive, albeit small, correlation between per capita HSW arising and weighted average age. Simply put, the test findings of this study show that higher rates of HSW arising are associated with advancements in age. The above outcomes are in line with the findings of [22] and also of [21].

Even so, it is important to note that all of the above outcomes were found to be insignificant from a statistical perspective. In other words, even though there were discernible differences were observed between the groups, at the 95% confidence interval the values exhibited along the DV did not differ sufficiently from each other to be statistically significant. As for the correlation test, higher HSW generation rates could be correlated to higher weighted average age amongst the sampled households; however, the relationship between the two is not substantial enough to be significant from a statistical standpoint.

Additionally, a key aspect that should be kept in mind regarding any correlational analysis is that correlation does not imply causation. It cannot be said with certainty that one variable necessarily causes a change in the other and vice versa, even when the outcomes show a strong positive correlational relationship. This is because, more often than not, causality could be attributed to more than just one factor; chief among them being the presence of another confounding variable, whether measured or unmeasured that impacts the results of the correlation analysis. In this instance, it is more accurate to say that higher rates of per capita HSW generation have been observed in households attributed with older weighted average ages.

4 Conclusion

A truly efficient, sustainable and integrated solid waste management system is predicated on having the most accurate data on solid waste generation available [30]. This is reiterated by [31], who stated that 'accurate information on the quantities of solid waste being generated is fundamental to all aspects of solid waste management'. Crucial planning and operational parameters such as collection frequency, type of waste treatment and final disposal capacity requires accurate data on how much refuse is produced.

Population ageing together with unrelenting urbanization are major obstacles to be overcome in the twenty-first century. As cities grow, the portion of their older population grows commensurately. This not only requires a truly sustainable city to adapt to cater for an ageing population, it also necessitates a more nuanced and holistic approach in their methods of solid waste planning and management in order to ensure that their impacts on the environment remains minimal both in severity and quantity.

With the information gleaned from the outcomes of this study at hand, policymakers and stakeholders could formulate protocols and conduct integrated and sustained efforts that would serve to target the increasing HSW arising from specific age groups. Educational and awareness raising programmes could be tailored to suit the needs of younger or older target audiences and/or participants. More importantly, adequate measures should be undertaken to provide sufficient resources and the necessary infrastructure to handle the burgeoning amounts of HSW that will be generated in the future. Indeed, this is made all the more imperative in a country that is approaching the ageing nation status at an accelerated pace such as Malaysia, where according to [32], 14.4% of the population is forecasted to be above 60-years old by the year 2030.

Acknowledgements The study authors express their sincere gratitude to the DCP-2017-006/3 research grant for its support and financial aid in the undertaking of this study. The Japanese Ministry of Environment (NIES) was also instrumental in the provision of financial wherewithal necessary to carry out this research, for which the study team also expresses their utmost appreciation.

Conflicts of Interest
The authors hereby declare that they have no conflicts of interest pertaining to this research and the writing of this manuscript.

References

1. United Nations: World Population Prospects: The 2012 Revision. New York (2013)
2. Pitkanen-Brunsberg, J., Kostakos, G., Shine, C.: Salzburg Global Seminar Session Report 515. Salzburg Global Seminar, Salzburg (2013)
3. Periathamby, A., Hamid, F.S., Khidzir, K.: Evolution of solid waste management in Malaysia: impacts and implications of the solid waste bill, 2007. J. Mater. Cycles Waste Manag. **11**, 96–103 (2009).https://doi.org/10.1007/s10163-008-0231-3
4. Tanaka, N., Tojo, Y., Matsuto, T.: Past, present, and future of MSW landfills in Japan. J. Mater. Cycles Waste Manag. **7**, 104–111 (2005). https://doi.org/10.1007/s10163-005-0133-6
5. Benítez, S.O., De, V.C.A., Ramirez-Barreto, M.E.: Characterization and quantification of household solid wastes in a Mexican city. Resour. Conserv. Recycl. **39**, 211–222 (2003). https://doi.org/10.1016/S0921-3449(03)00028-4
6. Murad, M.W., Raquib, M.A., Siwar, C.: Willingness of the poor to pay for improved access to solid waste collection and disposal services. J. Environ. Dev. **16**, 84–101 (2007). https://doi.org/10.1177/1070496506297006
7. Peavy, H.S., Rowe, D.R., Tchobanoglous, G.: Solid waste management. In: Salvato, N.L.N.J. (ed.) Environmental Engineering. Wiley, pp. 755–888 (2003)
8. Minghua, Z., Xiumin, F., Rovetta, A., et al.: Municipal solid waste management in Pudong New Area, China. Waste Manag. **29**, 1227–1233 (2009). https://doi.org/10.1016/j.wasman.2008.07.016
9. Medina, M.: Solid wastes, poverty and the environment in developing country cities: challenges and opportunities. Helsinki (2010)
10. Guerrero, L.A., Maas, G., Hogland, W.: Solid waste management challenges for cities in developing countries. Waste Manag. **33**, 220–232 (2013). https://doi.org/10.1016/j.wasman.2012.09.008
11. SWCorp: Kompendium Pengurusan Sisa Pepejal Malaysia 2016. SWCorp (2017)
12. European Commission: Municipal waste statistics. In: Eurostat (2011). http://epp.eurostat.ec.europa.eu/portal/page/portal/waste/key_waste_streams/municipal_waste
13. Grossman, D., Hudson, J.F., Marks, D.H.: Waste generation models for solid waste collection. J. Environ. Eng. Div. **100**, 1219–1230 (1974)
14. Wertz, K.L.: Economic factors influencing households' production of refuse. J. Environ. Econ. Manag. **2**, 263–272 (1976)
15. Medina, M.: The effect of income on municipal solid waste generation rates for countries of varying levels of economic development: a model. J. Solid Waste Technol. Manag. 24 (1997)
16. Li, R.S.: The influence of household income on waste disposal practices: a case study in Calgary, Alberta, Canada. In: SWANA Northern Lights Chapter 2013 Annual Conference. SWANA, Edmonton, pp. 1–13 (2013)
17. Orccsosupa, J.: Relationship among the per capita generation of household solid waste and socioeconomic variables. Santiago de Chile University, Provincia de Santiago de Chile, Santiago de Chile Province (2002)
18. Benítez, S.O., Lozano-Olvera, G., Morelos, R.A., De, V.C.A.: Mathematical modeling to predict residential solid waste generation. Waste Manag. **28**(Suppl 1), S7–S13 (2008). https://doi.org/10.1016/j.wasman.2008.03.020
19. Bandara, N.J.G.J., Hettiaratchi, J.P.A., Wirasinghe, S.C., Pilapiiya, S.: Relation of waste generation and composition to socio-economic factors: a case study. Environ. Monit. Assess **135**, 31–9 (2007).https://doi.org/10.1007/s10661-007-9705-3
20. Matsunaga, K.: Effects of affluence and population density on waste generation and disposal of municipal solid wastes. Earth Eng. Cent. Report 1–28 (2002)
21. Watanabe, K., Irwan, D., Basri, N.E.A.: Door-to-door measurement of household waste arisings in selected towns in Malaysia. J. Mater. Cycles Waste Manag. 1–10.https://doi.org/10.1007/s10163-015-0379-6

22. Irwan, D., Basri, N.E.A., Watanabe, K., Abushammala, M.F.M.: Influence of income level and age on per capita household solid waste generation in Putrajaya, Malaysia. J. Teknol (Sciences Eng) **65**, 21–28 (2013). https://doi.org/10.11113/jt.v65.2186
23. Department of Standards Malaysia: Malaysian Standard: Guidelines for sampling of household solid waste—Composition and characterisation analysis. Malaysia (2012)
24. Watanabe, K.: Chapter 6: What is in the bin? University of Cambridge (2003)
25. Afroz, R., Masud, M.M.: Using a contingent valuation approach for improved solid waste management facility: evidence from Kuala Lumpur, Malaysia. Waste Manag. **31**, 800–808 (2011). https://doi.org/10.1016/j.wasman.2010.10.028
26. Malaysian Department of Statistics: Abridged life tables Malaysia 2015–2017. Table 4.15 Frequency table for the variable "Recode Pax 4" (2017)
27. CIA: The World Factbook (2011). https://www.cia.gov/library/publications/the-world-factbook/geos/my.html. Accessed 23 Jan 2013
28. Cohen, J.W.: Statistical Power Analysis for the Behavioral Sciences, 2nd ed. Taylor & Francis Inc, Hillsdale, New Jersey (1988)
29. Pallant, J.: SPSS Survival Manual, 4th ed. McGraw-Hill, Berkshire (2010)
30. Chowdhury, M.: Searching quality data for municipal solid waste planning. Waste Manag. **29**, 2240–2247 (2009). https://doi.org/10.1016/j.wasman.2009.04.005
31. Tchobanoglous, G., Theisen, H., Vigil, S.: Integrated Solid Waste Management-Engineering Principles and Management Issues. McGraw-Hill, New York (1993)
32. The Star: No country for old and sick people. Star (2017)

Feasibility Study on Municipal Solid Waste (MSW) as Sustainable Engineering Material Using Suction Characteristics

M. V. Shah and A. J. Brahmbhatt

1 Introduction

In all around the globe, unsaturated soils are frequently found as surface soil as well as soils available at shallow depth especially arid and semi-arid regions. For unsaturated soils, compressibility and suction are very prominent parameters relating its moisture conditions. Soil suction formed due to capillarity effect occurs in the compressed unsaturated soil due to the presence of water and air in it. Only few literatures available on classification of municipal solid waste (MSW) due to the variability of minerals presented in it. Suction characteristics reflect the role of water in the compressed sample, and thus, suction of MSW is very much essential for carry out its re-usability study. The suction characteristics and compressibility of MSW at various moisture content are required to study for understanding its behavior as a compressed fill under different weather conditions.

In numerous study, it is shown that as per the IS: 1498-1970, MSW lies in the silty sand (SM) or poorly graded sand (SP) group of soils and in MSW almost 75–80% of granulated particles were observed. MSW used is well segregated and particles of larger size were removed from the sample. The index and engineering characteristics of the MSW show its similarity with silty sand. The suction and compressibility characteristics of MSW were compared to that of silty sand which having low-free swell index and of low plasticity to show the influence of minerals presented in MSW and check its suitability as engineering fill material. Suction, compressibility and microscopic study were carried out at optimum moisture content (OMC) and on dry side and wet side of optimum. The same characteristics of soil and MSW samples were also observed for submerged condition as the samples were soaked to water for 24 h.

M. V. Shah (✉) · A. J. Brahmbhatt
L.D. College of Engineering, Ahmedabad, Gujarat, India
e-mail: drmvs2212@gmail.com

I. Pal and S. Kolathayar (eds.), *Sustainable Cities and Resilience*, Lecture Notes in Civil Engineering 183, https://doi.org/10.1007/978-981-16-5543-2_24

Suction basically differentiated into total suction, matric suction and osmotic suction. For measuring all three components of suction characteristics, filter paper method using ASTM D5298-10 is very much effective as well as one of the methods which provides wide range of suction measurement. MSW samples were further observed for void space in it using a microscope having large magnification factor and capable of reflecting the 3D projection of sample to understand the role of moisture in void formation.

2 Experimental Programs and Discussion

MSW classification takes place through sieve analysis method as per IS 1498-1970, and it is classified as SP group of soil. Though it may contain some volatile materials in it, the energy dispersion X-ray (EDAX) analysis and proximate analysis on the waste is carried out using IS 1350 (Part-1) 1984. The waste used here contains very less amount of organic/inorganic content as over the years due to the compressibility waste materials converted into soil like material. Figure 1 shows the graphical representation of volatile matters as a result of EDAX analysis.

Further, direct shear test and standard proctor test was carried out to measure cohesion, angle of internal friction and optimum moisture content of the soil and solid waste. The engineering properties of waste and soil were shown in Table 1. For, reflecting the behavior of both materials in various weather condition, compressibility and suction was observed at moisture content below OMC (−2 and −4%) and above OMC (+2 and +4%) along with the one at OMC.

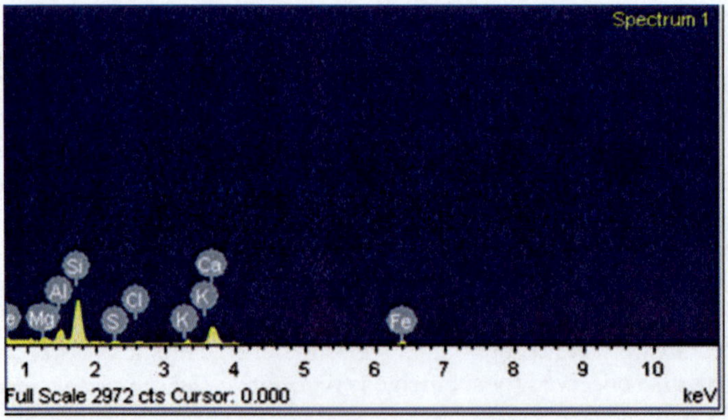

Fig. 1 Peak intensities of various elements present in MSW observed using EDAX analysis

Table 1 Properties of solid waste and soil

S. No.	Description	Indian standards	Solid waste	Soil
1	Particle size	IS:2720–4	$c_u = 3.43$, $c_c = 1.17$	$c_u = 4.63$, $c_c = 2.14$
2	Soil classification	IS:1498-1970	Poorly graded sand (SP)	Silty sand (SM)
3	Specific gravity	IS:2720-3	$G = 2.24$	$G = 2.65$
4	Compaction criteria	IS:2720-13	$C = 5.8$ kN/m^2, $\phi =$ 32°	$C = 19.6$kN/m^2, $\phi =$ 30°
5	Moisture and dry density	IS:2720-7	OMC = 18.5%, MDD = 15.76 kN/m^3	OMC = 12.5%, MDD = 18.74 kN/m^3

2.1 *Suction Characteristics and Strength Comparison*

Pirana landfills were about 10-years old, and the material extracted to study was beneath 1 m depth of the top surface of fill. Thus, in unsaturated landfills like Pirana suction characteristics play very important role in its compression and strength. Filter paper method allow to measure all three components of suction and support entire range of soil suction. Also, the range of suction for MSW is not previously known that is the basic reason to adopt this method for the study and many of the research study supports its effectivity of suction measurement.

For un-soaked sample, total suction for the MSW ranges from 310 to 537 kPa, which is similar to that of silty sand as total suction for silty sand is 340–480 kPa. However, there is a decrement observed in suction values for soaked condition. The range is declined to 53–129 kPa for soil and waste. The osmotic suction is higher for solid waste (ranges from 237 to 403 kPa) and plays major role to develop total suction. In soil, the matric suction covers the majority stretch of the suction range as it lies between 287 and 286 kPa. Matric suction ($u_a - u_w$) reflects effects of textures of material and capillary effect in soil due to water present in it and osmotic suction (π) shows the effect of disbanded ions and various salts in the samples.

Figures 2 and 3 show the suction values of soil and MSW for un-soaked and soaked sample, respectively, for different moisture content. The values clearly elaborate the behavior of the MSW with variation of moisture, and the suction shows decremented nature with increment in moisture. The nature of suction in the soil follows the same pattern. The suction reduces to considerable amount as the sample is soaked into water for 24 h that is due to the effect of submergence on compacted fill. On the dry side of optimum, the higher values of suction are observed for both soaked as well as un-soaked condition than that of on wet side of optimum.

The strength (S) of the MSW is further calculated using the compaction parameters and compared it to the suction characteristics. The results show that there is an increment in strength of the MSW with an increase in suction. Thus, the compaction on dry side of optimum will get higher strength than that of wet side of optimum. Thus, it is considered to compact the waste on dry side of optimum for practical

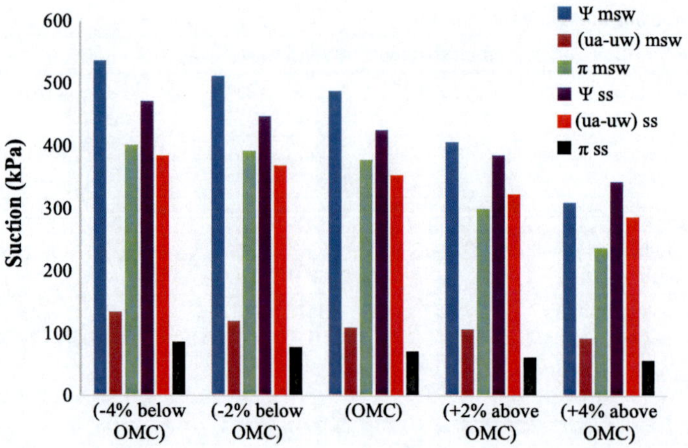

Fig. 2 Suction characteristics of soil and MSW for un-soaked sample at various moisture content

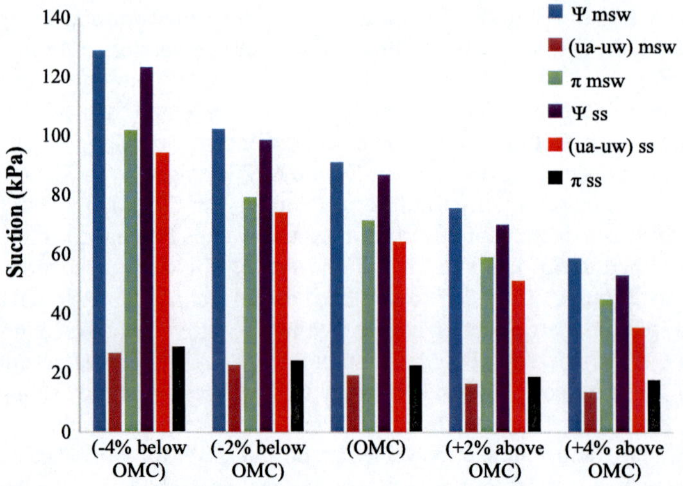

Fig. 3 Suction characteristics of soil and MSW for soaked sample at various moisture content

purposes. Figures 4 and 5 show the strength vs. suction representation of MSW for both the conditions. The considerable decrement in strength for soaked sample shows the influence of submergence on compressed fill.

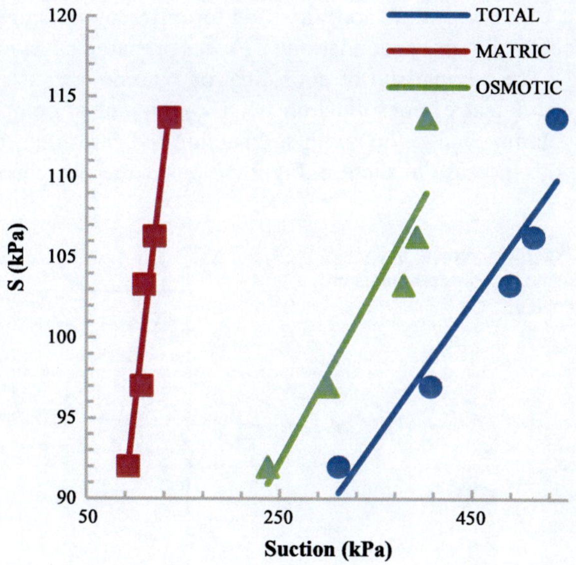

Fig. 4 Strength versus suction of MSW for un-soaked condition

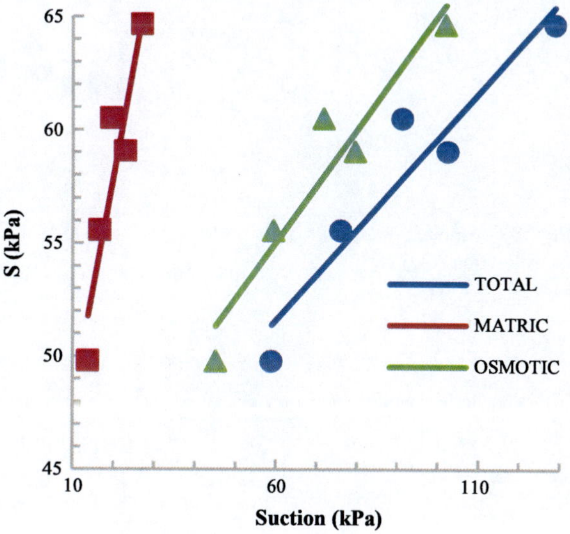

Fig. 5 Strength versus suction of MSW for soaked condition

2.2 *Compression and Suction Relation*

Compressibility of soil is essential component for the soil strength and to observe its relativity with suction the consolidation components were observed using IS:

2720-15 for MSW and silty sand for different moisture contents and the load versus coefficient of consolidation (C_V) was prepared as shown in Fig. 6.

The comparison of consolidation parameters such as compression index (C_C), coefficient of consolidation (C_V), coefficient of compressibility (a_v) coefficient of volume change (m_v) with soil suction indicates that the cc, cv, av and mv reduces with increase in suction. Figure 7 shows the decremental pattern of C_V as suction

Fig. 6 Increment in C_V values for different moisture content

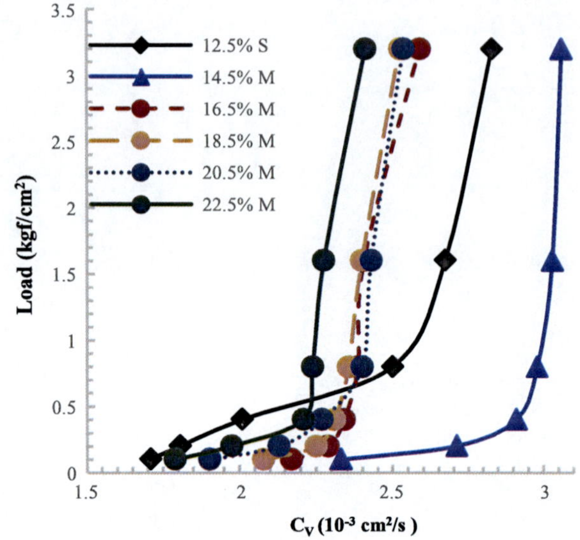

Fig. 7 Coefficient of consolidation (C_V) versus suction for MSW

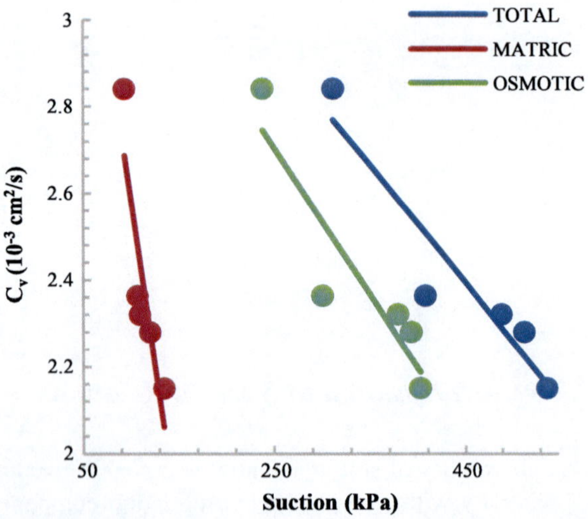

increases. This indicates that the compressibility behavior of soil is a function of soil suction (which depends on soil moisture content).

Hence, it can be concluded that unsaturated soil is expected to show lower compressibility as compared to saturated soil. Suction plays very important role in compressibility of solid waste and the relation of suction and compressibility shows the consolidation reduction with increment in suction. As the waste is compacted on dry side of optimum the higher suction and lesser consolidation is takes place due to the reduction in water availability.

2.3 Pore Size and Void Formation Observation

To understand the increment in strength of MSW as we move toward dry side, the void formation in the compressed structure is observed using microscope. Also, Laplace equation suggests that pore size has direct relation with matric suction (Ψm), the relation given as $\Psi m = 4\ \mathrm{T/d}$, the T is interfacial tension at air–water–solid interface = 0.072 N/m. So, by knowing the matric suction, the pore diameter can be easily carried out.

Figures 8 and 9 show the pattern of average pore diameter of compressed sample for different suction values for un-soaked and soaked conditions, respectively. The void formation in soaked sample considerably increases than that of un-soaked sample, and the same scenario is observed in silty sand sample. The reason behind this is the sample is soaked for 24 h which allows water to further penetrate into the small pores of compressed fill and widen the pores.

Fig. 8 Suction versus pore diameter of MSW for un-soaked condition

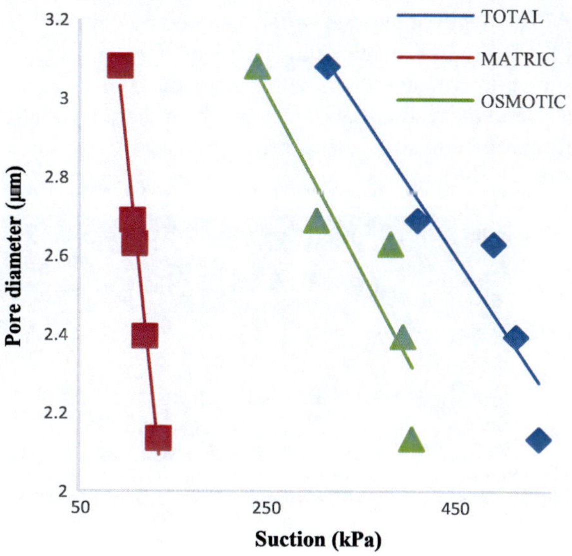

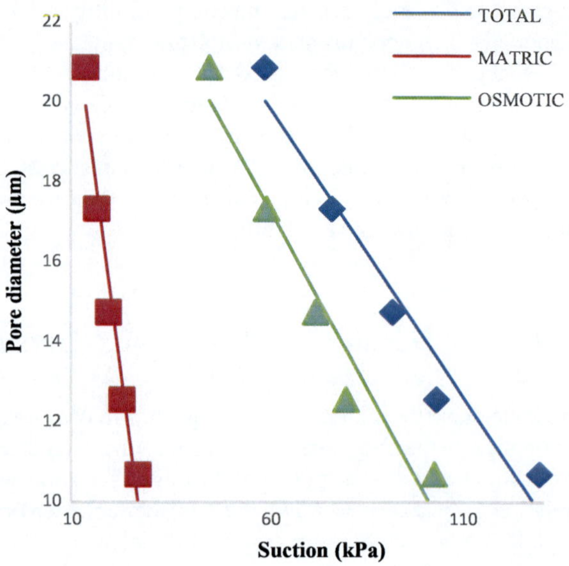

Fig. 9 Suction versus pore diameter of MSW for soaked condition

By using optical microscopy, it becomes easier to visualize the different ingredients such as soil particle, brick particle and pieces of plastic or glass, if any. The microscopical study was also carried out on MSW using microscope of high-magnification factor 50X. Figure 10 represents the void size variations in MSW sample. The dry side of optimum is showing more continuous structure than that of on wet side of optimum, and that must be the reason behind the relatively more strength on dry side of optimum than wet side.

The depth of various voids observed by converting the high-resolution images of MSW into 3D image using Helicon focus software. Relatively deeper voids were shown in wet side than that of dry side. The same study conducted on soil sample also and the results elaborates that the moisture variation plays important role in void formation and size of the pores.

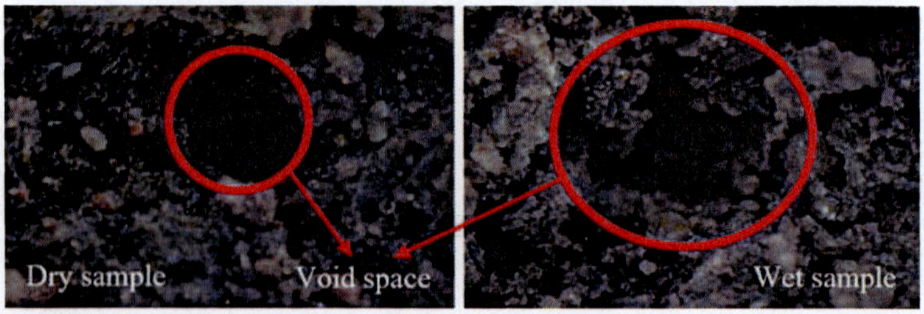

Fig. 10 Void space in MSW for dry and wet side of optimum

3 Conclusion

The study presents the importance of suction characteristics in municipal solid waste and the effect of suction characteristics of MSW in pores formation and attempt toward establishment of relationship of suction to compressibility and suction to strength, and the properties of MSW were further compared to silty sand for observing the suitability of MSW in absence of filling material. The following conclusions were carried out:

- Suction is very much essential component for understanding the behavior of compressed MSW as a filling material and has considerable impact on the strength and compressibility of the waste.
- For MSW, the suction is a reflection of the volatility present in it and the osmotic suction becomes very much important as compared to matric suction.
- The average pore size calculated in the study for silty sand and MSW at OMC, indicates higher average pore diameter for MSW as compared to silty sand. Hence, the corresponding capillary pressure causing matric suction is higher for silty sand as compared to MSW. Further, this also explains the higher dry density of silty sand at OMC as compared to dry density of MSW at its OMC.
- The deviation observed in compressibility characteristics and suction parameters were due to intra and inter structure moisture mobility, which is playing vital role and creating viscous and non-viscous film on particle because of organic contents present in this MSW.
- The soaked samples were also present the same pattern for suction characteristics; however, the values of suction are lesser compared to that of un-soaked samples of MSW and silty sand for different moisture content. The submergence reduces the strength of the soil and MSW.
- The variation in suction is clearly define the role of moisture mobility in unsaturated state. As the moisture variation plays major role toward the force required to extract water from pores and ultimately suction characteristics.

With proper segregation and by adding some important construction materials like lime, ash, etc., and by making it inert material, MSW can becomes reusable item. The suction has un avoidable impact on strength, compressibility and pore size of the compressed mass. The compaction on dry side of optimum is preferable for achieving higher strength and more continuous structure of compressed solid waste.

Acknowledgements The authors acknowledge the encouragement and facilities provided by the Prof.Dr. Rajul Gajjar, Principal, L D College of Engineering, Ahmedabad. Authors are also thankful to Dr. Kannan Iyer, Assistant Professor, IITRAM, Ahmedabad for necessary research guidance and to Research and Development Department, DDU, Nadiad for extending their lab facilities and allowing usage of microscope.

References

1. ASTM D5298-10: Suction measurement using filter paper. ASTM, American standards
2. Fredlund, D.G., Rahardjo, H.: Soil Mechanics for Unsaturated Soils. Wiley-Interscience Publications (1993)
3. Marinho, F.A.M., Stuermer, M.M.: The influence of compaction energy on the SWCC of residual soil. ASCE Adv. Unsaturated Geotech. 125–141 (2012)
4. Peron, H., Laloui, L., Hu, L.B., Hueckel, T.: Formation of drying crack patterns in soils: a deterministic approach. Acta Geotechnica **8**, 215–221 (2013)
5. Bray, J.D., Zekkos, D., Kavazanjian, E., Athanasopoulos, G.A., Riemer, M.F.: Shear strength of municipal solid waste. J. Geotech. Geoenviron. Eng. **135**, 709–722 (2009)
6. Singh, S.: A new approach for characterizing shear strength of municipal solid waste for land fill design. In: 18th International Conference on Soil Mechanics and Geotechnical Engineering, pp 3073–3075 (2013)
7. Stark, T.D., Huvaj-Sarihan, N., Li, G.: Shear strength of municipal solid waste for stability analyses. Environ. Geol. **57**, 1911–1923 (2009)
8. Zhan, T., Chen, Y.M., Ling, W.A.: Shear strength characterization of municipal solid waste at the Suzhou landfill. Eng. Geol. **97**, 97–111 (2008)

Urban Development and Sustainability

Smart Sustainable Cities: Principles and Future Trends

Bharani Alagirisamy and Poornima Ramesh

1 Introduction

The United Nations predicted that 66% of the world's population would live in urban areas by 2050 [64]. This indicates a substantial challenge in environmental and social sustainability issues [45]. Furthermore, the shape of the new cities was perceived as a cause of social and environmental issues. In addition to the inefficiency of the built environment, cities consuming about 70% of global capital, and thus, as a result of the urban population, the scale of the associated economic and social activities, are the main users of energy resources and are substantially contributing to greenhouse gas emissions (GHGs). Therefore, the current debates within urban and academic contexts continue to concentrate in response to significant problems emerging from the rapidly changing urbanisation as well as the unsustainability of current urban structures on the role of sustainability in urban planning and architecture.

Many nations worldwide are practising sustainable growth. This is apparent in the development of modern urban communities. The rationale behind this new boundary of urban planning is the revived awareness of the growth of human resources, safe living conditions and environmental issues [4, 58–59]. However, some essential factors must be present if a city is to be deemed sustainable. The sustainability dividends can be accomplished when paired with educated and ready people (sustainable education, green energies, energy conservation, sustainable transport, sustainable housing, waste management, etc.). A sustainable city uses fewer renewables than its rates of production and uses less than the rate at which renewable alternatives are produced, thus reducing their environmental effects [28]. In the future, the transformation of towns into permanent settlements will be one of society's biggest challenges [46]. Cities are centres for electricity usage but, thanks to their infrastructure and existing capital, having tremendous potential for incorporating green energy

B. Alagirisamy (✉) · P. Ramesh
Department of Environmental Sciences, Tamil Nadu Agricultural University, Coimbatore, Tamil Nadu 641003, India

I. Pal and S. Kolathayar (eds.), *Sustainable Cities and Resilience*, Lecture Notes in Civil Engineering 183, https://doi.org/10.1007/978-981-16-5543-2_25

[16]. There are various possibilities for certain sustainable energy. The emphasis on a green energy resource in particular, however, is itself a need of the hour [48]. In the sustainable development sector, it is all about the distinction between a sustainable building and a green building [19, 25], as is the real meaning of what constitutes a sustained building. If the environmental impact of a building can be reduced to reasonable standards, it can be defined as green, while sustainable buildings have to meet many more. There are many standards for a sustainable building, including human and environmental health, cultural heritage, social equity and social infrastructure [10] which define the sustainable construction of a building. The critical natural, social and economic impacts of construction have been well documented [34, 47, 49].

2 Sustainable Cities

Sustainable development undoubtedly has global coverage, but successful policy initiatives have to be targeted towards meso-orientation, for example by targeting individual economic sectors and regions, rapidly increasing. A geographic or urban angle is also definitely warranted in order to evaluate sustainability in an organisational context. In an urban setting, sustainable growth refers to a city's capacity for attaining a social, demographic, environmental and technical standard that in the long term strengthens the base of the urban structure itself. Only by initiating effective policy strategies is an environmental-friendly growth of a community feasible. The introduction and application of these ideas in various cities around the world is obviously different because each city has its own unique spatial, political and environmental context. Nevertheless, general integrative principles and protocols for assessment can be created and recommendations for the sustainability programmes of many cities. Sustainable cities require taking the entire life of buildings into consideration, the sustainability of the environment, the nature of functions and potential values. The cautious alignment of architecture with the electrical, mechanical and structural engineering tools therefore represents a sustainable structural concept. The design team must take into account the long-term expense of natural, financial and human capital, in addition to the conventional massing, orientation, proportion scale, texture, shadow and light aesthetics. Construction adds even more than we know to the quality of life. The climate surrounding the construction of the building promotes an alternative and cohesive solution. As sustainability continually influences the lives of businesses, people and the broader community, there are also growing opportunities for responsible and integrated thought. Biomimetic goods have four key goals for industrial sustainability: energy conservation and resource efficiency, toxic chemicals removal and regulation, the use of sustainable, biological resources and the addition of functions of materials and structures. Sustainable urban planning policies should also cover many sectors, for example urban rehabilitation, urban land use, urban transit networks, urban energy management, urban and conservation design, as well as urban cultural policies. In order to increase awareness of modern towns' sustained growth problems, quantitative metrics that include minimum production thresholds and vital threshold levels must then be identified, calculated and used as prediction methods. These critical thresholds will be shared by local authorities with all other urban space stakeholders (including the private sector). Sustainable urban planning is clearly a mechanism rippled with controversy and inconsistency. Commitment by key stakeholders in the city to strict environmentally balanced urban planning is crucial if sustainability policies are to be enforced effectively. This will also involve economic (market) incentives to improve productivity and cope with the negative facets of urban life in modern times. If a balanced strategy for urban growth is not established, urban spread will be strengthened and urban issues will become evident in a much larger region [55].

3 Sustainable Urban Development and Sustainable Cities

The term of 'sustainable settlements' is intriguing forever. Global and local commitments currently exist to make communities prosperous through separate 'sustainable urban growth' processes. Many actors engage in the academic and realistic facets of the film. For example, we see social scientists, experts on architecture and natural habitats, technicians and artists focusing on research and the development of policies and initiatives to fix sustainable urban elements. But it is problematic to know if we are really moving towards sustainable cities. In some ways, in raising the image of sustainable cities in the last 30 years, so much has been done that transformation is encouraging. The vocabulary of sustainability is infused with urban policy around the world and several exemplary projects can be found. However, we seem in some ways to go downwards to the point that hope is impossible to see. The rising population and lack of capital are escalating urban issues in developed countries. We see major new infrastructure and development schemes in the developing world, which defy any notion of sustainability but which are celebrated by the public and the industry. While enormous progress has been made in some fields of knowledge or some remarkable realistic measures, an impression of reform 'on the ground' is quite inconsistent or at least fractured. In the background of this dynamic image, often it is difficult to see the obstacles that common sustainable cities' face. However, in conceptualising progress centres, part of the issue lies in what we feel we want to see what improvements we want to make and what we value as 'progress'. The key obstacles are this conceptualization of sustainable cities. The following parts define (but not according to their significance) the ideals of sustainability, their key values and importance in creating strong sustainable cities [12]. The vital principles for a sustainable city are presented in Fig. 1.

Fig. 1 Sustainability principles for modern sustainable cities

4 Renewable Energy

The introduction of green energy is a key prerequisite to meet environmental targets and is the best and the simplest. In both developed and emerging countries, a movement towards sustainable strategies for the introduction and deployment of renewable energy can be studied in the metropolitan areas. Renewable electricity in rural and remote areas is also an outstanding instrument for sustainable growth and electrification. The need of electricity can be met with hybrid green energy sources in these remote off grid areas [32].

In scaling up the clean energy technology from laboratories to mass markets, government policies and state involvement are important [35]. Kilinc-Ata [39] argued that renewable energy integration is a welcoming improvement, but does not rely heavily on government policy, but on the form of policies and modes of implementation they are. In the international community, this claim is relevant because of the growing complexities of energy consumption, national security, environmental conservation and fossil price dynamics. In spite of comparatively high prices, a limited portion of the economy and a sluggish sector, renewable energy needs help policies compared with conventional energy sources [2]. At present, government interventions are re-defining green and clean energy pathways in the world. For example, the European environmental policies are focused more on renewables and the Union's response to environmental issues is renewable energy [17].

As of 2005, about 55 countries were committed to the adoption of the related measures, which will further meet diverse goals for renewable energy. By 2013, this pledge had grown to 144 nations [33]. In order to better appreciate the driving forces and processes behind their successes, Abdmouleh et al. [1] studied numerous policies and cases for clean energy deployment. This study classified renewable energy funding as ecological, fiscal, regulatory, regulation, technical and environmental initiatives. From multiple examples, the introduction of off-grid green energy networks through financial incentives such as lending, grants and subsidies has a larger effect and policy effects often have a greater influence on network structures. In a model to research their efficacy, Kilinc-Ata [39] used four well-known and widely used instruments of renewable energies, such as tariffs for feed-in (FIT), quota, tax and tender. FIT and quota-driven programmes are incentives based on clean energy production, and projects are based on tax and tender. The research was performed over a variety of time between 1990 and 2008 using available evidence. FIT, tax and tender incentives have been found to be more successful than the quotas policy instrument to promote clean energy technologies. The findings revealed that the usefulness of these policy instruments in many countries has been evaluated in a similar way [29, 31, 54].

Original initiatives around the world were directed at reducing the expense of green energy (with scientists' experiments and policymakers' incentives) so that fossil fuel rivals were reasonably affordable. Renewed measures to boost clean energy performance, and adoption are still directed after the cost considerations have been discussed and are still being addressed. The progress of sustainable energy production and adoption remains directly or indirectly correlated with the policy push of many decision-makers in the world from available literature on the current state of renewable energy. The developments of most EU and Chinese Member States over less than two decade have made it possible for renewables to enter, in particular to reduce the capital cost and to scale renewable energy for various purposes. Renewable energy technology's simple adoption and affordability, especially to incorporate into the grid, are not without sacrifices and challenges. It was defined how the effects of variable generation (VG) on electricity grids by wind and solar energy integrations could be minimised. A recent effort is being made to overcome these problems with various energy storage technologies [21].

4.1 Energy Efficiency

One significant aspect must be taken into account in energy conservation in order to reduce the environmental effect of the energy industry, in particular as regards climate change. Energy conservation is seen internationally as a core aspect of sustainability. A significant feature of renewable energy system, promoting processes of decreased emissions, enhancing energy stability and competition at regional level, is an improvement in energy quality. Information on the existing energy conservation situation and improvement in the area is required to enforce these steps [65]. Energy efficiency (EE) policies have a big fiscal, livelihood, emissions, climate change and energy security effect on a region [41]. Energy protection will improve environmental resilience through sustainable resource utilisation by implementing energy-efficient technology in sectors.

A holistic strategy must be adopted that can further improve energy efficiency in order to promote energy planning and management at regional level. Evaluating energy quality and reliability is a technique which is already available, and the knowledge needed for it for municipal energy services is difficult to obtain. To solve these EE-related problems, a new approach is therefore required. The solution should take into account the position and interests of operators and how the regional energy system is connected, which has historically not been taken into consideration in conventional models of energy [65]. Energy flows are explored in the framework and the selected metrics observe sustainability aspects. This strategy presupposes that the study of energy conservation is part of urban decision-making and related to community energy resources and priorities for sustainable growth. Related research has also been gathered to make energy service providers, and policymakers evaluate the impact of emerging energy technologies on energy quality, recycling, demand resilience and load changes, taking technology and environmental planning into account [65]. Ayres et al. [5] spoke about the economic growth position of energy efficiency. The study also clarified how critical the long-term sustainable global economy is to reduce GHG emissions. The factors for a smart sustainable city are illustrated in Fig. 2.

A modern, endogenous and quantitative economic theory integrates certain variables that the traditional theories did not recognise, such as oil supply or demand in energy services. The presumption that potential economic development relies on the continuing decrease of energy resources prices, as was previously stated, is an implication of this hypothesis not found in the conventional theory. This cannot be the case in the long term; discoveries like oil and gas and technical development have seen this downturn in the past. The generation of electricity and energy service that cannot proceed at a historic pace is an example of this. Indeed, energy prices and their suppliers are estimated to decline (e.g. electricity). It is predicted that economic growth will decrease, or fall, if energy cost rises by a significant proportion of GDP. Increasing economic activity does not then work to raise electricity prices by levying carbon taxes to limit environmental consumption, but facilitating energy saving at a negative expense by leveraging 'double-dividend' opportunities. On the other hand, the second argument suggests that emissions of GHG can be minimised as the technology-driven economic growth rises. To do this, we do not need modern technology, but energy controls or deregulation will work [5]. The latest solutions to sustainability are destructive. This makes it very difficult to introduce them. Proper business models (BMs) must be built to solve this problem. Global change and energy

Fig. 2 Factors influencing a smart sustainable city

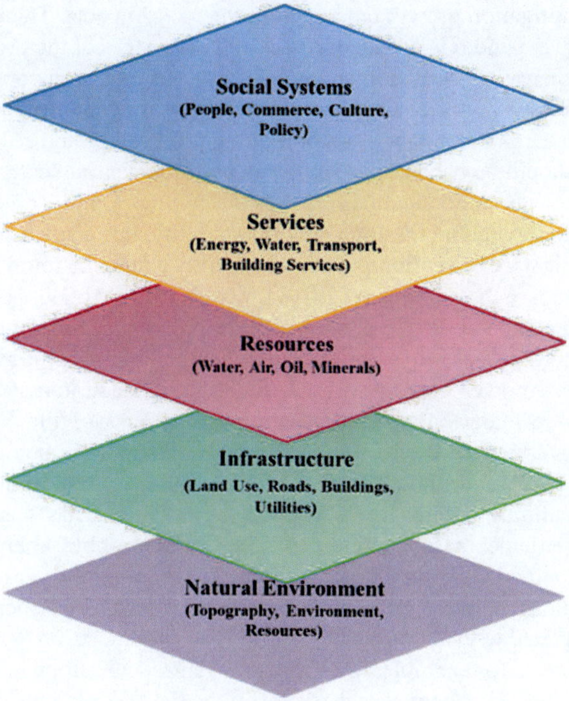

consequences can be discussed both for infrastructure and for control of the supply chain [26].

Industries have worked and increased their energy efficiency consistently in recent decades. Today, energy conservation is in the near term the most economic and efficient way to reduce greenhouse gas pollution in industries. This subsection also includes the potential contribution of energy-efficient manufacturing technology that will contribute to energy conservation and minimise greenhouse gas emissions by 2030 [67]. The use of energy-efficient systems and procedures to reduce the excess heat and materials is possible to achieve proposals for improving manufacturing energy efficiency [44]. In general, both developed and developing countries have not used entirely mitigation options. The explanation for this is obstacles, such as lack of access to resources, insufficient management focus, lack of expertise or unskilled service providers. It has proved to be difficult to quantify the ability to boost worldwide energy quality, and only a few research took this matter up. There are many risks with respect to mitigation and costs in 2030 that include the pace at which technology is developed and disseminated, the costs of future technology, emissions and future oil markets, manufacturing practises and the drivers of climate change and non-climate energy. Information gaps include baseline energy intensity for some industries, especially in transitional economies, potential for change in energy performance in non-energy-intensive industries, quantification of co-benefits, mitigation growth outcomes and market impacts. Therefore, more study is required to expand our knowledge base and awareness of the processes necessary to leverage energy conservation and greenhouse gas emissions opportunities [67]. Low energy and carbon system based on optimising fuel/energy cycle production and reducing incarnated energy or carbon through recycling and re-using of goods and materials should be encouraged [30]. The use of decentralised cogeneration (DCHP systems) can be economically useful when it comes to energy efficiency, reducing emissions of GHGs and increasing network reliability. It is also cheaper than central steam power plants. Conventional unified capability cannot be removed, but DCHP can be used to satisfy the anticipated 20-year demand rise [5]. New energy management solutions have the ability to achieve the desired energy-efficient mission. The application of heat-pumps, smart metres, micro-grids, air-conditioning building output systems, low-power lighting with LED, with thin isolation materials will do this.

Simulation of building efficiency is an evolving EE approach that involves an overview of the relationship between the energy demand for a building as regards HVAC and the availability of renewable and renewed energy supplies to make the building a zero-energy supply. Advances in BPS were tested for energy-efficient settlements for different sources of renewable energy (RES) [62]. Various CS methods and programmes for the design and reprogramming of existing buildings, along with the BPS method, to become cost- and energy-efficient using RES integrated with mixed energy sources and tri-generation to render renewable zero-energy areas, for example, in buildings, villages and settlements, have been created. Recent times have been very complex in the field of computing, data centres which operate the Internet, business, financial and business applications. The growth in demand for good networking, computing and storage is responsible. One data centre can have many servers that can be equivalent to a small city's electricity usage. Therefore, it is appropriate to operate these centres with tremendous computing capacity. Because of this large use of electricity, issues such as carbon gas emissions, high running costs, heavy fuel consumption, backup and recovery have to be tackled. The new IT infrastructure contribution to the overall CO2 footprint is 2%. This benefit would increase if no preventive action is taken. The green IT system will help strength and productivity of the data centre. This will add to energy efficiency and reduce greenhouse gas emissions, which could potentially reduce global warming. Cloud technologies, virtualization and green metrics are among the strategies [63]. Another modern technology which can reduce loss of electricity is micro-grid and intelligent grid. The EE micro-grid architecture project reveals that energy can be delivered to customers as consumers wish without risk of loss and higher prices [52]. This covers three areas of service, including performance, reliability and sustainability.

5 Sustainable Buildings

The critical impacts of construction have been well illustrated on the climate, culture and economy [34, 47, 49]. One of the main goals of sustainability is to reduce energy consumption in buildings [18]. Many countries now take into account the importance of sustainability in the building industry and play an important role, especially in reducing electricity. In Europe, for example, Germany has a big role to play in reaching 60 kWh/m^2/year or less HVAC plus lighting energy requirement, through the Energy Optimized Building (EnOB) Programme. This is 50% less than the norm for 2007. Based on their environmental impact at their development stage, the products used in building are an important player in environmental sustainability. Based on their environmental impacts at the manufacturing cycle, building materials are an important player in sustainability. Construction materials are valuable construction choices for eco-sustainable buildings that require low energy. The life cycle and material effect analyses, material movement studies and urban metabolism studies enhance the understanding of the usage of capital and inform the desire to create, build and maintain future infrastructure [6, 15, 23, 36]. The reuse of material is one thing, and it is complex and impossible to identify the optimal disassembly pathway for recycling particular construction materials, considering a variety of considerations, such as economic limitations, environmental requirements and building physical conditions. For sustainable buildings, the design process is an important period. Some important steps to take into account are as follows:

- The orientation, surface to volume ratio and thermal loss of the building
- Escalate the use of efficient and passive energy source
- Optimise high-performing envelope
- Choose high efficient, multipurpose energy equipments with effective control system.

5.1 *Sustainable Transportation*

In many metropolitan centres around the world, transportation concerns are one of the most critical topics [24]. Modern sustainable mobility methodologies require balanced steps that have impacts on better land use, development, management and utilisation of transport foundations and offices. It can also require actual capital investment and incorporate the environment into asset policies and decision-making as well as end-user expenditure. It can also boost the competitiveness of public transit with other contending metropolitan transport networks. The key issues of sustainability are upgrades on cars, quality of power and rewards for non-polluting fuel choices. It is primarily due to various core interests involved in diverse frameworks for simple governance in urban settings that the process towards a more efficient transport networks is uncertain [38]. In developed nations, more than half a million people die annually from transportation pollution, with a substantial proportion of automobile crashes [38]. Because the existing transport systems are unsustainable, it is critical that considerable simple speculations are made to achieve this aim. The arrangement would cover extraordinary long-term infrastructural initiatives, irrespective of whether it is a more constructive usage of the fewer land dangerous forms of transport or an adaptation of vehicle innovation [38]. Sustainable foundations of transport are governance, finance, utilities and districts. Each of the core elements is important: everything must be firmly developed to represent the needs of the society in a viable way.

The recommendations for strategies should concentrate on the elimination of carbon gases, the enhancement of air quality, the promotion of safety in transport, better access to road and wealth development. Innovation, efficient use of land, Renewable infrastructure, reliable data collection, competitive costs and better policy tools must all be included in the solution portfolios [42]. Three stages of mitigation take place: political/social, scientific and legal steps [50]. Several steps in transport to solve the issues of climate change include vehicle sharing, car parking-and-ride, parking space parking, smart cards, low-speed modes, telecommuting, smart growth and route assessment techniques, etc. The behavioural changes of passengers are also necessary in order to achieve sustainable transport, etc. Various behavioural improvements may be used to lead to sustainable travel. In the first place, people should follow more energy-efficient driving strategies, use cars for the right purpose, transform conventional fossil-fuelled vehicles into renewable energy saving or live close to their offices to minimise miles [56]. Steg and Gifford [57] recognised the inability to maintain the existing transportation system and emphasised the need to examine the difference between behavioural reform and technical innovations in sustainable transport. Behavioural reforms include reducing the amount of car use, while technological improvements concentrate on decreasing the negative impact per car per kilometre. Individuals are, therefore, oriented for technological advancement (e.g.) on the grounds that behaviour improvements—such as vehicle sharing, changes in destination, joint journeys, transition towards less polluting alternatives or less commuting—entail external efforts that might be challenging to adapt. Greater organisation, by supplying the travellers with a greater range of decisions and constant knowledge, may be needed to improve the beneficial outcomes. In the end, travellers' data will have the best travel time and energy benefits. These will affect the use of public transport or other alternatives or prolong their journey until the traffic is cleared [51]. Technological frameworks are emerging in the transport industry to fight the risks of climate change. These include developments in automobile engineering paired with fuel economy, impactful transport and simple accessibility networks, which can minimise private vehicles interest [51]. The mandate of noteworthy car miles is one of the better options for reducing GHG pollution from vehicles [51]. Smart transport networks with a smooth plan will have a huge impact on lowering emissions and fuel efficiency, smoothing and increasing speeds, decreasing travel, etc.

6 Food Wastes and Sustainability Issues

Food wasting is a concern worldwide and not limited to a single area, although the techniques used in food treatment which vary worldwide. The implementation of the sustainable development goals entails food waste as a big issue. In developed nations, in particular, in situations in which food waste management schemes are not as functional as in industrialised economies, food waste poses major environmental and health dangers. In terms of income, population development or public involvement in food waste management practises or projects between developing and developed countries, the discrepancies between food waste management systems may be related [61]. Disposal and discharge to sewers will accomplish food waste reduction. Food waste disposals can be used to distract food waste greatly from waste collection sites. While incineration dramatically reduces solid waste by 80–85% to deposits, a holistic sustainability approach is required for coping with fly ash and slag, incineration items. The behavioural approach to avoid food waste from touching waste deposits by means of home composting. It is necessary to prevent food waste. This is an example of how to use food waste to enrich and not destroy the environment. The best practises that take care of sustainable economic, environmental and social aspects are, however, accomplished by avoiding food being wasted or lost. The money used for planting, harvesting, refining, carriage and selling of foods is spared by this behavioural approach.

7 Sustainable Management of Natural Resources

Natural resources are not only protected and maintained for the non-renewable resources (such as fossil fuels) of the abiots, Crenna et al. [20] believes that the sustainable utilisation of biotic natural resources (renewable natural resources) similarly requires consideration. In order to successfully handle renewable energy, a change in order for their regeneration rates to be not smaller than their consumption rates has been proposed to a circular biotic economy. Economou and Mitoula [22] approach guidelines on natural gas are all-embracing for abiotic natural resources,those that encourage the use of natural gas instead of carbon-intensive fossil fuels encourage the use of trams in cities instead of individual cars, promote building automation in lieu of manual resourced energy-saving mode, encourage solar harvest for electricity production instead of fossil fuels.

8 Water Security

Water conservation is one of the best-discussed issues in the field of sustainability. The explanation is that it is the building block of almost all other recognised materials that is vital in existence. Ironically, water is one of the world's most available, but hard to reach, materials in most areas. The fresh water available is less than 1% of all water estimated to make up nearly 75% of the earth. In certain areas of water stress (e.g. parts of Australia and the Gulf countries), salt water is being reached and desalinated, however, at a high price both in cost and in sustainability. Any imaginable concept of sustainability has been extended to water; water shortage continues in most regions of the world from the reuse of water to recycling. As other materials, the use of circular economy to preserve the comparatively inaccessible water on the earth is essential. Application to wastewater by circular economy would produce deliverables; Ghosh and Mukherjee [27] figures worldwide of about 412 billion m3 (1 which is approximately 11% of the worldwide population) on the application of to wastewater by the circular economy in some underwater stressed regions (which include Ghana, California, North Asian Countries, Bangladesh, the United Arab Emirates and the Netherlands). For water sustainability, the storage and recycling of wastewater and other useful resources (e.g. electricity, fertiliser, nutrients, etc.) that can be reused on site for a number of urban, industrial and agriculture uses would need different technical solutions [37, 60, 66].

9 Future Prospects

The new surge of intelligent, progressive urban development has brought about significant shifts in urban and sustainable development. In the field smart sustainable cities, study and practise tend to be based-on the basis of broad and background data—on the identification of urban areas identified with sustainability aspects (such as transit, electricity, climate, mobility and connectivity, public and social services, as well as public safety). This encompasses how these spheres interrelate and influence one another in relation to unique physical entities and spatial organisations that are structured and coordinated [3, 9, 11]. In this respect, urban design (a community-specific way of life) has evolved in an equally integrated, coordinated and systematic manner of physical planning of the city and of its underlying infrastructural structures, procedures, functions and services for its management, planning and growth as a feature of sensed, stored, analysed, modelled, replicated and networked urban data [7–9, 13–14]. The definition and evolution of clever, sustainable cities thus require the thinking about and concept of urban landscapes as constellations of instruments on a multiple scale in order to provide continuous data from urban areas through the use of all-round systems for tracking, learning, processing and monitoring.

Therefore, the urban ICT that the modern computer wave enables is radically transforming the way towns can be built on a wide range of sizes and over a number of time periods [7]. This means that cities are making their attempts to reach the requisite degree of sustainability more intelligent. The technological aspects of smart sustainable urban planning include the use of land use practises, natural environments, physical arrangements, strategic organisations, natural capital, transport systems, socio-economic networks and municipal facilities of advanced ICTs as a collection of scientific and technological processes [3, 7, 12–13, 40, 43, 53]. Current research argues that integrating these urban growth strands with ICT will make these cities more sustainable and therefore more attractive and attractive. Smart planning is important for the accomplishment of the long-term objectives of urban growth and is a crucial strategic sustainable development. Furthermore, the functioning, administration and organisation for sustainable urban planning structures, procedures and practises involve diverse interdisciplinary knowledge of sustainability and high technology as well as high-performance computing and data processing capability.

10 Conclusion

The design of sustainable cities needs several initiatives, as illustrated in the preceding sections; new and refurbished sustainable cities must establish new norms in future cities, ranging from human capital growth and the information economy to the conservation of the environment. As the examined papers revealed, many elements are needed to create sustainable cities, but the main emphasis is on the threefold approach to sustainability (i.e. environment, economics and equity). Each aspect is linked with these principles of sustainability either directly or indirectly. There are actually over 400 major cities and 23 big cities around the world, with a population of between one million and 10 million. Cities cover nearly 2% of the earth's atmosphere, while 60–80% of global energy is absorbed. The global metropolitan population ranged from 220 million to about 2.8 billion in the twentieth century, with an expansion to approximately 70% of the world's population of 6.9 billion by 2050. There are, then, many variables correlated with communities: large population, intellectual and social dignity centre, high demand, etc., which may have an effect on sustainable initiatives as they were aimed at cities. A modern trend is moving for the circular economy to rule and operate future towns sustainably. Close energy and content loops, decreased capital intake, pollution, waste and energy leakages can be added to cities' economic values. In the end, the potential of our future cities to imitate the carbon and water cycles of nature is critical for regeneration and reuse, rather than the present mitigation conditions that prioritise recycling.

References

1. Abdmouleh, Z., Alammari, R.A., Gastli, A.: Review of policies encouraging renewable energy integration and best practices. Renew. Sustain. Energ. Rev. **45**, 249–262 (2015)
2. Aguirre, M., Ibikunle, G.: Determinants of renewable energy growth: a global sample analysis. Energ. Pol. **69**, 374–384 (2014)
3. Al Nuaimi, E., Al Neyadi, H., Mohamed, N., Al-Jaroodi, J.: Applications of big data to smart cities. J. Internet Serv. Appl. **6**(1), 25 (2015)
4. Aquilani, B., Silvestri, C., Ioppolo, G., Ruggieri, A.: The challenging transition to bio-economies: towards a new framework integrating corporate sustainability and value co-creation. J. Clean. Prod. **172**, 4001–4009 (2018)
5. Ayres, R.U., Turton, H., Casten, T.: Energy efficiency, sustainability and economic growth. Energy **32**(5), 634–648 (2007)
6. Baccini, P.: Understanding regional metabolism for a sustainable development of urban systems. Environ. Sci. Pollut. Res. **3**(2), 108–111 (1996)
7. Batty, M.: Big data, smart cities and city planning. Dialogues Human Geogr. **3**(3), 274–279 (2013)

8. Batty, M.: Urban informatics and big data. ESRC (The UK Economic and Social Research Council). Cities Expert Group (2013b)
9. Batty, M., Axhausen, K.W., Giannotti, F., Pozdnoukhov, A., Bazzani, A., Wachowicz, M., Portugali, Y., et al.: Smart cities of the future. Eur. Phys. J. Special Top. **214**(1), 481–518 (2012)
10. Berardi, U.: Clarifying the new interpretations of the concept of sustainable building. Sustain. Urban Areas **8**, 72–78 (2013)
11. Bibri, S.E.: Big data analytics and context-aware computing: core enabling technologies, techniques, processes, and systems. In: Smart Sustainable Cities of the Future, pp. 133–188. Springer, Cham (2018)
12. Bibri, S.E., Krogstie, J.: On the social shaping dimensions of smart sustainable cities: a study in science, technology, and society. Sustain. Urban Areas **29**, 219–246 (2017)
13. Bibri, S.E., Krogstie, J.: Smart sustainable cities of the future: an extensive interdisciplinary literature review. Sustain. Urban Areas **31**, 183–212 (2017)
14. Böhlen, M., Frei, H.: Ambient intelligence in the city overview and new perspectives. In: Handbook of Ambient Intelligence and Smart Environments, pp. 911–938. Springer, Boston, MA (2010)
15. Boyle, C., Mudd, G., Mihelcic, J.R., Anastas, P., Collins, T., Culligan, P., Edwards, M., Gabe, J., Gallagher, P., Handy, S., Kao, J.J.: Delivering sustainable infrastructure that supports the urban built environment (2010)
16. Byrne, J., Taminiau, J., Kurdgelashvili, L., Kim, K.N.: A review of the solar city concept and methods to assess rooftop solar electric potential, with an illustrative application to the city of Seoul. Renew. Sustain. Energ. Rev. **41**, 830–844 (2015)
17. Carrión, J.A., Estrella, A.E., Dols, F.A., Toro, M.Z., Rodríguez, M., Ridao, A.R.: Environmental decision-support systems for evaluating the carrying capacity of land areas: optimal site selection for grid-connected photovoltaic power plants. Renew. Sustain. Energ. Rev. **12**(9), 2358–2380 (2008)
18. Chenari, B., Carrilho, J.D., da Silva, M.G.: Towards sustainable, energy-efficient and healthy ventilation strategies in buildings: a review. Renew. Sustain. Energ. Rev. **59**, 1426–1447 (2016)
19. Cole, R.J.: Changing context for environmental knowledge. Build. Res. Inform. **32**(2), 91–109 (2004)
20. Crenna, E., Sozzo, S., Sala, S.: Natural biotic resources in LCA: towards an impact assessment model for sustainable supply chain management. J. Clean. Prod. **172**, 3669–3684 (2018)
21. Denholm, P., Mai, T.: Timescales of energy storage needed for reducing renewable energy curtailment. Renew. Energ. **130**, 388–399 (2019)
22. Economou, A., Mitoula, R.: Management of natural resources and protection of the coastal urban area of Glyfada. Land Use Policy **35**, 204–212 (2013)
23. Engel-Yan, J., Kennedy, C., Saiz, S., Pressnail, K.: Toward sustainable neighbourhoods: the need to consider infrastructure interactions. Can. J. Civ. Eng. **32**(1), 45–57 (2005)
24. Fedra, K.: Sustainable urban transportation: a model-based approach. Cybern. Syst. Int. J. **35**(5–6), 455–485 (2004)
25. Fowke, R., Prasad, D.K.: Sustainable development, cities and local government: dilemmas and definitions. Aus. Planner **33**(2), 61–66 (1996)
26. Gauthier, C., Gilomen, B.: Business models for sustainability: energy efficiency in urban districts. Organ. Environ. **29**(1), 124–144 (2016)
27. Ghosh, S.K., Mukherjee, T.: Circular economy through treatment and management of industrial wastewater. In: Waste Water Recycling and Management, pp. 1–13. Springer, Singapore (2019)
28. Goldman, T., Gorham, R.: Sustainable urban transport: four innovative directions. Technol. Soc. **28**(1–2), 261–273 (2006)
29. Haas, R., Panzer, C., Resch, G., Ragwitz, M., Reece, G., Held, A.: A historical review of promotion strategies for electricity from renewable energy sources in EU countries. Renew. Sustain. Energ. Rev. **15**(2), 1003–1034 (2011)
30. Hammond, G.P.: Towards sustainability: energy efficiency, thermodynamic analysis, and the 'two cultures.' Energ. Pol. **32**(16), 1789–1798 (2004)

31. Held, A., Ragwitz, M., Haas, R.: On the success of policy strategies for the promotion of electricity from renewable energy sources in the EU. Energ. Environ. **17**(6), 849–868 (2006)
32. Hossain, F.M., Hasanuzzaman, M., Rahim, N.A., Ping, H.W.: Impact of renewable energy on rural electrification in Malaysia: a review. Clean Technol. Environ. Pol. **17**(4), 859–871 (2015)
33. Hua, Y., Oliphant, M., Hu, E.J.: Development of renewable energy in Australia and China: a comparison of policies and status. Renew. Energ. **85**, 1044–1051 (2016)
34. Hwang, B.G., Zhao, X., See, Y.L., Zhong, Y.: Addressing risks in green retrofit projects: the case of Singapore. Proj. Manag. J. **46**(4), 76–89 (2015)
35. Jamil, M., Ahmad, F., Jeon, Y.J.: Renewable energy technologies adopted by the UAE: prospects and challenges—a comprehensive overview. Renew. Sustain. Energ. Rev. **55**, 1181–1194 (2016)
36. Junnila, S., Horvath, A., Guggemos, A.A.: Life-cycle assessment of office buildings in Europe and the United States. J. Infrastruct. Syst. **12**(1), 10–17 (2006)
37. Kefeni, K.K., Msagati, T.A., Mamba, B.B.: Acid mine drainage: prevention, treatment options, and resource recovery: a review. J. Clean. Prod. **151**, 475–493 (2017)
38. Kennedy, C., Miller, E., Shalaby, A., Maclean, H., Coleman, J.: The four pillars of sustainable urban transportation. Transp. Rev. **25**(4), 393–414 (2005)
39. Kilinc-Ata, N.: The evaluation of renewable energy policies across EU countries and US states: an econometric approach. Energ. Sustain. Dev. **31**, 83–90 (2016)
40. Kramers, A., Höjer, M., Lövehagen, N., Wangel, J.: Smart sustainable cities—exploring ICT solutions for reduced energy use in cities. Environ. Model. Softw. **56**, 52–62 (2014)
41. Lo, K.: A critical review of China's rapidly developing renewable energy and energy efficiency policies. Renew. Sustain. Energ. Rev. **29**, 508–516 (2014)
42. May, A.D.: Urban transport and sustainability: the key challenges. Int. J. Sustain. Transp. **7**(3), 170–185 (2013)
43. Neirotti, P., De Marco, A., Cagliano, A.C., Mangano, G., Scorrano, F.: Current trends in smart city initiatives: some stylised facts. Cities **38**, 25–36 (2014)
44. O'Rielly, K., Jeswiet, J.: Strategies to improve industrial energy efficiency. Procedia Cirp **15**, 325–330 (2014)
45. OECD: OECD environmental outlook to 2050. The consequences of inaction. OECD Publishing http://www.naturvardsverket.se/upload/miljoarbete-i-samhallet/internationelltmiljoarbete/multilateralt/oecd/outolook-2050-oecd.pdf. Accessed 21.12.2020
46. Petersen, J.P.: Energy concepts for self-supplying communities based on local and renewable energy sources: a case study from northern Germany. Sustain. Urban Areas **26**, 1–8 (2016)
47. Ravindu, S., Rameezdeen, R., Zuo, J., Zhou, Z., Chandratilake, R.: Indoor environment quality of green buildings: case study of an LEED platinum certified factory in a warm humid tropical climate. Build. Environ. **84**, 105–113 (2015)
48. Ribeiro, A.E.D., Arouca, M.C., Coelho, D.M.: Electric energy generation from small-scale solar and wind power in Brazil: the influence of location, area and shape. Renew. Energ. **85**, 554–563 (2016)
49. Roetzel, A., Tsangrassoulis, A., Dietrich, U.: Impact of building design and occupancy on office comfort and energy performance in different climates. Build. Environ. **71**, 165–175 (2014)
50. Samaras, Z., Vouitsis, I.: Transportation and Energy, pp. 183–205. Academic Press, Oxford (2013)
51. Shaheen, S.A., Lipman, T.E.: Reducing greenhouse emissions and fuel consumption: sustainable approaches for surface transportation. IATSS Res. **31**(1), 6–20 (2007)
52. Shahidehpour, M., Clair, J.F.: A functional microgrid for enhancing reliability, sustainability, and energy efficiency. Electr. J. **25**(8), 21–28 (2012)
53. Shahrokni, H., Årman, L., Lazarevic, D., Nilsson, A., Brandt, N.: Implementing smart urban metabolism in the Stockholm Royal Seaport: smart city SRS. J. Ind. Ecol. **19**(5), 917–929 (2015)
54. Smith, M.G., Urpelainen, J.: The effect of feed-in tariffs on renewable electricity generation: an instrumental variables approach. Environ. Resour. Econ. **57**(3), 367–392 (2014)
55. Sodiq, A., Baloch, A.A., Khan, S.A., Sezer, N., Mahmoud, S., Jama, M., Abdelaal, A.: Towards modern sustainable cities: review of sustainability principles and trends. J. Clean. Prod. **227**, 972–1001 (2019)

56. Steg, L.: Sustainable transportation: A psychological perspective. IATSS Res. **31**(2) (2007)
57. Steg, L., Gifford, R.: Sustainable transport and quality of life. In: Building Blocks for Sustainable Transport: Obstacles, Trends. Solutions. Emerald Group Publishing Limited (2007)
58. Szopik-Depczyńska, K., Cheba, K., Bąk, I., Stajniak, M., Simboli, A., Ioppolo, G.: The study of relationship in a hierarchical structure of EU sustainable development indicators. Ecol. Ind. **90**, 120–131 (2018)
59. Szopik-Depczyńska, K., Kędzierska-Szczepaniak, A., Szczepaniak, K., Cheba, K., Gajda, W., Ioppolo, G.: Innovation in sustainable development: an investigation of the EU context using 2030 agenda indicators. Land Use Pol. **79**, 251–262 (2018)
60. Theregowda, R.B., Vidic, R., Landis, A.E., Dzombak, D.A., Matthews, H.S.: Integrating external costs with life cycle costs of emissions from tertiary treatment of municipal wastewater for reuse in cooling systems. J. Clean. Prod. **112**, 4733–4740 (2016)
61. Thi, N.B.D., Kumar, G., Lin, C.Y.: An overview of food waste management in developing countries: current status and future perspective. J. Environ. Manag. **157**, 220–229 (2015)
62. Todorović, M.S.: BPS, energy efficiency and renewable energy sources for buildings greening and zero energy cities planning: harmony and ethics of sustainability. Energ. Build. **48**, 180–189 (2012)
63. Uddin, M., Rahman, A.A.: Energy efficiency and low carbon enabler green IT framework for data centers considering green metrics. Renew. Sustain. Energ. Rev. **16**(6), 4078–4094 (2012)
64. United Nations: World urbanization prospects. The 2014 revision. Department of Economic and Social Affairs, New York (2015). http://esa.un.org/unpd/wup/Publications/Files/WUP2014-Report.pdf. Accessed 22.1.2020
65. Viholainen, J., Luoranen, M., Väisänen, S., Niskanen, A., Horttanainen, M., Soukka, R.: Regional level approach for increasing energy efficiency. Appl. Energ. **163**, 295–303 (2016)
66. Wilcox, J., Nasiri, F., Bell, S., Rahaman, M.S.: Urban water reuse: a triple bottom line assessment framework and review. Sustain. Urban Areas **27**, 448–456 (2016)
67. Worrell, E., Bernstein, L., Roy, J., Price, L., Harnisch, J.: Industrial energy efficiency and climate change mitigation. Energ. Eff. **2**(2), 109 (2009)

Tire Derived Aggregate as a Sustainable Technique to Mitigate Transient Seismic Effect on Buried Concrete Pipes

Saif Alzabeebee

1 Introduction

Seismic waves propagation and permanent ground movement have been identified as the main reasons for buried pipelines failure during earthquakes [1]. Hence, there has been growing interest in the latest decade to enhance the understanding of the effect of earthquakes on the response of buried pipes, where previous studies examined the effect of waves propagation and faults movement induced due to seismic effect on the response of buried pipes [2–7]. These studies illustrated the significant impact of the seismic events on the behaviour of pipelines and showed how the pipe behaves when subjected to either transient seismic shake (propagating waves) or permanent ground movement induced by seismic effect. Consequently, these studies recommended to consider the effect of the earthquake in the design philosophy of buried pipes. However, this will remarkably increase the cost of buried pipes projects considering that the pipes are usually extended over many kilometres to transport water, wastewater, and oil. Hence, some studies suggested to use tire derived aggregate (TDA) as a technique to protect pipes against faults movement induced by seismic effect [1, 8]. The use of TDA has been proposed in these studies as this material is a cheap alternative, and it is also considered as a green material as it helps to reduce the problems associated with the discarding of used tires. Ni et al. [1] utilized finite element analysis to study the response of steel and high-density polyethylene pipes crossing faults, where the study simulated strike-slip fault movement induced due to seismic effect. The TDA is considered as a material surrounding the whole pipe. They found that the TDA enhances the structural performance of the pipes compared with loose sand or expanded polystyrene. However, this study did not consider the effect of the transient seismic shake associated with earthquakes. Demirci [8] also

S. Alzabeebee (✉)
Department of Roads and Transport Engineering, College of Engineering, University of Al-Qadisiyah, Al-Qadisiyah, Iraq
e-mail: Saif.Alzabeebee@qu.edu.iq

I. Pal and S. Kolathayar (eds.), *Sustainable Cities and Resilience*, Lecture Notes in Civil Engineering 183, https://doi.org/10.1007/978-981-16-5543-2_26

examined the use of TDA as a mitigation technique to reduce additional stresses applied on pipes crossing slip faults using experimental models and finite element analysis. He noticed that the TDA reduces bending stresses and axial strains induced due to fault movement. Demirci [8] also did not consider the effect of transient seismic shake in his experiments and numerical analyses.

Based on the discussed review, it is obvious that the TDA is a promising technique to mitigate earthquake effect. However, previous studies did not study the performance of the TDA as a mitigation technique for buried pipes subjected to transient seismic shake, although transient seismic shake was found to remarkably increase the stresses developed in the pipe wall [2, 3]. Hence, to fill this gap, this study investigates the efficiency of using tier derived aggregate (TDA) as a material supporting buried concrete pipe subjected to transient seismic shake. This is an important first step to develop techniques that reduce the impact of the transient seismic event on the behaviour of buried concrete pipes.

2 Cases Considered in This Study

A buried pipe with an external diameter of 0.8 m, a thickness of 0.1 m, and buried with a backfill height of 2.0 m has been modelled in this study. Two scenarios of pipe support have been modelled to address the aim of the study; these are shown in Fig. 1. It is clear from the figure that for scenario 1, the dense sand is employed as a soil surrounding the pipe at the haunch zone and bedding zone and also as a backfill soil. Scenario 1 has been modelled in compliance with the ASSHTO Type 2 installation requirements [2]. For scenario 2, the TDA is considered as a bedding material and also as a material supporting the pipe, while the dense sand is considered as a backfill soil. The use of TDA as a bedding and supporting material of the pipe also ensures good haunch support and fulfils the AASHTO requirements for Type 2 installation conditions based on the mechanism of pressure transfer around the pipe wall [9]. It

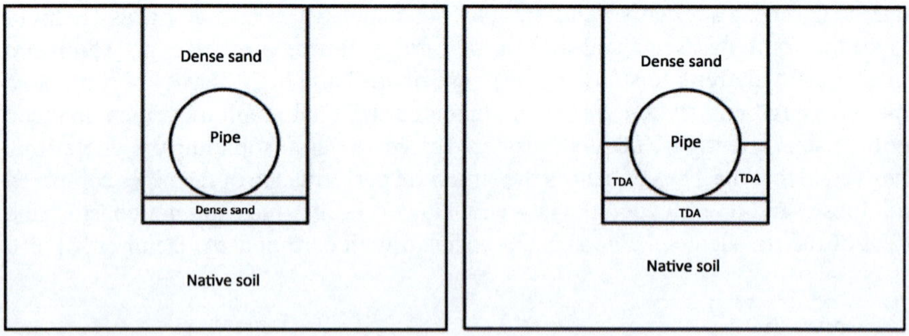

a) Scenario 1 b) Scenario 2

Fig. 1 Bedding and support conditions of the cases considered in this study is also necessary to state that dense sand is considered as a backfill soil above the haunch zone for both scenarios to allow a direct comparison between the two cases by ensuring that the pipe is carrying the same backfill weight on top of it.

3 Earthquakes Employed in This Study

Five earthquake records have been collected from previous studies and used in this research to ensure the consideration of different predominant frequencies. These records are the Parkfield earthquake 1966, Loma Prieta earthquake 1989, Upland earthquake 1990, Northridge earthquake 1994 and Hyogo-ken (Kobe) earthquake 1995. These records have been obtained from different references [10–13]. These records have been amended to ensure all the records have the same peak ground acceleration (PGA). The peak ground acceleration for all records is considered equal to 0.24 g, which is the PGA of the Upland earthquake. The amendment of each record has been done by scaling by the ratio of 0.24 g divided by the original PGA of each record. This has been done to enable easier comparison to unambiguously illustrate the impact of the predominant frequency. Same approach has also been considered in previous studies [3, 14, 15]. Figure 2 shows the scaled earthquake records.

In addition, Fourier transformation analysis, with the aid of SeismoSignal, has been conducted using the scaled records to obtain the predominant frequency. Based on the results (which have not been presented here for sake of briefing), it can be said that the predominant frequency of these records ranges between 0.66 and 2.90 Hz. Table 1 summarizes the predominant frequency for each record.

4 Constitutive Model and Soil Parameters

The hardening soil constitutive model is employed in this study to model the response of the soil and the TDA. The model is developed by Schanz et al. [16] based on the famous Duncan-Chang hyperbolic soils [17]. The model is able to simulate the impact of the applied load and the confining pressure on the stiffness of the soil, non-linearity of the soil stiffness in the elastic range and also the hysteresis damping. It is worthy to state that the model does not simulate the stiffness degradation and small strain stiffness like the more advanced hardening soil model with small strain stiffness. However, initial analyses comparing the hardening soil model and the hardening soil model with small strain stiffness showed that both models produce very close results for the shear force and bending moment developed in the pipe wall, where the percentage difference between both models ranged between 4 and 10%. Thus, the hardening soil model is considered a robust choice for this problem.

As it is clear from Fig. 1, three materials have been considered in the analyses: these are the natural soil, the backfill soil and the tire derived aggregate (TDA). The backfill soil is considered as a dense sandy soil with a relative density of 70%.

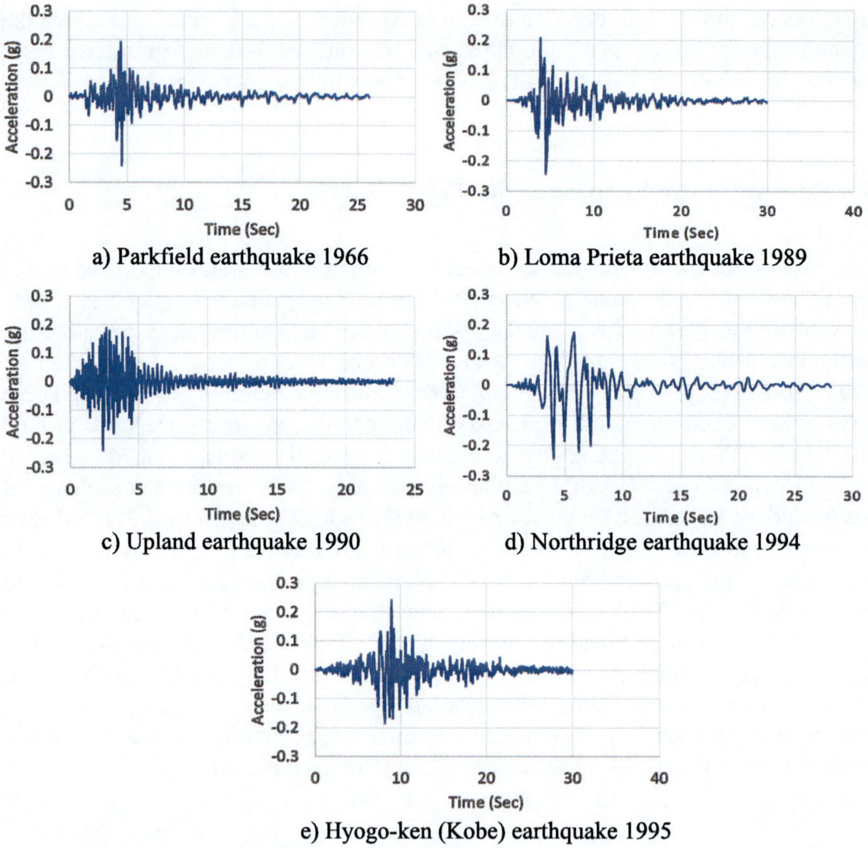

Fig. 2 Earthquake records used in this study

Table 1 Predominant frequency of earthquake records used in the study

Earthquake record	Predominant frequency (Hz)
Parkfield earthquake 1966	2.10
Loma Prieta earthquake 1989	0.66
Upland earthquake 1990	2.90
Northridge earthquake 1994	0.93
Hyogo-ken (Kobe) earthquake 1995	1.50

The properties of this soil are adopted from Al-Defae et al. [18]. The natural soil is considered to have properties similar to the backfill soil, but with a cohesion of 45 kPa to avoid numerical convergence problems associated with excavating and filling the trench. Also, the TDA properties are adopted from Mahgoub and El Naggar [19],

Table 2 Hardening soil model properties used in the analyses

Parameter	Natural soil	Dense sand	TDA
γ(kN/m^3)	16.60	16.60	7.00
E_{50}^{ref}(kPa)	47,150	47,150	2750
E_{oed}^{ref}(kPa)	37,720	37,720	2200
E_{ur}^{ref}(kPa)	113,160	113,160	8250
ϕ'(°)	43.0	43.0	26.5
ψ'(°)	13.5	13.5	0.0
c(kPa)	45.0	0.1	24.0
m	0.53	0.53	0.54
R_f	0.90	0.90	0.95
υ_{ur}	0.2	0.2	0.2
p^{ref}	100	100	25

who calibrated these properties using results of triaxial tests combined with three-dimensional finite element analysis and further proved the validity of the properties by numerically modelling a case study of steel pipe subjected to traffic load with a TDA layer placed on top of the crown of the pipe. It is worthy to add that the size of the TDA particles used was in the range of 13–63 mm [20]. Table 2 illustrates the adopted parameters for the natural soil, dense sand and TDA.

5 Details of the Numerical Model

The scenarios used in this study have been modelled using Plaxis 2D. This is because the problem involves only soil load and transient seismic shake, and these can be accurately simulated using the two-dimensional analysis. The models developed in this research are based on robust models developed by the author in previous research studies [2, 3, 15]. The model width is considered equal to 20 m and the model depth is considered equal to 20 m. Solid elements have been employed to model the soil and the TDA, while plate elements have been used to model the buried pipe. 15 nodes triangular elements have been used to model the soil and the TDA, while 5 nodes plate elements have been used to model the pipe. In addition, very fine mesh has been used in the area surrounding the pipe (i.e., the trench area) based on sensitivity analyses conducted by the author in previous research studies [2, 3]. Also, the elements sizes comply with the wave propagation requirements [21]. The pipe has been modelled using liner elastic constitutive model with a modulus of elasticity of and Poisson's ratio of 25,000 MPa and 0.15, respectively.

The boundary conditions used in the analyses are based on Plaxis recommendation, where the model is allowed to move only in the vertical direction at the sides and fixed in the vertical and horizontal directions at the bottom. In addition, Plaxis absorbent boundaries are employed in the dynamic analysis to

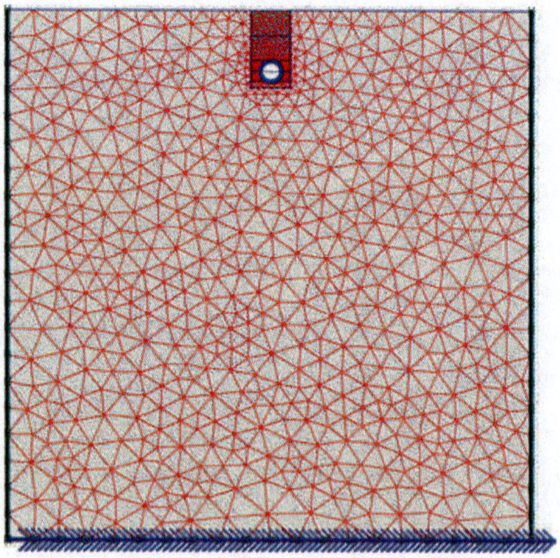

Fig. 3 Finite element mesh of the problem

avoid issues related to wave reflection. Also, the earthquake is modelled as a prescribed acceleration that is applied at the bottom of the model in a technique similar to previous studies in the literature [2, 3, 22–26]. The stages of analysis involved calculation of in-situ stress using at rest earth pressure coefficient (Stage 1). Excavating the trench in sub-steps (Stage 2), placing the bedding layer (Stage 3), placing the pipe (Stage 4), placing the soil surrounding the pipe in the haunch zone (Stage 5), placing the soil surrounding the pipe in the shoulder zone (Stage 6), adding the backfill on top of the pipe in sub-steps (Stage 7), and applying the earthquake shake using the time history analysis (Stage 8). The time history analysis is performed with a time step of 0.004 s, calculated based on the element size and also considering the Courant-Friedrichs-Lewy condition [21]. Figure 3 illustrates the finite element mesh of the problem.

6 Results and Discussion

Figure 4 compares the shear forces developed in the pipe wall for scenario 1 (referred to as soil in the figure) and scenario 2 (referred to as TDA in the figure) for static loading (after completing the trench backfilling) and for transient seismic shake. Figure 4 shows that using TDA as a support material produces similar distribution of the shear forces around the pipe compared to the traditional installation condition (i.e., using dense sand) for all the cases modelled in this paper. In addition, it is clear from the figure that the use of the TDA reduces the shear forces developed in the pipe wall. The maximum absolute shear force has been obtained from each case and the percentage decrease has been calculated, where it was found that the percentage decrease is equal to 37% for the static condition, while it ranges between 30 and 43% for the transient seismic cases depending on the frequency of the earthquake.

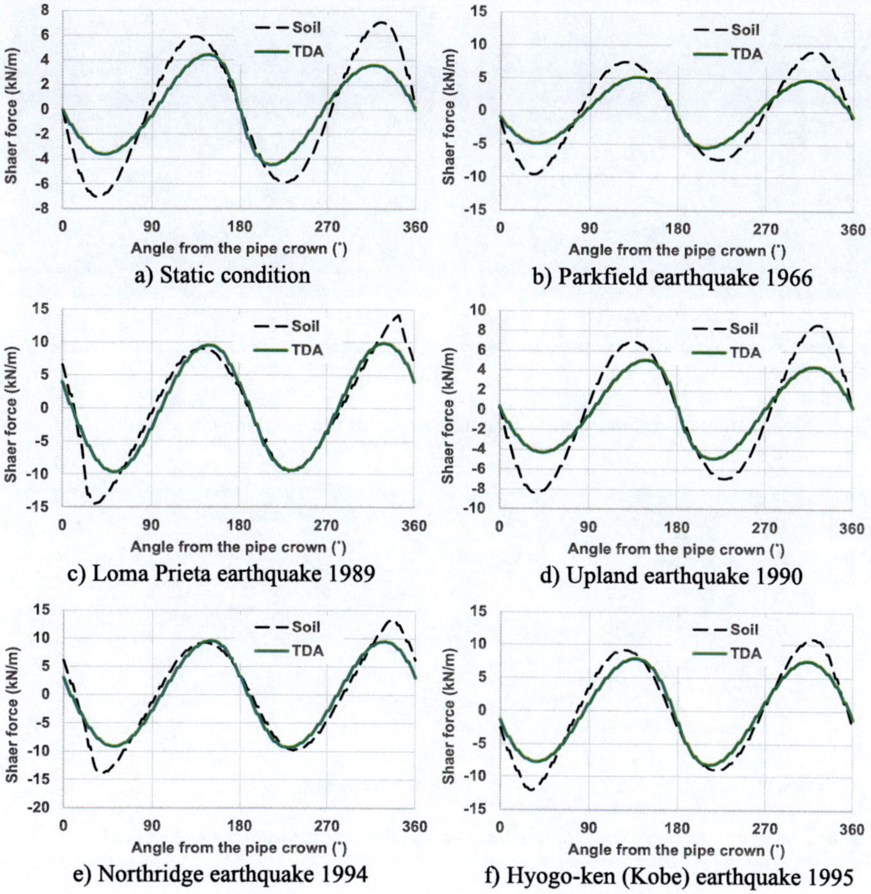

Fig. 4 Shear forces developed around the pipe for static and seismic conditions

Similar observations are also noted for the developed bending moment (figures have not been included due to page limitations). However, the percentage decrease of the maximum bending moment is slightly less compared with the reduction of the shear force, where it is equal to 33% for the static case and ranges between 22 and 38% for the seismic case. The reduction of the bending moment and shear forces is due to the decrease of the soil pressure applied on the pipe when using the TDA material as shown in the contour lines presented in Fig. 5. This is actually due to soil arching effect, where reducing the stiffness of the material supporting the pipe reduces the overall stiffness of the soil-pipe system and hence leads to shading the soil pressure away from the pipe crown to the trench as can be clearly noticed in the principal direction of the soil pressure around the pipe presented in Fig. 6.

To better understand the effect of trainset seismic shake for both scenario 1 and 2, the maximum absolute shear force and maximum absolute bending moment have been plotted against the predominant frequency of each earthquake. Figure 7 displays

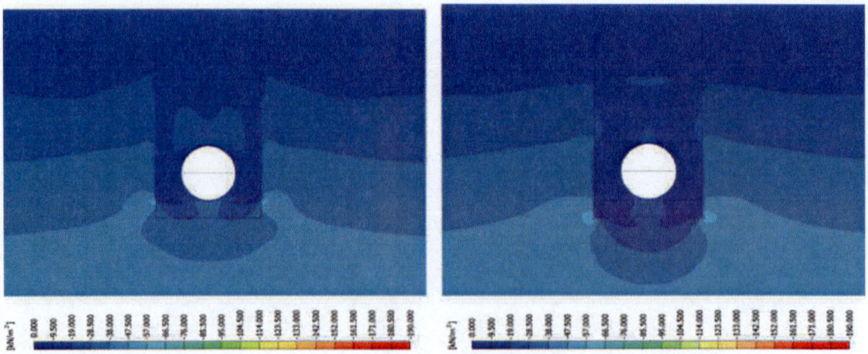

a) Scenario 1 b) Scenario 2

Fig. 5 Mean soil pressure around the pipe after completing the backfilling process

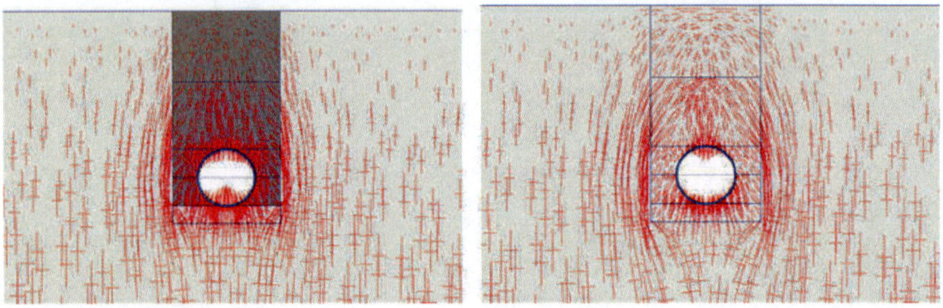

a) Scenario 1 b) Scenario 2

Fig. 6 Principal direction of the soil pressure around the pipe showing the soil arching

Fig. 7 Effect of the predominant frequency on the maximum shear force

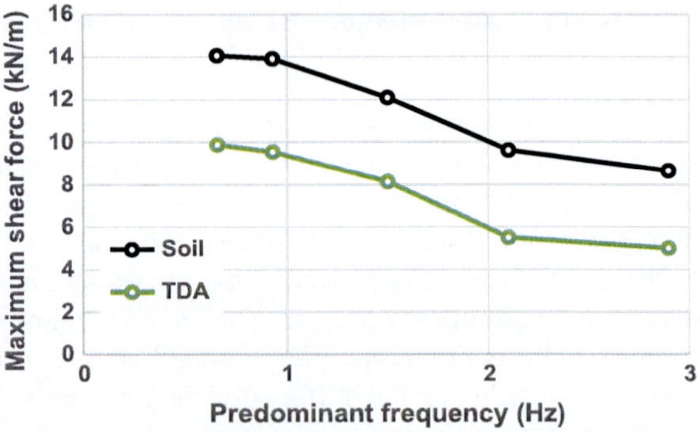

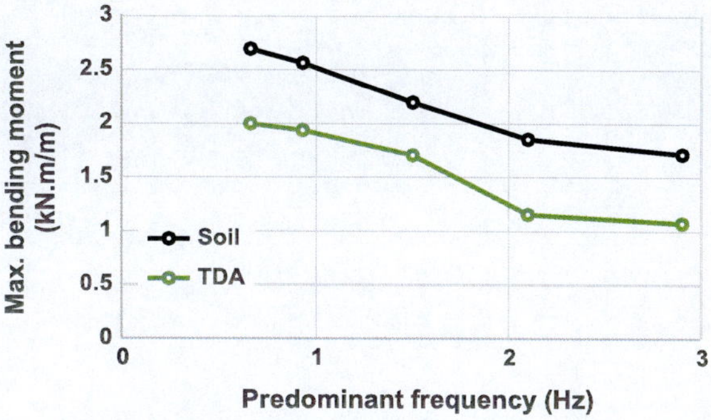

Fig. 8 Effect of the predominant frequency on the maximum bending moment

the relationship between the maximum absolute shear force and the predominant frequency for both scenarios. Furthermore, Fig. 8 displays the relationship between the maximum absolute bending moment and the predominant frequency for both scenarios. The figures clearly show the efficiency of the TDA in reducing the maximum shear force and bending moment regardless of the predominant frequency of the transient seismic shake. In addition, it is also clear from both figures that increasing the predominant frequency decreases the shear force and bending moment developed in the pipe wall. This can be justified by the densification of the soil surrounding the pipe as the predominant frequency increases, which ultimately improves the pipe support, leading eventually to decrease the soil arching [26, 27]. It is worthy to add that this finding is similar to a previous observation noted by the author for the case of flexible pipes, where increasing the predominant frequency was found to reduce the thrust and bending moment for uPVC pipes buried in sandy soil (more details on this can be found in [3]).

To understand the effectiveness of using the TDA for different peak ground accelerations, the Parkfield earthquake 1966 record (presented in Fig. 2) has been scaled to produce two additional seismic records with a PGA of 0.15 and 0.50 g. These records have been used in further analyses for both scenarios and the maximum absolute shear force and the maximum absolute bending moment have been obtained from these analyses. Figures 9 and 10 illustrate the obtained maximum shear force and bending moment as a function of the peak ground acceleration, respectively. These figures again show the effectiveness of the TDA compared with the normal installation for all of the PGA range utilized in the analyses. The percentage decrease of the maximum absolute shear force ranges between 32 and 43%, while the same percentage for the bending moment ranges between 23 and 39%. Based on these thorough analyses and discussions, it is obvious that using TDA even in a very limited area around the pipe (bedding layer and haunch zone) remarkably decreases the shear forces and bending moments induced in the pipe wall for both static and seismic conditions. It is also clear that the TDA is effective regardless of the predominant frequency or the PGA of the earthquake. Hence, the TDA can be considered as a promising mitigation technique that reduces the additional forces applied on the pipe due to transient seismic shake.

Fig. 9 Effect of the PGA on the maximum shear force

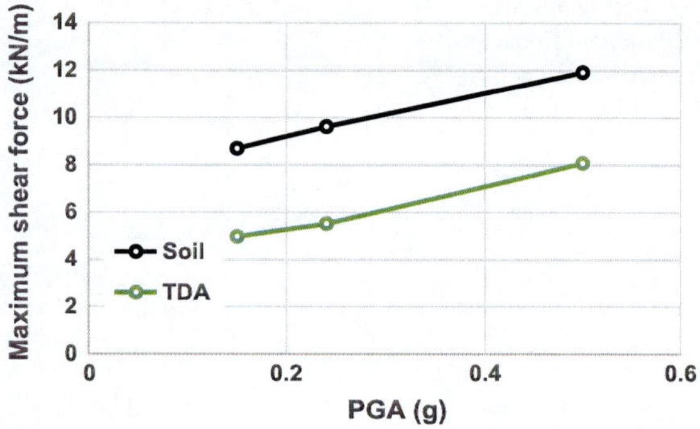

Fig. 10 Effect of the PGA on the maximum bending moment

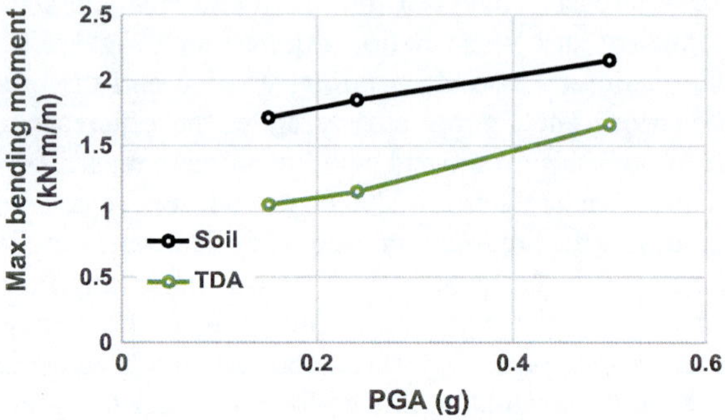

7 Conclusions

A robust two-dimensional finite element model has been developed to study the efficiency of tire derived aggregate (TDA) as a mitigation technique to reduce the additional forces applied on the pipe due to transient seismic shake. In addition, calibrated parameters have been employed in the analyses utilizing advanced soil constitutive model. Two installation scenarios have been modelled to compare the response. Scenario 1 simulated the tradition installation condition, in which the pipe is well supported at the haunch and invert zones by dense soil, while scenario 2 considered using TDA as a material supporting the pipe. The type of backfill soil on top of the haunch zone has been kept constant for both scenarios to allow straightforward judgement of the findings from both support conditions. Furthermore, wide range of transient seismic shakes has been considered by using earthquake records with a predominant frequency range of 0.63–2.90 Hz and a peak ground acceleration of 0.15–0.50 g. The results of the analyses illustrated the effectiveness of using the TDA as sustainable solution to reduce the shear forces and bending moments developed in the pipe wall due to earthquake effect. The percentage reduction of the maximum shear force is found to be equal to 37% for static case, while this percentage depends on the predominant frequency and the peak ground acceleration for the transient seismic cases and it ranges between 33 and 43%. Similarly, the percentage reduction of the maximum bending moment is found to be equal to 33% for the static case and it ranges between 22 and 38% for the transient seismic cases. In summary, the study illustrated the effectiveness of the TDA, and hence, it is hoped that the findings of this study will encourage practitioners to implement sustainable solutions in routine designs of buried pipes.

References

1. Ni, P., Qin, X., Yi, Y.: Use of tire-derived aggregate for seismic mitigation of buried pipelines under strike-slip faults. Soil Dyn. Earthq. Eng. **115**, 495–506 (2018)
2. Alzabeebee, S.: Seismic response and design of buried concrete pipes subjected to soil loads. Tunnel. Underground Space Technol. 93, 103084 (2019a)
3. Alzabeebee, S.: Response of buried uPVC pipes subjected to earthquake shake. Innov. Infrast. Solut. **4**(1), 52 (2019)
4. Xie, X., Symans, M.D., O'Rourke, M.J., Abdoun, T.H., O'Rourke, T.D., Palmer, M.C., Stewart, H.E.: Numerical modeling of buried HDPE pipelines subjected to strike-slip faulting. J. Earthquake Eng. **15**(8), 1273–1296 (2011)
5. Sim, W.W., Towhata, I., Yamada, S.: One-g shaking-table experiments on buried pipelines crossing a strike-slip fault. Geotechnique **62**(12), 1067–1079 (2012)
6. Sim, W.W., Towhata, I., Yamada, S., Moinet, G.M.: Shaking table tests modelling small diameter pipes crossing a vertical fault. Soil Dyn. Earthq. Eng. **35**, 59–71 (2012)
7. Demirci, H.E., Bhattacharya, S., Karamitros, D., Alexander, N.: Experimental and numerical modelling of buried pipelines crossing reverse faults. Soil Dyn. Earthq. Eng. **114**, 198–214 (2018)
8. Demirci, H.E.: Experimental and numerical modelling of buried continuous pipelines Crossing Active Faults (Doctoral dissertation, University of Surrey) (2019)
9. Alzabeebee, S., Chapman, D.N., Faramarzi, A.: Economical design of buried concrete pipes subjected to UK standard traffic loading. Proc. Instit. Civil Eng.-Struct. Build. **172**(2), 141–156 (2019)
10. Brinkgreve, R.B.J., Broere, W., Waterman, D.: Plaxis, finite element code for soil and rock analyses user's manual, The Netherlands (2006)
11. Bakr, J.A.: Displacement-based approach for seismic stability of retaining structures. PhD Thesis, The University of Manchester, United Kingdom (2018)
12. Xu, R., Fatahi, B.: Novel application of geosynthetics to reduce residual drifts of mid-rise buildings after earthquakes. Soil Dyn. Earthq. Eng. **116**, 331–344 (2019)
13. Midas, 2019. Manual of the Midas GTS/NX
14. Bakr, J., Ahmad, S.M.: A finite element performance-based approach to correlate movement of a rigid retaining wall with seismic earth pressure. Soil Dyn. Earthq. Eng. **114**, 460–479 (2018)
15. Alzabeebee, S.: Seismic settlement of a strip foundation resting on a dry sand. Natural Hazards: J. Int. Soc. Prevent. Mitigation Natural Hazards **103**, 2395–2425 (2020)
16. Schanz, T., Vermeer, P.A., Bonnier, P.G.: The hardening soil model: formulation and verification. Beyond 2000 in computational geotechnics, pp. 281–296 (1999)
17. Duncan, J.M., Chang, C.: Nonlinear analysis of stress and strain in soils. J. Soil Mech. Found. Division ASCE **96**(5), 1629–1653 (1970)
18. Al-Defae, A.H., Caucis, K., Knappett, J.A.: Aftershocks and the whole-life seismic performance of granular slopes. Géotechnique **63**(14), 1230–1244 (2013)
19. Mahgoub, A., El Naggar, H.: Using TDA as an engineered stress-reduction fill over preexisting buried pipes. J. Pipeline Syst. Eng. Practice **10**(1), 04018034 (2019)

20. Ashari, M., El Naggar, H., Martins, Y.: Evaluation of the physical properties of TDA-sand mixtures. In: GeoOttawa, the 70th Canadian Geotechnical Conference Ottawa, Canada. Canadian Geotechnical Society (2017)
21. Alzabeebee, S., Chapman, D.N., Faramarzi, A.: A comparative study of the response of buried pipes under static and moving loads. Transport. Geotech. **15**, 39–46 (2018)
22. Azzam, W.R.: Finite element analysis of skirted foundation adjacent to sand slope under earthquake loading. HBRC J. **11**(2), 231–239 (2015)
23. Bhatnagar, S., Kranthikumar, A., Sawant, V.A.: Seismic analysis of dam under different upstream water levels. Adv. Comput. Design **1**(3), 265–274 (2016)
24. Alzabeebee, S.: Influence of soil model complexity of the seismic response of shallow foundations. Geomech. Eng. **24**(2), 193–203 (2021)
25. Alzabeebee, S.: Dynamic response and design of a skirted strip foundation subjected to vertical vibration. Geomech. Eng. **20**(4), 345–358 (2020)
26. Alzabeebee, S., Chapman, D.N., and Faramarzi, A., 2017, Numerical investigation of the bedding factor of concrete pipes under deep soil fill. In: the Proceedings of the 2nd World Congress on Civil, Structural, and Environmental Engineering (CSEE'17) Barcelona, Spain, paper number 119.
27. Kang, J., Parker, F., Yoo, C.H.: Soil-structure interaction and imperfect trench installations for deeply buried concrete pipes. J. Geotech. Geoenviron. Eng. **133**(3), 277–285 (2007)

Temporal Variation of Land Surface Temperature in Response to Changes in Vegetation Index of Bhawal National Park, Bangladesh

Ha-mim Ebne Alam, Md. Yeasir Arafat, Kazi Tawkir Ahmed, and Md. Nizam Uddin

1 Introduction

Remotely sensed land surface temperature (LST) has become one of the major constituents in the study of the earth's environment and climate processes [1]. Land surface temperature refers to the temperature that is measured at the land-atmosphere interface spatially and temporally using satellite thermal infrared remote sensing and acts as a key index of thermal behavior of the surface. Several factors such as vegetation coverage, soil moisture, rainfall, and urbanization can affect and alter the distribution and variation of surface temperature at the local and global scale [2].

Protected forest areas bear an imperative role for a country by conserving biodiversity and providing several services like ecological, social, and economic [3]. Forests also have an immediate and convoluted relationship with pollution, global warming, and climate change [4]. According to the Forest Department of Bangladesh, there are 37 protected areas in Bangladesh with a total area of 265,981 ha among which there are 17 national parks with an area of 45,746 ha. Bhawal National Park is one of the 17 national parks in Bangladesh which is mainly a deciduous Sal forest situated very close to the capital city of Dhaka. It is the most threatened and declining protected area of Bangladesh with a threat score of 36% as the development of heavy industries, human intrusion and urbanization have been taking place very rapidly in this area [5]. This human disturbance is causing rapid deforestation over time in this protected area resulting in high land surface temperature and thus warming of the urban atmosphere. So, it is very important to know about the productive and deforested areas of this forest to study the temporal warming pattern in relation to the changes in vegetation, and remotely sensed vegetation indices can come in handy and be effective in this issue.

Normalized difference vegetation index (NDVI) is one of the most effective and employed vegetation indices to study the vegetation dynamics and monitor the

H. E. Alam · Md. Y. Arafat (✉) · K. T. Ahmed · Md. N. Uddin
Department of Oceanography, University of Chittagong, Chittagong 4331, Bangladesh

I. Pal and S. Kolathayar (eds.), *Sustainable Cities and Resilience*, Lecture Notes in Civil Engineering 183, https://doi.org/10.1007/978-981-16-5543-2_27

forestry and infrastructures. NDVI value is computed from multi-spectral information as the normalized ratio between the red and the near infrared bands [6]. The value of NDVI points out the healthiness and status of vegetation. A healthy forest area shows low red light reflectance and high near infrared reflectance yielding a higher value of NDVI. Again, the land portions with low vegetation such as barren soils, clouds, and water demonstrate NDVI value close to zero and negative values followed by NDVI FAQs [7].

This study targets to investigate the temporal and seasonal variation of land surface temperature (LST) in response to rapid urbanization and deforestation in the Bhawal National Park which is the most threatened deciduous forest of Bangladesh by analyzing the NDVI value for 30 years.

2 Methodology

2.1 Study Area

The Bhawal National Park is located on the Dhaka-Mymensingh Highway which is about 40 km north of the capital city of Dhaka and 13 km northwest of Gazipur district. Geographically, it is emplaced between 23°55′N to 24°00′N latitude and 90°20′E to 90°25′E longitude. This forest is declared as a national park on May 11, 1982, with an area of 5022 ha by the then Ministry of Agriculture and Forest of Government of Bangladesh to conserve the vegetation and wildlife. It is bordered by the Rajendrapur Cantonment on the east, Shakipur and Kaliakoir Upazilas of Tangail district on the west, Savar, Uttara, and Rupganj Thana of Dhaka district on the south, and Gaffargaon and Bhaluka Upazilas on the north (see Fig. 1). Bhawal National Park is classified as a category V protected area by the Government of Bangladesh according to the criteria set by IUCN.

The landscape of this forest is proportionately plane with a little elevation with a few valleys and depressions with a height ranging from 3 to 4.5 meters formed by strong acidic yellowish-red soils of sandy and clay loam [8]. It is linked with a bigger bio-ecological region of Madhupur Sal Tract [9], and this national park is classified as a tropical moist deciduous forest consisting primarily of Sal tree (*Shorea Robusta*) and occupying almost 90% of the canopy due to the highly favorable climatic condition. Except for Sal, there are 220 plant species can also be found in this area [10].

Human intrusion is increasing very rapidly in and near this protected area due to easy accessibility to the capital city of Dhaka, high population growth, and development of large and heavy industries such as readymade garments (RMG) and tannery plants [5].

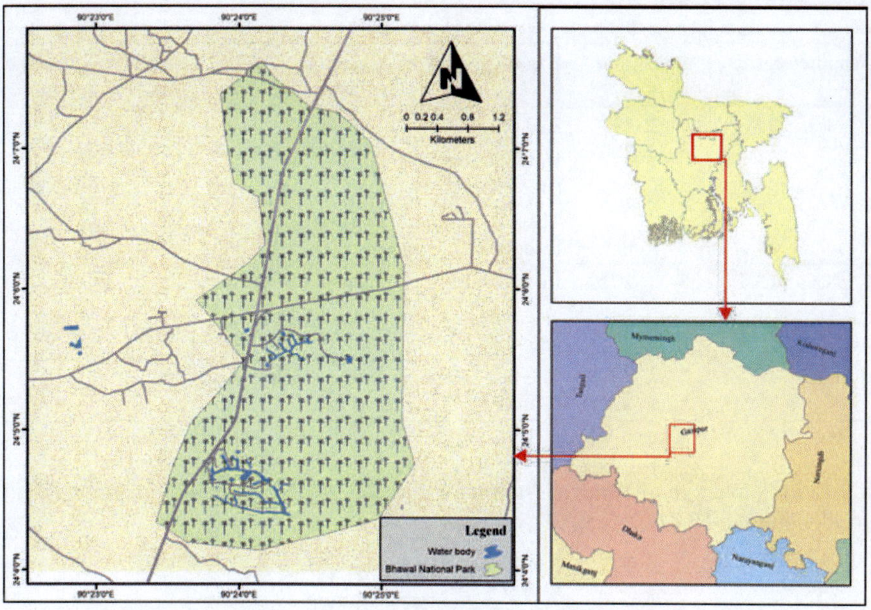

Fig. 1 Geographical location of Bhawal National Park, Bangladesh

2.2 *Data Collection and Pre-Processing*

This study uses Landsat 5 Thematic Mapper (TM), and Landsat 8 Operational Land Imager (OLI) multi-spectral data of winter season at 15-year interval for 30 years were acquired from Earth Explorer to analyze the temporal variation of surface temperature in Bhawal National Park. Images used in this study were obtained during November and December of each sampling year. The dates of data collection were free of clouds and had a clear atmospheric condition. The Landsat images were further rectified to a common Universal Transverse Mercator coordinate system using ArcGIS Pro for further analysis. The satellite images provided by Landsat also had no radiometric and geometrical distortions; hence, a little atmospheric and geo-correction was applied by processing the images in ENVI to improve the quality (Table 1).

2.3 *NDVI Analysis*

The normalized difference vegetation index (NDVI) is used to quantify the growth and health of the vegetation of an area by computing the spectral reflectance of the surface. Non-infrared (NIR) and red bands of Landsat images were used in this study to determine the value of NDVI. Total three images taken during the winter season in the years of 1990, 2005, and 2020 were analyzed to show the temporal variation in

Table 1 Landsat datasets that have been used in the study

Satellite/sensor	Acquisition date(s)	Bands used	Spectral wavelength (μm)	Spatial resolution (m)
Landsat-5/TM	09–12, 1990 16–11, 2005	Red (Band 3)	0.63–0.69	30
		Near infrared (Band 4)	0.76–0.90	30
		Thermal infrared (Band 6)	10.40–12.50	120
Landsat-8/OLI	25–11, 2020	Red (Band 3)	0.64–0.67	30
		Near infrared (Band 4)	0.85–0.88	30
		Thermal infrared (Band 6)	10.60–11.19	100

vegetation coverage of Bhawal National Park. The following equation was applied to calculate the NDVI value [11],

$$NDVI = \frac{\text{NIR} - \text{RED}}{\text{NIR} + \text{RED}} \tag{1}$$

where NIR and RED refer to the near infrared and red spectral reflectance values. The value of NDVI ranges from $+1.0$ to -1.0, and the area with a value of NDVI less than -1 or greater than $+1.0$ is considered a no data zone.

2.4 LST Analysis

The thermal band is used to convert the raw value into the black body temperature in degree Celsius using ArcGIS Pro software. The TM thermal infrared band 6 (10.4–12.5μm) data and OLI thermal infrared band 10 (10.6–11.19μm) were utilized to derive the LST.

The first step is to convert the digital number (DN) values of band 10 to at-sensor spectral radiance using the following equation [12],

$$L_\lambda = \left(\frac{L_{\text{max}_\lambda} - L_{\text{min}_\lambda}}{Q_{\text{cal}_{\text{max}}} - Q_{\text{cal}_{\text{min}}}}\right) \times \left(Q_{\text{cal}} - Q_{\text{cal}_{\text{min}}}\right) + L_{\text{min}_\lambda} \tag{2}$$

where

L_{max_λ} L_{max_λ} = Maximum spectral radiance scaled to $Q_{\text{cal}_{\text{max}}}$ in $[\text{watts}/(m^2 + sr + \mu m)]$.

L_{min_λ} Minimum spectral radiance scaled to $Q_{\text{cal}_{\text{min}}}$ in $[\text{watts}/(m^2 + sr + \mu m)]$

Table 2 Constants used for the calibration of thermal band

Satellite	Sensor	K_1	K_2
Landsat 5	TM	607.76	1260.56
Landsat 8	OLI	774.8853	1321.0789

After that, the conversion of spectral radiance to temperature in kelvin [12] is

$$T_{\text{Kelvin}} = \left[\frac{K_2}{\ln\left(\frac{K_1}{L_\lambda} + 1\right)} \right] \tag{3}$$

where

K_1 Calibration constant 1
K_2 Calibration constant 2
T_{Kelvin} Surface temperature in kelvin

The calibration constants K_1 and K_2 obtained from Landsat data user's manual are given in Table 2.

Conversion of Kelvin to Celsius,

$$T_{(^\circ\text{C})} = T_{\text{Kelvin}} - 273.15 \tag{4}$$

2.5 *Regression Analysis*

Statistical programming language R was used to estimate the regression equation between NDVI and LST of Bhawal National Park, Bangladesh. All the pixels of NDVI and LST were used in the regression analysis, where NDVI was considered as the independent and LST as the dependent variable, as LST has been found to be strongly determined by the value of vegetation index [13, 14]. In most studies, the LST has been found to be negatively correlated with NDVI [17–19].

3 Results and Discussion

The NDVI analysis performed in the study area of Bhawal National Park for the years of 1990, 2005, and 2020 is illustrated in Fig. 2.

The lowest value represents no vegetation coverage and refers to urban areas, infrastructures, water bodies such as lakes and ponds; in contrast, the highest value refers to healthy and dense vegetation in the reserved and protected region of the park.

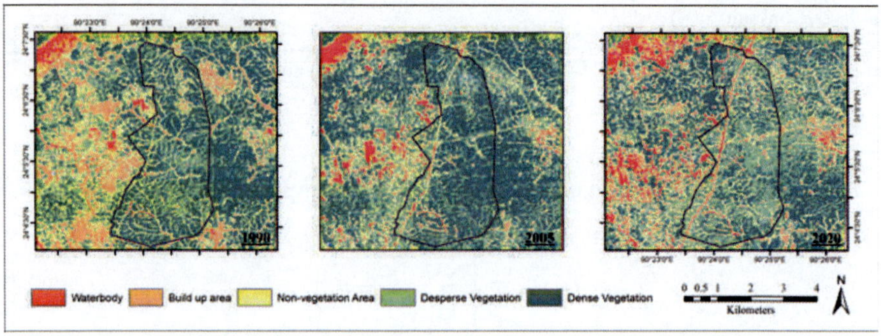

Fig. 2 NDVI computed from Landsat data for 1990, 2005, and 2020 of Bhawal National Park

The value of NDVI varied from −0.37 to 0.58 in 1990, −0.24 to 0.64 in 2005, and −0.16 to 0.52 in 2020. The analyzed images show dense and healthier vegetation in the year 1990 and 2005, respectively, where the year 2020 shows distortion in the vegetation index indicating a deterioration in the vegetation coverage, which can be linked to the enormous development of infrastructures and urbanization for the last 15 years. The factors responsible for forest degradation and biodiversity loss are extensive fuelwood collection for cooking purposes, industrial pressure, forest land encroachment,expansion of agricultural lands, and illicit removal of timber wood. It is confirmed that the number of floral species had declined one-third compared to the number of species recorded in previous decades [15]. Because of military activities, ground fires are destroying flora and killing medicinal herbs and shrubs as well, although they are still disappearing by human intervention and present practice [16].

The thermal variation of land surface over the study area from 1990 to 2020 with a 15 years gap has been shown in Fig. 3.

The change in temperature in each consecutive illustration after 1990 is ascending. The LST analysis presented here depicts a rise in the lower and upper range of the temperature where the minimum temperature ranged between 19.28 °C (in the year 1990) to 21.80 °C (in the year 2020) and the maximum limit of temperature varied between 23.68 °C (in the year 1990) to 34.63 °C (in the year 2020) showing a gradual rise of temperature in between the years of the study period. This increase of land surface temperature in the current year is due to the vegetation loss that has been proved by the decreased value of NDVI found in 2020 which can be connected to the rapid industrialization and urbanization near the study area.

The value of correlation coefficient between NDVI and LST is found −0.5908 and indicates the correlation between NDVI and LST, and the calculated regression coefficient from NDVI to LST is negative (see Fig. 4), which indicates the relation between the NDVI and LST is just opposite, where the higher land surface temperature for lower NDVI value and lower land surface temperature for higher NDVI value as higher LST increases the evapotranspiration and the dense vegetation (higher NDVI) with more evapotranspiration decreases the LST vice versa. The

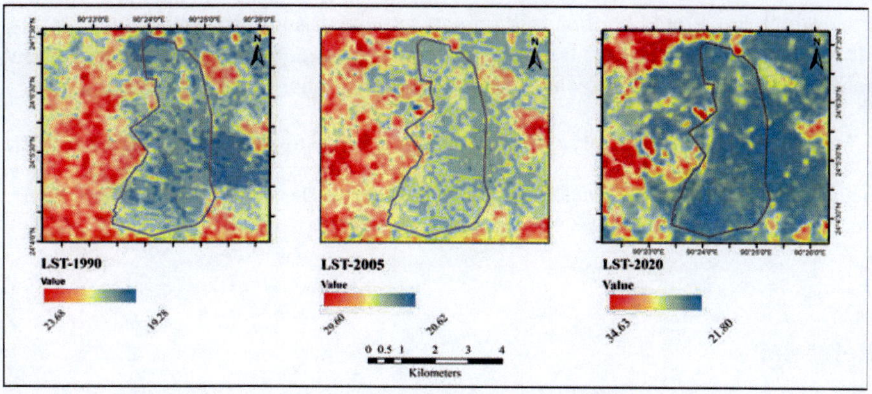

Fig. 3 Variation of land surface temperature estimated for 1990, 2005, and 2020 of Bhawal National Park

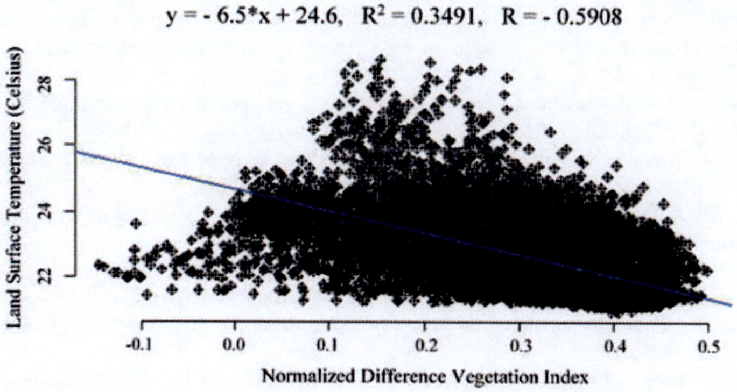

Fig. 4 Relationship between NDVI and LST for the study period

relationship between NDVI and LST is significant in according to the previous studies [17–19].

4 Conclusion

The rise in temperature in the recent year and the correlation coefficient found in this study support the progressive declination of vegetation coverage in the Bhawal National Park of Bangladesh. This study also confirms that the loss of vegetation and protected forest areas are responsible for the upward sweep of land surface temperature affecting the local weather and climate. The findings also indicate the negative effects of deforestation and the importance of the forest areas in the city areas which are altered by the rapid human settlement and urbanization day by day. Moreover, this attempt points to the necessity of strengthening the management policies to mitigate the obstacles and protect the national parks of the country.

Acknowledgements The authors express their deep gratitude to Md. Enamul Hoque, Assistant Professor, Department of Oceanography, University of Chittagong, for his valuable suggestions in preparing the manuscript. The authors are also thankful to the Department of Forest, Bangladesh, for providing data that helped to conduct the study smoothly.

References

1. Trigo, I.F., Monteiro, I.T., Olesen, F., Kabsch, E.: An assessment of remotely sensed land surface temperature. J. Geophys. Res.: Atmosp. **113**(D17) (2008)
2. Li, Z.L., Tang, B.H., Wu, H., Ren, H., Yan, G., Wan, Z., ... Sobrino, J.: A Satellite-derived land surface temperature: current status and perspectives. Remote Sens. Environ. **131**, 14–37 (2013)
3. Naughton-Treves, L., Holland, M.B., Brandon, K., et al.: The role of protected areas in conserving biodiversity and sustaining local livelihoods. Annu. Rev. Environ. Resour. **30**, 219–252 (2005)
4. Abrams, M.D.: Adaptations of forest ecosystems to air pollution and climate change. Tree Physiol. **31**(3), 258–261 (2011)
5. Ahsan, M.M., Aziz, N., Morshed, H.M.: Management effectiveness assessment of protected areas of Bangladesh. Bangladesh Forest Department, (2016)
6. Xue, J., Su, B.: Significant remote sensing vegetation indices: a review of developments and applications. J. Sensors (2017)
7. NDVI FAQs: Top 23 Frequently Asked Questions About NDVI. https://eos.com/blog/ndvi-faq-all-you-need-to-know-about-ndvi. Last Accessed 08 Nov 2020
8. Alam, M. K.: Diversity in the woody flora of sal forests of Bangladesh. Bangladesh J. Forest Sci. 41–51 (1995)
9. Nishat, A., Huq, S.I., Barua, S.P., Reza, A.H.M.A., Khan, A.M.: Bio-Ecological Zones of Bangladesh. The World Conservation Union (IUCN), Dhaka, Bangladesh (2002)
10. Rahman, M.M., Rahman, M.M., Guogang, Z., Islam, K.S.: A review of the present threats to tropical moist deciduous Sal (Shorea robusta) forest ecosystem of central Bangladesh. Tropical Conserv. Sci. **3**(1), 90–102 (2010)
11. Rouse, J.W., Haas, R.H., Schell, J.A., Deering, D.W.: Monitoring vegetation systems in the great plains with ERTS. In: Proceedings of 3rd ERTS Symposium, SP-351 I, pp. 309–317. NASA, USA (1973)
12. Amiri, R., Weng, Q., Alimohammadi, A., Alavipanah, S.K.: Spatial–temporal dynamics of land surface temperature in relation to fractional vegetation cover and land use/cover in the Tabriz urban area. Iran. Remote Sens. Environ. **113**(12), 2606–2617 (2009)
13. Zhang, X.X., Wu, P.F., Chen, B.: Relationship between vegetation greenness and urban heat island effect in Beijing City of China. Procedia Environ. Sci. **2**, 1438–1450 (2010)
14. Yue, W., Xu, J., Tan, W., Xu, L.: The relationship between land surface temperature and NDVI with remote sensing: application to Shanghai landsat 7 ETM+ data. Int. J. Remote Sens. **28**, 3205–3226 (2007)
15. Alauddin, M., Hossain, M.N., Islam, M.B., Islam, S., Islam, M.K.: Management strategies for sustainable forest biodiversity conservation in protected areas of Bangladesh: a study of Bhawal National Park. Gazipur. Grassroots J. Natural Resour. **3**(3), 56–72 (2020)
16. Chowdhury, R.I.: Attitudes towards co-management: is Satchari National Park a suitable model for Bhawal National Park? In: Fox, J., Mustafa, M.G., Bushley, B.R., Brennan, S.M., Durand, L.

(eds.) Connecting Communities and Conservation: Co-Management Initiatives Implemented by IPAC in Wetlands and Forests of Bangladesh, pp. 24–43. USAID, Dhaka (2013)

17. Grover, A., Singh, R.B.: Analysis of urban heat island (UHI) in relation to normalized difference vegetation index (NDVI): a comparative study of Delhi and Mumbai. Environments **2**(2), 125–138 (2015)
18. Karnieli, A., Agam, N., Pinker, R. T., Anderson, M., Imhoff, M. L., Gutman, G. G., Goldberg, A., et al.: Use of NDVI and land surface temperature for drought assessment: merits and limitations. J. Climate **23**(3), 618–633 (2010)
19. Schultz, N.M., Lawrence, P.J., Lee, X.: Global satellite data highlights the diurnal asymmetry of the surface temperature response to deforestation. J. Geophys. Res. Biogeosci. **122**(4), 903–917 (2017)

Use of Stream Power as a Tool for the Detection of Critical Reaches in Channeled Streams

Daniel Rios

1 Introduction

The stream power concept was introduced by Bagnold within its formulations for sediment transport [1] and has been analyzed on nature streams by Knighton [2], Barker et al. [3] among others. Furthermore, nowadays some authors like Bizzi and Lerner [4] have proposed using the concept to evaluate channel sensitivity to erosion and deposition processes.

In the present work, we use stream power concept for detection of critical reach in channeled streams. The need of a tool for prioritize intervention reach in the Aburrá River arises because there is a canalized reach of approximately 24 km along the valley, and additionally, there is a metro line and main roads parallel to the canal on great extent of its length.

2 Materials and Methods

2.1 *Study Area*

The Aburrá River (Fig. 1) is located on Colombia (South America) and over the Andes mountain range. The headwaters of the basin are located at the top of San Miguel (2400 m above the sea level). The region's climate is tropical in character, mainly determined by altimetric variations, the topography of the relief and the influence exerted by the movement of the Intertropical Confluence Zone (ITCZ), which generates two wet and two dry periods in its path that appear interspersed throughout the year.

D. Rios (✉)
Cooperative University of Colombia, Medellín, Colombia
e-mail: juan.riosar@campusucc.edu.co

I. Pal and S. Kolathayar (eds.), *Sustainable Cities and Resilience*, Lecture Notes in Civil Engineering 183, https://doi.org/10.1007/978-981-16-5543-2_28

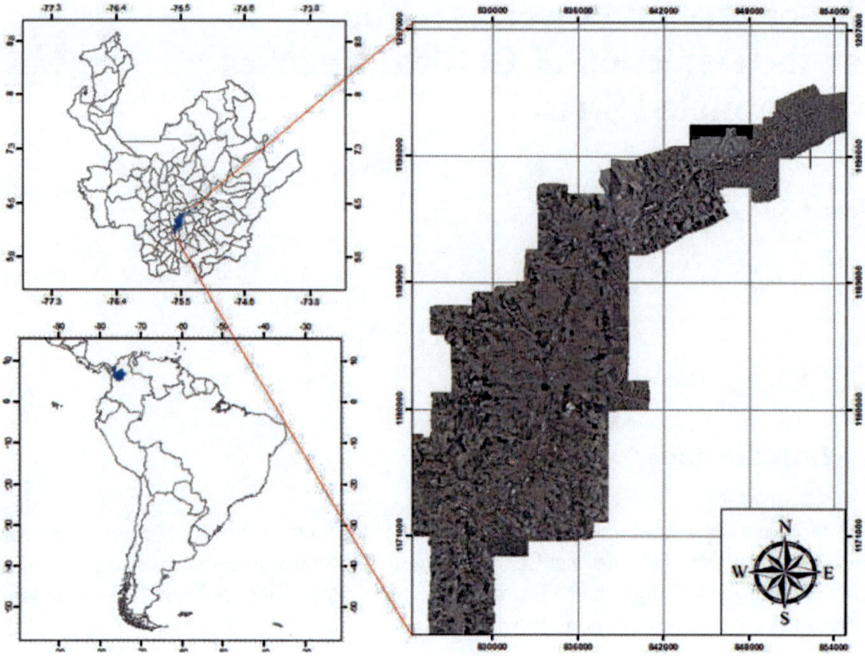

Fig. 1 Location of Aburrá River

2.2 Dataset

The dataset of discharges (Table 1), widths and friction slopes are available thanks to the network of environmental monitoring of the hydrographic basin "RED RIO."

Table 1 Maximum discharge (m^3/s) in the stations of analysis

Station	Area (km^2)	Return period (Years)					
		2.33	5	10	25	50	100
South nnickpoint	133.20	45.62	59.62	71.03	85.44	96.13	106.74
Before San Fernando	237.08	80.47	101.66	118.92	140.73	156.91	172.97
After San Fernando	273.94	92.77	116.28	135.43	159.63	177.58	195.4
Guayaquil bridge	333.54	112.61	139.69	161.76	189.63	210.31	230.84
Environmental classroom	474.38	159.27	194.16	222.59	258.49	285.13	311.58
Acevedo bridge	510.91	171.34	208.14	238.12	275.99	304.09	331.99

Table 2 Criterion of the task force stream power. (*Source* https://extension.umass.edu/riversmart/sites)

Risk	Erosion			Aggradation (Deposition)		
	Low	Moderate	High	Low	Moderate	High
Specific stream power (W/m^2)	0–60	60–300	300	300	60–300	0–60

Below are the flows estimated for different return periods at the six analysis sites (Station), In general, it is observed that the flow increases as the basin area increases and the friction slope decreases.

2.3 *Methods*

We use the definition of Bagnold [1] for calculate the total stream power and unitary stream power:

$$\Omega = \rho\, g\, Q\, S \tag{1}$$

where: Ω is the total stream power (W/m), ρ is the fluid (water) density, g is the acceleration of gravity, Q is the discharge (m^3/s) and S is the friction slope (m/m).

$$\omega = \frac{\rho\, g\, Q\, S}{w} \tag{2}$$

where: ω is the unitary stream power (W/m^2), ρ is the fluid (water) density, g is the acceleration of gravity, Q is the discharge (m^3/s), S is the friction slope (m/m) and w is the width of the stream.

Furthermore, we used the criterion of the Task Force Stream Power of United States Department of Agriculture (Table 2) for reclass the unit stream power and to find the reaches with risk of erosion and aggradation (Deposition).

3 Results and Discussion

The total stream power (Fig. 2) and unitary (specific) stream power (Fig. 3) calculated are shown below, on this it is observed that the stream power decreases as the basin area increases.

In addition, it is observed that the first three analysis sites present high risk of erosion for different periods of return and the last three analysis sites present on the contrary moderate to low risk of aggradation (Deposition) according to the criterion of the Task Force Stream Power of United States Department of Agriculture (Table 2).

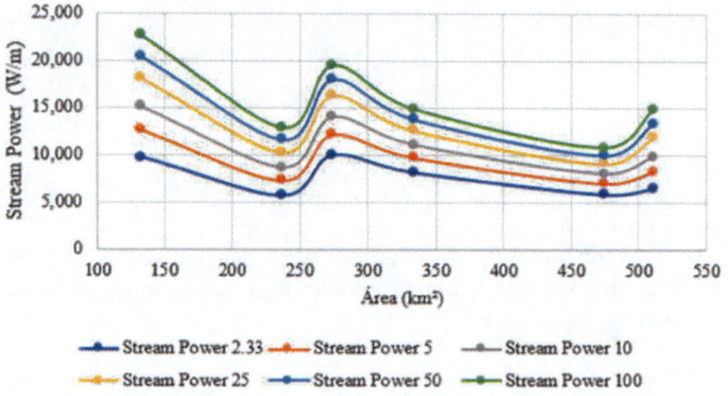

Fig. 2 Total stream power (W/m)

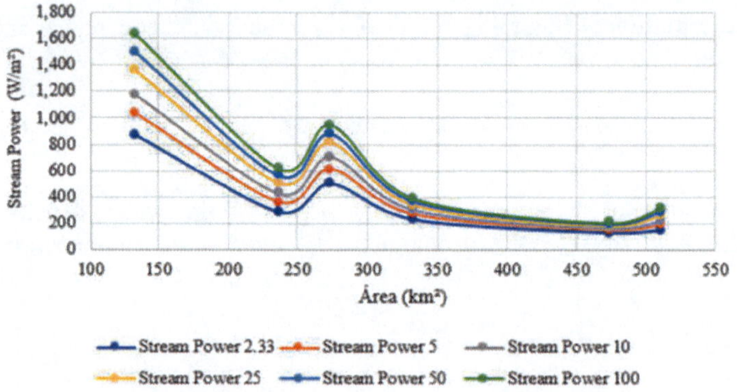

Fig. 3 Unitary stream power (W/m^2)

These results are consistent with the evidenced of field (Fig. 4), however, today the processes are in a state of quasi equilibrium due to the maintenance works carry out by Urban Environmental Authority.

4 Conclusion

The results show that the use of stream power in combination with the criterion of the Task Force Stream Power of United States Department of Agriculture (Table 2) is a useful tool for the detection of critical reaches in channeled streams and schedule your preventive maintenance.

Although the concept of power is old, it is possible to take it up again for the systematic analysis of river systems with rigid river training.

Fig. 4 Physical appearance of six analysis sites

References

1. Bagnold, R.A.: In: An Approach to the Sediment Transport Problem From General Physics. (1966)
2. Knighton, A.D.: Downstream variation in stream power. Geomorphology **29**(3), 293–306 (1999)

3. Barker, D.M., Lawler, D., Knight, D., Morris, D.G., Davies, H., Stewart, E.J.: Longitudinal distributions of river flood power: the combined automated flood, elevation and stream power (CAFES) methodology. Earth Surf. Process. Landforms **34**, 280–290 (2009)
4. Bizzi, S., Lerner, D.N.: The use of stream power as an indicator of channel sensitivity to erosion and deposition processes. River Res. Appl. **31**(1), 16–27 (2015)

Urban Facilities Management: A Way of Attaining Sustainable Cities in Sri Lanka

S. P. M. Dasandara, W. A. P. R. Weeratunga, and Piumi Dissanayake

1 Introduction

Facilities Management (FM) can be defined as the process of establishing the environment to carry out the organisation's primary operations, supplying customer satisfaction and adding value for money by taking an integrated view of the infrastructure services for an enhancement of the core business [11]. It simply deals with people, place, process and technology to support the core business of the organisation [7]. Although the concept of FM was originally limited to the built environment, it has been started evolving to the society level, as per [10]. In other words, nowadays, traditional FM, which is only limited to micro-level, is moving to macro-level, which is called as Urban Facilities Management (UFM). A recent study by Yuan [18] comprehensively disclosed that UFM is performing in a larger area than the traditional FM, and it manages the integration of services to drive the urban facilities with better operation and sustainability. In this context, it simply implies that UFM is an effective and novel approach, brought in to manage the space, place, people and technology at a macro-level. More importantly, UFM stresses on providing a dynamic platform to bring the government and private sectors together, as disclosed by Boyle and Michell [3].

Having an in-depth understanding about UFM is highly important in order to achieve urban sustainability [10]. He indicated that urban sustainability can be identified as performing innovative, equitable and economically feasible actions to meet aspirations and needs of the present, without compromising the capability of meeting aspirations and needs of future generation. Further, a close partnership between private, public sectors and government is required for the development of urban sustainability, as specified in the study by Michell [10]. In such a context, UFM

S. P. M. Dasandara (✉) · W. A. P. R. Weeratunga · P. Dissanayake
Department of Building Economics, University of Moratuwa, Moratuwa, Sri Lanka

P. Dissanayake
e-mail: piumid@uom.lk

I. Pal and S. Kolathayar (eds.), *Sustainable Cities and Resilience*, Lecture Notes in Civil Engineering 183, https://doi.org/10.1007/978-981-16-5543-2_29

offers a flexible and general platform to manage social, economic, ecological and institutional dimensions of sustainability to implement the urban sustainability [3].

Nevertheless, UFM, like the broader concept of urban sustainability, has many challenges, when it comes to implementation. Similarly, many plans for UFM have fallen short of their desired outcomes [3]. When moving to the Sri Lankan context, FM is currently performing in the individual buildings and stand-alone facilities [9]. It is where the FM is performing at the micro-level. However, with the emergence of the UFM concept, FM can now perform at the macro-level or in the urban level, as emphasised by the above authors. However, as stated by Boyle and Michell [3], adaptation of the concept to the Sri Lanka need to be studied well prior to the implementation. Although it is, adaptability of this concept to the Sri Lankan context has not been studied yet. Hence, it has become a worthy researchable area. This paper, therefore, aimed to analyse drivers and barriers to the successful adaptation of UFM to the Sri Lankan context.

2 Literature Review

The following subsections explore the relevant literature in the research arena with a major focus on the concept of UFM and its applicability, drivers and barriers affecting the successful implementation of UFM.

2.1 The Concept of UFM and Its Applicability

FM can be simply identified as the practice of managing people, places and process to accomplish the core business function of the organisation [14]. As further noted by the authors, FM can be identified as a multi-disciplinary field, which has a combination of different types of fields and it affects the organisations' fundamental strategic goals. Presently, FM concept has been moved from micro-level to macro-level, which means FM functions are now not limited for a single building but also can be applied to specific geographical area [12]. In this context, the concept of UFM, which manages the urban areas in a new, innovative and more sustainable way has been emerged from the concept of FM [10]. As pointed by Roberts [13], UFM is a comprehensive concept rather than FM, which mainly focuses on greater sustainability and resilience and it is paramount to support human activities for a standardised living style. A similar argument was brought forward by Michell [10] stating that UFM provides a flexible platform for both public and private sectors to uplift the whole society in an urban precinct scale. Accordingly, it can be argued that the notion of UFM plays a vital role in the way of creating sustainable cities by embracing the underlying concepts of space, place and people, which underpin the concept of FM.

Table 1 Drivers of UFM

Drivers of UFM	A	B	C	D	E	F	G	H	I	J
Economic and regulatory development	√	√	√							
Increased service integration				√						
Nationalisation and internationalisation of contract procurement		√			√					
Proper organisation structure				√						
Improved knowledge and skills of facility managers				√		√				
Good teamwork					√		√			
Proper change management		√				√		√	√	
Development of Information and Communication Technology (ICT)		√								
New competencies and education		√		√					√	√

A: [3], **B**: [12], **C**: [1], **D**: [13], **E**: [5] **F**: [8], **G**: [9], **H**: [15], **I**: [6], **J**: [14]

2.2 Drivers Influencing Adaptation of UFM

When performing UFM as an overarching construct for urban sustainability through the whole community scale, many drivers can be identified, which encourage the adaptation of the concept [12]. The drivers identified through the existing literature are presented in Table 1.

A recent study by Pearce [12] disclosed many drivers that influence the adaptation of UFM in the Sri Lankan context. These drivers include economic and regulatory development, nationalisation and internationalisation of contract procurement, proper change management, development of ICT and new competencies and education as shown in Table 1. Moreover, [13] identified in his study that increased service integration, proper organisation structure and improved knowledge and skills of facility managers as some main drivers of UFM. When mapping with Table 1, it is obvious that proper change management and new competencies and education are the two main drivers, which were mostly identified by the researchers.

In this context, aforementioned factors can be identified as the major driving forces for adaptation of UFM concept in a successful way.

2.3 Barriers Influencing Adaptation of UFM

However, UFM, like the broader concept of urban sustainability, has many challenges, when it comes to implementation. As noted by Drion et al. [4], these challenges can be arisen during the implementation, continuous improvement as well as maintenance of the concept of UFM. The barriers identified through the existing literature are presented in Table 2.

Table 2 Barriers of UFM

Barriers of UFM	A	B	C	D	E	F	G	H	I
High cost		√	√						
Conflicting stakeholder requirements		√	√	√					
Lack of information and knowledge	√	√	√			√			
Lack of understanding of contextual issues		√			√				
Lack of integration of stakeholder knowledge	√								√
Lack of funding for private owners		√	√						
Lack of communication among parties	√						√	√	

A: [13], **B**: [12], **C**: [1], **D**: [3], **E**: [16], **F**: [4], **G**: [2], **H**: [8], **I**: [11]

According to Table 2, many barriers, which influence the adaptation of UFM were identified in a recent study by Pearce [12], namely high cost, conflicting stakeholder requirements, lack of information and knowledge, lack of understanding of contextual issues and lack of funding for private owners. In addition to that lack of integration of stakeholder knowledge and lack of communication among parties were identified as by the other researchers as presented in Table 2.

Accordingly, these barriers can be identified as the challenges of UFM implementation, which discourage the way of implementing it within the current context.

3 Research Methodology

This study was designed to analyse the drivers and barriers to the successful adaptation of the concept of UFM into the Sri Lankan context. Accordingly, research question can be developed as,

RQ: *"How could drivers and barriers be influenced the adaptation of UFM in Sri Lanka?"*

Initially, an extensive literature review was conducted, which facilitate to strengthen the base of the study by reviewing the prevailing knowledge around the research area. Since there is a need of carrying out an in-depth investigation on the analysis of drivers and barriers to the successful adaptation of UFM into Sri Lankan context, a qualitative research approach with qualitative interview survey strategy was selected to conduct the study. Further, qualitative interview survey was carried out using semi-structured interviews with experts in the field who have wide knowledge regarding current FM practices, and also with the key stakeholders who are engaged in the management and development of sustainable cities to gather data. The literature review (refer Sect. 2.1) facilitates the researcher to identify the relevant interviewees for the study. Altogether, fifteen (15) respondents were selected with their familiarity with the urban context, profession and the year of experiences as tabulated in Table 3.

Table 3 Profile of the respondents

Respondent	Profession	Industry experience (years)
Respondent 1 (R1)	Town and transport planner	15
Respondent 2 (R2)	Facility manager	05
Respondent 3 (R3)	Chartered quantity surveyor	16
Respondent 4 (R4)	Town and country planner	07
Respondent 5 (R5)	Chartered architect	07
Respondent 6 (R6)	Chartered architect	15
Respondent 7 (R7)	Chartered quantity surveyor	25
Respondent 8 (R8)	Architect	15
Respondent 9 (R9)	Facility manager	09
Respondent 10 (R10)	Facility manager	05
Respondent 11 (R11)	Property manager	10
Respondent 12 (R12)	Architect	12
Respondent 13 (R13)	Facility manager	09
Respondent 14 (R14)	Quantity surveyor	18
Respondent 15 (R15)	Facility manager	06

When moving to data analysis, content analysis can be identified as the most applicable, flexible as well as commonly used technique, which can be used to analyse textual data, as suggested by White and Marsh [17]. Thus, it can be identified as the mostly preferred data analysis technique for this study. The drivers and barriers were extracted through analysis of research findings using an abductive analysis.

4 Research Findings and Analysis

A detailed description of analysis of the findings is presented under three subsections, namely understanding of the concept of UFM in Sri Lanka (Sect. 4.1), drivers (Sect. 4.2) and barriers (Sect. 4.3) to the successful adaptation of UFM to the Sri Lankan context.

4.1 The Concept of UFM in Sri Lanka

UFM is a novel concept to many stakeholders, who engaged in the management and development of cities in Sri Lanka, except facility managers. Being facility managers by profession, R2, R9, R10, R13 and R15 have already aware of the concept UFM as they are continuously seeking innovative and modern ways to carry out their FM practices. As further stated by R2 and R9, the principles and basics of UFM have been using in their professional role as they are mainly involved in urban development projects. They perceived UFM as a systematic process, which uses micro-level FM practices at a macro-level with the participation of both public and private sectors. A similar perception was shared by the other respondents as well, though the concept is new to them still. As stated by R8, UFM practices are currently performing in the field in Sri Lanka anyhow without having a proper idea about it. Furthermore, R6 and R12 explained that the country needs a balanced development with available resources through UFM to achieve urban sustainability. Otherwise, the country will fail to achieve the goals of the adaptation of UFM.

4.2 Drivers to the Successful Adaptation of UFM in Sri Lanka

According to the experts' opinions, driving forces, which encourage the relevant stakeholders to perform UFM as a concept in Sri Lanka are discussed as drivers of UFM under this section.

It was witnessed through the empirical findings that **economic and regulatory developments** has become a major driver of UFM today. Identifying the priorities of the country and establishing the required goals to achieve the major requirement is the basis of implementing a new concept within the country, as per R3 and R2. Furthermore, developing a proper regulatory framework has also become a major need, when adapting a novel concept within the country in a practical way. Therefore, as UFM is a major concept, which affects the lifestyle of the whole society, proper economic goals and regulations need to be established in an appropriate way, as stressed by R11. With the presence of such economic and regulatory developments, adapting UFM into the current practices will not be a huge challenge for the stakeholders. As per the views of many respondents, **increased service integration** was identified as another main driver of UFM. Most minor FM services can be linked with major FM services and better UFM services can be provided for the society due to service integration, which is the basis of performing UFM. As elaborated by R10 and supported by R4, if the service integration does not practice, performing UFM can be failed since services cannot provide from one single hub. As per the opinion of R7, this service integration needs to be done with a proper plan and proper communication between suppliers, by reducing the issues that can occur during implementing tasks. However, the attention of the stakeholders should be paid for integration of services for better application of UFM into the current practices.

When implementing a new concept into a country, there is a higher need of an improved procurement system without any corruption. As described by R4 and R5, these procurement systems need to have transparency procedures and also function with local as well as global standards. It was further justified by R7 stating that although nationalisation is enough for contract procurement, there is a major requirement of internationalisation since it will transfer funds from many countries and will provide opportunities to get involve human resources with many innovative ideas, when implementing UFM. Such internationalisation of contract procurement will be very easy today as FM services have a good international market, as per R11 and R15. Thus, it is obvious that both **nationalisation and internationalisation of contract procurement** can be considered as one of the major drivers of UFM. As the views of R3 as well as R15, when implementing UFM as a novel concept in Sri Lanka, a proper organisation structure need to be in advance indicating roles, responsibilities, authorities and communication procedures of all stakeholders, who involved in the UFM. It was strongly insisted by R12 mentioning that UFM implementation is definitely lead to its ultimate goals by objectives and characteristics of the organisational structure. According to the opinion of R10, as the objectives and characteristics can be changed with the time due to external environmental factors and business cycle, they need to be revised appropriately to achieve the aims of UFM. Further, awareness and knowledge regarding the concept of UFM are highly required for the parties in the organisational structure for better implementation of it, as mentioned by R2. Accordingly, it can be argued that **organisation structure** is very important during the implementation process of UFM into the current practices.

Another important factor, which was captured through the interviews is **knowledge and skills of facility managers**. Facility managers play a foremost role in the process of UFM implementation because it has emerged as a novel concept from the concept of FM, where facility managers have the basic knowledge and skills on FM practices. It was clearly witnessed through the discussions with R2, R9, R10, R13 and R15, who are working as Facility Managers. They further elaborated that currently facility managers are involving in managing individual buildings with requirements of FM and it can get as an exposure to perform FM practices in an urban scale. Accordingly, it is obvious that current performance is totally different from performing UFM as the scale is bigger than present, though the concept is same. Therefore, proper communication with the stakeholders is highly required for facility managers to get actions, which fulfil the requirements of the stakeholders and covering UFM objectives, as further stated by R10. Moreover, **teamwork** is a major requirement for every work to achieve the required goals. Thus, it can be applied for the process of implementing UFM as well. As mentioned by R4, interpersonal and interprofessional skills are need to be developed between all participants for implementing UFM. The implementation process will be highly encouraged by skilful, knowledgeable, experienced and professional teamwork as further explained by all respondents.

In the way of implementing UFM, different changes can be happened in many aspects positively as well as negatively, as shared by R12. These changes need to be properly identified and properly managed with appropriate solutions as unwanted situations may automatically occur with the absence of proper change management. With the results of such change management, the process of adapting UFM into current context will become more efficient and innovative, as stated by R2 and supported by R10. That is why **proper change management** can be identified as one of the major drivers of UFM. As per the views of majority of the respondents, new and advanced technologies are highly required for implementing and practicing UFM. As further elaborated by R5 and R3, there is a major need for lot of data and information to manage and operate urban practices in a standard way, which highlights the importance of ICT in adapting UFM. Although many new technologies can apply to the urban practices, unwanted technologies need to be rejected as they may be less advantaged for Sri Lanka, which is still a developing country with minor facilities and resources, as per R1 and R11. However, **development of ICT** makes the UFM implementation process more effective when it is used in its best way and appropriate situations.

When implementing the concept of UFM, professionals who involve in the implementation process should have a proper knowledge about it because being a new concept, UFM can make both positive as well as negative impacts on current practices, as explained by R6. With the experience in the field, he further elaborated that in such negative situations, professionals need to find solutions without failing the adaptation of the concept. Moreover, the society also needs to have a proper idea about how they can adapt to this new concept. Hence, **establishing new educational paths** for professionals as well as common society to get an idea about the UFM concept will be a major driver to its implementation. **Uplifting laws and ordinance** is also an enabler of UFM according to the views of R4 because current laws and ordinance are not properly functioned and some cannot be applicable today. Hence, uplifting them in an applicable manner for the present situation will be a main driver for UFM implementation.

4.3 Barriers to the Successful Adaptation of UFM in Sri Lanka

Despite the drivers, UFM, like the broader concept of urban sustainability, has many challenges, when it comes to implementation. Those barriers are comprehensively discussed with relevance to the views of the respondents under this section.

According to the opinions of R8 and R12, when implementing UFM as a new concept within the country, all the processes and systems need to be converted into a systematic way as it is trying to integrate all FM services in a macro-level. Hence, this implementation process can be costly, especially the initial cost. Though this cost can be covered by the savings of UFM adaptation, it may somewhat difficult due to the lack of resources in a county like Sri Lanka, which is still a developing country. Therefore, this **high cost** may be a barrier for implementing UFM. Many stakeholders have been involved during the implementation process of UFM. The requirements of these stakeholders of UFM can differ resulting conflicts among those requirements. These **conflicting stakeholder requirements** may hinder the UFM implementation process as stakeholders focus more to achieve their personal objectives rather than achieving the common goals of UFM, as stated by R5. As emphasised by R2, when implementing a new concept, the stakeholders should need to give their best support and priority to achieve the goals of the concept. Unfortunately, such support or priority is lacking in current situation of the country, as further stated by R2. Moreover, implementation of UFM needs to have every proper, clear and validated information as well as improved knowledge regarding the area in order to become successful. However, as mentioned by R8 and supported by R13, getting each and every information has become very difficult today due to poor communication between relevant parties. Further, R6 explained that existing knowledge, specially existing FM knowledge is not sufficient for UFM implementation as it deals with many broad areas rather than FM and this was endorsed by R15 as well. Thus, **lack of information and knowledge** creates another huge barrier for UFM implementation to achieve its great success.

Another important barrier, which was captured through the findings is **lack of understanding of contextual issues**. Presently, many urban issues have emerged in Sri Lanka and there is a higher need of overcoming those issues through proper adaptation of UFM, as mentioned by R7. Further, he stated that when doing this, stakeholders of UFM should have a proper understanding of the current context. However, as per the experiences of the interviewees, it was apparent from the findings that the understanding of the stakeholders on contextual issues is very poor and it has made a main barrier to implement UFM successfully. In line with the preceding barrier, another barrier can be identified as **lack of integration of stakeholder knowledge**, as per the views of the respondents. What they have elaborated in here is, proper integration needs to be developed among both theoretical and practical knowledge of the stakeholders from different fields regarding adapting UFM. However, currently, such integration can be seen at its infancy, which may result many issues and conflicts. As further disclosed by R4, it may difficult to provide the best, possible solutions for available urban issues with the absence of such integration among stakeholder knowledge, which creates a barrier for UFM implementation into Sri Lankan context.

Moreover, **lack of funding for private owners** has also become a huge challenge today for UFM implementation. As previously discussed, both public and private parties have to involve in the process of adapting UFM into current practices. As per the arguments of R8, although the public organisations are driving only for well-being of the society, private sector is also seeking it, while driving for profit earnings. Thus, proper funding should be done for the private investors by the public sector as private sector is also very essential in achieving goals of UFM. However, currently, it is poorly happened creating a barrier for UFM implementation. According to R2 and R6, current working patterns in government organisations have become more inefficient as majority of employees do not tend to fulfil their duties and responsibilities properly. Therefore, these **inefficient government organisations** create a huge barrier in the way of implementing UFM because, when implementing such new concept within the country, active participant of both public and private sectors is highly required.

5 Discussion

UFM has emerged as a novel concept based on the concept of FM to manage current urban practices in a modern, innovative and sustainable way [10]. Although FM is currently performing at the micro-level in Sri Lanka, with the emergence of the notion of UFM, it can be performed at the macro-level [9]. However, adaptability of UFM into Sri Lankan context needs to be studied well prior to the implementation. It is, therefore, this study endeavoured to analyse the drivers and barriers for proper adaptation of UFM in Sri Lanka. When mapping with the literature findings, [3] proved that living standards of the people will be highly increased through the adaptation of UFM by providing quick access to a wide range of services, which are managed in a macro-level by different stakeholders from different fields. This was further confirmed by the respondents stating that adaptation of UFM will facilitate the provision of urban facilities and services for the community in a comfortable and convenient way. However, adaptation of UFM is still at its infancy in Sri Lanka, where the many driving forces as well as challenges can be arisen during the path of UFM implementation.

When reviewing the literature, nine main drivers of UFM were identified based on the existing literature, namely economic and regulatory development, increased service integration, nationalisation and internationalisation of contract procurement, proper organisation structure, improved knowledge and skills of facility managers, good teamwork, proper change management, development of ICT and new competencies and education. Although these findings were in general, similar results can be seen in Sri Lankan context as well according to the respondents. Apart from these drivers, establishing new educational paths and uplifting laws and ordinance were captured as newly identified drivers through the views of the respondents. As elaborated by the respondents, professionals as well as common society can gain proper understanding on the concept of UFM to adapt to this new concept properly by establishing new educational paths. Moreover, uplifting the existing laws and ordinance will be a main driver for UFM implementation since they are not currently practiced properly and not applicable for today according to the respondents. At the same time, seven main barriers for adaptation of UFM were discovered through the prevailing literature. They include high cost, conflicting stakeholder requirements, lack of information and knowledge, lack of understanding on contextual issues, lack of integration of stakeholder knowledge, lack of funding for private owners and lack of communication among parties. Respondents confirmed that as same as the identified drivers through the literature, these barriers also can be applicable into Sri Lankan context, though they have discussed generally in the literature. Additionally, inefficient government organisations were identified as a new barrier for UFM adaptation through interview findings. Michell [10] disclosed in his study that active participation of both private and public sectors is highly acknowledged, when implementing a new concept for a country. Similarly, respondents argued that effective participation of private and public sectors is highly required for adaptation of UFM into Sri Lanka. However, current working patterns in government organisations in Sri Lanka have become more inefficient as further highlighted by them. Hence, it creates a major barrier towards the UFM implementation process. On the whole, these identified drivers and barriers can be summarised as presented in Fig. 1 (refer Annexure 1).

An exclusive role is played by all these drivers and barriers, when implementing a new concept like UFM into Sri Lankan context. As per the views of the respondents, all stakeholders should highly dedicate for encouraging these drivers as well as overcoming the barriers through appropriate solutions for a better adaptation of UFM, which ultimately focuses on urban sustainability in performing urban practices within the country as indicated in a study by Pearce [12].

6 Conclusions

Currently, in Sri Lanka, FM is performing in a micro-level. However, with the emergence of the concept of UFM, FM activities can be performed in a macro-level to manage the built environment in a modern, innovative and sustainable way for uplifting and standardising the living styles of the society. However, adaptability of this concept to the Sri Lankan context has not been studied yet. When adapting this novel concept into Sri Lanka, many drivers as well as barriers can be identified. Thus, this paper intended to analyse these drivers and barriers for proper adaptation of UFM to the Sri Lankan context. Altogether, eleven drivers for UFM implementation were identified through the empirical findings of the study. They include economic and regulatory development, increased service integration, nationalisation and internationalisation of contract procurement, proper organisation structure, improved knowledge and skills of facility managers, good teamwork, proper change management, development of ICT, new competencies and education, industrialisation and uplifting laws and ordinance. At the same time, eight barriers, which create many obstacles in the way of UFM implementation were identified as high cost, conflicting stakeholder requirements, lack of information and knowledge, lack of understanding on contextual issues, lack of integration of stakeholder knowledge, lack of funding for private owners, lack of communication among parties and inefficient government organisations. According to the findings, it was witnessed that the adaptation of UFM is at the nascent stage in Sri Lanka. Thus, these drivers need to be further encouraged and the barriers need to be overcome by applying appropriate strategies for the successful implementation of UFM into current urban practices in Sri Lanka. In fact, implementation of the concept of UFM definitely leads for a standardised living pattern for the whole society through a sustainable, innovative and modern urban management. However, all the stakeholders should highly contribute in achieving the goals of UFM by managing aforementioned drivers and barriers successfully.

The study provides a contribution to the literature on applicability of UFM into Sri Lankan context. Hence, the knowledge generated through this study can be assisted in many ways by the respective industry experts in Sri Lanka in implementing the concept of UFM successfully.

Annexure 1

See Fig. 1.

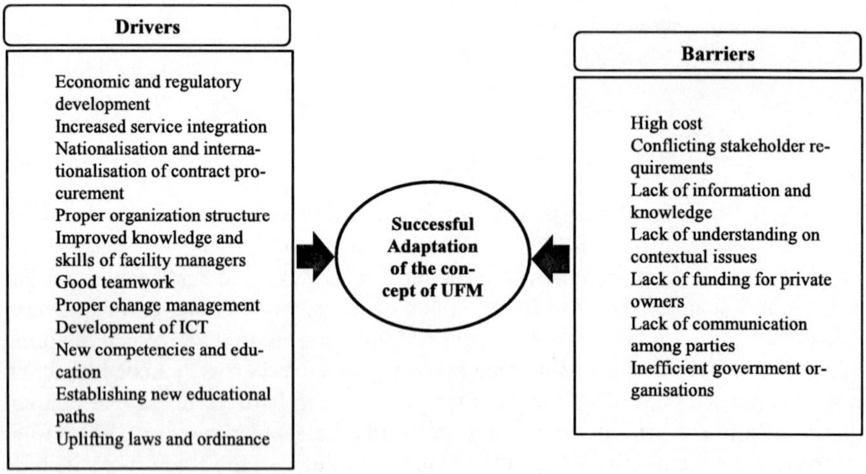

Fig. 1 Summary of the identified drivers and barriers to the successful adaptation of UFM in Sri Lanka

References

1. Alexander, K.: Community-based facilities management. Facilities **24**(7/8), 250–268 (2006). https://doi.org/10.1108/02632770610666116
2. Amaratunga, D. et al.: Process improvement in facilities management: the SPICE approach. Facilities 90–106 (2003)
3. Boyle, L., Michell, K.: Urban facilities management: a systemic process for achieving urban sustainability. Int. J. Sustain. Dev. Plan. **12**(3), 446–456 (2017). https://doi.org/10.2495/SDP-V12-N3-446-456
4. Drion, B., Melissen, F., Wood, R.: Facilities management: lost, or regained? Facilities **30**(5/6), 254–261 (2012). https://doi.org/10.1108/02632771211208512
5. Hallam, S.: 'FM research—the leading edge'. Facilities **18**(7) (2007)
6. Hosseini, M.N.: Cultural planning and urban planning. J. Urban Manage. **8**, 56–61 (2007)
7. International Facility Management Association.: Energy Efficiency Indicator Summary Report (2009). Accessed 12 December 2020
8. Larsen, J.L., et al.: Urbanising facilities management: the challenges in a creative age. Facilities **29**(1), 80–92 (2012). https://doi.org/10.1108/02632771111101340
9. Manjula, N.H.C., Senarathne, S.: Team work in facilities management. Global Challenges Construct Indus. 258–265 (2012)
10. Michell, K.: Urban facilities management: a means to the attainment of sustainable cities? J. Facil. Manag. **11**(3), 3–5 (2013). https://doi.org/10.1108/jfm.2013.30811caa.001
11. Noor, M.N.M., Pitt, M.: A critical review on innovation in facilities management service delivery. Facilities **27**(5), 211–228 (2009). https://doi.org/10.1108/02632770910944943
12. Pearce, A.R.: 'Sustainable urban facilities management', encyclopedia of sustainable technologies. Elsevier **2**, 351–363 (2017). https://doi.org/10.1016/B978-0-12-409548-9.10183-6
13. Roberts, P.: FM: new urban and community alignments. Facilities **22**(13), 349–352 (2004). https://doi.org/10.1108/02632770410563059
14. Roper, K.O., Payant, R.: In: The Facility Management Handbook. 4th edn. AMACOM (2014). Available at https://www.oreilly.com/library/view/the-facility-management/9780814432150/
15. Rosa, M.D.: Land use and land-use changes in life cycle assessment: green modelling or black boxing?'. Ecol. Econ. Elsevier B.V. **144**, 73–81 (2018). https://doi.org/10.1016/j.ecolecon.2017.07.017
16. Weerasinghe, N., Sandanayake, Y., Bandusena, P.: Collaborative facilities managementin urban regeneration projects: problems and potentials. In: Proceedings of the International Conference on 'Cities, People and Places', pp. 316–329. (2014)
17. White, M.D., Marsh, E.E.: Content analysis: a flexible methodology. Libr. Trends **55**(1), 22–45 (2006). https://doi.org/10.1353/lib.2006.0053
18. Yuan, J., et al.: Examining sustainability indicators of space management in elderly facilities—a case study in China. J. Clean. Prod. (2018). https://doi.org/10.1016/j.jclepro.2018.10.065

Fuzzy-AHP Based Design and Performance Indexing Model for Tall Buildings

Shubham Pandey, Prateek Roshan, and Shobha Ram

1 Introduction

Ever since the beginning of human civilisation, tall structures have always fascinated mankind. Right from ancient times, humans have always built structures like temples, cathedrals and pyramids, which had significant height, pointing to the sky, in their constant pursuit of reaching the sky. Nowadays, there seems to be a marathon between countries, where everybody wants to come out on top, be it, technological advancements, upgrading defence capabilities, reaching out in space, or constructing "the tallest building". Due to technological advancements and expanding interest for business and residential space, there is a critical requirement for high-rise structures.

A structure is known to be a tall building if 50% of the structure is habitable; otherwise, it is just a tall structure, not a "tall building". As per Bureau of Indian Standards, a building with a height greater than 50 m and less than 250 m is a tall building (IS 16700:2017) [1]. In this study, tall buildings within this specified height range are considered.

Tall buildings have various components, and designing a tall building requires a significant amount of knowledge about each component and their impact on the design of tall buildings. Therefore, the decision regarding design, construction and performance of a tall building in service life relies on various parameters and factors, making it a complex problem to deal with. Analytic Hierarchy Process (AHP) is quite possibly the most useful and valuable multiple criteria decision-making (MCDM)

S. Pandey (✉) · S. Ram
Gautam Buddha University, Uttar Pradesh, Greater Noida 201310, India

S. Ram
e-mail: shobharam@gbu.ac.in

P. Roshan
Delhi Technological University, Delhi 110042, India
e-mail: prateek_2k19phdce02@dtu.ac.in

I. Pal and S. Kolathayar (eds.), *Sustainable Cities and Resilience*, Lecture Notes in Civil Engineering 183, https://doi.org/10.1007/978-981-16-5543-2_30

strategy that can be used to manage such complex subjective and qualitative decision-making problems. Saaty initially developed AHP [2] which is a technique by which a complex decision-making problem can be broken down into simple and manageable components and represented as a hierarchical model. Based on the judgement of experts' opinions, priorities or weights of various components or factors are computed. After that, calculations are performed to arrive at the final result or to decide among a set of alternatives. The complete calculation process of conventional AHP is not explained here due to brevity. The AHP approach is often subjected to a certain amount of uncertainty and ambiguity because of the lack of apt decision making sometimes by experts. Due to the uncertainty that prevails in this method because of an expert's optimism and pessimism, the conventional AHP method is used with a concept of fuzzy embedded into it, making it a Fuzzy Analytic Hierarchy Process (FAHP).

The analysis of tall buildings is a very complex process; therefore, different aspects of a tall building are thoroughly reviewed to ease this. The problem is then broken down into small and manageable components, which are then incorporated into a hierarchical model by grouping these components and aspects of similar categories. The relative weightage of attributes or components at one level is determined through an exhaustive experts' judgement. In total, fifteen experts' opinions were taken into consideration by using a questionnaire survey form. The survey opinion was collected in the form of pairwise comparison tables. The α-cut method, as used by Chang [3], has been employed to calculate weights for each criterion, sub-criterion and alternatives in the present study. A design and performance assessment index for tall buildings is hence developed. This model was applied to a set of existing tall buildings, and individual scores were calculated for each building. Later on, the ranking of these buildings was carried out and also comparisons of the design and performance features are discussed. This model can efficiently and effectively accommodate both the qualitative and quantitative design, safety and performance aspects of a tall building.

2 Literature Review

2.1 Review on Analytic Hierarchy Process

The fuzzy modification of AHP was developed by Buckley [4] following Buckley's work, fuzzy-AHP methods have found many applications. Saaty [5] studied about relative technique making use of the pairwise comparisons among the elements to a common alternative. Chang [3] applied FAHP to develop an expert decision-making process. The main objective of the study was to tackle the uncertainties faced during evaluation stages, in order to avoid that the author represented expert's judgements as fuzzy triangular numbers thus the results obtained were represented in the form of decision matrices and then by using a fuzzy prioritisation method crisp values were obtained. Wang and Chin [6] found that the existing defuzzification techniques were very sophisticated and were also subjected to some significant drawbacks. Henceforth they proposed a logarithmic fuzzy preference programming method (LFPP). Chaphalkar and Shirke [7] applied MCDM for selecting the type of bridge to be constructed at a particular site. The hierarchical model was formed, and with the help of a panel of six experts, they arrived at the result. They used pairwise comparisons to arrive at the result using FAHP methodology.

2.2 *Review on Tall Buildings*

The Indian standard code IS 16700:2017 [1] is used in order to study the building configuration parameters, height limit, slenderness ratio, storey stiffness, plan geometry and the aspect ratio of the structural plan. Fu [8] well-researched floor system studies and lateral stability provisions were adopted as parameters in the study. The monograph published by the Council on Tall Buildings and Urban Habitat (CTBUH) [9] covers the topic of structural systems for tall buildings and how it has changed over the past decade. For wind provisions, the Indian standard code, IS 875 (Part 3) [10] was adopted. Mendis et al. [11] completely cover the primary and essential aspects associated with the design of tall buildings on the effects of wind loading.

Parameters covered under this study include wind speed, design wind loads, types of wind design, design criteria, along and crosswind loading directions. The concept of vortex shedding and how it can be avoided by doing some aerodynamic modifications to the building, and avoiding any vibrations which cause occupants inconvenience along with a brief introduction of purpose-designed damping systems are also studied in this paper. Li et al. [12] evaluated the factors that affect the fire resistance of high-rise buildings and how efficiently a building is designed to overcome the fire outbreak in the building. Fire safety in a tall building is a crucial safety concern that needs to be addressed at the design and planning stage. To take care of this aspect, fire safety factor has been included in this study which consists of parameters like fire safety hardware facilities, e.g. availability of a proper fire extinguishing system, a proper smoke control system, availability of a fire lift and a properly functioning alarm system.

3 Methodology

In the present study, Saaty's fundamental scale of absolute numbers (Saaty 1980) [2], which uses the scale of 1, 3, 5, 7 and 9 for relative comparison of the importance of two criteria over one another was used for procuring the experts' judgement. The typical AHP procedure for calculation of weight matrices is carried out as used by Roshan et al. [13, 14] and Sharma et al. [15]. AHP is very useful for checking the consistency of the decision-maker's evaluations or the experts' judgement. The consistency of

Table 1 Random index (R.I.) values given by Saaty [2]

n	1	2	3	4	5	6	7	8	9	10
R.I	0.00	0.00	0.58	0.90	1.12	1.24	1.32	1.41	1.45	1.49

the judgemental matrix can be measured by the term known as consistency ratio (C.R.).

C.R. is defined as:

$$\text{Consistency ratio(CR)} = \text{Consistency Index(CI)}/\text{Random Index(RI)} \quad (1)$$

The consistency index (C.I.) can be given as

$$\text{C.I.} = \frac{\lambda_{\max} - n}{n - 1} \quad (2)$$

where "n" is the number of evaluation criteria considered, and $\lambda_{\max}$ is the Eigen value. A perfectly consistent decision gives C.I = 0, but small values of inconsistency may be tolerated. If the consistency ratio exceeds 0.1, then the set of judgements are not consistent and hence are not reliable. R.I. is the random index, i.e., the consistency index when the entries of matrix "A" are completely random. The values of R.I. for different values of "n" are shown in Table 1.

3.1 *The Fuzzy-AHP Methodology*

In normal AHP method, developed by Saaty, there were some ambiguities and uncertainties. So to improve the concept of fuzzy, first introduced by Zadeh [16], was incorporated to AHP, to form a new technique called Fuzzy Analytic Hierarchy Process (FAHP) methodology. The earliest published work on FAHP was by Laarhoven [17]. In fuzzy-AHP, the experts' opinions are considered fuzzy values that spread around a crisp central value with different degrees. This degree of association is called the membership function. There are three major elements of FAHP methodology, namely, hierarchy construction, priority analysis and consistency verification. The utilisation of fuzzy logic in an MCDM problem has made the process for comparing multiple criteria and attributes has become free of any uncertainties and ambiguity. This process of representing the parameters into linguistic terms is known as fuzzification.

Defuzzification is a process of breaking the linguistic term of the problem into crisp numbers. There are numerous defuzzification techniques, like the geometric mean method, fuzzy logarithmic least square method, the centre of gravity method, lambda-max method and α-Cut method [18–21]. In this study, the α-cut method has been implemented for the defuzzification of the data by considering a triangular membership function.

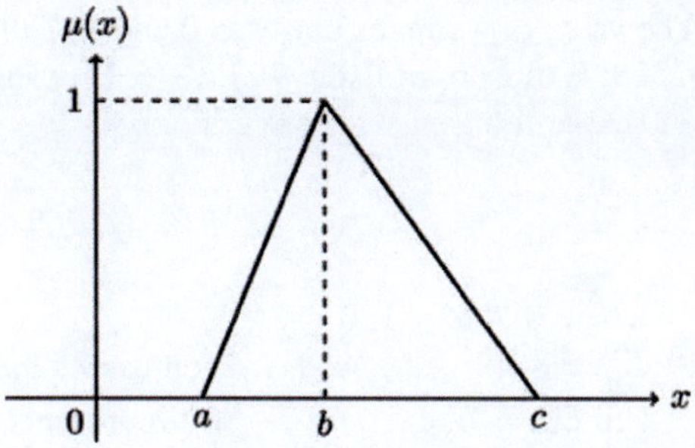

Fig. 1 Graphical representation of triangular fuzzy number

α-cut Method

In the α-cut method, the opinions of the expert are demonstrated as a triangular area between L and U values representing the lower and upper limits of the membership function, respectively, and M is the geometric mean of experts' opinion representing the major value of the shape function. Using the values of L, M and U a fuzzy comparison matrix is obtained.

$$\overline{a}_{ij} = \left(L_{ij}, M_{ij}, U_{ij}\right) \text{ where } L_{ij} \le M_{ij} \le U_{ij}$$
$$\text{and } L_{ij}, M_{ij}, U_{ij} \in \left[\tfrac{1}{9}, 1\right] \cup [1, 9]$$

$$L_{ij} = \min(B_{ijk}),\ M_{ij} = \sqrt[n]{\prod_{k=1}^{n} B_{ijk}} \quad (k = 1, 2, 3, \ldots, n)\ U_{ij} = \max(B_{ijk}) \tag{3}$$

where B_{ijk} represent opinions of expert k for the relative comparisons of two criteria i and j. So, the expert opinions obtained are demonstrated in a triangular area. A graphical representation is shown in Fig. 1. $\overline{a}_{ij}$ is an element of fuzzy comparison matrix as represented by Eq. (4), where, U_{ij},M_{ij} and L_{ij} denotes the figures of the highest, geometric mean, and the lowest values of the experts' opinion, respectively.

$$\left[A\right] = \overline{a}_{ij} = \begin{pmatrix} 1 & \overline{a}_{12} & \ldots & \overline{a}_{1n} \\ \frac{1}{\overline{a}_{12}} & 1 & \ldots & \overline{a}_{2n} \\ \vdots & \vdots & \vdots & \vdots \\ \frac{1}{\overline{a}_{1n}} & \frac{1}{\overline{a}_{2n}} & \ldots & 1 \end{pmatrix} \tag{4}$$

The α-cut method tends to show the uncertainty (α) and risk tolerance (λ) in the fuzzy opinion of the decision-maker. The degree of uncertainty is maximum and minimum for the two values of α at 1 and 0, respectively. The value of α can vary from 0 to 1. It can also be showed as a degree of stability or fluctuation. The increase in value of decision making means increase in the stability and reduction in variance of decision making. Risk tolerance (λ) is reviewed as degree of expert's pessimism.

The value of λ ranges between 0 and 1. The value of $\lambda = 0$ means that the decision-maker is most optimistic, while $\lambda = 1$ means the decision-maker is most pessimistic and expert consensus is lower bound.

$$\left(a_{ij}^{\alpha}\right)^{\lambda} = \left(\lambda \cdot L_{ij}^{\alpha} + (1 - \lambda)U_{ij}^{\alpha}\right) \tag{5}$$

where $0 \leq \alpha \leq 1, 0 \leq \lambda \leq 1$,
$L_{ij}^{\alpha} = \left(M_{ij} - L_{ij}\right)\alpha + L_{ij}$ represents the left end value of the α-cup for a_{ij},
and $U_{ij}^{\alpha} = U_{ij} - \alpha\left(U_{ij} - M\right)$ represents the right end value of the α- cup for a_{ij}.
Finally, the resultant defuzzified single pairwise comparison matrix is generated as given in Eq. (6).

$$\left[(A^{\alpha})\right]^{\lambda} = (a_{ij}^{\alpha})^{\lambda} = \begin{pmatrix} 1 & \left(a_{12}^{\alpha}\right)^{\lambda} & \ldots & \left(a_{1n}^{\alpha}\right)^{\lambda} \\ \left(a_{21}^{\alpha}\right)^{\lambda} & 1 & \ldots & \left(a_{2n}^{\alpha}\right)^{\lambda} \\ \vdots & \vdots & \vdots & \vdots \\ (a_{n1}^{\alpha})^{\lambda} & (a_{n2}^{\alpha})^{\lambda} & \ldots & 1 \end{pmatrix} \tag{6}$$

4 Proposed Model for Performance Assessment Indexing of Tall Buildings

The performance assessment index model of tall buildings is divided into four sub-categories, design, lateral stability provisions, structural systems and fire safety. The design of tall buildings is further sub-divided into building configuration, foundation type, floor systems and soil type. The flow chart of the proposed study is shown on Fig. 2.

Building configuration is divided into elevation and plan. Elevation has been further sub-divided into height limit, slenderness ratio and storey stiffness. The plan of a tall building is then further divided into plan geometry and plan aspect ratio. The plan geometry of a tall building is sub-divided into rectangular, circular, square and elliptical shapes. Floor systems used in a typical tall building can be broadly divided into concrete floor systems and composite floor systems. The concrete floor system can be divided into three categories based on slabs used: solid slab, post-tensioned slab and flat slabs. Composite floor system used in a tall building is divided into solid R.C., precast slab and composite truss. The type of foundation used in tall buildings is sub-divided into pile, raft and composite pile and raft foundation. The soil type of the site is divided into soft, medium and hard.

Lateral load resisting provisions of tall buildings includes wind and seismic provisions. Wind provisions further include topography, aerodynamic modifications and dampers. The topography is further divided into four main terrains based on the terrain of the particular building site. Various engineering enhancements like

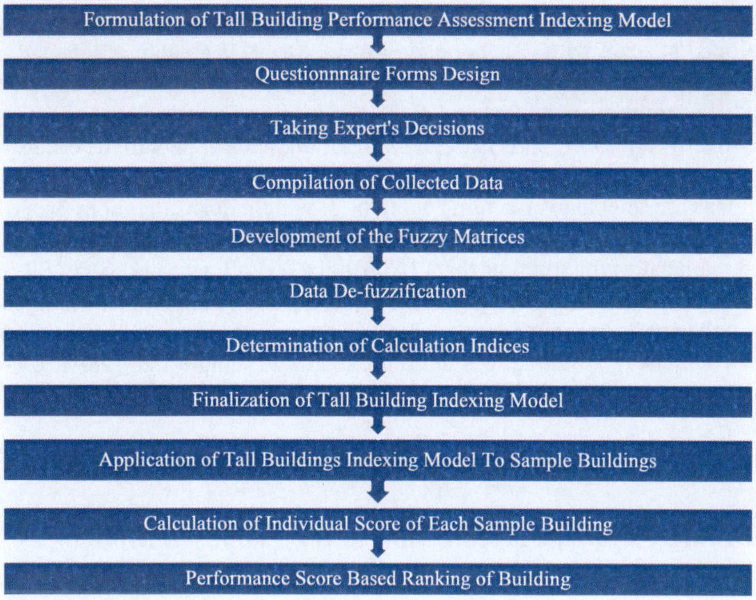

Fig. 2 Flow chart of the proposed study

tapering, twisting, porosity and corner shortening may be done to tackle the high wind loading in the tall buildings.

Dampers used in tall buildings are broadly divided into three categories: active, passive and semi-active. To deal with seismic loading in a tall building, seismic provision is taken at the designing level. These can be further divided into base isolation, shock absorber, damping systems and provisions for the shear wall. Structural systems used in constructing tall buildings are broadly divided into the rigid frame, braced and shear-walled frame, outrigger and tubular frame. Fire safety is divided into two sub-criteria: warning systems and emergency evacuation. The warning system is further sub-divided into the fire alarm and smoke control. The emergency evacuation includes staircase width, staircase per floor, evacuation guidance and fire lift.

To facilitate the development of the hierarchical model and to identify the critical factors that influence the design of tall buildings, an exhaustive review of past literature was carried [8–12, 22–27].

4.1 *Questionnaire Formation and Collection of Data*

After the development and finalisation of the tall buildings indexing hierarchical model, a detailed questionnaire in the form of tables (Appendix I) for comparing parameters at each hierarchical level, a feature of AHP, was crafted to gather the experts' opinions. The comparative assessment between the two criteria "*a*" and "*b*" is developed at scale with values ranging from 1 to 9. If a value is selected on the left side of the scale, that would mean that "*a*" has more weightage than "*b*" which means "*a*" is more important than "*b*" and vice-versa. A value of 9 towards right implies that the criterion "*b*" is extremely important with respect to the criterion "*a*". Similarly, a value of 7 towards left means criterion "*a*" is stronger than "*b*" and so on. Suppose the selected value turns out to be 1, which means that both the criterion is equally important. With the questionnaire survey form, the experts were given a detailed explanatory instruction sheet (Appendix II) on how to properly fill out the survey forms and do the scaling part. These questionnaire survey forms (Appendix I) were sent to experts engaged in the academics, construction and designing industries. AHP is a subjective method, does not require a large sample size [28–30]. For this reason, data were collected from a total of fifteen experts through the questionnaire form.

4.2 *Buildings for the Application of the Proposed Model*

After the successful distribution of questionnaire survey forms to the experts for their judgement, data collection was done. The data collection was done by visiting the site location and talking to the respective construction authorities to get all the relevant data for the study. The buildings which were taken for consideration in this study are given in Table 2. The names of the five buildings were given as A-96, B-106, C-155, D-130 and E-55 with the numbers after alphabets representing the respective height of the buildings.

5 Calculations Involved

Fuzzy matrices were developed from the compiled data obtained from survey opinion forms by using values of L_{ij}, M_{ij} and U_{ij}. Calculation of L_{ij}, M_{ij}, U_{ij} is done using Eq. (3). Hence, the table obtained after the calculation is shown in Table 3. The values obtained from the experts' opinion through the pairwise comparisons were shown as

Table 2 Buildings for the proposed tall buildings performance assessment index model application

S.No	Buildings	Location
1	A-96	Ghaziabad
2	B-106	Noida city centre
3	C-155	Noida
4	D-130	Gurugram
5	E-55	Gurugram

Table 3 Experts' judgement with the values of L_{ij}, M_{ij}, U_{ij}

Criteria "A"	Criteria "B"	L_{ij}	M_{ij}	U_{ij}
Height limit	Slenderness ratio	1/5	0.890	7
Height limit	Storey stiffness	1/5	2.629	7
Slenderness ratio	Storey stiffness	1/5	1.817	6

Table 4 Fuzzified matrix for different parameters with respect to: elevation

Criteria "a"/"b"	Height limit	Slenderness ratio	Storey stiffness
Height limit	1,1,1	1/5, 0.890, 7	1/5, 2.629, 7
Slenderness ratio	1/7, 1.124, 5	1,1,1	1/5, 1.817, 6
Storey stiffness	1/7, 0.380, 5	1/6, 0.550, 5	1,1,1

an arrangement of fuzzy matrices by means of the values of L_{ij}, M_{ij} and U_{ij}. Hence, a fuzzified matrix is obtained which is shown in Table 4.

Fuzzy matrices values are converted into a crisp value by using Eq. (5). These values are carried out by taking the central value (0.5) of uncertainty (α) and risk tolerance (λ) factors in the fuzzy opinion of the decision-makers from 0.1 to 0.9. Hence, the obtained values of $L_{ij}{}^{\alpha}$, $U_{ij}{}^{\alpha}$ and $(a_{ij}{}^{\alpha})^{\lambda}$ are given in Table 5. Hence, the resultant defuzzified matrix obtained is given in Table 6. The principal Eigenvalue and Eigenvector of these defuzzified matrices are evaluated. Eigenvalues of defuzzified matrix obtained are shown as "λ_{max}", whereas normalised Eigenvector of the same is represented as "w". For checking, the consistency of the comparison matrix of performance model of tall buildings Eqs. (1) and (2) were used. The whole procedure

Table 5 Calculation of $L_{ij}{}^{\alpha}$, $U_{ij}{}^{\alpha}$ and $(a_{ij}{}^{\alpha})^{\lambda}$ for performance assessment index of tall buildings

α	λ	L_{ij}	M_{ij}	U_{ij}	$L_{ij}{}^{\alpha}$	$U_{ij}{}^{\alpha}$	$(a_{ij}{}^{\alpha})^{\lambda}$
0.5	0.5	1/5	0.890	7	0.5448	3.9448	2.2448
0.5	0.5	1/5	2.629	7	1.4147	4.8147	3.1147
0.5	0.5	1/5	1.817	6	1.0086	3.9086	2.4586

Table 6 Defuzzified matrix for different parameters with respect to: elevation

Criteria 'a'/'b'	Height limit	Slenderness ratio	Storey stiffness
Height limit	1	2.2448	3.1147
Slenderness ratio	0.445478	1	2.4586
Storey stiffness	0.321062	0.406742	1

Table 7 Values of different calculation indices for the sample matrix

n	Random Index	Eigen value, λ_{max}	Eigen vectors, w	Consistency Index, CI	Consistency Ratio, CR
3	0.58	3.0366	0.5516, 0.3000, 0.1484	0.0183	0.0315

is repeated for all the comparison matrices obtained in the study. Hence, the obtained values of all calculation indices for the matrix have been summarised in Table 7.

5.1 *Scoring for the Performance Assessment Indexing Model of Tall Building*

For the calculation of the scores, a simple and systematic approach has been evolved in this study. All the criteria related to the performance assessment indexing of tall buildings are placed in the first columns with their respective weights placed in the next column. Columns next to alternatives contain the values of measurable items and observed items. The measurable items at the very end of the hierarchical model have a value equal to 1, and the corresponding complementary value is equal to 0. For the calculation of scores, a weighted sum is used for any given option with respect to each criterion. Similarly, the score at the topmost level is determined. A particular building's maximum score is achieved when all the performance assessment index model criteria meet their maximum weights obtained. The maximum score thus found for performance assessment indexing of tall buildings is 0.6048. Similarly, the index value for each building is calculated and the final scores thus obtained are shown in Table 8.

Table 8 Total performance assessment weighted score of building with obtained relative normalised scores (NS) and their corresponding ranks

S. no	Buildings	Weighted score	Normalised Score (NS) of the building $NS = \frac{WS}{WS_{max}} \times 100$	Rank
1	A-96	0.3456	57.1429	2
2	B-106	0.3169	52.3975	3
3	C-155	0.3080	50.9259	4
4	D-130	0.2842	46.9907	5
5	E-55	0.3870	63.9881	1

5.2 *Relative Performance Assessment Index of Tall Building*

Once the weighted score of the sample buildings is calculated, they are further normalised at the base of 100. The scaling procedure taken up for the study is explained below.

For example, let us assume *WS* is the obtained "weighted score" for a sample tall building and WS_{max} (i.e., 0.6048) is the extreme likely score possible in this indexing study. The calculation of the "normalized score" at base of 100 can be represented as "*NS* 100" and is evaluated as shown in Eq. (7):

$$NS = \frac{WS}{WS_{max}} \times 100 \tag{7}$$

The relative *NS* scores hence obtained with their corresponding ranking based on performance score are represented in Table 8.

6 Results and Discussions

The performance assessment index model for tall buildings developed in this study was applied to five sample tall buildings. The scores obtained for each building were then normalised on the scale of 100 to obtain Relative NS score for a particular building. The sample tall buildings were ranked according to their Relative NS scores.

E-55, with a score of 63.9881 was ranked first. E-55 is a residential building having 16 floors. The main reason behind the building's better performance score is the introduction of an aerodynamic modification in the building; the corner shortening technique (corner chamfering) is adopted in the building to overcome any wind-induced vibrations. In contrast, no such modifications were present in any other buildings taken in the study. In the past, researchers [31] have reported that chamfered corners effectively reduce the dynamic along-wind and crosswind responses in comparison with the basic rectangular shape of a tall building.

The building with the second-highest relative NS score was found to be A-96 with a score of 57.1429. Due to the lack of aerodynamic modifications in the buildings, it was ranked slightly below the E-55.

A score of 52.3975 was obtained in B-106. It performed well in all the design parameters but lacked in fire safety features as there were no smoke control devices installed in the building and also no proper evacuation guidance system was installed for the safety of the occupants.

C-155 was ranked fourth in this study with an NS score of 50.9259. C-155 has two identical towers. Both have 45 storeys. The height limit criteria, adopted from IS: 16,700 (2017) [1], was not found to be in "In range" criteria. This caused the relative score of building to fall short.

D-130 was the least performing building, among others. Due to not being in an "In range" category of height limit according to IS: 16,700 (2017) [1], D-130 performed poorly in this assessment. D-130 had the best safety features among other studied buildings, owing to the presence of a proper evacuation guidance system installed in the building. Hence, safety features give a slight upper hand to this building.

6.1 General Findings

Some of the typical findings for the tall buildings studied are discussed below:

1. In most of the studied buildings, there were no aerodynamic modifications to counter wind vibrations except in E-55 where the corner shortening technique was adopted.
2. All the studied buildings considered had a concrete floor system, making use of one-way and two-way slabs. There was no use of a flat slab or any post-tensioned slab in any of the buildings.
3. There was no single building making use of any dampers to tackle the wind and seismic induced vibrations. In order to avoid the vibrations, heavy foundations were used rather than using a damper.
4. Provision for the shear wall was adopted in each of the buildings to reduce seismic damages. Techniques like base isolation, using shock absorbers and damping systems were not found in any of the building considered in the study. Generally, tall buildings are more flexible and have a larger mass; therefore, they have a longer natural period. Past researchers [32] have reported that the incorporation of base isolation may reduce inter-storey drifts and floor level accelerations considerably in buildings.
5. In order to protect the building and the its occupants during any fire outbreak, fire lift was not used in any of the buildings, although all the essential fire safety features like smoke control and fire alarm system were properly installed.

6.2 Suggestions

General findings of the buildings were analysed, and their effect on each of the criteria was studied. Some suggestions to improve the behaviour of tall building and consequently the relative score of the building which is discussed below:

Designers and clients must prioritise occupants' safety by making it necessary to incorporate the techniques like base isolation and damping systems wherever required. Cost-cutting should not be done at the expense of the safety of the occupants because this can prove to be disastrous in case of any major hazard. From the observations, it was also seen that influence of nearby structures in the construction of a new building is normally assessed, but the reverse is often sidelined. Designers must also pay sufficient attention to the aerodynamic interference effect on the surrounding structures by constructing a newly proposed tall building in an area, as proposed by Quan et al. [33].

Fire lift is a lift mainly installed to enable the fire service personnel to reach the upper floor with minimum delay, they are recommended by the Model Building By-Laws [34]. The speed of such a lift should be such that firefighters can reach the top floor from the ground floor within a minute. To avoid any electric failure, the electric supply for this lift must be kept separate from the building's electric supply mains. The building bye-laws also recommend using fire escapes or exterior staircases, whose entrance must be separate and remote from the building's internal staircases.

7 Conclusions

In the past few decades, the rapid urbanisation and population growth have forced the designers to build tall buildings to fulfil residential and commercial requirements. Design, safety and performance aspect of a tall building are different from the regular structures. This study is an effort to encompass all such parameters into a simple, easy to understand hierarchical unit, which is very simple and can highlight the crucial parameters and make it possible to assess the comparative need and requirement of one factor and criterion over the other. FAHP has been utilised for carrying out the same. The generated model is applied to a set of buildings, and the individual scores are obtained, and hence, a detailed study is carried out to investigate the reasons and factors responsible for the high and low scores of all the studied buildings. In the end, the prevailing design practices and norms are reviewed and some suggestions have been summarised considerations of which can further enhance and improve the likely performance of the tall buildings.

More detailed structural investigation and inclusion of practical factors for a better analysis can be incorporated for a more refined result as a future scope.

Acknowledgements The authors would like to thank Mr. Nishant Bhardwaj, Structural Engineer, for his valuable guidance and sharing his structural design expertise in refining the present model. The authors also thank Mr. Siraj Husain, Structural Engineer, for his timely help throughout the work.

Appendix I—Questionnaire Form

***Refer to Appendix II: Explanatory Sample Sheet for Filling the Questionnaire Form*

Table 1.	Compare the relative preference with respect to : Parameters governing the design of Tall Buildings																	
		Extreme		Very Strong		Strong		Moderate		Equal		Moderate		Strong		Very Strong		Extreme	
S.No.	Criteria 'a'	9	8	7	6	5	4	3	2	1	2	3	4	5	6	7	8	9	Criteria 'b'
		Rating Scale																	
1	Design																		Lateral Stability Provisions
2	Design																		Structural Systems
3	Design																		Fire Safety
4	Lateral Stability Provisions																		Structural Systems
5	Lateral Stability Provisions																		Fire Safety
6	Structural Systems																		Fire Safety

Table 2.	Compare the relative preference with respect to : Design																		
		Extreme		Very Strong		Strong		Moderate		Equal		Moderate		Strong		Very Strong		Extreme	
S.No.	Criteria 'a'	9	8	7	6	5	4	3	2	1	2	3	4	5	6	7	8	9	Criteria 'b'
		Rating Scale																	
1	Building configuration																		Floor Systems
2	Building configuration																		Foundation Type
3	Building configuration																		Soil Type
4	Floor Systems																		Foundation Type
5	Floor Systems																		Soil Type
6	Foundation Type																		Soil Type

Table 3.	Compare the relative preference with respect to : Building Configuration																	
		Extreme		Very Strong		Strong		Moderate		Equal		Moderate		Strong		Very Strong		Extreme	
S.No.	Criteria 'a'	9	8	7	6	5	4	3	2	1	2	3	4	5	6	7	8	9	Criteria 'b'
		Rating Scale																	
1	Elevation																		Plan

Table 4.	Compare the relative preference with respect to : Elevation																	
		Extreme		Very Strong		Strong		Moderate		Equal		Moderate		Strong		Very Strong		Extreme	
S.No.	Criteria 'a'	9	8	7	6	5	4	3	2	1	2	3	4	5	6	7	8	9	Criteria 'b'
		Rating Scale																	
1	Height Limit																		Slenderness Ratio
2	Height Limit																		Storey Stiffness
3	Slenderness Ratio																		Storey Stiffness

Table 5.	Compare the relative preference with respect to : Plan																	
		Extreme		Very Strong		Strong		Moderate		Equal		Moderate		Strong		Very Strong		Extreme	
S.No.	Criteria 'a'	9	8	7	6	5	4	3	2	1	2	3	4	5	6	7	8	9	Criteria 'b'
		Rating Scale																	
1	Plan Geometry																		Plan Aspect Ratio

Table 6.	Compare the relative preference with respect to : Plan Geometry																		
		Extreme		Very Strong		Strong		Moderate		Equal		Moderate		Strong		Very Strong		Extreme	
S.No.	Criteria 'a'	9	8	7	6	5	4	3	2	1	2	3	4	5	6	7	8	9	Criteria 'b'
		Rating Scale																	
1	Rectangular																		Circular
2	Rectangular																		Square
3	Rectangular																		Elliptical
4	Circular																		Square
5	Circular																		Elliptical
6	Square																		Elliptical

Table 7.	Compare the relative preference with respect to : Floor Systems																		
		Extreme		Very Strong		Strong		Moderate		Equal		Moderate		Strong		Very Strong		Extreme	
S.No.	Criteria 'a'	9	8	7	6	5	4	3	2	1	2	3	4	5	6	7	8	9	Criteria 'b'
		Rating Scale																	
1	Concrete Floor Systems																		Composite Floor System

Table 8.	Compare the relative preference with respect to : Concrete Floor Systems																		
		Extreme		Very Strong		Strong		Moderate		Equal		Moderate		Strong		Very Strong		Extreme	
S.No.	Criteria 'a'	9	8	7	6	5	4	3	2	1	2	3	4	5	6	7	8	9	Criteria 'b'
		Rating Scale																	
1	Solid Slab																		Flat Slab
2	Solid Slab																		Post-tensioned Slab
3	Flat Slab																		Post-tensioned Slab

Table 9.	Compare the relative preference with respect to : Composite Floor System																		
		Extreme		Very Strong		Strong		Moderate		Equal		Moderate		Strong		Very Strong		Extreme	
S.No.	Criteria 'a'	9	8	7	6	5	4	3	2	1	2	3	4	5	6	7	8	9	Criteria 'b'
		Rating Scale																	
1	Solid R.C.																		Precast Slab
2	Solid R.C.																		Composite Truss
3	Precast Slab																		Composite Truss

Table 10.	Compare the relative preference with respect to : Foundation Types																		
		Extreme		Very Strong		Strong		Moderate		Equal		Moderate		Strong		Very Strong		Extreme	
S.No.	Criteria 'a'	9	8	7	6	5	4	3	2	1	2	3	4	5	6	7	8	9	Criteria 'b'
		Rating Scale																	
1	Pile																		Raft
2	Pile																		Composite Pile and Raft
3	Raft																		Composite Pile and Raft

Table 11. Compare the relative preference with respect to : Soil Type

		Extreme		Very Strong		Strong		Moderate		Equal		Moderate		Strong		Very Strong		Extreme	
S.No.	Criteria 'a'	9	8	7	6	5	4	3	2	1	2	3	4	5	6	7	8	9	Criteria 'b'
		Rating Scale																	
1	Soft																		Medium
2	Soft																		Hard
3	Medium																		Hard

Table 12. Compare the relative preference with respect to : Lateral Load resisting provisions

		Extreme		Very Strong		Strong		Moderate		Equal		Moderate		Strong		Very Strong		Extreme	
S.No.	Criteria 'a'	9	8	7	6	5	4	3	2	1	2	3	4	5	6	7	8	9	Criteria 'b'
		Rating Scale																	
1	Wind Provisions																		Seismic Provisions

Table 13. Compare the relative preference with respect to : Wind Provisions

		Extreme		Very Strong		Strong		Moderate		Equal		Moderate		Strong		Very Strong		Extreme	
S.No.	Criteria 'a'	9	8	7	6	5	4	3	2	1	2	3	4	5	6	7	8	9	Criteria 'b'
		Rating Scale																	
1	Topography																		Aerodynamic modifications
2	Topography																		Dampers
3	Aerodynamic modifications																		Dampers

Table 14. Compare the relative preference with respect to : Topography

S.No.	Criteria 'a'	9	8	7	6	5	4	3	2	1	2	3	4	5	6	7	8	9	Criteria 'b'
		Extreme		Very Strong		Strong		Moderate		Equal		Moderate		Strong		Very Strong		Extreme	
		Rating Scale																	
1	Terrain 1																		Terrain 2
2	Terrain 1																		Terrain 3
3	Terrain 1																		Terrain 4
4	Terrain 2																		Terrain 3
5	Terrain 2																		Terrain 4
6	Terrain 3																		Terrain 4

Table 15. Compare the relative preference with respect to : Aerodynamic Modifications

S.No.	Criteria 'a'	9	8	7	6	5	4	3	2	1	2	3	4	5	6	7	8	9	Criteria 'b'
		Extreme		Very Strong		Strong		Moderate		Equal		Moderate		Strong		Very Strong		Extreme	
		Rating Scale																	
1	Tapering																		Twisting
2	Tapering																		Porosity
3	Tapering																		Corner Shortening
4	Twisting																		Porosity
5	Twisting																		Corner Shortening
6	Porosity																		Corner Shortening

Table 16.	Compare the relative preference with respect to : Dampers																		
		Extreme		Very Strong		Strong		Moderate		Equal		Moderate		Strong		Very Strong		Extreme	
S.No.	Criteria 'a'	9	8	7	6	5	4	3	2	1	2	3	4	5	6	7	8	9	Criteria 'b'
		Rating Scale																	
1	Active																		Passive
2	Active																		Semi-active
3	Passive																		Semi-active

Table 17.	Compare the relative preference with respect to : Seismic Provisions																		
		Extreme		Very Strong		Strong		Moderate		Equal		Moderate		Strong		Very Strong		Extreme	
S.No.	Criteria 'a'	9	8	7	6	5	4	3	2	1	2	3	4	5	6	7	8	9	Criteria 'b'
		Rating Scale																	
1	Base Isolation																		Shock Absorber
2	Base Isolation																		Damping Systems
3	Base Isolation																		Provision for Shear Wall
4	Shock Absorber																		Damping Systems
5	Shock Absorber																		Provision for Shear Wall
6	Damping Systems																		Provision for Shear Wall

Table 18.	Compare the relative preference with respect to : Structural Systems																		
		Extreme		Very Strong		Strong		Moderate		Equal		Moderate		Strong		Very Strong		Extreme	
S.No.	Criteria 'a'	9	8	7	6	5	4	3	2	1	2	3	4	5	6	7	8	9	Criteria 'b'
		Rating Scale																	
1	Rigid Frame																		Braced and Shear-walled Frame
2	Rigid Frame																		Outrigger
3	Rigid Frame																		Tubular
4	Braced and Shear-walled Frame																		Outrigger
5	Rigid Frame																		Tubular
6	Outrigger																		Tubular

Table 19.	Compare the relative preference with respect to : Fire Safety																		
		Extreme		Very Strong		Strong		Moderate		Equal		Moderate		Strong		Very Strong		Extreme	
S.No.	Criteria 'a'	9	8	7	6	5	4	3	2	1	2	3	4	5	6	7	8	9	Criteria 'b'
		Rating Scale																	
1	Warning Systems																		Emergency Evacuation

Table 20.	Compare the relative preference with respect to : Warning Systems																		
		Extreme		Very Strong		Strong		Moderate		Equal		Moderate		Strong		Very Strong		Extreme	
S.No.	Criteria 'a'	9	8	7	6	5	4	3	2	1	2	3	4	5	6	7	8	9	Criteria 'b'
		Rating Scale																	
1	Fire Alarm																		Smoke Control

Table 21.	Compare the relative preference with respect to : Emergency Evacuation																		
		Extreme		Very Strong		Strong		Moderate		Equal		Moderate		Strong		Very Strong		Extreme	
S.No.	Criteria 'a'	9	8	7	6	5	4	3	2	1	2	3	4	5	6	7	8	9	Criteria 'b'
		Rating Scale																	
1	Staircase Width																		Staircase per floor
2	Staircase Width																		Evacuation guidance
3	Staircase Width																		Fire Lift
4	Staircase per floor																		Evacuation guidance
5	Staircase per floor																		Fire Lift
6	Evacuation guidance																		Fire Lift

Comments and reviews (optional):
Kindly provide some details about you

Name:
Maximum Qualification:
Expertise in:
Organisation and post:
Experience (in years):

Thank You for Your Precious Time and Effort.

Appendix II: Explanatory Sample Sheet

Mark your opinion about the relative importance of the factors given on the two sides of the scales of category 'a' over category 'b' (refer the scoring pattern). Please put tick marks on the number of your choice on each scale.

If category 'a' is more important, move towards left hand side (LHS) of the scale.

If category 'a' and 'b' are equally important, put tick mark on center portion (MIDDLE) of the scale.

If category 'b' is more important, move towards right hand side (RHS) of the scale.

Scoring Pattern

Relative importance	Score
Equal importance	1
Slightly more important	3
Strongly more important	5
Very strongly important	7
Absolutely more important	9
Intermediate values	2, 4, 6, 8

Sample: If relative preference of the following sample table is to be filled.

Table 1.	**Compare the relative preference with respect to: Performance Index Assessment of Building**																		
		Extreme		Very strong		Strong		Moderate		Equal		Moderate		Strong		Very strong		Extreme	
S.No.	**Criteria ‘a’**	9	8	7	6	5	4	3	2	1	2	3	4	5	6	7	8	9	**Criteria ‘b’**
		Rating Scale																	
1.	Height Limit																		Slenderness Ratio
2.	Height Limit																		Storey Stiffness
3.	Slenderness Ratio																		Storey Stiffness

For example,

- If you think “Height Limit” (element of criteria ‘a’) is very strongly important over “Slenderness Ratio (element of criteria ‘b’), then, write “**R**” under column 7 towards criteria ‘a’ (i.e. towards LHS).
- If you think “Storey Stiffness” (element of criteria ‘b’) is strongly important over “Height Limit” (element of criteria ‘a’), then, write “**R**” under column 5 towards criteria ‘b’ (i.e. towards RHS).
- If you think “Slenderness Ratio” (element of criteria ‘a’) is equally important to “Storey Stiffness” (element of criteria ‘b’), then, write “**R**” under column 1 (i.e. in the middle).

Table 1.	Compare the relative preference with respect to: Performance Index Assessment of Building																		
		Extreme		Very strong		Strong		Moderate		Equal		Moderate		Strong		Very strong		Extreme	
S.No.	Criteria 'a'	9	8	7	6	5	4	3	2	1	2	3	4	5	6	7	8	9	Criteria 'b'
		Rating Scale																	
1.	Height Limit			R															Slenderness Ratio
2.	Height Limit													R					Storey Stiffness
3.	Slenderness Ratio									R									Storey Stiffness

Likewise, I would like to express your opinion as expert to select amongst the alternatives (or elements) for comparing their relative weightage for all the tables in the questionnaire.

References

1. IS 16700:2017: Criteria for structural safety of tall buildings, special structures Sectional committee CED-38 (2007)
2. Saaty, T.L.: In: Analytical Hierarchical Process, McGraw hill. N. Y (1980)
3. Chang, C.-W.: Applying fuzzy hierarchy multiple attributes to construct an expert decision making process. Expert Syst. Appl. **36**, 7363–7368 (2009)
4. Buckley, J.J.: Fuzzy hierarchical analysis. Fuzzy Sets Syst. (1985)
5. Saaty, T.L.: Analytical Hierarchy Process. University of Pittsburgh, Pittsburgh, USA (1986)
6. Wang, Y.M., Chin, K.-S.: Fuzzy analytic hierarchy process: a logarithmic fuzzy preference programming methodology. Int. J. Approximate Reason. **52**, 541–553 (2011)
7. Chaphalkar, N.B., Shirke, P.P.: Application of multi-criteria decision making techniques for bridge construction. Int. J. Innov. Res. Sci. Eng. Technol. **2** (2013)
8. Fu, F.: Design and Analysis of Tall and Complex Structures. Elsevier Science, United States (2018)
9. Council on Tall Buildings and Urban Habitat (CTBUH).: Structural systems for tall buildings. McGraw-Hill, Inc. (1995)
10. IS 875 (Part 3): 2015: Design loads (other than earthquake) for buildings and structures –code of practice, Part 3 wind loads, Special Structures Sectional committee CED-37 (2015)
11. Mendis, P. et al.: Wind loading on tall buildings. EJSE Special Issue: Loading on Structures (2007)
12. Li, S.Y., et al.: Fire risk assessment of high-rise buildings based on gray-FAHP mathematical model. Proc. Eng. **211**, 395–402 (2018)
13. Roshan, P., Pal, S., Kumar, R.: Seismic damage assessment index of buildings using fuzzy-AHP approach. Int. J. Tech. Innov. Mod. Eng. Sci. **4**(08) (2018)
14. Roshan, P., Pal, S., Kumar, R.: Performance assessment indexing of buildings through fuzzy AHP methodology. In: Ahmed, S., Abbas, S., Zia, H. (eds.) Smart Cities—Opportunities and Challenges. Lecture Notes in Civil Engineering, vol 58. Springer, Singapore (2020)
15. Sharma, R., Roshan, P., Ram S.: Fuzzy-AHP-Based indexing model for performance assessment of highways and expressways. In: Singh, D., Awasthi, A.K., Zelinka, I., Deep, K. (eds) Proceedings of International Conference on Scientific and Natural Computing. Algorithms

for Intelligent Systems. Springer, Singapore (2021). https://doi.org/10.1007/978-981-16-1528-3_14
16. Zadeh, L.: Fuzzy sets. Inform. Control **8**, 338–353 (1965)
17. Laarhoven, V.: A fuzzy extension of Saaty's priority theory. Fuzzy Sets Syst. **11**, 229–241 (1983)
18. Crawford, G.B.: The geometric mean procedure for estimating the scale of a judgment matrix. Math/ Modelling. **9**(3–5), 327–334 (1987)
19. Wang, Y-M. et al: A modified fuzzy logarithmic least squares method for fuzzy analytic hierarchy process. Fuzzy Sets Syst. **157**, 3055–3071 (2006)
20. Weck, M., et al.: Evaluating alternative production cycles using the extended fuzzy AHP method. Eur. J. Oper. Res. **100**, 351–366 (1997)
21. Csutora, R., Buckley, J.J.: Fuzzy hierarchical analysis: the lambda-max method. Fuzzy Sets Syst. **120**, 181–195 (2001)
22. Taranath, B.S.: In: Structural Analysis and Design of Tall Buildings, McGraw-Hill book company (1988)
23. Kwok, K.C.S.: Effect of building shape on wind-induced response of tall building. J. Wind Eng. Ind. Aerodyn. **28**, 381–390 (1988)
24. Haldera, L., Duttab, S.C.: Wind effects on multi-storied buildings: a critical review of Indian codal provisions with special reference to American standard. Asian J. Civil Eng. (Building and Housing) **11**, 345–370 (2010)
25. Dutton, R., Isyumov, N.: Reduction of tall building motion by aerodynamic treatments. J. Wind Eng. Ind. Aerodyn. **36**, 739–747 (1990)
26. Davenport, A.G.: In: The Treatment of Wind Loading on Tall Buildings. (1967)
27. Cowlard, A. et al.: Fire safety design for tall buildings. Proc. Eng. **62**, 169–181 (2013); Jignesh, A., Ashok, A.: Aerodynamic modifications to the shape of the buildings: a review of the state-of-the-art. Asian J. Civil Eng. **11**(4), 433–450 (2010)
28. Lam, K., Zhao, X.: An application of quality function deployment to improve the quality of teaching. Int. J. Qual. Reliab. Manage. **15**, 389–413 (1998)
29. Cheng, E.W.L., Li, H.: Construction partnering process and associated critical success factors: quantitative investigation. J. Manage. Eng. **18**, 194–202 (2002)
30. Hassanzadeh, R., Honarmand, M., Hossienjani Zadeh, M., Naseri, F.: New approaches to modelling of local seismic amplification susceptibility using direct characteristics of influencing criteria: case study of Bam City, Iran. Nat. Hazards Earth Syst. Sci. **19**, 1989–2009 (2019). https://doi.org/10.5194/nhess-19-1989-2019
31. Amin, J., Ahuja, A.: Aerodynamic modifications to the shape of the buildings: a review of the state-of-the-art. Asian J. Civil Eng. **11**(4), 433–450 (2010)
32. Becker, T.C., Yamamoto, S., Hamaguchi, H., Higashino, M., Nakashima, M.: Application of isolation to high-rise buildings: a Japanese design case study through a U.S. design code lens. Earthquake Spectra. **31**(3), 1451–1470 (2015). https://doi.org/10.1193/052813EQS136M
33. Quan, Y et al.: Aerodynamic interference effects of a proposed taller high-rise building on wind pressures on existing tall buildings, Struct. Design Tall Spec. Build. (2019)
34. Model Building By-Laws-2016: In: Town and Country Planning Organisation, Ministry of Urban Development, Government of India (2016)

Carbon Neutral Communities: Model for Integrating Climate Action into Development Planning

Sajan Sharon Maria, S. Lakshmy, Davis K. Nidhin, and Nair K. Shibu

1 Introduction

1.1 Climate Change as a Social and Political Issue

Climate change is the most profound threat faced by humanity today which has become the axis of development around the globe. The 'climate crisis' has revealed the need to replace the business-as-usual models with alternate climate resilient development trajectories. Current conventional development programmes are not equipped to address this 'climate emergency', and the impacts of climate change exacerbates vulnerabilities that already exist in society. The underprivileged communities, in terms of social, political and economic aspects, are exposed to the most adverse effects of climate change like floods, droughts and heat waves, escalating social unrest. The growing problem of climate change and global warming has created a need to reduce anthropogenic greenhouse gas (GHG) emissions and its atmospheric concentration, which is the prime cause of climate change. As the current economy is carbon-intensive, there arises an urgent need for a radical transformation in society to shift into a low carbon economy. This transition makes climate change

S. Sharon Maria (✉)
Assistant Professor at Government Engineering College, Thrissur, Volunteer of Zero Waste and Climate Action, Thanal, Thiruvananthapuram, Kerala, India
e-mail: climate@thanal.co.in

S. Lakshmy
Program Officer, Zero Waste and Climate Action, Thanal, Thiruvananthapuram, Kerala, India

D. K. Nidhin
Junior Technical Expert, GIZ India, Volunteer of Zero Waste and Climate Action, Thanal, Thiruvananthapuram, Kerala, India

N. K. Shibu
India Cordinator at Global Alliance for Incinerator Alternatives, Volunteer of Zero Waste and Climate Action, Thanal, Thiruvananthapuram, Kerala, India

I. Pal and S. Kolathayar (eds.), *Sustainable Cities and Resilience*, Lecture Notes in Civil Engineering 183, https://doi.org/10.1007/978-981-16-5543-2_31

not only an environmental problem, but a socioeconomic and political one. This climate constraint can only be overcome by a paradigm shift in the economy and technologies to one that corresponds to the present challenges.

Climate change gained attention in social, political and economic decision-making in the international platforms since the Earth Summit in Rio de Janeiro (Brazil) in 1992, where global campaigns began which culminated with the formation of the United Nations Framework Convention on Climate Change (UNFCCC), an apex body to deal with climate change internationally. UNFCCC and the Kyoto Protocol (1997), which defined precise targets for GHG emission reductions, laid the foundation for future regimes globally.

1.2 Greenhouse Gases and Climate Change—A Reality

Climate change is driven by greenhouse gases (GHGs) emitted from anthropogenic activities such as deforestation, industrial farming, waste generation and burning of fossil fuels. GHGs trap heat energy from the sun in the atmosphere and prevent it from radiating from the earths' surface, thereby creating a greenhouse effect and increasing global temperature. Carbon dioxide (CO_2), methane (CH_4), nitrous oxide (N_2O), hydrofluorocarbons, perfluorocarbons, sulphur hexafluoride, etc. are the major GHGs that remain in the atmosphere and oceans for hundreds of years trapping heat energy causing 'global warming'. Carbon dioxide is the major GHG that has the largest share in global emissions and is responsible for 64% of human-induced global warming. Methane is the most intense GHG which traps up to 100 times more heat in the atmosphere than carbon dioxide within a five-year period. This increase in global mean temperature alters the natural climate pattern in a short time, rendering humans and other living beings unable to adapt.

1.3 Impacts of Climate Change and the Need for Action

The global average temperature has increased by 1.1 degree Celsius in 2018 since the pre-industrial level which corresponds to the increase in GHG concentration of about 147% of pre-industrial level [1]. Climate change, which is triggered by global warming, has direct impacts such as extreme weather events like floods, droughts, heat waves and cold storms, rising sea-levels and melting of glaciers. It also has indirect impacts that affect the things we depend on and value—air, water, ecosystem, wildlife, transportation, energy, agriculture and human health. Climate change is a global issue, and its impacts are transboundary in nature. Due to the rising levels of anthropogenic emissions, global institutions like the UN, World Bank and European Union have taken efforts to prevent and reverse dangerous human interference with the climate systems. These institutions encourage national, regional and local levels of climate actions under global co-operations. 'Think global, Act local' is a concept introduced during the Rio Earth Summit which called for actions at municipal level. The impacts of the climate crisis are extreme to the underprivileged groups and communities, especially in developing and underdeveloped nations. These severe impacts on their livelihoods, health, social security, quality of life and resource accessibility call for community level initiatives for ensuring sustainable development.

2 Framework for Global Sustainable Development

2.1 Rethinking Human-Centric Development—Mainstreaming Climate Change and Disaster Management in the Development Process

Development can mean many things to many, but fundamentally, it revolves around the economic prosperity of people. It is considered that economic growth, building human capabilities and improving people's well-being and quality of life are the important measures of development. People are placed at the centre of development in the present-day economy. But, this type of development enhances over exploitation and increased inequalities and lacks effective resource management which in turn makes it vulnerable to environmental and climate challenges. Hence, it is high time to address this current global threat by placing people and environment/climate at the centre of development for a sustainable future. The mainstream notion that environmental protection hinders development or economic growth destroys the environment has to be demolished to pay way for a new economy that puts climate change at the forefront.

2.2 2030 Agenda for Sustainable Development and Sustainable Development Goals

The term sustainability was introduced in the beginning of the eighteenth century, and the concept gained global attention in the second half of twentieth century. Sustainable development was officially defined in 1987 in the Brundtland report as 'development that meets the needs of the present without compromising the ability of future generations to meet their own needs' [2].

The 2030 Agenda for Sustainable Development was released in 2015 by the UN which came into action on 1st January 2016 as a historic decision to build a better future for all with a promise of leaving no one behind. It focuses on five critical pillars of sustainability (5p's) as **People, Prosperity, Planet, Partnership and Peace**. 17 Sustainable Development Goals (SDGs) with 169 associated targets were announced and pledged by world leaders from 193 countries for common action and endeavour across a broad and universal agenda for the first time. 'We are the first generation that can end poverty, the last that can end climate change' [3].

The sustainable development goals and targets are integrated and indivisible, global in nature and universally applicable, taking into account different national realities, capacities and levels of development and respecting national policies and priorities [4]. The UN urges all countries to work to implement this agenda in regional and global levels by incorporating global commitments while considering the realities, capacities, level of development and their priorities as a nation. The 2030 Agenda is considered as universal, integrated, transformative, nationally owned and right-based and requires global partnership. The 2030 Agenda and SDGs, along with its targets and indicators, help to translate the core values and principles underlying the agenda to concrete and measurable results.

2.3 Climate Action and Sustainable Development

Climate action is the thirteenth goal set in the Sustainable Development Goals of 2030 Agenda—'take urgent action to combat climate change and its impacts'—to address climate change, and it is also intrinsically connected with other SDGs. This synergy helps in achieving sustainable development in different dimensions by global climate action. Climate action in energy efficiency can increase the access to energy and reduce energy expenditure (SDG 1, 7), can improve health (SDG 3), reduce pollution (SDG 11, 14, 12) and also improve economic productivity (SDG 8, 9) (Fig. 1). The reduced usage of fossil fuels in vehicles, energy industries and households helps to cut down emissions and provide us cleaner and fresher air. This can ensure improvement in quality of life (SDG 3). Global climate action also ensures the conservation and protection of forest ecosystems, oceans, peat lands and

Fig. 1 Sustainable development goals cross cutting with climate action

other fragile land areas ensuring healthy life under water and on land. Achieving SDGs ensures that future generations will be able to enjoy the resources and lead a sustainable life.

2.4 *National and State Level Actions to Achieve SDGs*

As a rapidly developing economy which is home to 17.7% of the world population, India has made strong commitments to achieve SDGs as the policies adopted by the nation can play a huge role in leading the world to a sustainable future. The National Action Plan on Climate Change (NAPCC) released in 2008 created a movement in India to address climate change. As a response to this, each state developed their own state action plans to incorporate climate change in their development policies. In October 2015, India pledged to reduce the emission intensity of its gross domestic product (GDP) by 20–25% from its 2005 levels by 2020 and by 33–35% by 2030 [5]. India's motto for development 'Sabka Saath, Sabka Vikaas' (collective efforts, inclusive growth) reflects the SDGs, and several flagship programmes of the government like the Swachh Bharat Mission and Make In India are driven by SDGs. Niti Ayog is the principal body in India that coordinates programmes and policies that helps to achieve SDGs in the country.

The state and local self-governments play a major role in achieving SDGs in India. Kerala is the leading state in achieving SDGs in India in 2019 as per the SDG India Index developed by Niti Ayog which tracks the progress of policies related to SDGs in the country. The state's strong decentralized governance system enabled local self-governments to carry out projects at the grass-roots level and achieve this top position.

3 Climate Responsive Planning at Local Level

3.1 *Local Level Development Planning for Carbon Neutral Communities*

Countries around the globe have framed ambitious national level development policies to achieve SDGs. But, these policies can be accomplished successfully when they are executed at sub-national/regional or local level as the participation of citizens and all stakeholders in the society plays key roles in achieving the goals. While low carbon or carbon neutral development has become the path to attain sustainable development, community-level initiatives have gained the spotlight with success stories flowing from across the globe. Community-level carbon neutral development has received traction in many countries as involving the community, people, governments, private sector and other important actors in the society seems to be the best way to achieve SDGs and carry out climate action.

3.2 *Meenangadi as a Carbon Neutral Community*

'Carbon Neutral Meenangadi' is a pilot project under the 'Carbon Neutral Wayanad' project of the state government of Kerala initiated by Dr. T. M. Thomas Isaac, Former Finance Minister of Kerala and a well-known economist, as a commitment in response to the Paris Agreement on Climate Change. This project was created to set an example in the state for the rest of the nation to follow in its low carbon and climate resilient footsteps. Meenangadi is one of the 25 Grama Panchayats (local self-government) in Wayanad district of Kerala which is situated at an elevation of 700–2100 m above sea level on the Western Ghats, highly vulnerable to the adverse impacts of climate change. The panchayat is situated between two major urban centres, Kalpetta and Sulthan Bathery. It has a geographical area of 53.52 sq km. with rocky hills, valleys and plains and has a population of 34,601 distributed across 8199 households. It is a primarily rural panchayat with an urban area stretched along the National Highway-766 that passes through Meenangadi which connects Kozhikode in Kerala to Mysore in Karnataka. Providing better livelihood opportunities and socioeconomic resilience through nature-based adaptation strategies is the backbone of this project which makes this attempt unique.

3.3 *Methodology*

The scientific methodology used for GHG emission estimation of Meenangadi is based on the 2006 IPCC[1] guidelines for National Greenhouse Gas Inventories. The GHG inventory has been prepared for the baseline year 2016–17 for four sectors—transportation, energy, waste and Agriculture, Forests and Other Land Use (AFOLU). Carbon dioxide (CO_2), nitrous oxide (N_2O) and methane (CH_4) emissions were considered for this particular inventory since the emissions of other GHGs are negligible. CO_2 being the most abundant GHG contributing almost 75–80% of total GHG emissions, quantities of N_2O and CH_4 are converted in terms of CO_2 equivalent (CO_2 eq.) by multiplying it with corresponding global warming potential (GWP) values. The amount of carbon sequestered in Meenangadi is calculated for forests, plantations, homestead trees and organic carbon in soil. The equations for GHG emission and carbon sequestration calculations are given below.

[1] IPCC—Intergovernmental Panel on Climate Change—the scientific body of the UN that deals with climate science.

$$\text{Total GHG emissions} = \Sigma \text{ Activity Data X Emission Factor} \quad (1)$$

$$\text{Total carbon sequestration} = \Sigma \text{ Sectoral Data X Sequestration Factor} \quad (2)$$

where
Activity data—is the data on the magnitude of human activity resulting in emissions or removals taking place during a given period of time (e.g. tonnes of coal mined nationally in a given year).
Emission factor—is the average emission rate of a given GHG for a given source, relative to units of activity (e.g. CH_4 emitted per tonnes of coal mined) [6].

Emission factors for various sectors and activities were considered according to 2006 IPCC guidelines with TIER 1[2] precision.

Activity data pertaining to various sectors were collected through surveys with the help of students and volunteers and secondary data from respective government departments. Emission for transportation, energy and waste sectors was calculated using emission factors from internationally approved toolkits relevant for India, while secondary data were used for estimating emissions from AFOLU sector.

3.4 Results and Discussions

The total GHG emission of Meenangadi is estimated at 33,375 tonnes of CO_2 eq., and net GHG emissions (surplus of total emissions over total sequestration in terms CO_2 eq.) was 11,412.57 tonnes of CO_2 eq. in the base year 2016–17 [7]. The following figures explain the GHG emission profile of Meenangadi. The estimate shows that transportation (45%) and energy (39%) sectors are the largest contributors of GHG emissions in Meenangadi. Similarly, the carbon sequestration profile explains how carbon is sequestered naturally and stored in different forms. There is a significant share of emission removal by forests (38%) and homestead trees (34%) in Meenangadi (Figs. 2, 3 and 4).

To achieve carbon neutrality, the surplus of GHG emission over the annual carbon removal has to be brought down to zero. As there will be change in annual emissions every year, it is very important to understand how the emission rates are going to be in future for setting targets for emission reductions. An emission profile of 2008 and 2012 was also estimated for plotting the trend line of future emissions of Meenangadi under business-as-usual scenario. The emission projection is as shown below (Figs. 5 and 6):

[2] Tier 1: uses default data and simple equation for emission estimation.

Tier 2: uses country-specific emission factors; Tier 3: uses region-specific emission factors.

Tiers 2 and 3 are more demanding than Tier 1 in terms of complexity and data requirements.

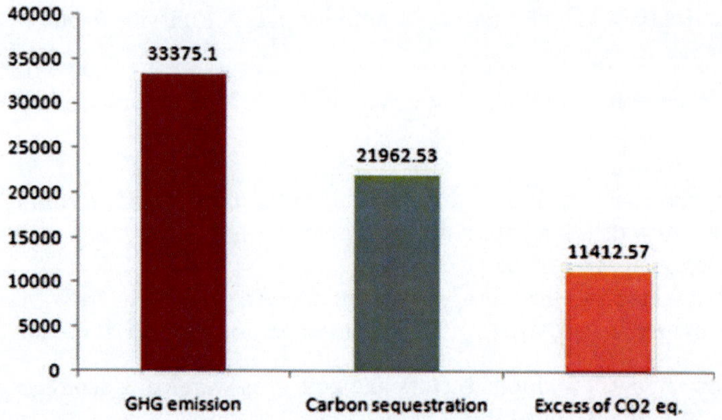

Fig. 2 Total GHG emission, sequestration and excess emission graph of Meenangadi

Fig. 3 GHG emission contribution from various sectors

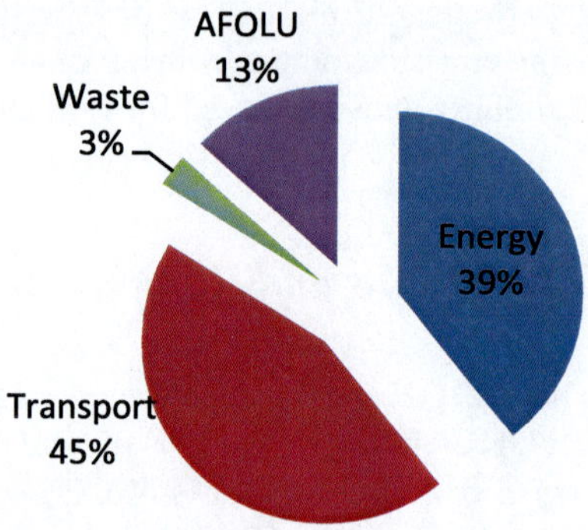

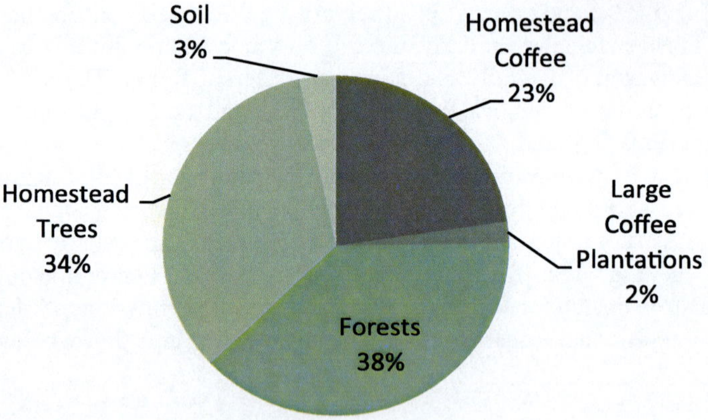

Fig. 4 Carbon sequestration profile in Meenangadi

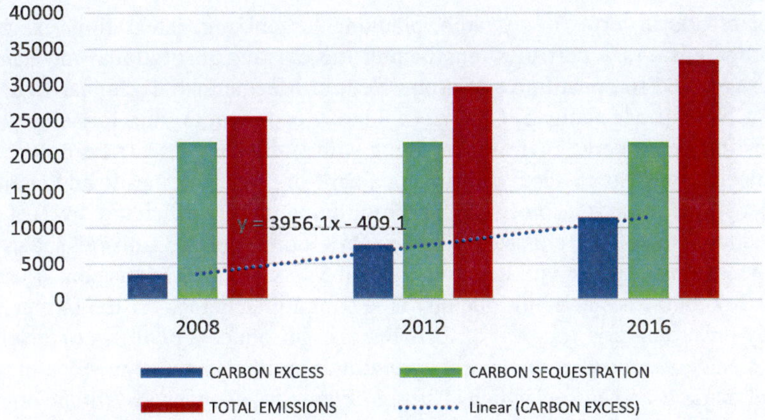

Fig. 5 Trend line projection of CO_2 excess in Meenangadi Grama Panchayat

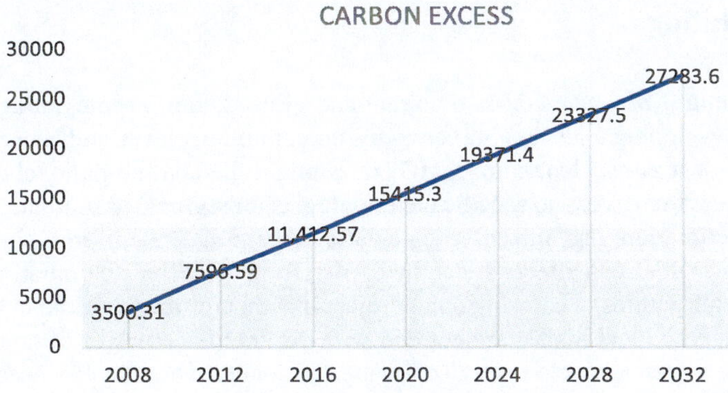

Fig. 6 Carbon excess projections for Meenangadi Grama Panchayat

3.5 *Recommendations for Low Carbon Development at Community Level*

The recommended strategies for achieving and maintaining carbon neutrality are i. *reducing emissions*, ii. *improving carbon sinks* and iii. *creating carbon credits/reserves through nature-based solutions.* Apart from quantitative estimation of GHG emissions and carbon sequestration, a qualitative assessment of the livelihood needs of the people was a crucial element for the success of this project. To know more about the needs, issues and potentials in Meenangadi, a ward-level and panchayat-level stakeholder consultations were conducted in a non-conventional manner. For implementing the 'Carbon Neutral Meenangadi' project, it required a lot of efforts in terms of research, planning, organizing, integration, execution and campaigns. As a part of strengthening the existing institutional mechanism, it was suggested to constitute a 'Carbon Neutral Meenangadi Technical Cell' and 'Carbon Neutral Meenangadi Campaign Team' for managing the implementation and creating a conducive environment along with embedding an ownership status of the project among the residents. Targeting a shift of 30% of domestic and commercial energy consumption to off-grid and improving energy efficiency by 10% will bring about 10% of net emission savings. Promoting electric auto-rickshaws and implementing zero waste will help to cut down 8% of net GHG emissions. Planting around 2,00,000 trees and introducing 'Tree banking' schemes for the farmers will not only ensure natural removal of CO2 but also introduce a platform to develop a local green economy. Carbon neutral agricultural products have high value in international markets and coffee, which is the major crop in Meenangadi, will be branded as 'Carbon Neutral Coffee' which will generate more income for the farmers.

4 Limitations

'Carbon Neutral Meenangadi' is a unique and pilot attempt in integrating climate action for developing a sustainable economy at a community level, and hence, it faced numerous constraints. Generating GHG emission profile at a sub-national level was challenging as there were no specific methodologies for regional or sub-national level emission estimation, and the entire calculations were dependent on national-level determinants which can affect its accuracy. Apart from the technical limitations, there were other difficulties in carrying out a project like this for the first time in the state. Even though the local bodies are entitled with decentralized power of governance, the existing systems have a centralized nature in reality which makes execution of such unique and outstanding initiatives from the bottom level of governance more laborious. The communication gap between stakeholders, institutions, government departments, etc. also creates some order of limitations in the implementation stages. The lack of human resources such as technical support, researchers and experts at the local level also calls for capacity-building programmes for strengthening the same.

5 Conclusion

5.1 Low Carbon Development as a Universal Model

As countries around the globe have realized the importance of reducing anthropogenic emissions to tackle the current climate crisis, carbon neutral development has become the sustainable option for a green future ahead. The society today, which is highly dependent on fossil fuels, is in direct need of radical transformation into a low emission economy. Many countries have come forward with promises to phase out coal and other fossil fuels and instead promote technologies that use renewable resources. Developing into a low carbon economy not only helps to combat climate change but also helps to create a sustainable, self-sufficient, healthy and green future for all.

Achieving carbon neutral status is an arduous but worthy task as the resulting sustainable economy will benefit all. By adopting policies and schemes and setting short-term and long-term projects, a community can easily become carbon neutral with the support of all actors in the community. As each and every community across the globe attains carbon neutrality and takes steps to maintain it; the planet will become a better place for the present and future generations. Mainstreaming low carbon development is the new path to achieve SDGs, combat climate change and make a sustainable future a reality.

6 Discussion

6.1 Implementation Status of Recommendation Including Achievements in GHG Reductions or Sequestration Capacity

During the period of 2017–19, Meenangadi panchayat has planted about 4 lakhs trees in private land. Tree banking schemes introduced as part of low carbon development has become a source of income for farmers who planted trees under the project. Bamboo is also being planted extensively in the panchayat due to its high sequestration rates, especially in riverbanks which also helps in conservation and soil preservation. The panchayat also started implementing energy-efficient measures such as energy auditing, promoting LED bulbs, waste management plants, vermi-composting and supply of kitchen composting equipment to households. The project anticipates improving the livelihood of people, adapting to climate change and emerging as a sustainable community. The methodology developed for Meenangadi panchayat in Kerala can be adapted for any region or local government for planning a sustainable low carbon economy.

6.2 Low Carbon Development in Costa Rica

Costa Rica is the first country in the world to sign a National Pact to the Advancement of the SDGs in 2016 to achieve the 2030 Agenda. This pact involves multiple stakeholders in the country including the government, civic society, private sector, religious organizations and citizens to join hands for sustainable development. In 2019,

Costa Rica released the National Decarbonization Plan 2018–2050 to achieve net-zero emissions by 2050, which is more ambitious than its Paris Agreement targets. The country adopted new policies to electrify its transportation sector, which is the largest source of its GHG emissions, by providing incentives and tax concessions. Costa Rica also promotes carbon neutral coffee, banana, pineapple and cattle produces. With the help of all these sustainable policies, Costa Rica is close to achieve its 2030 Nationally Determined Contributions (NDC) goals.

References

1. WMO. World Meterorological Organization: World Meterorological Organization. [Cited: January 14, 2021] https://public.wmo.int/en/media/press-release/greenhouse-gas-concentrations-atmosphere-reach-yet-another-high#:~:text=Carbon%20dioxide%20is%20the%20main,rate%20over%20the%20last%20decade
2. United Nations: Report of the World Commission on Environment and Development : Our Common Future (1987)
3. —: UN Meeting Coverages and Press Releases. https://www.un.org/press/en/2015/sgsm16800.doc.htm. Last accessed 14 Jan 2021
4. UN—: Transforming our world: 2030 Agenda for Sustainable Development. https://sdgs.un.org/2030agenda. Last accessed 2021/01/14
5. Subramanian, K.: Down To Earth. https://www.downtoearth.org.in/blog/climate-change/is-india-on-track-to-meet-its-paris-commitments-67345. Last accessed 14 Jan 2021
6. IPCC.: 2006 IPCC Guidelines on National Greenhouse Gas Inventories. IGES, Hayama, Japan
7. Thanal: Carbon Neutral Meenangadi. Thiruvananthapuram : Thanal (2018)

Cross-Cutting Issues

Review of Free Vibration Response of Spar-Supported Wind Turbine with Tuned Mass Damper

K. K. Akheel, Muhamed Safeer Pandikkadavath, and A. P. Shashikala

1 Introduction to the Investigation

The effort for renewable and clean energy is pushing the world to focus more on wind energy [1]. Onshore and near offshore wind farms near populated areas are wreaked with noise and visual difficulty disturbances [2]. Additionally, onshore wind farms may consume costly land area to a greater extend [3]. Deepwater floating wind turbines (such as spar-type) can eliminate these issues considerably [4]. The aforementioned deep water floating-type offshore wind turbines have increased power production efficiency compared to the onshore or near offshore wind turbines, mainly owing to smaller wind shear and superior wind speed [5]. Such wind turbines also have an option to go for larger capacity (in terms of size well) to have a higher quantum of power generation [6]. Thus, floating offshore wind turbines can be a reasonable choice to meet the increasing energy demands for sustainable growth [4]. However, these types of wind turbines are highly vulnerable to extreme marine environmental conditions such as wave, wind, and current loadings. This situation makes the floating-type wind turbines more sensitive to instability and fatigue-related difficulties to both the structure and mooring ties [6]. This may affect power production efficiency, service life of the wind turbines (adversely) and increases the maintenance cost. To mitigate the extreme marine condition, stimulated vibration responses of floating wind turbines passive control devices such as tuned mass dampers (at nacelle) can be utilized [7–10]. The present study aims to review the free vibration responses and its control on an offshore wind turbine (spar supported) using a tuned mass damper. The natural frequencies corresponding to various degrees of freedom are found in the beginning. Subsequently, free vibration analyses corresponding to fore-aft initial displacement are carried out. For the free vibration response control,

K. K. Akheel · M. S. Pandikkadavath (✉) · A. P. Shashikala
National Institute of Technology Calicut, Kozhikode, India
e-mail: msafeerpk@nitc.ac.in

I. Pal and S. Kolathayar (eds.), *Sustainable Cities and Resilience*, Lecture Notes in Civil Engineering 183, https://doi.org/10.1007/978-981-16-5543-2_32

different mass ratios have been adopted and corresponding response reductions are analyzed.

2 Details of Study Wind Turbine

For the study, a 5-MW rated baseline wind turbine (OC3-Hywind type platform) used in previous investigations is adopted [6, 11]. The rotor diameter, hub height, blade length, hub diameter, and tower height above ground dimensions are 126, 90, 61.5, 3, and 87.6 m, respectively. The corresponding blade mass, nacelle mass, hub mass, and overall mass are 17,740, 2,40,000, 56,780, and 2,49,718 kg, respectively. Similarly, the spar (having mass = 7,466,330 kg) diameter above taper and spar diameter below taper are 6.5 and 9.4 m, respectively. The mass (per unit length), diameter, unstretched length, fairlead radius, fairlead depth below sea water level, anchor radius, and anchor depth below the seawater level of mooring are 77.71 kg/m, 0.09, 902.2, 5.2, 70, 853.87, and 320 m, respectively (The properties of OC3-Hywind platform is summarized in Table 1 and a typical schematic sketch of the platform is shown

Table 1 Properties of OC3-Hywind [6, 11]

Rated power	5 MW
Cut-in, rated, cut-out wind speed	3 m/s, 11.4 m/s, 25 m/s
Depth of water	320 m
Total draft	120 m
Rotor diameter	126 m
Hub height	90 m
Blade length	61.5 m
Blade mass	17,740 kg
Nacelle mass	240,000 kg
Hub mass	56,780 kg
Hub diameter	3 m
Tower height above MSL	87.6 m
Tower mass	249,718 kg
Platform mass	7,466,330 kg
Spar diameter above taper	6.5 m
Spar diameter below taper	9.4 m
Mass of mooring lines	77.7 kg/m
Diameter of mooring lines	0.09 m
Unstretched length	902.2 m
Fair lead radius	5.2 m
Depth of fair lead from MSL	70 m
Radius of anchor	853.87 m

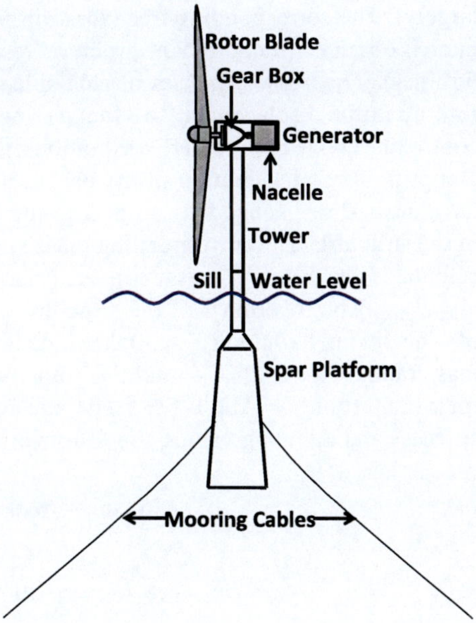

Fig. 1 Typical schematic plot of spar-supported OC3-Hywind wind turbine

Table 2 Natural frequencies with various degrees of freedom

Degrees of freedom (Platform)	Natural frequency (Hz)	Degrees of freedom (Tower)	Natural frequency (Hz)
Surge/sway	0.008	Fore-aft	0.476
Heave/roll/pitch	0.032	Side-side	0.471
Yaw	0.120		

in Fig. 1). This wind turbine incorporated OpenFAST framework is used to evaluate the response under free vibration scenario [12]. The natural frequencies corresponding to the different tower and platform degrees of freedom are obtained. The tower response-related degrees of freedom; fore-aft and side-side natural frequencies were 0.476 and 0.470 Hz, respectively. Similarly platform response-related degrees of freedom; surge, sway, heave, roll, pitch, and yaw natural frequencies are 0.008, 0.008, 0.032, 0.032, 0.032, and 0.120 Hz, respectively. These details are given in Table 2.

3 Free Vibration Response Evaluations

For the free vibration analysis, an initial displacement of 2 m is applied along the fore-aft direction (Since the fore-aft response governs the power generation efficiency largely). The corresponding free vibration response of fore-aft, side-side, surge, and pitch is obtained (predominant structural responses for deepwater floating-type wind turbines). The peak responses of side-side, surge, and pitch (corresponding to 2 m fore-aft initial displacement) are found to be 0.21, 0.74, and 0.52 m, respectively. To control the above described free vibrations, tuned mass damper, TMD is incorporated along the fore-aft direction in the nacelle [7–10]. Tuned mass damper is tuned to have desired resonance with respect to the fore-aft vibration frequency (0.476 Hz) of the structure. The corresponding mass and stiffness of the tuned mass damper are calculated for different mass ratios, μ (mass of TMD, m_{TMD}/ mass of the structure, $m_{\text{Structure}}$). The total mass of the structure (mass of nacelle + mass of hub + mass of 3 blades) is found to be 3, 50,000 kg. Corresponding to this structural mass, five mass ratios, i.e., 1, 2, 3, 4, and 5% (mostly the mass ratio of tuned mass dampers for wind turbines will be 1–5% range) are considered. To find out the corresponding stiffness and damping values, the following equations are used [13].

$$K_{\text{TMD}} = m_{\text{TMD}}\omega_{\text{Fore-Aft}} \tag{1}$$

$$\xi_{\text{TMD}} = \sqrt{3\mu/8(1+\mu)} \tag{2}$$

where K_{TMD} and ξ_{TMD} are the stiffness and damping ratio of TMD for a given mass ratio, μ; $\omega_{\text{Fore-Aft}}$ is the natural frequency in the fore-aft direction (0.476 Hz); m_{TMD} is the mass of tuned mass damper. The calculated values of stiffness and damping corresponding to 2% mass ratio are obtained as 62,664 N/m and 3592 Ns/m (damping ratio, $\xi = 8.6\%$), respectively. The relevant values of tuned mass damper corresponding to different mass ratios, μ, are given in Table 3.

The fore-aft 2 m initial displacement without tuned mass damper condition takes around 100 s to die out the fore-aft free vibration response. Once the TMD is applied, the same response dies out within (around) 40 s. As the mass ratio changes (increases), the dying out nature of fore-aft vibrations intensifies. For the same initial displacement-induced free vibration without tuned mass damper, side-side response takes around 120 s to die out completely, with a peak response of 0.214 m at around 30 s. With the enabling of tuned mass damper, the vibration tendency remains the same but the vibration magnitude depletes out faster with the increase of tuned mass

Table 3 Properties TMD corresponding to respective mass ratios

Mass ratio, μ	TMD mass m_{TMD} (kg)	Stiffness, K_{TMD} (N/m)	Damping ratio, ξ_{TMD}	Damping (Ns/m)
0.01	3500	31,332	0.061	1276
0.02	7000	62,664	0.086	3592
0.03	10,500	93,996	0.105	6567
0.04	14,000	125,328	0.120	10,061
0.05	17,500	156,660	0.134	13,994

Table 4 Percentage reductions in side-side responses with respect to different tuned mass damper mass ratios

Mass ratio, μ (%)	Maximum side-side response (m)		
	Without TMD	With TMD	% Reduction
1	0.214	0.177	17
2	0.214	0.168	21
3	0.214	0.161	25
4	0.214	0.155	27
5	0.214	0.148	30

damper mass ratio. The percentage reduction in side-side response with different mass ratios of tuned mass damper is given in Table 4 (as the reductions are very evident compared to other responses). Corresponding to the 2% mass ratio, the side-side response is decreased to 0.168 m from 0.214 m (without tuned mass damper) to have a percentage reduction in the level of 21%. Similarly, corresponding to 5% of mass ratio, the same response is reduced to 0.148 m to have percentage reduction in the range of 30%.

The surge and pitch vibration responses under 2 m fore-aft displacement have a larger wavelength compared to fore-aft and side-side vibrations and it takes more than 200 s to die out completely. Once the tuned mass damper is applied, the magnitude of surge and pitch vibration response decreases with respect to increase of mass ratio. Still, it takes more than 200 s to die out this response completely possibly owing to large wavelength characteristics. The fore-aft, side-side, surge, and pitch vibration responses under the 2 m fore-aft initial displacement with and without tuned mass damper cases are shown in Fig. 2.

4 Concluding Comments

A brief review of the necessity and advantages of offshore floating-type wind turbines compared to the onshore wind turbines are carried out. Successively a floating-type spar-supported OC3-Hywind wind turbine is selected for the investigation. The structural modeling details are studied and OpenFAST software framework is used for the analysis. Initially natural frequencies of fore-aft, side-side (both for the tower), surge, sway, heave, roll, pitch, and yaw (all for the platform) are found by respective free vibration trials. Later free vibration analysis corresponding to 2 m initial fore-aft displacement is carried out. Corresponding to this initial displacement, predominant responses fore-aft, side-side, surge, and pitch are studied. In the next step, the mass, stiffness, and damping of the tuned mass damper are calculated for different values of mass ratio (damper mass to nacelle mass). The tuned mass damper is enabled in the nacelle and the free vibration test (2 m fore-aft initial displacement) is repeated. The corresponding changes in the considered structural degrees of freedom vibrations are

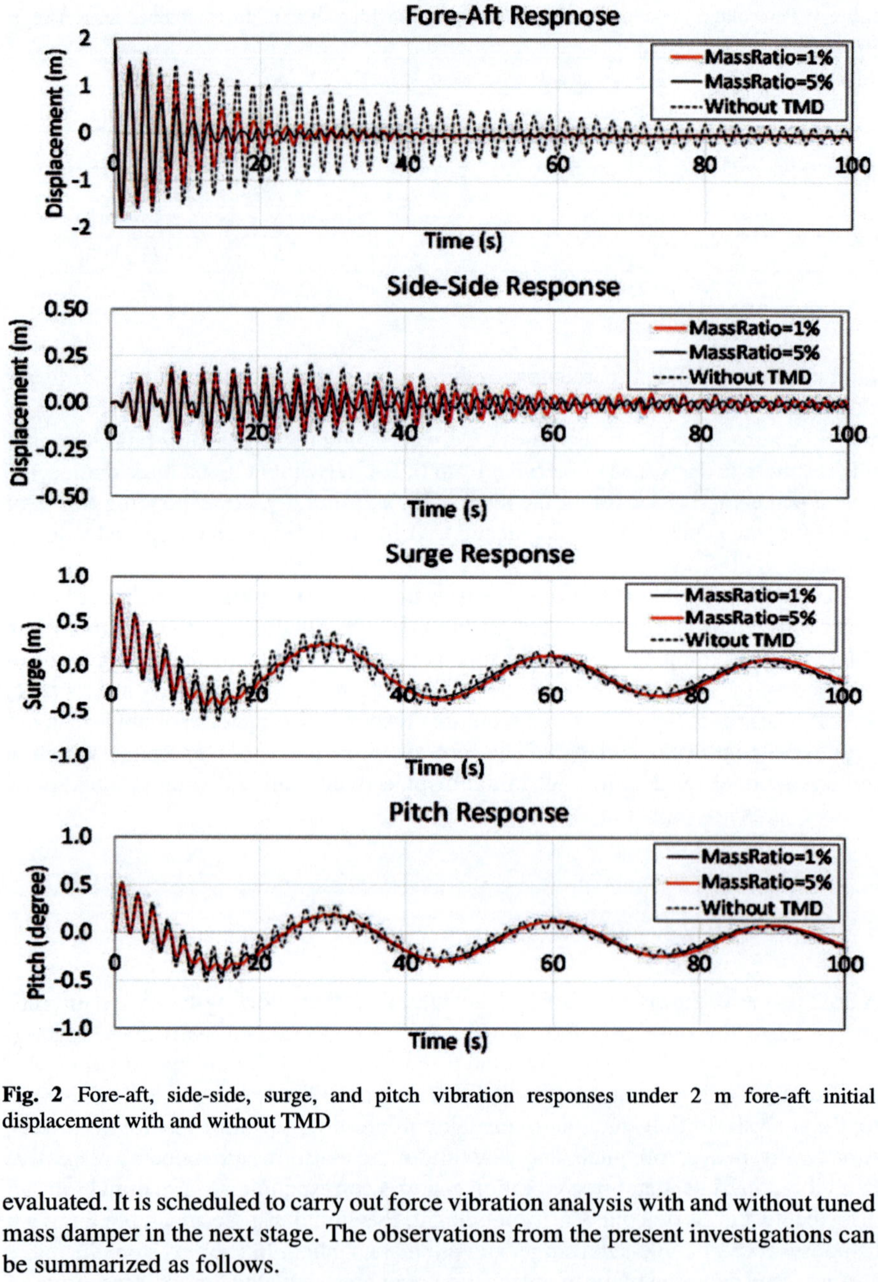

Fig. 2 Fore-aft, side-side, surge, and pitch vibration responses under 2 m fore-aft initial displacement with and without TMD

evaluated. It is scheduled to carry out force vibration analysis with and without tuned mass damper in the next stage. The observations from the present investigations can be summarized as follows.

- When tuned mass damper with 2% mass ratio (practical) is incorporated, the maximum fore-aft response had a reduction of 21% (0.214–0.168 m) from its peak value (without tuned mass damper).

- Corresponding to this instant, the fore-aft, surge, and pitch responses had reductions of 80% (0.664–0.130 m), 53% (0.124–0.058 m), and 99% (0.1266–0.0001°), respectively.
- It is found that the response control improves (further) positively with the increase of tuned mass damper mass ratio.

References

1. Leung, D.Y.C., Yang, Y.: Wind energy development and its environmental impact. Renew. Sustain. Energy Rev. **16**(1), 1031–1039 (2012)
2. Kaldellis, J.K., Kapsali, M.: Shifting towards offshore wind energy-recent activity and future development. Energy Policy **53**, 136–148 (2013)
3. Si, Y., Karimi, H.R., Gao, H.: Modeling and optimization of a passive structural control design for a spar-type floating wind turbine. Eng. Struct. **69**, 168–182 (2014)
4. Yang, J.J., He, E.M.: Coupled modeling and structural vibration control for floating offshore wind turbine. Renew. Energy **157**, 678–694 (2020)
5. Esteban, M.D., Diez, J.J., Lopez, J.S., Negro, V.: Why offshore wind energy? Renew. Energy **36**(2), 444–450 (2011)
6. Jahangiri, V., Sun, C.: Three-dimensional vibration control of offshore floating wind turbines using multiple tuned mass dampers. Ocean Eng. **206**, 107196 (2020)
7. Lackner, M., Rotea, M.: Passive structural control of offshore wind turbines. Wind Energy **14**(3), 373–388 (2011)
8. Stewart, G.M., Lackner, M.A.: The impact of passive tuned mass dampers and wind-wave misalignment on offshore wind turbine loads. Eng. Struct. **73**, 54–61 (2014)
9. Dinh, V.-N., Basu, B.: Passive control of floating offshore wind turbine nacelle and spar vibrations by multiple tuned mass dampers. Struct. Control. Health Monit. **22**, 152–176 (2015)
10. Jin, X., Xie, S., He, J., Lin, Y., Wnag, Y., Wang, N.: Optimization of tuned mass damper parameters for floating wind turbine by using the artificial fish swarm algorithm. Ocean Eng. **167**, 130–141 (2018)
11. Jonkman, J., Butterfield, S., Musial, W., Scott, G.: Definition of a 5MW reference wind turbine for offshore system development. NREL Technical Report (2009)
12. OpenFAST: Documentation. National Renewable Energy Laboratory, Golden, CO, USA (2021)
13. Den hartog, J. P.: Mechanical Vibrations. Dover, New York, USA (1985)

A Review on Methods for Analysis of Laterally Loaded Piles

Abitha Babu and Sitaram Nayak

1 Introduction

1.1 General

Pile foundations are generally used to transmit loads from the superstructure to the surrounding soil, when the adequate bearing capacity is not available at shallow depths. In addition to the vertical loads, piles are often subjected to lateral loads and moments due to the forces such as wind, waves and earthquake. Such foundations need to be analyzed for lateral loads since the pile behavior under lateral loads are more critical than that under axial loads. High lateral loads are usually experienced by piles which are used under tall chimneys, television towers, high rise buildings, high retaining walls, offshore structures, etc.

Piles subjected to lateral load act as a transversely loaded beam and they transmit the lateral load to surrounding soil by means of lateral resistance of the soil.

Under static or dynamic loading conditions, the ultimate lateral resistance of a pile and the deflection of pile as the load builds up are complex matters which involve soil-structure interaction. The pile presses in the direction of applied load against the soil in front of it and develops stress and strain in the soil which offers resistance to movement of the pile. The shallow surface layers are of greatest importance for the study of laterally loaded piles since they gain most of their support from top portion of the soil. After a certain length pile length loses its significance (termed as critical length), such piles with length greater than the critical value are called as long flexible piles. Short piles behave as a rigid member, which loses significance of its flexural stiffness (EI). In practice, depending on the relative stiffness of the pile and pile cap and the type of connections, there may be two pile head fixity conditions which are free head and fixed head. Based on pile head conditions and length, the

A. Babu (✉) · S. Nayak
National Institute of Technology Karnataka, Surathkal, India
e-mail: abitha.192gt002@nitk.edu.in

I. Pal and S. Kolathayar (eds.), *Sustainable Cities and Resilience*, Lecture Notes in Civil Engineering 183, https://doi.org/10.1007/978-981-16-5543-2_33

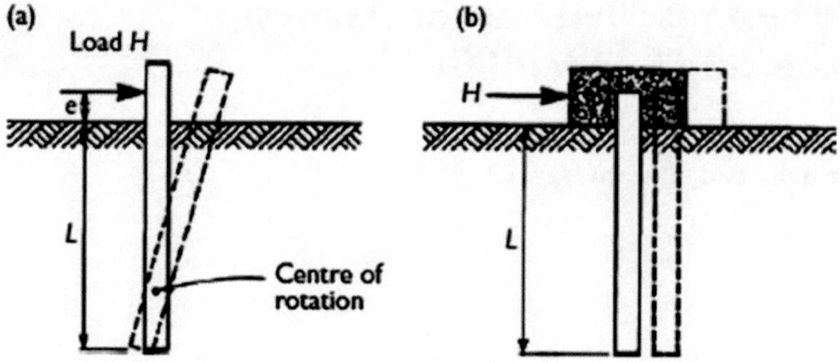

Fig. 1 Short vertical pile under horizontal load. **a** Free head. **b** Fixed head [1]

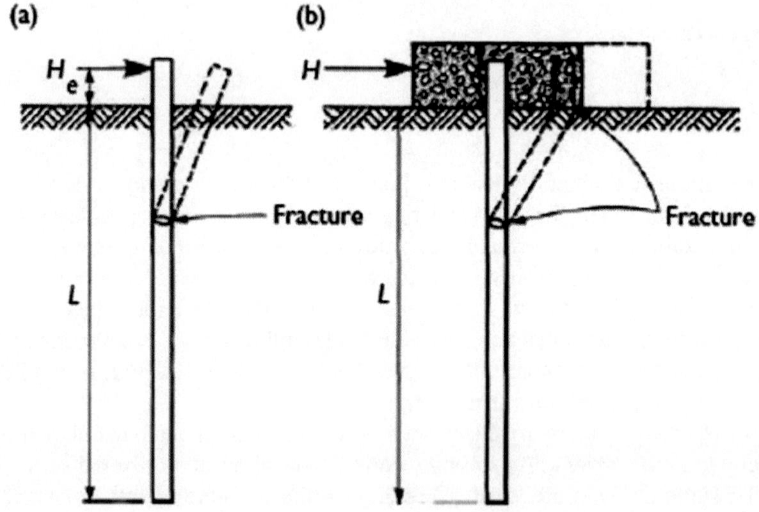

Fig. 2 Long vertical pile under horizontal load. **a** Free head. **b** Fixed head [1]

pile as a part or whole tries to shift horizontally in the direction of applied lateral load which leads to its failure either by rotation or translation or fracture as shown in Figs. 1 and 2.

1.2 Analysis of Laterally Loaded Piles

In the case of laterally loaded pile, even under the working load the lateral movement of a pile is significant, which is the major difference between the design and analysis of an axially and laterally loaded pile. The determination of pile resistance under lateral loads can be done either by calculating ultimate lateral resistance or by calculating permissible deflection at working load [2]. Before the pile undergoes failure, the pile head may move horizontally over a considerable distance to such an extent that the displacement of the structure supported by the pile exceeds tolerable limits [1]. Therefore, while designing laterally loaded pile foundation, the major concern of engineers is the allowable deflection of pile rather than ultimate lateral load. Reviewing the available literatures, it was observed that the commonly used methods of calculating lateral resistance of piles can be divided into two categories in which allowable lateral load is taken as follows:

(1) Load obtained by dividing ultimate lateral load with an appropriate safety factor.
(2) Load corresponding to permissible deflection.

The effect of vertical loads on the pile have been very well established in the past years, but researches on vertical piles subjected to lateral load are limited and the researchers are still working on it to obtain a well defined method of analysis considering all the influencing factors. This paper gives an overview of different methods which are used for the analysis of lateral load capacity of piles. Also an attempt is made to include various studies carried out by researchers to employ different software in analyzing the piles.

2 Ultimate Lateral Resistance of Piles

To determine the ultimate lateral load resistance of a single pile two methods proposed by Brinch Hansen and Broms [3] are commonly used. The basic concept used in these two approaches is explained here. The mechanism of mobilization of ultimate soil resistance to resist combination of lateral force Q (or H) and moment M is illustrated in the Fig. 3.

By considering equilibrium conditions $\sum F_x = 0$ and $\sum M = 0$, the ultimate lateral resistance (Q_u) and corresponding moment ($M = Q_u e$) are related with the ultimate soil resistance p_u as following.

$$Q_u - \int_{x=0}^{x=x_r} p_{xu} B dx + \int_{x=x_r}^{x=L} p_{xu} B dx = 0 \quad (1)$$

$$Q_u e + \int_{x=0}^{x=x_r} p_{xu} B x dx - \int_{x=x_r}^{x=L} p_{xu} B x dx = 0 \quad (2)$$

In Eqs. (1) and (2), x_r is depth of point of rotation and Q_u is ultimate lareal resistance. The values of x_r and Q_u can be obtained using above equations, if the distribution of ultimate unit soil resistance p_{xu} along pile length is known.

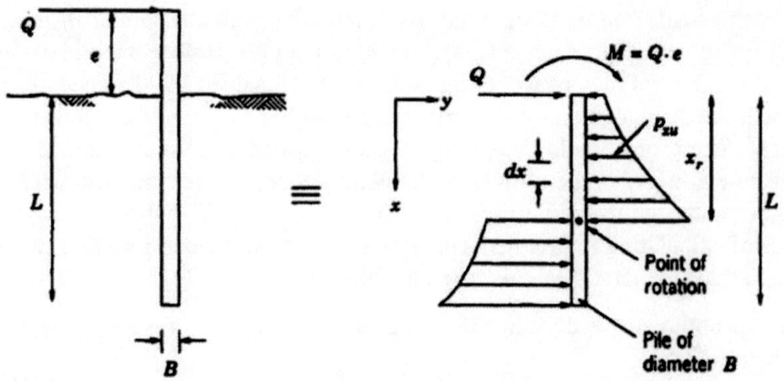

Fig. 3 Mobilization of lateral resistance for a free head laterally loaded rigid pile [2]

2.1 Brinch Hansen's Method

Brinch Hansen suggested a method for obtaining distribution of soil resistance for short piles [4]. This method of analysis based on earth pressure theory is applicable to *c*-Φ soil, uniform soils and layered system. The center of rotation is determined by taking of moment of all forces as zero at the point of application of load. It is possible to measure the ultimate lateral resistance on the basis of Eq. (1), so the relation for ultimate soil resistance at any depth is given as follows:

$$P_{xu} = \sigma'_{vx} K_q + c K_c \tag{3}$$

In Eq. (3), σ'_{vx} is vertical effective overburden pressure, c is cohesion of soil, K_c and K_q are factors that are function of ϕ and *x/B*. The undrained strength c_u and ϕ can be taken as zero for short term loading such as wave forces and the drained effective strength values (*c*', ϕ ') are used for long term sustained loading conditions. However the applicability of method is limited to analysis of short piles and to locate point of rotation trial and error solution is required.

2.2 Brom's Method

Broms [3, 5] proposed a method for determining lateral resistance of vertical pile based on earth pressure theory like the Brinch Hansen method, but with few simplifying assumptions. This method can be used for the analysis of short and long piles with both the pile head fixity conditions in either purely cohesive or cohesionless soil. However the method is not applicable to c-Φ soil and layered system. For short piles in cohesionless soil, ultimate resistance of pile is given by:

$$p = 3B\sigma_v' K_p = 3\gamma' \mathrm{LBK}_p \tag{4}$$

In Eq. (4), p is unit soil pressure, σ'_v is effective overburden pressure at any depth, ϒ'is effective unit weight of soil, L and B are embedded length and width of pile, K_p is coefficient of Rankine's passive earth pressure. For short piles in cohesive soil, ultimate resistance of pile is assumed to be zero from the ground surface up to a depth of 1.5B and a constant value of $9c_uB$ afterwards. In the case of long piles in cohesionless soil, L is being replaced by x_o (depth below which soil reaction decreases) in the above equation. For long piles in cohesive soil, soil pressure decreases beyond a depth of $(1.5\mathrm{B} + x_o)$.

3 Permissible Lateral Deflection of Piles

The modulus of subgrade reaction approach and elastic continuum approach are two methods mainly used for calculating lateral deflection of piles. In addition to these two methods, this section discusses extensions and modifications of subgrade reaction method carried out by several researchers.

3.1 Modulus of Subgrade Reaction Approach

Reese and Matlock [6] developed non-dimensional solutions for pile subjected to horizontal force Q (or H) and moment M, with soil modulus assumed proportional to depth. They considered laterally loaded piles as a beam on elastic foundation [7]. Soil was assumed to be Winkler soil model, according to which a series of infinitely closely spaced independent elastic springs replace the elastic soil medium. For normally consolidated clay and cohesionless soils, Reese and Matlock have established a series of curves. The deformed shape of the pile and the corresponding bending moments, shearing forces, and soil reactions are depicted in Fig. 4.

The behavior of pile can be evaluated by using equation of an elastic beam supported on an elastic foundation, given as follows:

$$\mathrm{EI}\frac{\mathrm{d}^4 y}{\mathrm{d}x^4} + p = 0 \Rightarrow \frac{\mathrm{d}^4 y}{\mathrm{d}x^4} + \frac{K_h y}{\mathrm{EI}} = 0 \tag{5}$$

In general, the solution of Eq. (5) can be expressed as:

$$y = f\,(x, T, L, K_h, \mathrm{EI}, H, M) \tag{6}$$

From this solution, Matlock and Reese [8] developed equations for the determination of deflection, slope, moment, shear and soil reaction along the pile. Equations for the free headed condition are given below:

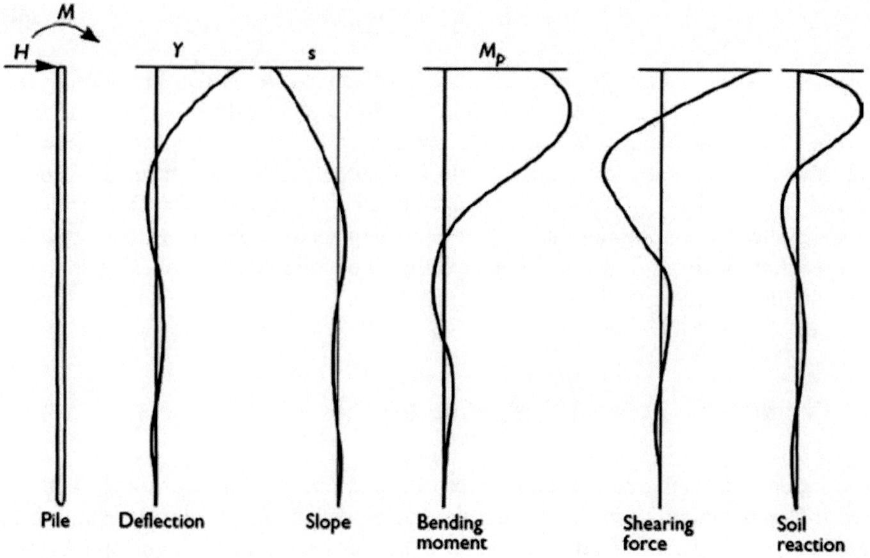

Fig. 4 Deflection, slope, bending moments, shearing forces and soil reactions for elastic conditions [6]

$$\text{Deflection, } y = y_A + y_B = \left[\frac{\text{HT}^3}{\text{EI}}\right] A_y + \left[\frac{\text{MT}^2}{\text{EI}}\right] B_y \tag{7}$$

In Eq. (7), y is total deflection of the pile at any point, y_A is deflection due to horizontal load H and y_B is deflection due to moment load M.

$$\text{Similarly; Slope, } S = S_A + S_B = \left[\frac{\text{HT}^2}{\text{EI}}\right] A_s + \left[\frac{\text{MT}}{\text{EI}}\right] B_s \tag{8}$$

$$\text{Moment, } M_p = M_A + M_B = (\text{HT}) A_m + (M)\, B_m \tag{9}$$

$$\text{Shear, } V = V_A + V_B = (H) A_V + \left[\frac{M}{T}\right] B_V \tag{10}$$

$$\text{Soil reaction, } p = p_A + p_B = \left[\frac{H}{T}\right] A_p + \left[\frac{M}{T^2}\right] B_p \tag{11}$$

In Eqs. (7–11), A_y and B_y are the deflection coefficients, A_s and B_s be the slope coefficients, A_m and B_m the moment coefficients, A_v and B_v the shear coefficients and A_p and B_p the soil reaction coefficients respectively for horizontal load H (or Q) and moment load M acting on the pile at the ground level. These coefficients for long piles are given in tabular form and for intermediate piles it can be taken from graphs given by Matlock & Reese. This method of analysis has gained significant experience in applying the theory to practical situations but the continuity of soil mass is not taken into consideration in the analysis.

3.2 Elastic Continuum Approach

Poulos [9] analyzed the behavior of laterally loaded piles by considering soil as an elastic continuum, which makes this approach theoretically more realistic than subgrade reaction method. However the realistic determination of soil modulus E_s (K_h and E_s are sometimes used interchangeably) was an obstacle to its practical application. And the approach requires more field verification before theoretical concept could apply to practical problems. Using Mindlin equation for horizontal loads within a semi-infinite mass, soil displacement was calculated and the pile displacements have been evaluated using beam on elastic foundation Eq. (5). Then the solutions for lateral deflection and maximum moment were obtained by assuming that soil modulus increases linearly with depth.

3.3 P-y Curve Method

Reese and Matlock's [6] analytical approach is applicable only for the pile deflection which are within the range of the elastic compression of the soil. However, this analytical method can be extended beyond the elastic range to analyze movements where the soil yields plastically up to and beyond the stage of shear failure. This could be made possible by employing the p–y curves [8], which represent the deformation of the soil at any given depth below the ground surface as shown in Fig. 5. Since the *p-y* curves employed in the method were obtained from full scale field test, this method has been considered as most convenient and commonly used procedure for analyzing laterally loaded piles. In the analysis, however some significant factors such as soil continuity and few pile properties are not taken into account. Therefore various researchers made attempt to improve efficiency of *p y* curve method and some of them have introduced new methods.

3.4 Characteristic Load Method

Duncan et al. [10] modified *p-y* curve method and referred it as the Characteristic Load Method (CLM) which was simpler than *p-y* method of analyses, but closely approximates results of *p-y* analysis. In this method, dimensional analysis has been employed to model the nonlinear behavior of laterally loaded piles and drilled shafts. In order to determine the degree of approximation involved in CLM, the results acquired from both CLM and *p-y* analyses was compared, which indicates a close

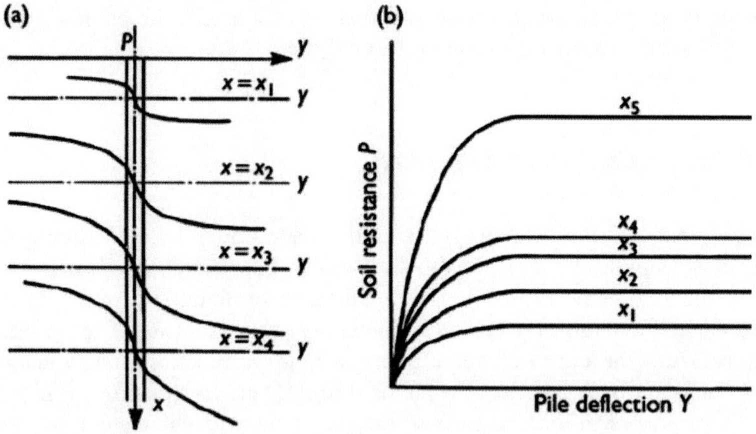

Fig. 5 *P-y* curves for laterally loaded piles. **a** Shape of curves at various depths x below soil surface. **b** Curves plotted on common axes [8]

approximation for static loading and fair approximation for cyclic loading on piles or drilled shafts in soft clay. The major drawback of the CLM is that its application is limited to piles and drilled shafts that are long enough so that the pile length loses its significance on pile behavior. In the case of shorter piles and drilled shafts in uniform soil the CLM overestimates the lateral deflection while underestimating bending moments. The results of the study show a reasonable agreement with the results of field load tests performed by Reese et al. and Cox et al. on single piles. Later Brown et al. [11] suggested an advanced analytical method of interpreting lateral load test on piles and drilled shafts using inclinometer data [11].

3.5 Strain Wedge Model

Ashour et al. described the application of the strain wedge (SW) model in the analysis of lateral response of single long piles, drilled shafts, and pile groups in layered soil (either sand or clay) and rock deposits [12]. The capabilities of the SW model over other procedures and approaches are highlighted in this study. The SW model analysis is based on a representative soil-pile interaction that includes pile and soil properties. In the analysis, prediction of related *p-y* curve at any point along the deflected part of pile, liquefaction potential of soil, interference among adjacent piles are considered. Through several comparative studies with model tests and full scale lateral load tests, the SW model has been validated and confirmed. The SW model has the potential to determine the response of a laterally loaded pile group in layered soil on the basis of more realistic assumptions of pile interference as compared to currently employed or proposed methods and practices.

4 Numerical Analysis of Laterally Loaded Piles

Howe [13], Matlock and Reese [8], and Bowles described their attempts to use the finite-difference method (FDM) to analyze a laterally loaded piles. Since 1960's computers have been available with computer programs for analyzing laterally loaded pile, which are obviously better tools for these analyses. Bowles [14] initially used the FDM for analyses but soon it became evident that the finite element method (FEM) offered a considerable improvement. FEM has both node translation and rotation, while the FDM has only translation. The elastic curve can be defined better using both translation and rotation. Most of the recently published numerical analyses were based on the finite element method using software such as PLAXIS Foundation, PLAXIS 2D, PLAXIS 3D, ALP OASYS, ABAQUS, ANSYS, FLAC, LPILE etc. Some of those studies which are performed with the aid of software are discussed here.

Kok et al. [15] presented a numerical modeling of laterally loaded pile based on some analytical results and case history. The single pile response obtained was then compared with the published results. A two dimensional finite element program PLAXIS 2D has been used to simulate lateral response of two single piles, one near an excavation and the other one used to stabilize sliding slope. The major limitation of the whole study is that, since the pile was modeled as an infinitely long wall in this 2D finite element analysis, the shear flow around the pile seems to be ignored, thereby underestimating the maximum moment along the pile.

Ahmari et al. introduced a three dimensional finite element model using the ANSYS computer program, to analyze laterally loaded piles in clay [16]. Two field measured full scale case studies, one conducted by Matlock in soft clay at Lake Austin and the other by Reese in stiff clay, have been analyzed using the developed finite element model. The comparison of results shows a good agreement between the results of finite element analysis and field measurements. By applying the developed model in the first case study, the *p*-*y* curve was obtained and then compared with the traditional *p*-*y* curves suggested by Matlock & Reese and the hyperbolic *p*-*y* curves proposed by Wu et al.

Jayantha et al. [17] proposed an alternative theoretical method for predicting the *p*-*y* curves for laterally loaded single piles placed in undrained clay, which directly takes into account governing variable groups. This method is based on finite difference analysis using commercial FLAC software. The ultimate lateral displacement predicted using the proposed closed form solutions were compared with that of the Matlock (1970). Results obtained for the tension failure condition was much closer to the value obtained using the empirical method of Matlock.

Kim et al. have done a non linear finite element modeling using PLAXIS 3D Foundation to investigate numerically the deflection of pile, bending moment and *p*-*y* curve along the pile length [18]. The results from finite element analyses were validated with results of lateral field load tests performed on the steel pile and drilled shaft embedded in soil profiles in South Korean offshore deposits. The validation of the lateral response of the pile has been presented in terms of lateral deflection,

Table 1 Comparison of field test and two numerical analyses [19, 20]

Method	Displacement (mm)
Field test	4.12
PLAXIS 2D analysis [19]	5.51
ALP OASYS analysis [20]	5.83

bending moment distribution and *p-y* curves at various depths. The comparison shows a good agreement of the study with general trends observed in field measurement. They also made an attempt to carryout parametric studies for verifying the *p-y* characteristics, which shows that the modulus of subgrade reaction (K) and the ultimate soil resistance (p_u) increase linearly with pile diameter, whereas pile elastic modulus (E_p), interface property (R_{inter}) and pile-toe condition have no significant influence.

Naveen et al. [19] presented the results of a field test carried out on a laterally loaded bored cast in-situ piles of 1 m diameter and 17 m length and of M35 grade concrete, embedded in residual soil. They made an attempt to simulate these lateral load test on larger diameter piles in residual soil using finite element model in PLAXIS 2D. Naveen et al. [20] simulated the same lateral load test using ALP OASYS software. Final results of finite element modeling were validated with results of field test on laterally loaded pile. The curve obtained from numerical simulation using PLAXIS 2D and ALP OASYS shows good agreement with field test results. Displacement obtained from field test, PLAXIS 2D and ALP OASYS were compared in the paper [20]. In the field test, simulation of displacement such as 12 mm or 10% of pile diameter (IS code requirement) are not possible using existing techniques, especially for larger diameter piles. Accurate numerical methods can be adopted under such situation to simulate displacement as per code provisions (Table 1).

Fath et al. [21] performed finite element analyses of laterally loaded piles in medium stiff clay using ABAQUS program. In terms of ultimate lateral load, maximum bending moment and the *p-y* curves, the lateral response of pile is presented. The effects of various factors such as L/d ratio, pile dimensions and soil shear strength on the performance of pile have been studied. The conclusion drawn from the study is that with the increase in soil shear strength, the L/d ratio and the pile diameter, the ultimate lateral pile capacity increases. In the study, the maximum bending moment occurs in between a depth of 0.12–0.18 times the length of pile.

Thadapaneni et al. [22] performed the analysis of piles subjected to lateral and vertical loads by using various methods. In the analysis, piles were modeled as linear elements and the soil-structure interaction was taken into consideration. Lateral load analysis of piles was carried out by using empirical equations and finite element methods. Empirical methods used were Vesic's method using subgrade reaction and Brom's method and the finite element analysis were done with the aid of STAAD Pro and LPILE software. The comparison of results obtained from various methods of analysis has been presented. When software is not available for the analysis, Brom's method which holds good in computing pile head deflection under lateral loads could be used in small scale projects. Though the analysis using LPILE software provides conservative results, the soil-pile behavior such as bending moment of pile, soil reactions, *p-y* curves etc. can be analyzed with less effort within a short period of time. The STAAD Pro analysis can be used effectively to determine pile head deflections with a certain multiplication factor due to its significantly lower deflection values.

5 Conclusions

The effects of vertical load on the pile have been very well established in the past years, but researches on piles under lateral load are still going on to obtain a well defined method of analysis. This paper has presented a detailed literature review of various methods generally used for the analysis of laterally loaded piles. The applicability, advantages, limitations and comparison of different methods are discussed in this study. The modifications and extensions suggested by various researchers on the existing procedures and approaches are also reviewed.

The lateral load resistance of pile depends on soil properties, pile material, pile diameter, pile length, loading type etc. Even though it has been found that the analytical solutions are complicated and tedious, the implementation of computer-based programs has solved the problem to a great extent. With the increased computational power of the computers and the availability of appropriate software, numerical simulation and analysis of the problem has become the most cost effective and accurate method. Real soil-pile interaction is three dimensional which makes it difficult to model highly accurate two dimensional model of pile. However some researchers managed to simulate lateral response of pile in 2D finite element program by simplifying the geometry of model. For the better understanding of different methods and their comparison some figures and table are included. There is still need to explore the available features in various numerical modeling software, in order to perform much more efficient and quick analysis of laterally loaded piles under various loading conditions and in different types of soil.

References

1. Tomlinson, M., Woodward, J.: Pile Design and Construction Practice, 5th edn. E & FN Spoon Publishers (2008)
2. Prakash, S., Sharma, H.D.: Pile Foundations in Engineering Practice. A Wiley-Interscience Publication. John Wiley & Sons Inc., New York (1990)
3. Broms, B.: Lateral resistance of piles in cohesionless soils. J. Soil Mech. Found. Eng., ASCE **90**(3), 123–156 (1964)
4. Brinch, H.J., Christensen, N.H.: The Ultimate Resistance of Rigid Piles Against Transversal Forces, pp. 5–9. Danish Geotechnical Institute Copenhagen, Denmark (1961)
5. Broms, B.: Lateral Resistance of piles in cohesive soils. J. Soil Mech. Found. Eng., ASCE **90**(2), 27–63 (1964)

6. Reese, L.C., Matlock, H.: Non-dimensional Solutions for Laterally Loaded Piles with Soil Modulus Assumed Proportional to Depth. In: Proceedings 8th Texas Conference on Soil Mechanics and Foundation Engineering, pp.235–250. Austin, TX (1956)
7. Punmia, B.C., Jain, A.K., Jain, A.K.: Soil Mechanics and Foundations, 17th edn. Laxmi Publications (2010)
8. Matlock, H., Reese, L.C.: Generalized solutions for laterally loaded piles. JSMFD, ASCE **86**(5), 63–91 (1960)
9. Poulos, H.G.: Behavior of laterally loaded piles: 1-single piles. J. Soil Mech. Found., ASCE **97**(5), 711–731 (1971)
10. Duncan, M.J., Evans, L.T., Phillip, S.K.: Lateral load analysis of single piles and drilled shafts. J. Geotech. Eng., ASCE **120**(6), 1018–1033 (1994). https://doi.org/10.1061/(ASCE)0733-9410(1994)120:6(1018)
11. Brown, D.A., Hidden, S.A., Zhang, S.: Determination of p–y curves using inclinometer data. Geotech. Test. J. **17**(2), 150–158 (1994)
12. Ashour, M., Noris, G., Pilling, P.: Strain wedge model capability of analyzing behavior of laterally loaded isolated piles, drilled shafts and pile groups. J. Bridg. Eng. **7**(4), 245–254 (2002). https://doi.org/10.1061/(ASCE)1084-0702(2002)7:4(245)
13. Howe, R.J: A Numerical method for predicting the behavior of laterally loaded piling. In: Exploration and Production Research Division Publ. No.412, Shell Oil Co., Houston, TX (1955)
14. Bowles, J.E.: Analytical and Computer Methods in Foundation Engineering. McGraw-Hill, NewYork (1974)
15. Kok, S.T., Huat, B.B.K.: Numerical modeling of laterally loaded piles. Am. J. Appl. Sci. **5**(10), 1403–1408 (2008). https://doi.org/10.3844/ajassp.2008.1403.1408
16. Ahmadi, M.M., Ahmari, S.: Finite element modeling of laterally loaded piles in caly. Proc. Inst. Civ. Eng., Geotech. Eng. **162**(3), 151–163 (2009). https://doi.org/10.1680/geng.2009.162.3.151
17. Kodikara, J., Haque, A., Lee, K.Y.: Theoretical p-y curve for laterally loaded single pile in undrained clay using bezier curves. J. Geotech. Geoenviron. Eng., ASCE **136**(1), 265–268 (2010)
18. Kim, Y., Jeong, S.: Analysis of soil resistance on laterally loaded piles based on 3D soil–pile interaction. J. Comput. Geotech. **38**(2), 248–257 (2011). https://doi.org/10.1016/j.compgeo.2010.12.001
19. Naveen B.P., Sitharam T.G., Vishruth S.: Numerical simulations of laterally loaded piles. In: Proceedings of the International Conference on Ground Improvement and Ground Control. https://doi.org/10.3850/978-981-07-3560-9_09-0907 (2012). ISBN: 978-981-07-3560-9
20. Naveen B.P., Kontoni, D.P.N. (2016) Comparison of field test and numerical analysis for laterally loaded piles. In: 7th International Conference from Scientific Computing to Computational Engineering. Vol. 1, pp. 117–124
21. Eldayem, F.E.E.N., Mohamedzein. Y.E.A.: Finite element analysis of laterally loaded piles in clays. In: Proceedings of the 4th Geo-China International Conference, pp. 175–182. Geotechnical Special Publications No. 259, ASCE, China (2016)
22. Kanakeswararao, T., Ganesh, B.: Analysis of pile foundation subjected to lateral and vertical loads. Int. J. Eng. Trends Technol. **46**(2),113–127. https://doi.org/10.14445/22315381/IJETT-V46P219 (2017)

Effect of Silica Fume and Steel Fiber on Mechanical Characteristics of High-Strength Concrete

A. Sumathi and K. Saravana Raja Mohan

1 Introduction

The chief binder part of concrete is ordinary portland cement (OPC), but high CO_2 emissions are increasingly affecting the infrastructure. As a consequence, it is deemed environmentally unacceptable. A variety of works have been carried out on alternative cement binders or replacements aimed at reducing the environmental effects. Minimum works were, however, published to examine HSFRC's structural behavior, especially with regard to its ultra-high ductility. This is one of the explanations for the analysis mentioned here, with an emphasis on silica fume (SF) and HSSFRC. The uses of SFRC over the past thirty years have been so varied and so wide spread, that it is difficult to categorize them. The most common applications are pavements, tunnel linings, airport pavements, bridge deck slab repairs, and so on. SF used is a by-product of silicon and ferrosilicon. The fineness and the specific surface area of SF are very high, about 2000 m^2/kg, whereas this value is only about 350 m^2/kg for OPC. Because of its very efficient pozzolanic action, this has led to the evolution of the use of microsilica as a cement substitute, but as more laboratory and field results are obtained, the material becomes an additional cementitious material giving increased strength and durability in fresh and hardened states [1–3]. Silica fume does not only influence in the rate of hydration but also reacts with the hydration products. For improved concrete performance based on strength and durability, use of SF as cement supplement has been tried with success. Different researchers have achieved different optimal values. However, most research studies agree on the replacement level ranging between 5 to 10% depending on the type of desired concrete properties [4–13]. Higher silica fume content led to enormous weight loss and strength loss

A. Sumathi (✉) · K. Saravana Raja Mohan
School of Civil Engineering, SASTRA Deemed To Be University, Thanjavur, India
e-mail: sumathi@civil.sastra.edu

K. Saravana Raja Mohan
e-mail: srm@civil.sastra.edu

I. Pal and S. Kolathayar (eds.), *Sustainable Cities and Resilience*, Lecture Notes in Civil Engineering 183, https://doi.org/10.1007/978-981-16-5543-2_34

under aggressive environment [14–16]. Mahmoud Nili [17] on their study inferred that the SF and steel fiber enhance the concrete properties such as long-term compressive strength (CS), dynamic frequency, and less water absorption for 1% steel fiber on SF concrete. Al sakiny [18] studied the performance of steel fibers content and high-range water reducer (HRWRA) in UHSFRC. From the tests, control concrete with 2 to 3% with 0.5% increment of fibers in volume fractions of steel fibers showed a minimum reduction in CS at early period of curing. Soulioti [19] carried out an experimental work to evaluate the mechanical properties of SFRC and hybrid fibers with regression model. The result showed also that the increase in fiber volume lead to an improvement in the first and final peak strengths, residual strength as well as flexural strength of the tested beams with hooked-end fiber exhibiting higher flexural toughness behavior. Addition of carbon steel fibers, viz., low and high to concrete enhanced the compressive and flexural strength at 28 days [20]. Baran [21] investigated the pull-out behavior of prestressing elements embedded in SFRC. The variables considered were steel fibers with two different lengths of 30 and 60 mm each of 0, 0.2, 0.4, 0.6, and 0.8% content in volume. Based on the results, there was an improvement in the tensile behavior and flexural post-cracking response of SFRC due to crack control by the fibers. [22, 23] via experimental studies, different mechanical properties of HSC forta-ferro and steel fibers with and without nanosilica and SF found that pozzolan fiber-reinforced concrete improved the different properties of HSC compared to plain concrete. This paper deals with the basic materials properties, mix design, and testing methods. It also consists of experimental results and discussions on mechanical properties of HSC and HSSFRC like cube compression strength, indirect tensile strength, modulus of rupture, elastic modulus, and stress–strain curve under compression for concrete along with experimental and statistical interpretations.

2 Experimental Program

2.1 Materials Used

Ordinary portland cement (OPC) of 53 grade conforms to IS 12269 [24]. Specific gravity (SG) is 3.23, 55, and 355 min are the initial and final setting times, respectively. Elkem microsilica 920 D was used as cementitious material of SG of 2.26. Maximum size of 4.75 mm river sand and coarse aggregates of size 12.5 and 20 mm were used. The fineness modulus of fine and coarse aggregate was 3.39 and 5.25, respectively [25]. Superplasticizer (SP) with the SG in the range of 1.220–1.225 and pH greater than 6 was used. Hooked-end steel fibers of 80 aspect ratio and density 7850 kg/m^3 were used.

Table 1 Mix proportions (kg/m^3)

Concrete mix	M60
Cement (kg/m^3)	450
Silica fume 10% (kg/m^3)	31.5
Fine aggregate (kg/m^3)	780
Coarse aggregate (kg/m^3)	1130
Superplasticizer	0.8–1.2%

2.2 *Mix Design and Methods*

The mix configuration for the M60 grade was performed in the present study in compliance with IS 10262 [26], and the proportion of the mix used for the study was 1:1.73:2.51:0.3. SF dosages are 5, 10, 15, 20, 25, and 30%, replaced by cement percentages. To study the SF and steel fiber effect on HSC material features, steel fibers are added to this by 0.5, 1, 1.5, and 2% by volume. The water-to-binder ratio arrived from the slump test is having water-to-binder ratio 0.27 with the addition of superplasticizer. Specimens were prepared based on the exact quantity of material for each mix as presented in Table 1.

2.3 *Details of Specimens*

Cubical specimens of size 150 mm were used for compressive strength (CS), cylindrical specimens of 150 mm diameter and 300 mm long specimens were used for indirect tensile strength (TS), elastic modulus (E) and stress–strain curve, prism specimens of size 100 × 100 × 500 mm were used for modulus of rupture (MOR). Mixtures containing 0, 5, 10, 15, 20, 25, and 30% of SF as cement replacement and 0 to 2% steel fibers with an increment of 0.5% have been used. Six samples for each mix were casted for various water-to-binder ratio, the percentage of cement replacement level, percentage of steel fiber content in volume fractions, and period of curing, respectively, for testing compressive strength and three samples for other properties. The casted and cured specimens were tested on 3000 kN capacity compression-testing machine in accordance with IS 516 [27] to obtain the CS and TS of concrete at the curing days of 7, 28, 90, and 180 days. The MOR is carried out on prisms to estimate the flexural strength. For each concrete batch, tests were conducted on three prisms in flexural strength-testing machine of capacity 100 kN in accordance with ASTM C78 [28]. The center distance between supporting rollers for beams (prism) is 400 mm, and the center distance between loading rollers is 133 mm. The specimens were tested for 28, 90, and 180 days. The stress–strain curve was studied in 150 Ø × 300 mm long cylinder in a servo-hydraulic closed-loop CTM of capacity 3000 kN with a displacement rate of 0.05 mm/min at the age of 28 days. Before testing, the extreme ends of the specimens were made horizontal by grinding and constant length for all cylinder specimens.

3 Results and Discussion

3.1 Silica Fume Effect on CS

The average CS studied in six specimens for each mix with replacing cement with six different percentage of SF for different curing days is shown in Table 2. The results showed an increasing trend up to replacement of 10% of silica fume by cement. Similar trends were observed by other researchers [7, 10, 30]. The SF contains amorphous silica, and pozzolanic reaction gives the better strength results. When the replacement ratio is greater than 10%, the intensity is found to be decreased. Compared with cement, the CaO content in the silica fume is comparatively lower. Even though the strength decreases after silica fume replacement of 10%, the characteristic strength of 60 MPa was achieved for SF replacement more than 10%. Since silica fume is cheaper than cement, 10% SF as cement replacement is considered as optimum replacement for HSC.

3.2 Silica Fume Effect on TS

In each mix, three cylindrical specimens were tested to calculate the split TS of high-strength concrete. The variations of average tensile strength with respect to silica fume content are presented in Table 2. The maximum split TS obtained is 8.16 MPa, which is about 28% more than the control concrete specimen. The maximum strength is obtained for 10% silica fume at 28 days.

3.3 Silica Fume Effect on MOR

The results of MOR of control and SF concrete are presented in Table 2. The maximum MOR concrete at the age of 28, 90, and 180 days is 6.92, 8.15, and 8.39 MPa, respectively. The maximum increase in MOR for 10% silica fume compared to control concrete is observed as 17.48, 19.5, and 18.84% at 28, 90, and 180 days.

Table 2 CS, split TS, and MOR test results

SF (%)	Fiber (%)	CS (MPa)					TS (MPa)			MOR (MPa)		
		7	28	90	180	7	28	90	180	28	90	180
0	0	35.96	60.08	64.45	68.92	4.58	5.98	6.78	7.28	5.89	6.82	7.06
	0.5	37.18	64.62	68.72	73.35	4.83	6.77	7.83	8.54	6.62	7.81	8.08
	1	41.29	67.21	70.66	75.65	5.08	7.19	8.31	8.99	7.72	9.06	9.53
	1.5	42.73	64.79	68.83	73.61	5.67	7.98	8.09	8.66	7.38	8.65	9.08
	2	38.5	63.60	67.56	72.36	5.12	7.87	8.93	8.50	7.07	8.29	8.65
5	0	36.5	65.37	69.62	75.06	4.92	6.13	7.02	7.53	6.33	7.40	7.72
	0.5	40.18	69.53	74.32	80.09	5.05	6.97	8.13	8.80	7.13	8.50	8.87
	1	44.81	72.62	76.55	82.56	5.89	7.39	8.63	9.42	8.34	9.89	10.52
	1.5	47.99	70.16	74.62	80.38	5.74	7.21	8.39	9.08	7.96	9.43	9.97
	2	45.83	68.87	73.22	78.87	5.45	7.08	8.22	9.11	7.63	9.02	9.50
10	0	37.18	68.02	74.52	78.46	4.97	7.63	7.58	8.16	6.92	8.15	8.39
	0.5	45.97	72.62	79.71	83.89	5.07	7.74	8.88	9.76	7.87	9.44	9.73
	1	46.28	75.59	82.93	87.47	5.68	8.07	9.39	10.36	9.22	10.92	11.55
	1.5	45.1	73.42	80.60	84.79	5.13	7.90	9.15	9.96	8.79	10.47	10.94
	2	43.89	71.70	79.01	83.10	5.08	7.76	8.94	9.76	8.43	10.04	10.41
15	0	34.81	66.38	71.80	76.30	4.82	5.63	6.08	6.62	6.37	7.46	7.72
	0.5	43.66	70.68	76.50	81.22	5.12	6.34	7.00	7.51	7.18	8.55	8.87
	1	45.29	73.01	78.89	83.87	5.76	6.76	7.41	7.96	8.38	9.95	10.52
	1.5	43.72	71.12	76.83	81.58	5.54	6.55	7.22	7.64	8.01	9.49	9.96
	2	42.51	69.68	75.31	78.99	5.15	6.46	7.23	7.47	7.68	9.10	9.49

(continued)

Table 2 (continued)

SF (%)	Fiber (%)	CS (MPa)				TS (MPa)				MOR (MPa)		
		7	28	90	180	7	28	90	180	28	90	180
20	0	33.62	63.08	64.98	71.04	3.91	5.23	5.85	6.27	5.59	6.52	6.72
	0.5	35.22	66.98	68.91	75.37	4.39	5.90	6.73	7.27	6.27	7.45	7.69
	1	36.97	68.94	71.14	77.94	4.95	6.20	7.12	7.57	7.31	8.65	9.04
	1.5	34.89	67.10	69.31	75.92	4.41	6.05	6.93	7.31	6.97	8.25	8.63
	2	34.1	65.76	68.14	73.56	4.21	5.90	6.80	7.16	6.69	7.91	8.22
25	0	32.39	55.98	58.20	63.33	2.92	4.88	5.51	5.90	5.32	6.20	6.39
	0.5	33.66	59.43	61.69	67.17	3.47	5.49	6.32	6.77	5.96	6.34	7.39
	1	34.91	61.10	63.65	69.25	4.68	5.77	6.68	7.11	6.93	8.20	8.52
	1.5	33.18	59.57	61.98	67.47	4.08	5.62	6.47	6.82	6.62	7.82	8.18
	2	32.05	58.19	61.93	65.50	3.97	5.55	6.33	6.67	6.36	7.51	7.80
30	0	24.81	51.66	54.68	62.86	2.65	4.30	4.82	5.14	5.15	6.06	6.21
	0.5	27.45	54.65	57.85	66.56	3.39	4.80	5.48	5.84	5.76	6.90	7.08
	1	29.08	56.35	59.63	68.72	4.08	5.02	5.77	6.13	6.69	8.00	8.21
	1.5	28.1	54.85	58.14	66.90	3.83	4.90	5.61	5.87	6.40	7.63	7.94
	2	27.83	53.75	57.14	65.45	3.67	4.79	5.49	5.72	6.15	7.33	7.56

3.4 Steel Fiber Effect on CS

The CS of cubes with various steel fiber content at all ages are detailed in Table 2. There is an increase in strength up to 1% steel fiber content in volume fractions, and there is a reduction in strength beyond 1%. The enhanced strength was observed for steel fiber specimens than control for 3, 7, 28, 90, and 180 days of curing. The percentage strength increase is 11.86, 19.27, and 25.91% compared to 28 days control concrete.

3.5 Steel Fiber Effect on TS

In each mix, three specimens were casted and tested to estimate the split TS with (0, 0.5, 1, 1.5, and 2%) fiber content in volume fractions. From Table 2, the increase in development of strength was found when the fiber content increases. HSC with fiber 1% tends to maximum increase in strength compared to control specimen at different days. At 28, 90, and 180 days, the percentage increase was 33.44, 38.96, and 50.33%, respectively, compared to 28 days control concrete.

3.6 Steel Fiber Effect on MOR

The flexural strength (MOR) studied at different curing ages for concrete with different percentage of SF and steel fiber is presented in Table 2. The highest value of flexural strength is attained for mixes with 1% of fibers at all ages. The 28, 90, and 180 days flexural strength was 7.72, 9.06, and 9.53 MPa. It showed an increase of 31% at 28 days and 53.82% at 90 days and 61.8% at 180 days compared to 28 days control concrete.

3.7 Silica Fume and Steel Fiber on CS

The CS for 10% SF with different percentage of steel fiber content were tested for 7, 28, 90, and 180 days curing. Based on the variations of strength with respect to fiber content and silica fume, it is observed that the maximum strength was 87.47 MPa, obtained for S10F1 mix, i.e., 10% silica fume and 1% fiber content in volume fractions. Maximum increase in strength over control concrete was found to be about 45.59% at 180 days. The general inference is from the results, all mixes show similar trend of steep changes between 7 and 28 days but flatten out for additional duration. This is due to the bonding action of the fibers in concrete matrix that is effective initially up to 28 days [6, 11]. Hence, the fiber content increases, the strength of the concrete also increases. It is found that the reduction in compressive strength due to introduction of silica fume could be easily compensated through the addition of fibers. At 28 days, silica fume increases the pozzolanic reaction, and it was sufficient to contribute to the strength under compression. Thus, the efficiency of the silica fume to act as cementitious material has increased substantially at the age of 28, 90, and 180 days. Compressive strength for an optimum mix (S10F1) varied from 75.59 to 82.93 MPa at 90 days and 82.93 to 87.47 MPa at 180 days. There was a 9.71% increase in strength at 90 days and 15.72% at 180 days compared to 28 days strength.

3.8 Silica Fume and Steel Fiber on TS

Splitting TS for silica fume and fiber content was observed. The maximum tensile strength for 3, 7, 28, 90, and 180 days was 3.97, 5.68, 8.07, 9.39, and 10.36 MPa which is about 17.10, 24.017, 34.95, 38.49, and 42.30% than control concrete. From Fig. 2, the highest strength is obtained for mixes containing 10% SF with 1% of fiber at all ages. Beyond 10% replacement of cement by SF and 1% of fibers, there is a decrease in strength. For this optimum range, the percentage increase in split tensile strength is 16.36% at 90 days and 28.38% at 180 days when compared to 28 days.

3.9 Silica Fume and Steel Fiber on MOR

MOR studied at different curing ages for various percentage of SF and steel fiber and the results are shown in Table 2. The maximum strength is obtained for mixes with 10% silica fume and 1% of fibers at all ages. The 28, 90, and 180-days flexural strength is 9.22, 10.92, and 11.55 MPa. It showed an increase of 18.44% at 90 days and 25.27% at 180 days compared to 28 days.

3.10 Stress–Strain Curve Under Compression

An experimental stress–strain curve under compression for optimum silica fume 10% with different steel fiber content is shown in Fig. 1. The addition of fibers to HSC enhances the curve at the post-peak segment and increased the strain at peak stress. It can be observed that the elastic portion of the curve increases when the strength increases in ascending portion and decreases the drops in the softening portion of the stress–strain curve for control concrete and gradually flatters for steel fiber-reinforced concrete [33]. The ultimate strains at failure for control specimen were 0.00697, and steel fiber-reinforced specimens were found to be in the range of 0.012 to 0.019. Earlier findings noticed that steel fibers with hooked-end are much

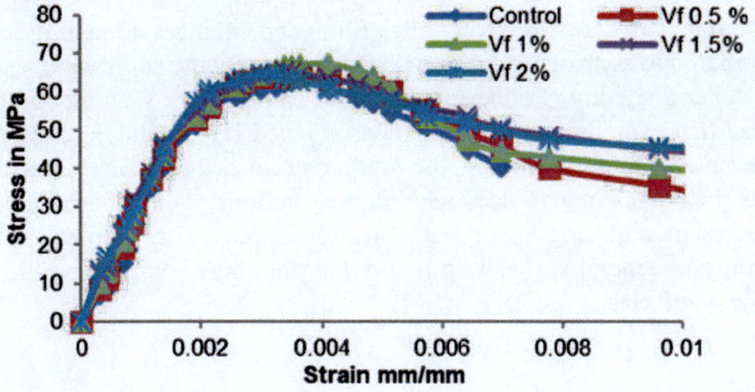

Fig. 1 Stress–strain curve

Fig. 2 Modulus of elasticity for SF-based fiber in V_f (%)

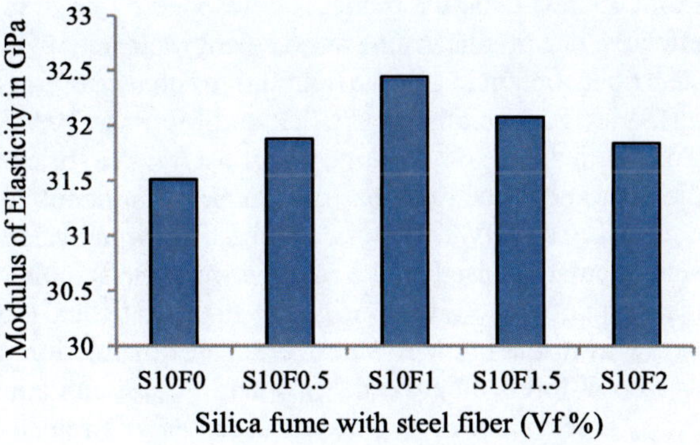

effective in mechanical properties and energy absorption at post-peak load carrying capacity [31].

3.11 Elastic Modulus

Elastic modulus (secant modulus) was found from the cylinder specimen for the optimum percentage of silica fume with steel fibers in different volume fractions at 28 days. According to ACI 318 [29], the elastic modulus is estimated from the slope of the straight line between the 40% compressive stress and the strain point. The variation of the elastic modulus for optimum silica fume 10% by weight of cement, various fiber content in volume fractions of each mix shows higher order than control concrete, but the modulus of elasticity seems to change very little over the test range are shown in Fig. 2. The percentage increase in elastic modulus of fiber concrete relative to silica fume concrete (control) was 1.2, 2.96, 1.8, and 1.05% for 0.5, 1, 1.5, and 2%, respectively [11, 23]. From the results, the maximum improvement in elastic modulus is obtained for 10% silica fume and 1% fiber content (S10F1) was 2.96% compared to control concrete [23]. This means by the addition of steel fibers, the beam flexural rigidity is rather unaffected. From the stress–strain curve, it may be observed that even though the specimens showed less modulus of elasticity for larger volume of steel fiber content, their larger strain sustainability without sudden failure was the notable one to use high volume fractions of steel fibers, i.e., 1.5 and 2% in specimens requiring larger flexibility for increased rotation or ductility for large deflection. Generally, it can be found that the fibers addition to HSC slightly enhance the static elastic modulus.

3.12 Statistical Method

From the test data, the model is developed by graphical approach to enumerate the effective use of silica fume as a cement replacement to develop a HSC and effect of steel fiber content in volume fractions to enhance the mechanical properties to develop a HSSFRC especially on CS, TS, and MOR, at 10% SF with steel fiber content (0, 0.5, 1, 1.5, and 2%) in volume fractions. It was to be checked whether the variation in the fiber and silica fume has statistically significant effect on the strength. This was done using ANOVA. Tables 3, 4, and 5 show the values of F_{actual} and $F_{critical}$. F_{actual} was calculated based on the ratio between the sample mean square to within sample mean square. $F_{critical}$ was verified from the statistical tables, and it was compared with F_{actual} from the known values of degrees of freedom and significance level. There was no difference between F_{actual} and $F_{critical}$ and considered as null hypothesis, if F_{actual} is less than $F_{critical}$ means accepted and rejected when F_{actual} is greater than $F_{critical}$. Tables 3, 4, and 5 show the variance ratio for various strengths for different

Table 3 Variance ratio values for compressive strength

Parameters	Actual variance ratio, F_{actual}				Critical variance ratio, $F_{critical}$
	7 days	28 days	90 days	180 days	
SF (%)	17.83	35.52	51.63	38.25	2.69
Fiber (%)	7.97	6.71	5.27	6.08	2.87
SF and fiber	39.77	41.92	58.58	46.48	2.42

Table 4 Variance ratio values for split tensile strength

Parameters	Actual variance ratio, F_{actual}				Critical variance ratio, $F_{critical}$
	7 days	28 days	90 days	180 days	
SF (%)	0.95	1.141	0.904	1.070	2.69
Fiber (%)	0.16	0.680	0.621	0.425	2.87
SF and fiber	0.68	1.162	1.485	1.856	2.42

Table 5 Variance ratio values for modulus of rupture

Parameters	Actual variance ratio, *F* actual			Critical variance ratio, *F* critical
	28 days	90 days	180 days	
SF (%)	0.405	0.576	0.640	3.55
Fiber (%)	0.506	0.745	0.917	3.88
SF and fiber	0.991	1.514	1.719	3.08

parameters, viz., silica fume content, fiber content, and both silica fume and fiber content. Table values are verified and concluded that fiber content and silica fume are much effective and statistically significant.

4 Conclusions

The important conclusions drawn from the experimental results are summarized below.

- Among different percentages of SF and steel fiber used in concrete, 10% SF and 1% steel fiber shows better strength and was considered as suitable percentage in terms of CS, split TS and MOR, and elastic modulus.
- The CS increases from 1% steel fiber to SF concrete ranging from 25 to 30%, at different curing days than control concrete. The incorporation of SF in the reinforced fiber concrete enhances the compressive strength due to an improvement in the bond of cement paste with aggregates and reduction of pores through pozzolanic reactions in cement paste, and concrete fibers provide resistance to crack propagation and control the rate of crack growth.
- The addition of steel fibers to HSC containing silica fume changes the complete basic trend of stress–strain response. The descending branch slope enhances based on the increase of fiber in volume fractions.
- The strength model was developed using experimental findings on the basis of regression analysis. In estimating the different intensities at 28, 90, and 180 days on the basis of estimation more or less similar with measurements, the proposed regression linear model was found to be accurate.

Acknowledgements Authors would like to thank SASTRA Deemed University for the support to perform the experimental works in the laboratory.

References

1. Selvaraj, R., Muralidharan, S., Srinivasan, S.: The influence of silica fume on the factors affecting corrosion of reinforcement in concrete a review. Struct. Concr. **4**, 4–19 (2003)
2. Berke, N.S., Pfeifer, D.W., Weil, T.G.: Protection against chloride induced corrosion. Concrete. Int. Des. Constr. **10**(12), 45–55 (1988)
3. Ozyildirim, C.: Rapid chloride permeability testing of silica fume concrete. Cement Concr. Aggregates **16**(1), 53–56 (1994)
4. Yogendran, V., Langan, B.W., Haque, M.A., Ward, M.A.: Silica fume in high strength concrete. ACI Mater. J. **84**(2), 124–129 (1982)
5. Abdullah, A.A., Beshr, H., Maslenuddin, M., Al-Amoud, O.S.B.: Effect of silica fume on the mechanical properties of low quality coarse aggregate concrete. Cement Concr. Compos. **26**, 891–900 (2004)
6. Mazloom, M., Ramezanianpour, A.A., Brooks, J.J.: Effect of silica fume on mechanical properties of high strength concrete. Cement Concr. Compos. **26**, 347–357 (2004)
7. Khedr, S.A., Abou-Zeid, M.N.: Characteristics of silica fume-concrete. J. Mater. Civ. Eng. ASCE **6**(3), 357–375 (1994)
8. Sabir, B.B.: High strength condensed silica fume concrete. Mag. Concr. Res. **47**(17), 219–226 (1995)
9. Bayasi, Z., Zhou, J.: Properties of silica fume concrete and mortar. ACI Mater. J. **90**(4), 349–356 (1993)
10. Bhanja, S., Sengupta, B.: Influence of silica fume on the tensile strength of concrete. Cem. Concr. Res. **35**, 743–747 (2005)
11. Mazloom, M., Ramezanianpour, A.A., Brooks, J.J.: Effect of silica fume on mechanical properties of high-strength concrete. Cement Concr. Compos. **26**, 347–357 (2004)
12. Gonen, T., Yazicioglu, S.: The influence of mineral admixtures on the short and long-term performance of concrete. Build. Environ. **42**, 3080–3085 (2006)
13. Thomas, M.D.A.: Using silica fume to combat ASR in concrete. Indian Concr. J. **75**, 671–676 (2001)
14. Asrar, N., Malik, A., Shahreer: A Corrosion protection performance of micro silica added concretes in NaCl and seawater environments. Construc. Build. Mater. **13**, 213–218 (1999)
15. Sancak, E., DursunSari, Y., Simseki, O.: Effect of elevated temperature on compressive strength and weight loss of the light weight concrete with silica fume and super plasticizer. Cement Concr. Compos. **30**, 715–721 (2008)
16. Behnood, A., Ziari, H.: Effects of silica fume addition and water to cement ratio on the properties of high strength concrete after exposure to high temperatures. Cement Concr. Compos. **30**, 106–112 (2008)
17. Mahmoud, N., Afroughsabet, V.: The long-term compressive strength and durability properties of silica fume fiber-reinforced concrete. Mater. Sci. Eng. **531**, 107–111 (2011)
18. Al-Sakiny, Z.H.: Engineering Properties of Ultra High Strength Fiber Reinforced Concrete. M.Sc. Thesis, University of Technology, Baghdad, (2002)
19. Soulioti, D.V., Barkoula, N.M., Paipetis, A., Matikas, T.E.: Effect of fiber geometry and volume fraction on the flexural behavior of steel fiber reinforced concrete. Int. J. Exp. Mech. **47**, 535–541 (2009)
20. Gonen, T.: Mechanical and fresh properties of fiber reinforced self-compacting lightweight concrete. Scientia Iranica Trans. A: Civ. Eng. **22**(2), 313–318 (2015)
21. Baran, E., Akis, T., Yesilmen, S.: Pull-out behavior prestressing strands in steel fiber-reinforced concrete. Constr. Build. Mater. **28**, 362–371 (2012)
22. Le, H.T.N., Poh, L.H., Wang, S., Zhang, M.H.: Critical parameters for the compressive strength of high-strength concrete. Cement Concr. Compos. **82**, 202–216 (2017)
23. Hasan-Nattaj, F., Nematzadeh, M.: The effect of forta-ferro and steel fibers on mechanical properties of high-strength concrete with and without silica fume and nano-silica. Constr. Build. Mater. **137**, 557–572 (2017)

24. IS 12269: Specifications for 53 grade Ordinary Portland Cement. Bureau of Indian Standards, New Delhi (1997)
25. IS 383: Specifications for coarse and fine aggregates from natural sources for concrete. Bureau of Indian Standards, New Delhi (1970)
26. IS 10262: Recommended guidelines for concrete mix design. Bureau of Indian standards, New Delhi (2009)
27. IS 516: Methods of Tests for Strength of Concrete. Bureau of Indian Standards, New Delhi (1959)
28. ASTM C78/C78M-16: Standard Test Method for Flexural Strength of Concrete using simple beam with Third-Point Loading
29. ACI 318: Building code requirements for Structural Concrete (2014)
30. Ranjbar, N., Mehrali, M.: High tensile strength fly ash based geopolymer composite using copper coated micro steel fiber. Constr. Build. Mater. **112**, 629–638 (2016)
31. Neves, R.D., Fernandes de Almeida, J.C.O.: Compressive behavior of steel fibre reinforced concrete. Struct. Concr. **6**(1), 1–7 (2005)
32. Afroughsabet, V., Ozbakkaloglu, T.: Mechanical and durability properties of high-strength concrete containing steel and polypropylene fibers. Constr. Build. Mater. **94**, 73–82 (2015)
33. Bencardino, F., Rizzuti, L., Spadea, G., Swamy, R.N.: Stress-strain behavior of steel fiber-reinforced concrete in compression. J. Mater. Civ. Eng. **20**(3), 255–263 (2008)

Particle Deposition Analysis Using DPM-DEM

Nurhanani A. Aziz, M. H. Zawawi, N. M. Zahari, Aizat Mazlan, Aizat Abas, Aqil Azman, and Muhammad Naqib Nashrudin

1 Introduction

Coastal bank is the most imperative financial locale that can support the satisfaction of human nature. Still, this region is always beneath frequent threat from various natural and man-actuated threats, including coastal erosion and shoreline scouring [1]. Coastal erosion is a natural phenomenon that influences a large number of coastal bank sites worldwide [1–10]. Storminess and endless sea level rise has become a major problem toward coastal erosion along with a worldwide increment in coastal advancement of numerous nations [10]. Coastal authority dealing with erosion difficulty and long-term shoreline stability depends essentially on balance between the rate of sediment supply and transport [11].

Due to wave movement, beach and shallow water sediments are continually responding to the erosion-causing action and affecting the dynamic equilibrium of the coastal bank [12]. This event is not only induced by unpredictable and extraordinary sea storms. It is, however, the result of the natural behavior and persistent impact of the waves on the coast [2]. Besides, the wave's impact toward the shoreline and the resultant erosion rely on a few aspects such as the strength, the action time, the type of soil, and the direction of propagation [2]. Numerous coastal preservation and protection measures have been applied for the improvement and rehabilitation of the coastal bank, but the destructive aftermath of erosion continues [8]. In order to build or create new coastal protection, it is essential to investigate more about the behavior of waves toward sand sediment to understand the nature of this erosion [13–15]. This paper will further discuss the simulation of sediment erosion using the DPM-DEM method.

N. A. Aziz (✉) · M. H. Zawawi · N. M. Zahari · A. Mazlan
Department of Civil Engineering, Universiti Tenaga Nasional, Kajang, 43000 Selangor, Malaysia

A. Abas · A. Azman · M. N. Nashrudin
School of Mechanical Engineering, Universiti Sains Malaysia, Parit Buntar, 14300 Pulau Pinang, Malaysia

I. Pal and S. Kolathayar (eds.), *Sustainable Cities and Resilience*, Lecture Notes in Civil Engineering 183, https://doi.org/10.1007/978-981-16-5543-2_35

2 Numerical Model

The equation for the conservation of the continuity equation can be composed as shown [16]:

$$\frac{\partial \rho}{\partial t} + \nabla \cdot (\rho \vec{v}) = S_m \tag{1}$$

Equation (1) is the general form of the mass conservation equation and can be utilized for compressible and incompressible flows. S_m is the mass included in the continuous phase of the dispersed second phase and any user-defined sources.

Conservation of momentum can be described by

$$\frac{\partial}{\partial t}(\rho \vec{v}) + \nabla \cdot (\rho \vec{v}\vec{v}) = -\nabla p + \rho \vec{g} + \vec{F} \tag{2}$$

where p is the static pressure and $\rho \vec{g}$ and $\vec{F}$ are the gravitational body force and external body forces, respectively. $\vec{F}$ also contains other model-dependent source terms such as porous-media and user-defined sources.

Energy conservation is expressed by

$$\frac{\partial}{\partial t}(\rho E) + \nabla \cdot \left[\vec{v}(\rho E + p)\right] = -\nabla \cdot \left(\sum_j h_j J_j\right) + S_h \tag{3}$$

2.1 Volume of Fluid (VOF)

Momentum equation for the volume of fluid [16] is given as

$$\frac{\partial}{\partial t}(\rho \vec{v}) + \nabla \cdot (\rho \vec{v}\vec{v}) = -\nabla p + \nabla \cdot \left[\mu\left(\nabla \vec{v} + \vec{v}^T\right)\right] + \rho \vec{g} + \vec{F} \tag{4}$$

The energy equation is shown below.

$$\frac{\partial}{\partial t}(\rho E) + \nabla \cdot (\vec{v}(\rho E + p)) = \nabla \cdot (k_{\text{eff}} \nabla T) + S_h \tag{5}$$

The VOF model treats energy, E, and temperature, T, as mass-averaged variables:

$$E = \frac{\sum_{q=1}^{n} \alpha_q \rho_q E_q}{\sum_{q=1}^{n} \alpha_q \rho_q} \tag{6}$$

where E_q is based on the specific heat of the phase and the shared temperature for each phase.

2.1.1 Air Wave Theory

The wave profile for the linear wave is shown below

$$\zeta(X, t) = A \cos \alpha \quad (7)$$

where

$$\alpha = k_x x + k_y y - \omega_e t + \varepsilon \quad (8)$$

x and y are the space coordinates in the $\hat{x}$ and $\hat{y}$ directions, respectively, while ε is the phase difference, and t is the time.

For both shallow and intermediate waves, the wave frequency ω is defined as given below:

$$\omega = \sqrt{gk \tanh(kh)} \quad (9)$$

And for short gravity waves:

$$\omega = \sqrt{gk} \quad (10)$$

where h is the height of the liquid, k is the number of waves, and g is the gravity magnitude.

2.1.2 Stokes Wave Theory

Based on the past study of John D. Fenton (1985), the Stokes wave theories in fluent is formulated [16]. These theories can be used for the large steepness of finite amplitudes of waves operating in the intermediate to deep liquid depth range.

For higher-order Stokes theories (second to fifth order), the generalized expression for wave profiles is shown as:

$$\zeta(X, t) = \frac{1}{k} \sum_{i=1}^{n} \sum_{j=1}^{i} b_{ij\xi^i \cos(j\alpha)} \quad (11)$$

The generalized expression of the associated velocity potential is shown as:

$$\Phi(X,t) = \frac{1}{k}\sqrt{\frac{g}{k}\tanh(kh)}\sum_{i=1}^{n}\xi^{i}\sum_{j=1}^{i}a_{ij}\cosh(\text{jk}(\text{z}+\text{h}))\cos(j\alpha) \tag{12}$$

where

$$\xi = \frac{kH}{2} \tag{13}$$

ξ is referred to wave steepness.

n = wave theory index (2–5: from second-order Stokes to fifth-order Stokes, respectively).

Wave celerity, c, is given as;

$$c = \sqrt{\frac{g}{k}\tanh(kh)\left(1 + c_3\xi^2 + c_5\xi^4\right)} \tag{14}$$

For second-order Stokes, $c_3 = c_5 = 0$, while for third and fourth-order Stokes, $c_5 = 0$.

Wave frequency, ω, is defined as

$$\omega = kc \tag{15}$$

The velocity components for surface gravity waves are derived from the velocity potential velocity function.

$$u = \frac{\partial\Phi}{\partial x}\cos\theta \tag{16}$$

$$v = \frac{\partial\Phi}{\partial x}\sin\theta \tag{17}$$

$$w = \frac{\partial\Phi}{\partial x}\tan\theta \tag{18}$$

2.2 *Discrete Particle Model*

Each particle is tracked individually in an Euler-Langrangian model by using Newton's second law of motion [17], given by

$$m_p\frac{\text{d}\vec{v}_p}{\text{d}t} = m_p\vec{g} + \frac{V_p\beta}{1-\epsilon}\left(\vec{u} - \vec{v}_p\right) - V_p\nabla p \tag{19}$$

Fluid phase hydrodynamics is described in the continuity equation and the volume-averaged Navier–Stokes equations [17]:

$$\frac{\partial(\in \rho_f)}{\partial t} + \nabla \cdot \in \rho_f \vec{u} = 0 \tag{20}$$

$$\frac{\partial(\in \rho_f \vec{u})}{\partial t} + \nabla \cdot \in \rho_f \vec{u}\vec{u} = - \in \nabla p - \nabla \cdot \in \vec{\tau}_f - \vec{S}_p + \in \rho_f \vec{g} \tag{21}$$

where $\vec{S}_p$ is the source term that accounts for the exchange of momentum between the fluid phase and the particle phase. $\vec{S}_p$ can be calculated as shown below:

$$\vec{S}_p = \frac{1}{V} \int \sum_{k=0}^{N_p} \frac{V_{p,k}\beta}{1 - \in} (\vec{u} - \vec{v}_{p,k}) \delta(\vec{r} - \vec{r}_{p,k}) dV \tag{22}$$

The δ function is to ensure that the reaction force acts as a point force at the position of the particle in the system.

2.3 *Discrete Element Model (DEM)*

The implementation of DEM is based on [18] and accounts for the forces that result from the particles colliding. The discrete element method (DEM) is a common computational approach introduced by [18] for modeling the dynamics of granular matter such as gravel, coal, and beads of any substance. It involves trajectories and spins of all particles and objects in the system and is capable of predicting their interactions with other particles and with their environment [16]. DEM simulation can be described by a high volume fraction of the particles, where interaction between the two particles is necessary [16].

2.3.1 The Spring-dashpot Collision Law

In spring-dashpot collision law, the spring constant K can be defined as in the spring collision law, together with a coefficient of restitution for the dashpot term (η) which $0 < \eta \leq 1$.

In preparation for the force calculations, the following equation is evaluated:

$$f_{\text{loss}} = \sqrt{\pi^2 + \ln^2 \eta} \tag{23}$$

$$m_{12} = \frac{m_1 m_2}{m_1 + m_2} \tag{24}$$

$$t_{\text{coll}} = f_{\text{loss}}\sqrt{\frac{m_{12}}{K}} \tag{25}$$

$$\gamma = -2\frac{m_{12}\ln\eta}{t_{\text{coll}}} \tag{26}$$

$$\vec{v}_{12} = \vec{v}_2 - \vec{v}_1 \tag{27}$$

where f_{loss} is a factor of loss, m_1 and m_2 are Particles 1 and 2 masses, respectively, while m_{12} is the "reduce mass." Then, t_{coll} is the time scale of the collision, $\vec{v}_1$ and $\vec{v}_2$ are the velocities of Particles 1 and 2, respectively, while $\vec{v}_{12}$ is the relative velocity. Here, γ is the damping coefficient, and $\gamma \geq 0$ because $\ln\eta \leq 0$.

The force on Particle 1 can be calculated as shown below;

$$\vec{F}_1 = (K\delta + \gamma(\vec{v}_{12} \cdot \vec{e}_{12}))\vec{e}_{12} \tag{28}$$

By the third law of Newton, the force of Particle 2 is given as

$$\vec{F}_2 = -\vec{F}_1 \tag{29}$$

3 Results and Discussion

3.1 Volume Fraction

Figure 1 illustrates the volume fraction of the seawater together with sand particles at various time steps. It can be seen that the seawater at its initial position is stable as the seawater just fills the model, and after few seconds, the wave generation can be observed in Fig. 1(b–e) as the seawater is not stable as it was at the initial position. Whenever these phenomena occur, the sand particles begin to switch their initial state and are transported to other locations. With the help of the drag force, the particles move slowly toward the direction of the incoming wave, leading to the erosion of the particles.

At the beginning of generation, sand particles are full within the sediment boundary before waves as shown in Fig. 1(a). Sand particles at the top of the sediment boundary begin to erode and have a gap after a few seconds. This phenomenon occurs because of the wave behavior that acts toward the sediment boundary and changes the sand formation. Based on Fig. 1(a–e), the gaps between sand particles at the top of the sediment boundary increase as time increases.

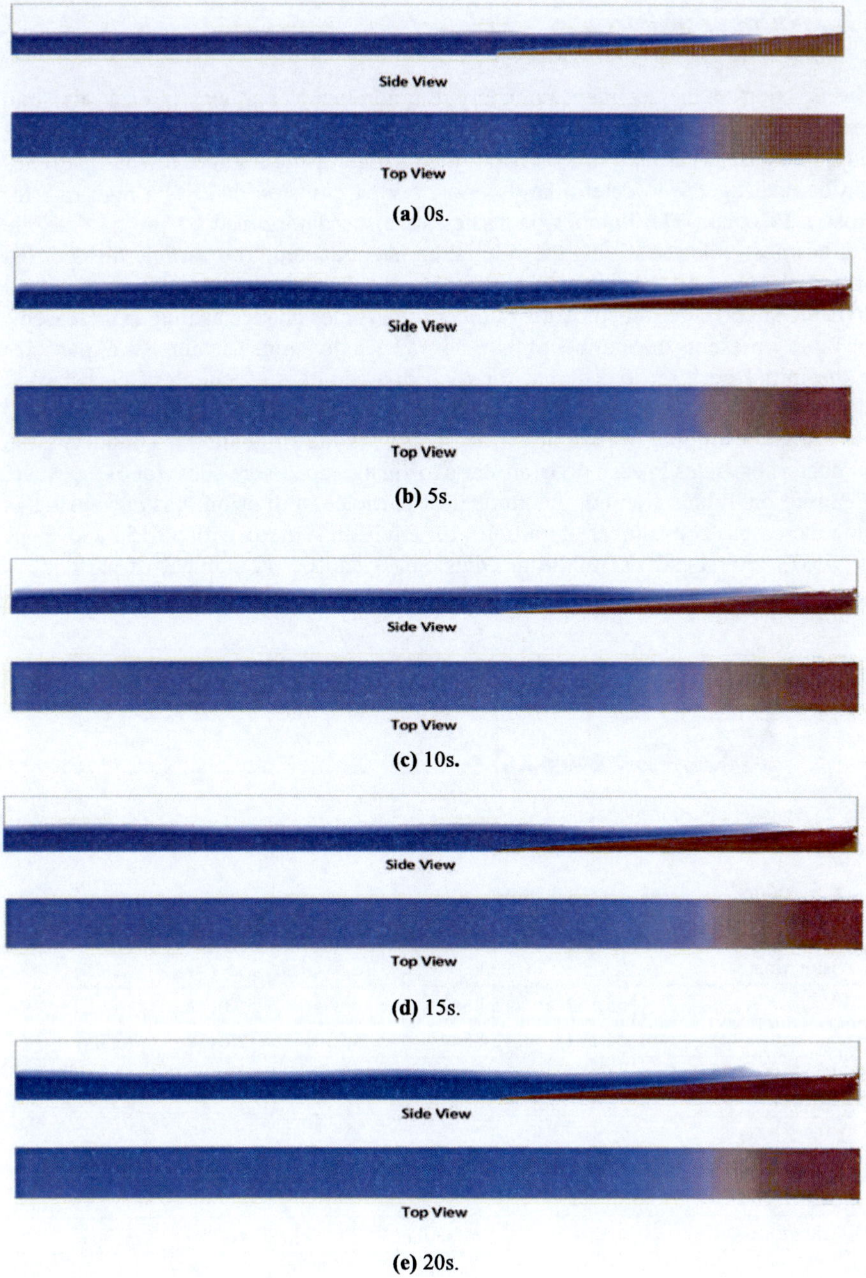

Fig. 1 Movement of waves and sand

3.2 Number of Particles

The division of the sediment boundary, the number of particles in each division, and the changes of particles over time are shown in Fig. 2 and Tables 1 and 2 simultaneously. Through this division, the change of the number of sand particles can be investigated to determine the most critical division that has a high risk for erosion to occur. The number of particles can be determined by the particle file that has been injected during the simulation is conducted. The particle files consist of particle size and the gap or distance between particles. Hence, the most critical division can be determined by the number of particles change as time is increased.

Table 1 presents the number of particles in each division. The number of particles at the initial position is different for each division as the volume of each part is different. Division A has a bigger number of particles with 55,152, while division C has a smaller number of particles with 14,472. As the simulation is conducted, the number of particles in each division starts to change due to the behavior of the wave.

Based on Tables 1 and 2, the number of particles in division A is decreasing as time increases. The number of particles for Division A starts with 55,152 and drops to 52,673 after the 20 s. While in Divisions B and C, the number of particles is increasing when the time increases. The number of particles for Divisions B and C

Fig. 2 Division of sediment boundary

Table.1 Number of particles in each division

Number of particles					
Division/time	0 s	5 s	10 s	15 s	20 s
A	55,152	54,342	53,694	53,142	52,673
B	34,272	34,844	35,210	35,457	35,517
C	14,472	14,642	14,888	15,159	15,542

Table 2 Changes in the number of particles over time

Change of particles				
Division/time	T_0–T_5	T_5–T_{10}	T_{10}–T_{15}	T_{15}–T_{20}
A	−810	−648	−552	−469
B	572	366	247	60
C	170	246	271	383

starts with 34,272 and 14,472, then increases to 35,517 and 15,542 simultaneously after the 20 s.

This phenomenon happens because of the drag force that acts toward the sediment boundary. The particles from the upper part (Division A) had been dragged by the waves toward the Divisions B and C. Therefore, the number of particles in Division A is decreasing, while Divisions B and C are increasing. Based on Tables 1 and 2, we can conclude that the most critical division between these three divisions is Division A since this division lost a lot of the number of particles compared to other divisions.

3.3 Validation

To investigate the efficiency and the accuracy of the numerical analysis of DPM-DEM, the validation with another method which is smooth particle hydrodynamics (SPH) should be made. The same model is used for both methods to compare and validate the results. However, the results of these two methods are slightly different due to the mesh used. DPM-DEM is a numerical analysis that needed mesh to analyze, while SPH is mesh-free.

The validation of the two methods is shown in Fig. 3. Based on the figure, the pattern of the particle erode is slightly the same between these two methods. For both methods, the particle erodes occur in the same area which is at the upper part of the sediment boundary (Division A). The particle erodes from the upper part because the particle is dragged by the drag force of the wave and move downward to the lower part followed by the direction of the incoming wave.

4 Conclusion

The sediment erosion has been successfully simulated numerically using a DPM-DEM method. Accordingly, we presumed that the findings were validated based on the SPH method as the pattern of the particle erode is the same for both methods. It is also proved that the length of time would generally increase the number of particles erosion and change the formation of sand that would lead to erosion. For instance, the number of particles for the top of the sediment boundary (Part A) keeps on decreasing as the time increases, while in other parts (Parts B and C), the number of particles keeps on increasing due to the loss of particles in Part A. This will, therefore, indicate that Part A has a high potential for erosion due to wave behavior.

Fig. 3 Validation between two methods

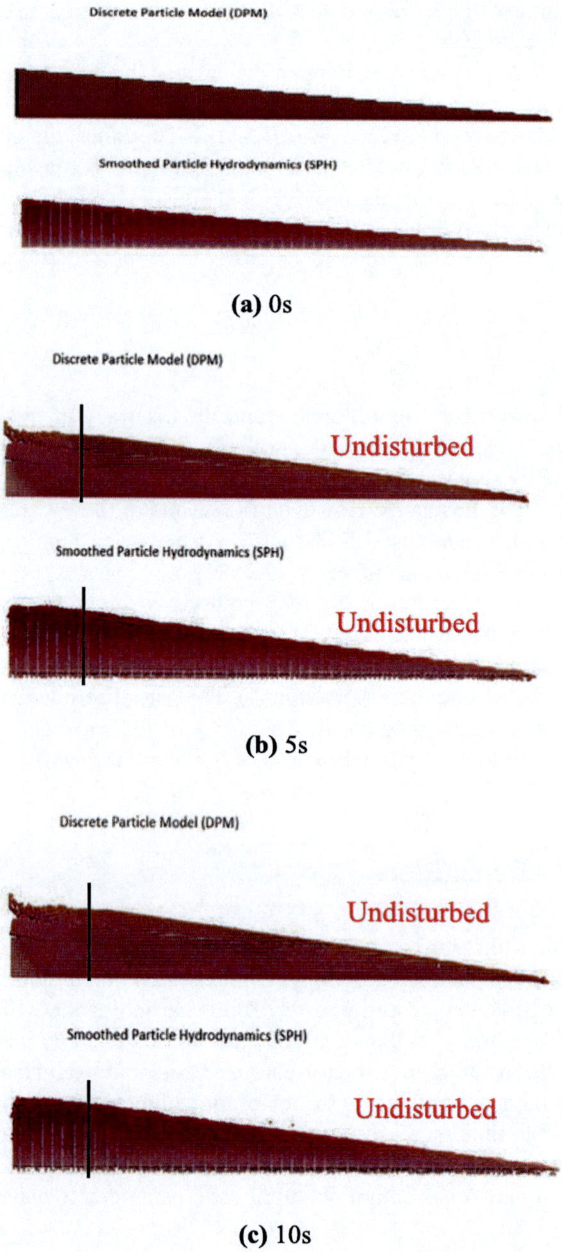

Acknowledgements This research project was funded by Tenaga Nasional Berhad seeding fund under grant no. U-TG-RD-18-05.

References

1. Hegde, A.V., Akshaya, B.J.: Shoreline transformation study of Karnataka Coast: geospatial approach. Aquat. Procedia **4**, 51–156 (2015)
2. Adamo, F., De Capua, C., Filianoti, P., Lanzolla, A.M.L., Morello, R.: A coastal erosion model to predict shoreline changes. Meas. J. Int. Meas. Confed. **47**(1), 734–740 (2014)
3. Zainol, M.R.R.M.A., Kamaruddin, M.A., Zawawi, M.H., Wahab, K.A.: Numerical analysis study of sarawak barrage river bed erosion and scouring by using Smooth Particle Hydrodynamic (SPH). Mater. Sci. Eng. **267** (2017)
4. Sedigh, M., Tomlinson, R., Cartwright, N., Etemad-Shahidi, A.: Numerical modelling of the Gold Coast Seaway area hydrodynamics and littoral drift. Ocean Eng. **121**, 47–61 (2016)
5. Besset, M., Anthony, E.J., Dussouillez, P., Goichot, M.: The impact of Cyclone Nargis on the Ayeyarwady (Irrawaddy) River delta shoreline and nearshore zone (Myanmar): towards degraded delta resilience? Comptes Rendus—Geosci. **349**(6–7), 238–247 (2017)
6. Zhang, J.M., Zhang, J.H., Wang, G., Chen, Y.: Safety evaluation of breakwaters based on physical and numerical modelling. Ocean Eng. **36**(11), 852–862 (2009)
7. Mohanty, P.K., Barik, S.K., Kar, P.K., Behera, B.: Impacts of ports on shoreline change along Odisha coast. Procedia Eng. **116**(Apac), 647–654 (2015)
8. Noujas, V., Thomas, K.V., Ajeesh, N.R.: Shoreline management plan for a protected but eroding coast along the southwest coast of India. Int. J. Sediment Res. **32**(4), 495–505 (2017)
9. Burningham, H., French, J.: Understanding coastal change using shoreline trend analysis supported by cluster-based segmentation. Geomorphology **282**, 131–149 (2017)
10. Adriana Gracia, C., Rangel-Buitrago, N., Oakley, J.A., Williams, A.: Use of ecosystems in coastal erosion management. Ocean Coast. Manag. **156**, 277–289 (2017)
11. Bahgat, M.: Optimum use of dredged materials for sustainable shoreline management in Nile Delta. Water Sci. 1–14 (2017)
12. Jyothi, K., Mani, J.S., Pranesh, M.R.: Numerical modelling of flow around coastal structures and scour prediction. Ocean Eng. **29**(4), 417–444 (2001)
13. Aziz, N.A., Zawawi, M.H., Zahari, N.M., Abas, A., Azman, A.: Simulation of Homogeneous Particle Size in Fluid Flow by Using DPM-DEM, vol. 1, Springer Singapore (2020)
14. Zawawi, M.H., Azman, A., Radzi, M.R.M., Abas, A., Aziz, N.A.: Coastal sediment transport study along shoreline by using smooth particle hydrodynamics. In: IOP Conference Series Material Science Engineering, vol. 815, no. 1 (2020)
15. Azman, A., Zawawi, M.H., Radzi, M.R.M., Abas, A., Aziz, N.A.: Smooth particle hydrodynamics study of sediment profile erosion. In: IOP Conference Series Material Science Engineering, vol. 815, no. 1 (2020)
16. Hurley, J., Stokes, N., Cleary, P.W.: Efficient collision detection for three dimensional super-ellipsoid particles. In: Proceeding 8th International Computer Technique Application Conference, no. January 1997, pp. 1–7, (1997)
17. Bokkers, G.A., Van Sint Annaland, M., Kuipers, J.A.M.: Mixing and segregation in a bidisperse gas-solid fluidised bed: a numerical and experimental study. Powder Technol. **140**(3), 176–186 (2004)
18. Cundall, P.A., Strack, O.D.L.: A discrete numerical model for granular assemblies. Géotechnique **29**(1), 47–65 (1979)
19. ANSYS Inc.: ANSYS Fluent Theory Guide. ANSYS Inc., USA, vol. 15317, no. August, pp. 724–746 (2016)

Recent Trends in IOT-Enabled Composter for Organic Wastes

P. Balaganesh, M. Vasudevan, R. Rameswari, and N. Natarajan

1 Introduction

The contemporary world is undergoing a silent smart revolution with the rising technological developments touching all aspects of life. The smart technologies located as the epicenter of digital transformations has drastically lead the innovations towards design of Internet of things (IoT) as the key pillar of latest Industrial Revolution (Industry 4.0) [1]. Waste management has always been a tough task for the government and local administrators despite spending huge budgets on collection, separation, treatment and disposal [2]. A few critical issues reported in the case of solid waste management in developing urban communities are, (i) ever-increasing volume surpassing all design capacities of handling; (ii) difficulties in actual separation; (iii) inefficient treatment facilities; (iv) poor financial support system; (v) NIMBYism (not in my backyard syndrome); (vi) lack of awareness and communication; and (vii) lack of coordination and participation. On a fair-base analysis, we can observe that the points (i.), (ii.) and (iii.) are more related with technological aspects, (iv.) with economic background and (v.), (vi.) and (vii.) with social elements. On the technological aspects, selection and success of proper treatment technology depend on the efficient separation of materials and suitable pre-treatments (if necessary) [3].

Composting is reported to have huge environmental benefits in terms of reduced pollution (reduction in open dumping, reduced emission of greenhouse gases and reduced subsurface leaching) as well as increased agricultural benefits (crop productivity, soil nutrient balance and soil reclamation) [4, 5]. In general, composting practices are conventional in small-scale in rural areas while urban composting practices are generally more innovative and in larger scales. The major difference is, however,

P. Balaganesh (✉) · M. Vasudevan · R. Rameswari
Bannari Amman Institute of Technology, Sathyamangalam, Tamil Nadu, India
e-mail: balaganesh@bitsathy.ac.in

N. Natarajan
Dr. Mahalingam College of Engineering and Technology, Pollachi, Tamil Nadu, India

I. Pal and S. Kolathayar (eds.), *Sustainable Cities and Resilience*, Lecture Notes in Civil Engineering 183, https://doi.org/10.1007/978-981-16-5543-2_36

in the successful operation through proper monitoring scheme which is not well developed so far in the rural communities. As the compost quality eventually determines its stability and application, there is a large scope of advanced technologies in large-scale compost monitoring.

Henceforth, 'smart composting' is the need of the hour which enables the IoT application and automation that can potentially eliminates the inebrieties of physical monitoring and other above-mentioned issues. Particularly, aerobic in-vessel composting devices (composters) are preferred to be the suitable reactors to prepare compost in urban areas that avoids space constraints and produces quick stable bio-manure. Qiu and Wang [5] developed a kitchen waste composting machine and identified the significant expectations of people as optimum size, aesthetic look, interactive panel, odorless operation and cost effectiveness. Zhou and Wang [6] analyzed various design elements of an automatic domestic food-waste composter based on perceptual engineering method and reported the significance of selection of feasible materials, surface texture, color and models based on a questionnaire survey.

Being an interdisciplinary area, a collaborative approach is necessary to break the natural hindrances in earmarking the required research components by demystifying the specific technical attributes from the physical science of composting and the digital science of IoT. Thus, the present study aims to provide a comprehensive review on the recent trends in the development of IoT-enabled composter by highlighting the technological aspects under various sections such as (i) selection of feed and suitable hardware components; (ii) significance of data acquisition and storage systems; (iii) aspects of IoT gateway selection; (iv) available standard protocols and (v) prospective data mining. According to the author's knowledge, this is the first of its kind study on the investigation for selection of digital technology on a conventional physical system.

2 Advances in Smart Composter Design

2.1 Feed Selection and Optimization

Suitable and optimum feedstock ensures efficient and rapid conversion of the organic constituents irrespective of their source resulting in good quality compost. The carbon-rich composting materials are normally called as browns, and nitrogen-rich materials are known to be greens (Fig. 1). The selection of feedstock proportion is based on optimizing the carbon and nitrogen ratio (*C*/*N*) of the feedstock in order to achieve stabilized compost (optimum range 15–30:1). In case of mixed feedstock, the overall *C*/*N* ratio can be obtained by taking their weighted average as shown in Eq. 1.

$$[C/N]_{\mathrm{mix}} = \frac{\sum_i [C/N]_i W_i}{\sum_i W_i} \tag{1}$$

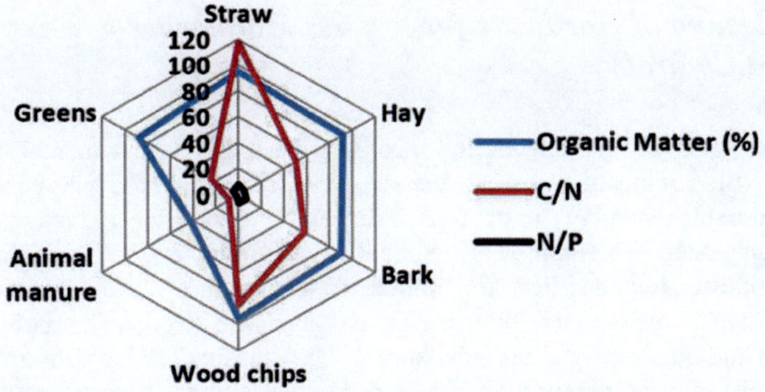

Fig. 1 Stability parameter ranges of the typical composting substrates

where $(C/N)_m$ is the C/N ratio of feedstock mixture, $(C/N)_i$ is the C/N ratio of individual feed types and W_i is the corresponding weight (kg) [7].

Regular and reliable measurement of environmental parameters such as temperature and moisture content are the preliminary requirement for the smart composting process [8, 9]. The organic wastes tend to degrade due to microbial activity, and thus it discharges energy in the heat form. It changes the temperature of the pile/bin to increase from the initial phase (mesophilic) to a range of 40–60 ° Celsius (C) (thermophilic phase). During the final stage of maturation (cooling phase), the approximate pile temperature fall in the range of 25–40 °C. The overall moisture content of feed mixture needs to be maintained in the range of 50 to 60% (v/v) for maintaining sufficient thermal comforts for effective decomposition. It is observed that provision for free air space (FAS) through regular mixing controls the entire aerobic reactions by optimizing the environmental conditions. A measure of aeration or oxygen uptake can bring more light to the efficiency of mixing and health of active biomass in the system. Proper aeration also eliminates the accumulation of odorous gases and decreases the composting period [7].

Considering the chemical transformations during composting, pH serves as a good indicator for the change in microbial activity. The optimum pH range for the composting is 6.5–8.0. When the pH increases above 8.5, the possibility of ammonia emission is high due to the biochemical conversion. The redistribution of common ionic species in the compost can be easily monitored by measuring the electrical conductivity which gives an overview of the amount of nutrient contents and salts present in the compost. In the real-time operation scenario, most of these aspects are seldom captured with rapt accuracy [10, 11]. Hence, it is important to realize that provisions for real-time monitoring and control of composter can aid in better performance. Now, we will consider the technology-related requirements for an IoT-enabled composting unit to work in the smart mode.

2.2 Selection of Hardware for Process Monitoring and Control

As there are plenty of parameters (both constitutional and environmental) involved in defining the composting process, the selection of suitable hardware components to obtain reliable data plays a vital role in successful monitoring and controlling for a smart composter. The common array of hardware components to be employed in a smart composter includes, but not limited to, temperature sensor/ thermocouple, humidity sensor, pH sensor, flow meter, gas analyzer (oxygen, carbon-dioxide, ammonia) and quartz crystal microbalance (QCM) sensors [12–15]. Pump steering system, flow adjuster, aeration fans, solenoid valves and hot wire anemometer are some of the common actuators that performs operation based on the user set point for automation. The hardware components for the processor and output generally consist of Arduino UNO, Node MCU, Raspberry Pi, microcontroller, track recorder, wireless communication module, Siretta antenna, etc. In addition, the data transfer system for an IoT-enabled composter needs uninterrupted power supply, provision for USB communication port and flexibility to operate with analog inputs while generating digital outputs (Table 1).

2.3 Data Acquisition and Storage Facilities

The data acquisition system varies with respect to reliability and as per user requirements. As mentioned in Table 1, there are a few automated composters deployed with Arduino, MATLAB, LABVIEW, PLC and Colony V software to access the data (Table 1). However, selection of suitable software is based on the user requirements for their intended result with accuracy. Similarly, hardware for data acquisition includes, Arduino UNO, Node MCU, Raspberry Pi, etc. Since data storage provision is a critical requirement for data analysis, the SD card module is widely used as low-cost technology to store the monitored data. In recent studies, a server-based cloud platform to store data finds its wide application with remote access. Wireless technologies such as WiFi, LoRa are also bridged with input devices for the efficient data acquisition and storage. The functioning of an IoT-enabled composter integrates a whole range of activities including data management system. The common peripherals include input–output pins, analog–digital converter, real-time clock, pulse-width modulator and universal asynchronous receiver–transmitter unit.

Table 1 Details of IoT-enabled/smart composter

Composter/composting type	Feed/composting details	Hardware for data collection	Data acquisition and storage	Communication	Significance	Ref
Windrow composting	Any compost pile	Temperature sensors, microcontroller, power input of 5 to 12 VDC and analog inputs	Arduino IDETM based on C/C + + , Digital quartz clock, 2-GB SD-Card attached to the Data Logger, IP 67 hermetic box, rechargeable-sealed battery	USB communication port, digital outputs	Economical automation system to monitor compost and ambient air temperature	[12]
Bin composting	Olive pomace, sewage sludge, dairy manure, tomato stalks, (32 days)	K type thermocouple, aeration fans, solenoid valves, hot wire anemometer, air pump,	Programmable logic controller(PLC)	(human machine interface) HMI	Automated composting process. Particularly, when the temperature set value varies using fan composting temperature controlled using PLC	[10]
Bin composting	sewage sludge, dairy manure, tomato stalks, (32 days)	K type thermocouple, aeration fans, solenoid valves, hot wire anemometer, air pump	PLC	HMI	Automated sewage sludge composting process using temperature controlled using PLC	[11]
Bin composting	Wheat-straw, sawdust	Multiple sensors (temperature, gas), pump steering, track recorder, flow meter, flow adjuster, light-box	PIAO IT SYSTEM	Direct transfer	Automated compost maturity assessment based on color	[13]

(continued)

Table 1 (continued)

Composter/composting type	Feed/composting details	Hardware for data collection	Data acquisition and storage	Communication	Significance	Ref
Bin composting	Organic solid wastes, 38 days composting	Temperature sensor, humidity sensor, Wheatstone bridge, 12 V electric pump, hot air circulatory system	MATLAB, PID controller, telemetric, human machine interface via Internet	HMI	Automated control reduced the composting time up to 30% with stabilized physico-chemical parameters of compost	[14]
Pile composting	Cow dung, kitchen waste, chicken droppings	Light emitting diode light source, charge coupled device camera, fluorescent microscopy	Colony V	Direct transfer	Quick and automatic enumerated system for microbial colonies in compost	[15]
Bin composting	Cellulosic substrate	Electromagnetic relay board, solenoid pinch valve, liquid level sensors, pump control, automated sampler	LABVIEW	Direct transfer	Automatic repetitive batch system to cultivate thermophilic bacteria using CO2 analysis	[16]
Bin composting	Food wastes, yard trimmings	Gas analyzer, flow meter, pump, water heater	LABVIEW 8.6, environmental scanning electron microscope	Direct transfer	Automatic multi-unit compost system can able to optimize and control the composting process of biodegradable polymers	[17]

(continued)

Table 1 (continued)

Composter/composting type	Feed/composting details	Hardware for data collection	Data acquisition and storage	Communication	Significance	Ref
Bin composting	Bananas, leaves, rice straw, dried leaves, 17 days of composting	Multiple sensors (temperature, humidity, pH, gas), microcontroller, actuators (heater, water pump, stirrer)	Arduino nano, Node MCU, Raspberry Pi, MATLAB, fuzzy logic	Direct transfer	Automated control system 2.35 times increased the rate of composting time compared with traditional system	[18]
Bin composting	Organic wastes	Quartz crystal microbalance (QCM) sensors array	MATLAB,	Monitor/online	Online monitoring of compost to attain stable yield	[19]
Bin composting	Organic wastes	Temperature sensor, humidity sensor, microcontroller, stirrer	PLX-DAQ, USB TTL	USB TTL	Automatic stirring for composter based on temperature control	[20]
Bin composting	Organic waste, 4 weeks composting	Multiple sensors (temperature, humidity, pH, gas), microcontroller	Server-based cloud platform to store data	Iot gateway and remote control	Remote management of compost as smart composter	[21]
Pile composting	Organic wastes	Sensors, microcontroller,	Wireless communication module, Raspberry Pi B, siretta antenna	IoT gateway, 4G mifi modem	Automated composting system designed to work remotely and eliminate manual work	[22]

3 Developments in System Architecture for IoT-Enabled Composter

3.1 Selection of IoT Gateways

Considering the scenario of multiple composting units located in a locality, an IoT-based integrated data management system can handle the entire data received from the multiple sensor nodes from multiple smart composting bins/piles and can process the same through suitable system architecture models (Fig. 2). The gateways integrate the protocols for networking, data storage for further analytics and ensure safe data transfer between edge devices. Recent studies advocate for the multi-task performance for the IoT gateway through the cloud environment. The gateway system comprises of a suitable hardware with sufficient memory (RAM and storage) and software with supporting cloud platform integration and datasheets. The common means of communication include 3G, 4G, GPS, GPRS, GSM, Zigbee/Thread, Ethernet, Bluetooth or Wifi depending on the preference of the service providers. If the cluster of composters is separated wide apart, it is quite reasonable to assign the gateway with a booster near high elevated points so as to minimize the power consumption

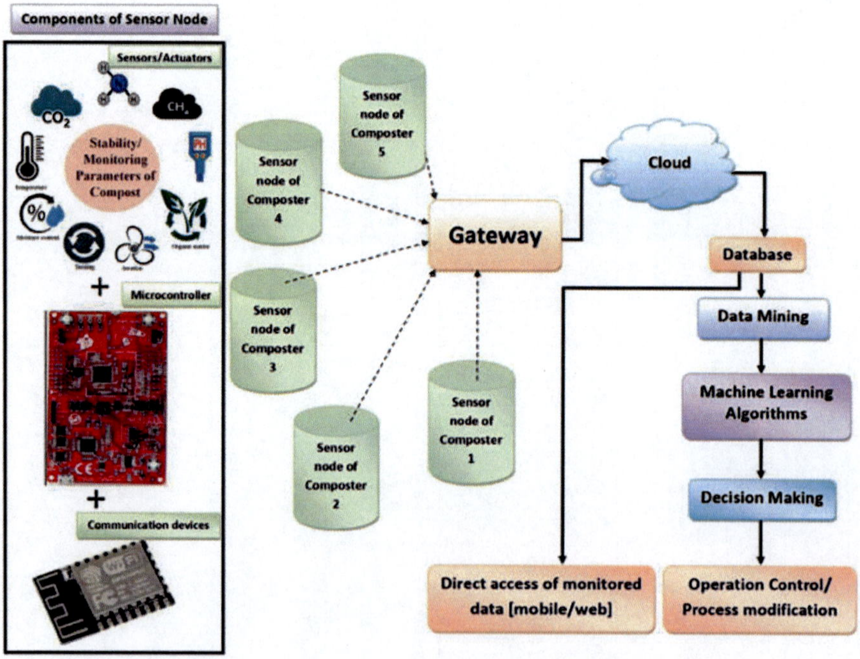

Fig. 2 Schematic diagram for smart composting system

and loss of data. The security of data transfer is ensured by limiting targeted connections to the Internet only through the gateway. Jo et al. [22] deployed an IoT gateway technology that contains Raspberry Pi B, HTTP (hypertext transfer protocol), REST (representational state transfer), LoRa wireless communication module and 4G MiFi modem to remotely monitor the composting operation in a farmland. In another study, Elalami et al. [21] used rotary composting having portable sensors to measure pH, temperature, humidity and ammonia concentration. The measured data were bridged to the cloud platform by integrating the gateway with 4G Mifi device for enabling remote control.

3.2 Standard Protocols for Data Transfer

The IoT application requires various standard protocols that have to be aligned with the operation of different components. extensible messaging presence protocol (XMPP), advanced massage queuing protocol (AMQP), constrained application protocol (COAP), data distribution service (DDS) and message queue telemetry transport (MQTT) are some of the different application layer protocols generally required to communicate with sublayers. Multicast DNS (mDNS) and DNS service discovery (DNS-SD) are the major service discovery protocols for the enriched (appropriate functioning) IoT services. Infrastructure protocols bridge various systems and devices with networks. Routing protocol, network adaptation layer protocol, link layer protocol and physical layer protocols are the available infrastructure protocols [23]. It is observed that in the case of wireless sensor networks, the frequent on–off operations can delay in network latency and delay in communications. A novel routing called lightweight non-increasing delivery-latency interval routing (LNDIR) was reported to solve this issue by changing individual time slots into non-increasing delivery-latency interval. In addition, the selection of operating system resembling an embedded system can be made suitable for long-time applications owing to their minimum power consumption and cheaper price. Such interventions can possibly deliver smooth and continuous data transactions for the successful operation of integrated smart composer systems.

3.3 Data Mining-Based Optimization

In order to optimize the composting operation, the monitored data from various sensors need to be transmitted to appropriate storage devices as raw data, which are further analyzed by suitable software for their reliability and applicability. Chakraborty et al. [23] optimized compost monitoring parameters by comparing the visible and near-IR spectroscopy data. They scanned around 55 (dairy manure) compost samples for the observation and used regression analysis to predict the relationship between organic matter and other monitoring (pH, electrical conductivity, moisture %) parameters. They also took around 100 compost samples and used visible near-infrared (VisNIR) and diffuse reflectance spectroscopy (DRS) to envisage enzyme activities of compost. Their method was considered to be economical and rapid based on the selected artificial neural network tool usage. Usage of machine learning and data mining algorithms can eventually result in better prediction of composter performance.

4 Futuristic Scope

4.1 Design of Smart Composter Bins

Adapting/merging the traditional physico-chemical process with advanced technology led to the creation of many innovative smart devices for waste management. In this regard, Sevik et al. [10] fabricated a smart composter using 100 L capacity stainless steel reactors with automated design for temperature (K type thermocouple) and gas measurements. Boniecki et al. [13] identified the maturity class of the compost samples by introducing an artificial neural network module which can predict based on the visual observations of the color of compost samples. Ahmadi et al. [14] fabricated a smart composter consisting of temperature and humidity sensors, Wheatstone bridge circuit, 12 V electric pump and hot air circulatory system. They automated the composter with MATLAB, PID controller, telemetric and human machine interface via Internet. In essence, many innovations are happening in the world of smart compost design by incorporating the fundamental concepts of embedded systems to reach rational technological breakthrough.

4.2 Advanced Data Analytics on Human Perception

Technological implementation enhances the quality of output by reducing the manpower and increasing the accuracy. However, human perception on choosing the right methodologies and product specifications can still be a vital component in the progress of automated systems. Qiu and Wang [5] studied the kitchen waste composting machine and classified the most critical expectations of people in terms of optimum size, aesthetic look, interactive panel, odorless operation and cost effectiveness. Similarly, Zhou and Wang [6] analyzed various design elements of the automatic domestic food waste composter based on perceptual engineering method and reported the significance of feasible materials, surface texture, color and models based on a questionnaire survey. It is also evident from the literature that IoT-enabled composters can pull huge demands in the market by adapting to a dynamic and people-centered approach in their design, implementation and operations. Considering the futuristic vision with increased connectivity, individual perceptions are bound to be influenced by the way the world looks around, and the data mining and analytics can also influence the development of perceptions through the increased human activity in the digital world.

4.3 *Prospective on Smart and Rapid Composters*

As reported in many studies, the optimum feed proportion for specific feedstock, appropriate amendments and design-based process modification can lead to increment the composting rate. Particularly, adiabatic composting systems are found to increase the rate of composting by optimizing the environmental parameters, resulting in the reduction of the composting time. Enhancement in the speed and quality of food-waste compost through addition of biochar and natural zeolite was reported by Waqas et al. [2]. The significance of microbial treatment in improving the rate of composting of domestic sewage sludge (bio-solids) was reported in many studies [3, 5, 6]. Considering the prospects of rapid composter in delivering quick solution to the growing issue of waste management, the incorporation of smart technologies in making a smart-rapid composter can be a better solution [3]. Though it is technically feasible, the associated economic liabilities certainly undermine the sustainability of large-scale application of smart-rapid composters as a favorable low-cost solution. To a greater extent, this issue can be handled by promoting federal incentives to the prospecting vendors and customers.

5 Conclusion

The review highlights the recent trends in technological innovations involved in the design and implementation of a smart composter enabled with IoT. The common hardware peripherals include sensors for measuring temperature, humidity, flow rate and pH while specific sensors are also available for special applications such as gas analyzer (oxygen, carbon-dioxide, ammonia). In addition, the pump steering system, flow adjuster, aeration fans, solenoid valves, hot wire anemometer are some of the common actuators that perform operations based on the user set point for automation. The control and accuracy of results mainly depend on the selection of the major processing units having suitable output-related hardware such as Arduino UNO, Node MCU, Raspberry Pi model 3, track recorder, Siretta antenna and wireless communication module. It is also observed that selection of proper Internet gateway is essential to address the challenges in data transmission, data safety and power consumption as in the case of a community-based cluster of smart composters. The study makes design-oriented prospects towards development of a smart-rapid composter involving provisions for advanced process control and quality checking. Apart from the technological innovation, the feasibility of fabricating a smart composter also depends on human perception and expectation such as optimum size, aesthetic look, interactive panel, odorless operation and cost effectiveness. It is envisaged that integration of smart technologies in conventional waste management operations can transform the entire sector by creating more avenues for customized design and implementation of smart composting units in varying environmental conditions.

References

1. Shimizu, N.: Process Optimization of Composting Systems. Robot. Mechatronics Agriculture, pp. 1–22 (2018). https://doi.org/10.1201/9781315203638-1.
2. Waqas, M., Nizami, A.S., Aburiazaiza, A.S., Barakat, M.A., Rashid, M.I., Ismail, I.M.I.: Optimizing the process of food waste compost and valorizing its applications: a case study of Saudi Arabia. J. Clean. Prod. **176**, 426–438 (2018). https://doi.org/10.1016/j.jclepro.2017.12.165
3. Kumar, S., Negi, S., Mandpe, A., Singh, R.V., Hussain, A.: Rapid composting techniques in Indian context and utilization of black soldier fly for enhanced decomposition of biodegradable wastes—a comprehensive review. J. Environ. Manage. **227**, 189–199 (2018). https://doi.org/10.1016/j.jenvman.2018.08.096
4. Chronicle, L., Report, T.: Res. Hyperspectral Prediction Mode. Soil Organic Matter Based Full Spectrum Data Min. Technol. **32**, 1–3 (2020)
5. Qiu, J., Wang, H.: Improved design of household kitchen waste composting machine based on human factors engineering. Springer International Publishing (2020). https://doi.org/10.1007/978-3-030-51549-2_27
6. Zhao, C., Wang, H.: Analysis of design elements of automatic household food waste composting machine based on perceptual engineering. Springer International Publishing (2020). https://doi.org/10.1007/978-3-030-51194-4_45
7. Hemidat, S., Jaar, M., Nassour, A., Nelles, M.: Monitoring of composting process parameters: a case study in Jordan. Waste Biomass Valorization **9**, 2257–2274 (2018). https://doi.org/10.1007/s12649-018-0197-x
8. Vasudevan, M., Karthika, K., Gowthaman, S., Karthick, K., Balaganesh, P., Suneeeth Kumar, S.M., Natarajan, N.: Aerobic in-vessel co-composting of dewatered sewage sludge with mixed municipal wastes under subhumid and semiarid atmospheric conditions. Energy Sources, Part A Recover. Util. Environ. Eff. (2019). https://doi.org/10.1080/15567036.2019.1624888
9. Balaganesh, P., Vasudevan, M., Natarajan, N., Suneeth Kumar, S.M.: Improving soil fertility and nutrient dynamics with leachate attributes from sewage sludge by impoundment and co-composting. Clean—Soil, Air, Water. (2020). https://doi.org/10.1002/clen.202000125.
10. Şevik, F., Tosun, Ekinci, K.: Composting of olive processing wastes and tomato stalks together with sewage sludge or dairy manure. Int. J. Environ. Sci. Technol. **13**, 1207–1218 (2016). https://doi.org/10.1007/s13762-016-0946-y
11. Şevik, F., Tosun, İ, Ekinci, K.: The effect of FAS and C/N ratios on co-composting of sewage sludge, dairy manure and tomato stalks. Waste Manag. **80**, 450–456 (2018). https://doi.org/10.1016/j.wasman.2018.07.051
12. Jordão, M.D.L., de Paiva, K., Firmo, H.T., Inácio, C.T., Rotunno Filho, O.C., e Lima, T. de A.: Low-cost automatic station for compost temperature monitoring. Rev. Bras. Eng. Agric. e Ambient **21**, 809–813 (2017). https://doi.org/10.1590/1807-1929/agriambi.v21n11p809-813.
13. Boniecki, P., Idzior-Haufa, M., Pilarska, A.A., Pilarski, K., Kolasa-Wiecek, A.: Neural classification of compost maturity by means of the Self-organising Feature Map artificial neural network and Learning vector Quantization algorithm. Int. J. Environ. Res. Public Health. **16** (2019). https://doi.org/10.3390/ijerph16183294

14. Ahmadi, T., Casas, C.A., Escobar, N., García, Y.E.: Municipal organic solid waste composting: development of a tele-monitoring and automation control system. Agron. Res. **18**, 1911–1925 (2020). https://doi.org/10.15159/AR.20.212
15. Wang, X., Yamaguchi, N., Someya, T., Nasu, M.: Rapid and automated enumeration of viable bacteria in compost using a micro-colony auto counting system. J. Microbiol. Meth. **71**, 1–6 (2007). https://doi.org/10.1016/j.mimet.2007.06.019
16. Reed, P.T., Izquierdo, J.A., Lynd, L.R.: Cellulose fermentation by Clostridium thermocellum and a mixed consortium in an automated repetitive batch reactor. Bioresour. Technol. **155**, 50–56 (2014). https://doi.org/10.1016/j.biortech.2013.12.051
17. Dagnon, K., Pickens, M., Vaidyanathan, V., D'Souza, N.: Validation of an automated multiunit composting system. J. Polym. Environ. **22**, 9–16 (2014). https://doi.org/10.1007/s10924-013-0596-9
18. Pratama, Y.F., Ariyanto, E., Karimah, S.A.: Prototype of automation of organic fertilizer manufacturing systems based on internet of things. In: 2019 7th International Conference Information Communication Technology ICoICT 2019, pp. 1–6 (2019). https://doi.org/10.1109/ICoICT.2019.8835299
19. Dickert, F.L., Lieberzeit, P.A., Achatz, P., Palfinger, C., Fassnauer, M., Schmid, E., Werther, W., Horner, G.: QCM array for on-line-monitoring of composting procedures. Analyst. **129**, 432–437 (2004). https://doi.org/10.1039/b315356h
20. Nasution, T.I., Banurea, R., Putra, A., Apriza, A.Y., Azis, P.F.A.: An automatic stirring system based on temperature control on composter for the production of organic liquid fertilizer. In: AIP Conference Proceeding, vol. 2221 (2020). https://doi.org/10.1063/5.0003248
21. Elalami, M., Baskoun, Y., Zahra Beraich, F., Arouch, M., Taouzari, M., Qanadli, S.D.: Design and test of the smart composter controlled by sensors. In: Proceeding 2019 7th International Renewable Sustainable Energy Conference IRSEC 2019, pp. 1–6 (2019). https://doi.org/10.1109/IRSEC48032.2019.9078197
22. Siswoyo Jo, R., Lu, M., Raman, V., Hanghui Then, P.: Design and implementation of iot-enabled compost monitoring system. In: ISCAIE 2019—2019 IEEE Symposium Computer Application Industrial Electronics, pp. 23–28 (2019). https://doi.org/10.1109/ISCAIE.2019.8743981
23. Chakraborty, S., Weindorf, D.C., Ali, M.N., Li, B., Ge, Y., Darilek, J.L.: Spectral data mining for rapid measurement of organic matter in unsieved moist compost. Appl. Opt. **52**, B82 (2013). https://doi.org/10.1364/ao.52.000b82

Study on the Effects of CNT and Nano-graphene in Clayey Soil of Aligarh City of Northern India

Jibran Qadri, Mohd. Aleem Farshori, Mohd Asim Khan, M. Nizamuddin, and Ibadur Rahman

1 Introduction

At present, a very limited literature related to nanotechnology is available in geotechnical engineering. A new intensive research methodology is systematically required, which can define new dimensions, and aspects for the improvement of the soil in a traditional engineering problem. In the case of fine-grained soil, nanoparticles can easily inflict and disperse through the grain pores, because nano-size nanoparticles compared to clay particles are smaller. Intensive analysis is required into the substance to enhance the soil's properties. The physicochemical relationship between nanomaterials and soil is worth remembering. Better understanding of the surface and interface interaction between nanomaterial and naturally occurring material is of great importance for new-generation nanomaterials used in geotechnical engineering that can improve the nanotechnology outcome. With the reduced size of the nanomaterial, a specific surface region increases, indicating an increase in the surface area in the molecule, which results in stronger contact with other natural matter particles. Soil with nanoparticles typically has a higher plastic and liquid limit, since it is apparent that more water molecules in its surroundings are in the specific region, resulting in improved liquefaction resistance. Nanotechnology involves an atomic

J. Qadri (✉)
Department of Civil Engineering, Aligarh Muslim University, Aligarh, U.P, India
e-mail: lncs@springer.com

Mohd. A. Farshori · M. Nizamuddin
Department of Electronics Engineering, Faculty of Engineering and Technology, JMI, New Delhi, India

M. A. Khan
School of Civil Engineering, Tianjin University, Tianjin, PR China

I. Rahman
Department of Civil Engineering, Faculty of Engineering and Technology, JMI, New Delhi, India

I. Pal and S. Kolathayar (eds.), *Sustainable Cities and Resilience*, Lecture Notes in Civil Engineering 183, https://doi.org/10.1007/978-981-16-5543-2_37

analysis, which includes atomic motion and atomic shifts. Nanotechnology investigates and manipulates on a molecular scale, and so the new substance with new functions and new fundamental characteristics can be obtained. Nanotechnology can be defined as a facilitator that in almost any area offers a new horizon. Nanotechnology reaches beyond discipline and field. In this regard, various studies [1–4] and [5] suggest that nanomaterials and numerous research analysis applications will soon be utilized. [6–9], many experimental research suggest promising outcomes, which are fascinating in the near future of nanotechnology and nanomaterial [10–12]. Carbon nanotubes can indeed be described simply as tubular-folded graphite sheets. CNTs exhibit excellent physical and chemical properties and researchers' curiosity in these nanomaterials in power storage, pharmaceutical goods, substance strengthening, microwave absorption, etc., is tremendous because of these extraordinary properties [13]. Stabilization of soil is a customary technique for civil construction to accommodate diverse types of engineering and construction specifications [14]. Traditional materials such as lime, which are in use since millennia back, and new materials such as cement are presently in practice, as materials such as fly ash derived from power stations, and goods such as silica fume and other conventional materials have been used continuously.

2 Experimental Program and Methodology

This research delineates laboratory testing and experimental testing conducted on soil samples in order to determine the effect on clayey soil on the complement of various CNTs and nano-graphenes (0gm, 0.2gm, and 0.6gm). The limits of Attenberg for the different amounts of nanomaterials were determined, and the plasticity index from test results was acquired. The study of the clay samples was performed by means of scanning electron microscopy (SEM). The basic sample analysis was also obtained for X-ray diffraction (XRD).

(1) Plastic limit determination: Plastic limits have been checked in compliance with IS: 2720(Part-6)-1972.
(2) Liquid limit determination: A measurement of the liquid limit was carried out in a laboratory using Casagrande equipment as provided in IS: 2720(Part5)-1985.
(3) Shear strength parameter (s) of the triaxial shear compression, unconsolidated and undrained, carried out according to IS 272(Part 11):1993; Shear strength parameter (s) of the triaxial shear compression

2.1 Materials

Nano-graphene and CNTs were bought from Sisco Research Laboratories Pvt. Ltd. (SRL)—India

Table 1 Synthesis details of clay nanocomposites

Sample I.D	Clay (gm)	Nanoparticles (gm)
Clay	300	NIL
Clay/CNT	300	02, 0.4 and 0.6
Clay/nano-graphene	300	0.2, 0.4 and 0.6

2.2 *Synthesis of Clay Nanocomposites with Nano-Graphene and CNTs*

Clay nanocomposites with nano-graphene and CNTs were prepared by mechanical mixing method in the presence of different amounts nanoparticles. The details of the synthesis are given in Table 1.

2.3 *Soil Characterization and Properties*

Clay samples were taken from dry pond nearby area of Aligarh district (India) and dug out using shovel from depth of near about 1 m from excavated pit obtained sample was disturbed sample. Figure 2 displays the XRD spectrum of the soil sample. The XRD pattern indicates that the soil primarily consists of quartz containing some small concentrations of montmorillonite and calcite. Figure 1 shows the SEM micrograph of pure clay having flakes-like structure.

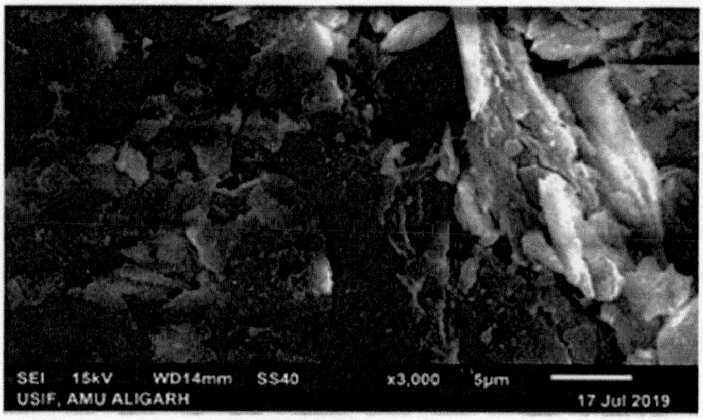

Fig. 1 SEM images of clay at 5 μm are shown

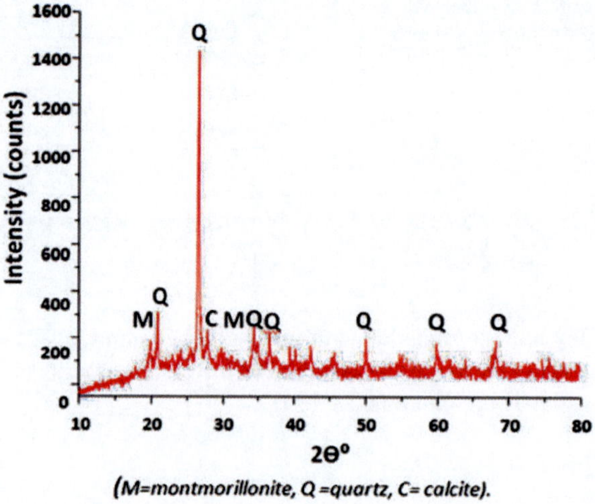

Fig. 2 XRD of plain clay soil

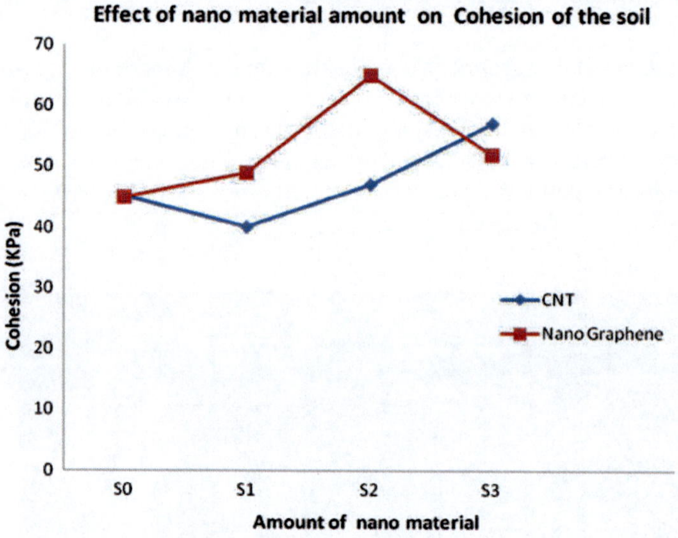

Fig. 3 Effect of nanomaterial on cohesion (KPa) of the soil

Table 2 Properties of the soil sample

Characteristics	Soil
Specific gravity	2.55
Soil classification	CL
Liquid limit	35
Plasticity index	14
Optimum water content	19%
Maximum dry unit weight (g/cm^3)	1.64

Table 3 Variation of water content with different amounts of CNT

S. No	Plastic limit	Liquid limit	Plasticity index
SO	21	35	14
S1	20	34	14
S2	20	32	12
S3	18	30	12

Table 4 Variation of water content with different amounts of nano-graphene

S. No	Plastic limit	Liquid limit	Plasticity index
SO	21	35	14
S1	21	34	13
S2	20	32	12
S3	20	31	11

3 Result and Discussion

3.1 Effect of Different Nanomaterials and Water Content on Engineering Properties of the Soil

The effect of nanoparticles on clayey soil is shown in this section. With the inclusion of nanoparticles in the soil, the liquid limit, plastic limit, and plasticity index decrease, lower plasticity indexes show dramatically improved soil properties. Thus, even in very small amounts, nanoparticles will improve their strength and enhance the properties as shown in Tables 3 and 4. The use of CNT and nano-graphene in clay soil leads to reduced plastic limits by 14 and 5%, and liquid limit by 14 and 11%, and plasticity index by 14 and 21.4%, respectively.

3.2 Effect of Nanomaterials on Cohesion and Angle of Friction of the Soil

A cohesive variation and the friction angles with the inclusion of nanoparticles are shown in Figs. 3 and 4. Friction angle and cohesion are the parameters which check the shear strength of soils. The cohesion is the molecular force of bonding. The frictional force can commonly be separated into two: Occlusion and sliding frictional intensity are highly determined by the friction angle. In clay, the cohesion value should be greater than the friction angle, and vice versa in sand. In order to find a relation between the nanomaterial content and the shear strength parameters, the cohesion (C) and the friction angle (Φ) shear strength parameters were correlated with the substance of nanomaterials, and the graphs were used to determine the association between the two parameters. Figures 3 and 4 demonstrate the increase

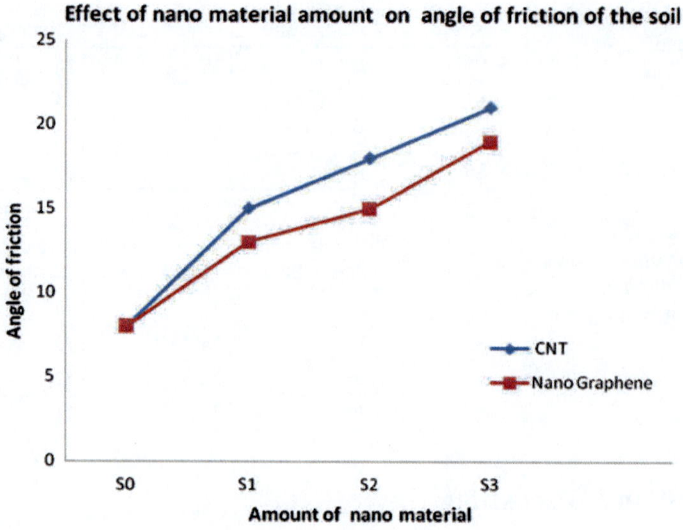

Fig. 4 Effect of nanomaterials on angle of friction (degree) of the soil

in nanomaterial content results in improvement in angle of friction and cohesion as nanomaterial CNTs and nano-graphenes increase, respectively. It indicates a growing trend of nanoparticles fill the gap between soil and continue to cause soil interlocking behavior, with more amounts of nanomaterials incorporated into soil. This describes the cause of high cohesion in the soils as the content of nanomaterials is increased.

4 Conclusion

The nanocomposites of clay, CNT and nano-graphene were successfully synthesized with the help of mechanical mixing process and characterized using SEM and XRD analysis. When the nanoparticles were applied to the liquid limit, the plastic limit and the plasticity index decreased. With the inclusion of nanoparticles exceeding the optimal quantity, the state of a mass cluster is associated, thereby influencing the mechanical features of soils. Thus, even in very small quantities the inclusion of nanoparticles in the soil will exquisitely increase strength will strengthen the properties of the soil. Cohesion and the angle of friction have increased, which suggests an increase in shear strength. The findings will lead to improving soil quality and other properties for researchers and field engineers.

Acknowledgements The authors acknowledge the help received from the USIF, AMU, and testing facilities at the Department of Civil Engineering, AMU, Aligarh.

Conflict of Interest: On behalf of all authors, there is no conflict of interest.

References

1. Abedi, M., Yasrobi, S.S.: Effects of plastic fines on the instability of sand. Soil DynEarthqEng **30**, 61–67 (2010)
2. Huang, Yu., Lin, W.: Experimental studies on nanomaterials for soil improvement: a review. Environ. Earth Sci. **75**, 497 (2016). https://doi.org/10.1007/s12665-015-5118-8
3. Majeed, Z.H., Taha, M.R., Jawad, I.T.: Stabilization of soft soil using nanomaterials. Res. J. Appl. Sci. Eng. Technol. **8**(4), 503–509 (2014)
4. Aroraa, A., Birpal, S., Parampreet, K.: Performance of Nano-particles in stabilization of soil: a comprehensive review. In: International Conference on Advanced Materials, Energy & Environmental Sustainability, ICAMEES (2018)
5. Taha, M.R.: Geotechnical properties of soil–ball milled soil mixtures. In: Proceeding 3rd Symposium Nanotechnology in Construction, pp. 377–382. Springer-Verlag (2009)
6. Alireza, S., Mohammad, M.Z., Hasan, B.M.: Application of Nanomaterial to Stabilize a Weak Soil, Seventh International Conference on Case Histories in Geotechnical Engineering (2013)
7. Zhang, G.: Soil nanoparticles and their influence on engineering properties of soils. In: Advanced in Measurement and Modelling of Soil Behaviour, ASCE (2007)
8. MajeedZH, T.M.R.: A review of stabilization of soils by using nanomaterials. Aust. J. Basic Appl. Sci. **7**(2), 576–581 (2013)
9. Sharma, D., Naval, S., Chandan, K.: Stabilisation of expansive soil using nanomaterials. In: International Interdisciplinary Conference on Science Technology Engineering Management Pharmacy and Humanities (2017)
10. Ghazavi, M., Bolhasani, M.: Unconfined compression strength of clay improvement with lime and nano- silica. In: Proceeding of 6th International Congress on Environmental Geotechnics, pp. 1490–1495. New Delhi, India (2010)
11. Majeed, Z.H., Taha, M.R.: An analysis on geotechnical properties of soil with different nanomaterials, Saudi. J. Eng. Technol. (2016)
12. Firoozi, A.A., Olgun, C.G., Firoozi, A.A., Baghini, M.S.: Fundamentals of soil stabilization. In: Firoozi et al. (eds.) Geo-Engineering, vol. 8, pp. 26 (2017). https://doi.org/10.1186/s40703-017-0064-9
13. Nigar, A., Rahil, H., Manshi, T., Neelam, Y., Jagriti, N.: Sens. Int. **1**, 100003 (2020)
14. Kolias, S., Rigopoulou, V.K., Karahalios, A.: Stabilisation of clayey soils with high calcium fly ash and cement. Cement Concr. Comp. **27**(2), 301–313 (2005)

Highway Development-Related Gully Erosion: The Case of the Okigwe-Isuikwuato Highway, Southeastern Nigeria

Site Onyejekwe, Jeremiah C. Obi, and Elamin Ismail

1 Introduction

In order to provide for an ever-growing population, humans have had to modify the natural environment to provide infrastructure required by this population. This is practically true in such areas as agriculture, housing, utilities, transportation, and such like where the natural environment has to be modified to suit the scheme as appropriate. These modifications typically involve a large-scale removal of vegetative cover, modification to the natural topography, alteration of the natural drainage system, etc., which mostly result in an increase in the volume of runoff and the attendant flooding and erosion when implemented improperly.

When done right, with all environmental safeguards in place, these modifications have a minimal impact on the environment and hence are generally considered to be beneficial. When executed haphazardly, the result is often severe environmental issues such as flooding and erosion. More often than not, the environmental safeguards required to remediate the effect of modifications to the natural environment are not put in place. Where they are put in place in the inception of a scheme, haphazard implementation and follow-through coupled with the lack of maintenance culture mean that these measures fall into distress shortly afterward resulting in severe environmental conditions like flooding and erosion, including gully erosion.

The causes of erosion (soil/gully) have often been traced to factors such as geology/soil type, topography, climate, vegetation, and anthropogenic actions. Erosion has been an issue in southeastern Nigeria for a long while. Several researchers

S. Onyejekwe (✉)
Road Sector Development Team, Federal Ministry of Works, Abuja, Nigeria

J. C. Obi
Missouri Department of Transportation, Chesterfield, MO 63017, USA

E. Ismail
Bara Geophysical Services, Rolla, MO 65401, USA

I. Pal and S. Kolathayar (eds.), *Sustainable Cities and Resilience*, Lecture Notes in Civil Engineering 183, https://doi.org/10.1007/978-981-16-5543-2_38

have studied the problem of soil erosion in southeastern Nigeria from various perspectives, the earliest being the review by Stamp [1], followed by those by Grove [2, 3]. Other researchers have studied soil erosion in southeastern Nigeria from various perspectives: geomorphology [4–7]; geology [8–10]; hydrology/hydrogeology [11, 12]; engineering–geological properties of the soil [13–18]; and anthropogenic activities and prevention and control measures [6, 19–22].

Almost all severe erosion gullies in southeastern Nigeria are located on moderate to very gently dipping, poorly consolidated sandstone, usually associated with local or regional highlands, among which the Udi-Orlu and the Okigwe-Ohafia-Arochukwu highlands are the most prominent [23, 24]. The major highlands, plateau, and their precipitous escarpments are formed by sandstone bedrocks (Ajalli sandstones and Nanka sands) while the lower slopes and plains are underlain by mainly shaley units (Imo, Mamu, Nsukka, and Bende-Ameki [Ameke] formations) [17, 23, 25]. The gentle slopes of undulating plains are covered by thick and highly sandy overburden. The highest point in the region (343 m) occurs around Okigwe while the southern plains of the region stand at about 61 m above sea level [17, 23, 25]. Gully processes are most prevalent in the fine- to medium-grained coastal plain sands (Pliocene to recent) and Nanka Sands (Eocene) and the medium- to coarse-grained Nsukka Sandstone and Ajalli Sandstone (Cretaceous) of the Anambra-Imo basin region [17, 25].

Various studies on erosion in southeastern Nigeria [14–16, 18] have shown that the initiation and propagation of gully erosion are influenced by three major groups of factors: rock/soil type and properties, topography and rainfall, and human activities. Combinations of all three factors have resulted in severe gully erosion issues on the Okigwe-Isuikwuato highway.

Constructed on a largely rugged terrain on the scarp slope of the 250 km north–south trending Arochukwu-Ohafia-Agwu-Udi-Nsukka cuesta in the late nineteen-seventies (1976–1978), a number of gullies have developed on the Okigwe-Isuikwuato highway mainly at points where inappropriately sized culverts have obstructed the natural drainage path.

This paper reports on an investigation into the factors (anthropogenic, hydraulic, hydrologic, etc.) that have led to the development of these gullies. Recommendations as to the mitigation of the effect of these gullies are also provided.

2 Physiography, Climate, and Geology of Study Area

The 25 km Okigwe-Isuikwuato highway is located within the old Isuikwuato/Okigwe Local Government Area (LGA) of Imo State. It originates at the Ugba (ABSU) Junction, Uturu (geographical coordinates: 5.828874 N, 7.395468 E) on the Okigwe-Afikpo highway and ends at the Akara (Ahaba) Junction (geographical coordinates: 5.722539 N, 7.560681 E) on the Umuahia-Uzuakoli-Ohafia-Arochukwu highway. The study area is located between Ohafia and Awgu within the scarp slope of the 250 km long north–south trending Arochukwu-Ohafia-Awgu-Udi-Nsukka cuesta.

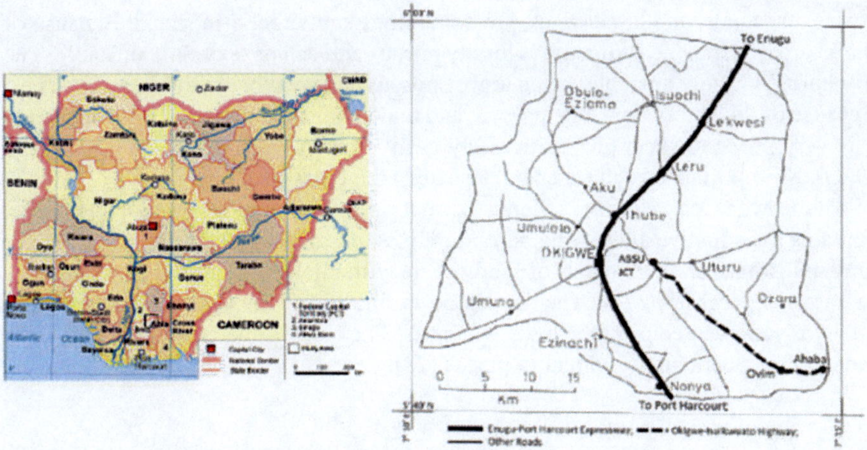

(a) (b)

Fig. 1 **a** Study area (adapted from Wikipedia [57]). **b** Highway alignment

The study area is shown in Fig. 1a. The road is shown in Fig. 1b. The former Isuikwuato/Okigwe Local Government Area is located within latitude 5° 40′–6° 03′ N and longitude 7° 10′–7° 35′ E, which locates in the tropical monsoon (AM) climatic zone based on the Köppen Climate Classification System [26, 27]. At an average of about 130–200 m above sea level, the region lies on a relatively higher terrain relative to the other parts of Imo and Abia states.

Typical of the rain forest regions of the tropics, rainfall is heavy with a range of 2250–2500 mm and accompanied by frequent storms of very high intensity. There are two seasons in a year: the rainy season (March–October) and the dry season (November–February) with the characteristic dry winds of the Harmattan period dominating most of December and January. The annual average monthly maximum and minimum temperatures range from 28 to 35 °C (during dry season) and 19–24 °C (during peak rainy season and the dry Harmattan season), respectively. Mean relative humidity ranges from 60 to 70% and 80 to 85% for the dry and rainy seasons, respectively.

Geologically, the study area falls within the southeastern Nigeria sedimentary basin. The rocks, rock systems, and geological history of the Isuikwuato/Okigwe region are derived from geological events that span the Cretaceous and Tertiary periods of the Mesozoic and Cenozoic eras, respectively [28]. The oldest rocks in the Isuikwuato-Okigwe region are of lower Cretaceous age, while the youngest rocks are still being formed. About ten main geological formations can be distinguished in the region. Following in order of stratification is a brief discussion on the geological formations through which the highway courses [28, 29].

The False-Bedded Sandstones (Ajalli Formation) consist of thick, friable, poorly sorted sandstones typically white in color, but sometimes iron-stained. The formation is often marked by repetitive banding of coarse and fine-grained layers [30]. The sand grains, especially the larger ones, are sometimes sub-angular in shape. Because of this characteristic, the formation is highly porous and can be texturally unstable. The formation is highly unstable, being highly porous and weakly consolidated in places. Because of the highly unstable, porous, and unconsolidated nature of this formation, it is very prone to sheet and particularly gully erosion when disturbed mindlessly. The most important structural characteristic is the cross-bedding.

The Lower Coal Measure (Mamu Formation), unlike the Upper Coal Measure, contains a distinctive alternating admixture of sandstones, dark shales, dark blue or gray mudstones and thin bands of sandy shale with thicker lignite and coal materials seams at several horizons. The sandstone is fine to medium-grained and may be white or yellow in color in the lower horizons. They are normally well-bedded, even though cross-bedding is evident in places [28].

3 Genesis of Present Problems

The 23.6 km Okigwe-Isuikwuato highway (a single lane two-direction highway) in southeastern Nigeria was constructed in the late nineteen-seventies (1976–1978) to link the Okigwe-Afikpo highway and the Umuahia-Uzuakoli-Ohafia-Arochukwu highway. The highway has its origin at the Ugba (Abia State University [ABSU]) Junction on the Okigwe-Afikpo highway and terminates at the Ahaba (Akara) Junction on the Umuahia-Uzuakoli-Ohafia-Arochukwu highway. The highway, located mainly in Isuikwuato and, to a lesser extent Okigwe (Uturu), was constructed on a largely rugged terrain on the scarp slope of the 250 km north–south trending Arochukwu-Ohafia-Agwu-Udi-Nsukka cuesta, one of the major cuestas in southeastern Nigeria (the other cuesta being the 75 km Northwest–Southwest trending Awgu-Awka Uplands cuesta).

Construction of highways through largely rugged terrains such as the Arochukwu-Ohafia-Agwu-Udi-Nsukka cuesta inevitably involves massive earthworks involving both cuts and fills that have the added effect of altering the natural drainage pattern and the hydraulics of the affected areas. Additionally, any earthworks or schemes constructed on geologically sensitive areas that are known to be highly susceptible to erosion are to be executed with utmost care, with special emphasis paid to the performance of the scheme throughout its service live. The Okigwe-Isuikwuato highway is mainly located within the medium- to coarse-grained Ajalli Sandstone (Cretaceous) of the False-Bedded Sandstone (Ajalli Formation) in the Anambra-Imo basin region which is known to be highly susceptible to the gullying process [16–18, 23, 25, 31–33]. The lesser proportion of the highway is located within the Lower Coal Measures (Mamu Formation) which is less susceptible to the gullying process.

Over the years, a number of gullies have developed on the Okigwe-Isuikwuato highway due to changes in the land use pattern that have resulted in an increase in the volume of runoff. The major gullies that have developed on the Okigwe-Isuikwuato highway are presented in Table 1. The first major gully developed in Umunnekwu, Isuikwuato at about Kilometer 11 sometime in 1981 or 1982. These gullies have

Table 1 Major erosion occurrences on the highway

Site*	Location	Distance* (km)	Status
1	Uturu	2.5	Active
2	Umunnekwu, Isuikwuato I	7.2	Active
3	Umunnekwu, Isuikwuato II	9.2	Active
4	Umunnekwu, Isuikwuato III	11.0	Very Active
5	Amokwe Amaiyi, Isuikwuato	13.0	Inactive
6	Amaba, Isuikwuato	17.0	Inactive

Note * = Relative to origin at Start [Ugba (ABSU) Junction] on Okigwe-Afikpo Highway

primarily developed at points where inappropriately sized culvert structures have obstructed the natural drainage path. The combination of the undersized culvert, pressure flow through the undersized culverts, topography, and vulnerable geologic formation has led to the development of massive gullies in some locations.

The following is a discussion of the various factors related to land use (deforestation, fire, grazing, cropping, and cultural practices) and factors related to the environment that have contributed to the increase in the quantity and the rate of runoff and hence to the development of gullies of various magnitudes along the highway. The hydrologic and hydraulic factors that have contributed to soil erosion and the development of these gullies are also discussed.

3.1 Land-Use Related Factors

Over the years since the construction of the road in the late 1970s, vast areas of forested land in the watershed have been converted to various uses in an attempt to meet the demand of food for humans and animals, fuel, housing, and general infrastructure. The change in land use and the range of cultural practice of soil and crop management employed have drastic effects on erosion. The following is a discussion of some land-use factors that affect the rate of erosion.

3.1.1 Deforestation

Vegetation offers general protection to the soil surface. Vegetation protects the soil surface from the erosive forces of trampling, raindrop impact, overland flow, and wind [18, 34]. Vegetation also produces the litter layer with which it also combines to protect the soil from compaction. The destruction of vegetation and possibly the litter layer, by way of cutting of trees, and burning of bushes exposes the soil to erosive forces [18, 34]. Since the construction of the highway in 1976–78, more areas of land have been cleared for agriculture, infrastructure, and other purposes, and this has resulted in increased exposure of the soil to erosive forces in terms of increased volume and velocity of runoff as evidenced by the hydraulic structures (line drains and culverts) on the highways that are now undersized.

3.1.2 Quarrying

The excavation of soil, sand, laterite, and other natural construction materials from hills or pits or sides of a river or stream causes erosion, especially in hilly areas [18]. The quarrying process also involves large-scale removal of vegetation and the development of earth roads and footpaths all of which are known to contribute to soil erosion [17, 35, 36]. Since the construction of the highway, several quarrying sites have been established along the highway corridor. The most prominent of them being the sand quarry at Umunnekwu, Isuikwuato, which is near one of the earliest locations where gully erosion breached the highway in about 1986 (Fig. 2).

Fig. 2 Active quarry site at Km 6.8, Umunnekwu, Isuikwuato

3.1.3 Crop Farming

Generally, due to the downturn in the economy and deliberate government policies that encourage agricultural production since the completion of the highway, increasingly more land has been put under cultivation for both commercial and subsistence purposes. This has led to an increase in erosion in the region. Traditionally, agriculture in the region is rain-fed and crop farming is by shifting cultivation. The farming cycle is such that crops are typically planted at the outset of the rainy season in March/April [29, 37], and they are not well established before the onset of heavy rains in the months of June to July [29, 37]. Hence, the farms are most vulnerable to soil loss due to the lack of vegetative cover during heavy rains [38–40]. This has led to an increase in erosion in the region.

3.1.4 Grazing

Erosion has been shown to increase with grazing, particularly overgrazing [41, 42]. Overgrazing, trampling of turf/breaking up of turf, and multiple animal tracks are some of the effects of livestock (cattle, and to a lesser extent, sheep and goats) that increase erosion by creating channelized and accelerated-velocity surface runoff [43]. The variation in soil structure wrought by overgrazing is not favorable for the establishment of the continuous vegetation cover required to limit soil erosion [18, 44]. Uncontrolled grazing has increased in the areas adjoining the Okigwe-Isuikwuato highway following the construction of the highway in 1976–78. This has significantly contributed to the increase in the forces of erosion as can be seen in the gullies that have developed within the highway right-of-way.

3.1.5 Footpaths and Earth Roads

Communities typically create footpaths and earth roads for a variety of purposes: access to rivers and streams, farms, homes, civic facilities (schools, health centers, etc.), major highways, and the like. Pressure from traffic (human and, or, motorized) on footpaths and earth roads creates an enabling environment for differential erosion of soil materials and soil erosion. Additionally, since they are neither properly designed nor planned in terms of their orientation relative to the ground contours, they mostly function to promote the concentration of runoff and erosion [17, 35, 36, 45, 46]. Figure 3a, b shows some instances of the erosive effect of footpaths and earth roads, respectively.

Fig. 3 a Earth road as erosion channel at Km 0.25, beside ABSU fence, Uturu. b Footpath as erosion channel at Km 13.7, NNPC Pipeline, Amiyi, Isuikwuato

3.2 *Environment-Related Factors*

Accelerated erosion is the increased rate of erosion that often arises when humans alter the natural system by various land-use practices and engineering constructions [4, 18, 47]. Accelerated or anthropogenic erosion is a manifestation of environmental imbalance [18, 38]. It is fast and very dangerous. The problem of soil and gully erosion is largely the problem of man-accelerated erosion. Accelerated erosion is a multifaceted and complex process that is affected by a multitude of interacting environmental parameters [48]. The following is a discussion on some of these environmental factors as they relate to the erosion on the Isuikwuato-Okigwe highway.

3.2.1 Rainfall

The former Isuikwuato-Okigwe LGA, within which the Okigwe-Isuikwuato highway is situated, is within the tropical monsoon (AM) climatic zone, based on the Köppen Climate Classification System [26, 27]. The annual rainfall ranges from 2250 to 2500 mm [29, 49]. Rainfall intensities exceeding 100 mm hr^{-1} sustained for 15 to 30 min are frequently observed in this region [49]. The high intensity of tropical rainfall is partly attributed to relatively large drop sizes. Rains with median drop size in excess of 2.5 mm are common [50, 51]. Rains with an energy load of 100 J m^{-2} mm^{-1} of rain are often received at a time when the protective vegetation cover is poor [50]. High-intensity rains are particularly damaging when vegetative cover is poor [38–40]. Crops are typically planted in Region at the outset of the rainy season in March/April and are not well established before the onset of heavy rains in June/July. Consequently, the farms are most vulnerable to soil loss during the months of rainfall due to the lack of effective vegetative cover.

3.2.2 Soils

The Okigwe-Isuikwuato Highway is mainly located within the medium- to coarse-grained Ajalli Sandstone (Cretaceous) of the False-Bedded Sandstone (Ajalli Formation) in the Anambra-Imo basin, region which is known to be highly susceptible to gullying process ([16–18, 23, 25, 31–33]. The lesser proportion of the highway is located within the Lower Coal Measures (Mamu Formation), which is less susceptible to the gullying process. Several studies have been conducted on the properties of soils in erosion-prone areas of southeastern Nigeria. Several researchers have studied the soils in the areas that are prone to severe gully erosion and have found that they are underlain by lateritic coastal sands that have very specific geotechnical properties (such as grain size and grain size distribution, uniformity, and dispersion characteristics) that are amenable to gully erosion [14–18]. The geotechnical properties of the soils in the highway right-of-way are within the range as detailed by Hudec et al. [17, 18].

3.2.3 Landform/Topography

The Okigwe-Isuikwuato Highway is situated on a largely rugged terrain on the scarp slope of the 250 km north–south trending Arochukwu-Ohafia-Agwu-Udi-Nsukka cuesta. Erosive forces are known to have greater effect on steep slopes than on flat land, with slope length and the amount of soil erosion being proportional to the steepness of the slope [5, 52]. The topography of the Okigwe-Isuikwuato region is classified into four broad relief units [29, 37]: (a) the lowland plains and lowlands including all the river valleys; (b) the low plateau; (c) the upland plains; and (d) the highland zone. It is observed that the uplands that are made up of highly friable sandstones (False-Bedded Sandstones) yield easily to erosion and induce gullying even on slopes of about 5%. The gradients over this landscape range from 5 to 10%, reaching 15% on valley sides [29, 37]. While the highlands have considerably stable lithology and hence are able to resist erosive forces, particularly gullying, they provide a massive amount of fast-flowing runoff which goes on to devastate the lowland areas particularly at the toe slopes and the valleys where many of the headwaters of the rivers in the area are located [29, 37].

3.3 Hydrology/Hydraulics-Related Factors

Highways alter the hydrology of the catchment basins they traverse. Hence, to avoid negative environmental outcomes, adequate attention should be paid to issues relating to the hydrology and hydraulics in highway development. Best environmental management practice requires that prime consideration is given to the hydrology of areas traversed by highways and the hydraulics of the drainage structures installed in the highway. This is particularly important in the determination of the appropriate structures required to transfer runoff from the catchment basin safely and at the termination of these hydraulic structures, where in order not to trigger gully erosion, geotechnical and geological factors are critical. Following is a discussion on some of these hydraulic and hydrological factors as they relate to the erosion on the Isuikwuato-Okigwe highway.

3.3.1 Hydraulics/Hydrology

Highways have a great impact on the hydrology of the catchment basins they traverse. Highways and associated structures serve either to retain or drain/pass runoff, often concentrating runoff where it did not exist before. The prevailing engineering practice relating to drainage is to protect the road and evacuate runoff from the road as soon as possible; highway authorities are responsible for the safe disposal of runoff about 20 m from the highway [53], leaving the disposal of runoff after the first 20 m to no one in particular. This practice is fraught with very severe shortcomings and has been one of the major causes of gully erosion on the Okigwe-Isuikwuato highway. Additionally, there is a serious lack of consideration for hydrologic and hydraulic factors in the state-of-the-practice of highway development in Nigeria as the most emphasis is placed on the geometric design aspects of highway development to the detriment of other aspects, particularly drainage and its effect on the immediate and wider environment. This has resulted in inappropriately sized and sometimes inappropriately located drainage structures that act as trigger points for erosion. These usually undersized line drains (trapezoidal, rectangular) often overflow under normal conditions and certainly overflow when partially clogged by debris, thereby exposing the soil at the point of overflow to the erosive power of surface runoff. The same applies to the undersized culvert which increases the erosive power of runoff via pressure flow conditions. Absent suitable outlet structures, there is continuous erosion of the soil at the outlet due to the high momentum of flow and its attendant erosive power. This high erosive power runoff eventually leads to the formation of gullies through head-cutting of the drainage channel and the collapse of sidewalls. Figure 4a, b shows active gullying at Km 11 while under remediation.

4 Suggestions for Mitigation

The need to arrest the wasteful trend in soil loss due to erosion has always been recognized in southeastern Nigeria. There is a long tradition of soil conservation being practiced in southeastern Nigeria. Indigenous techniques from the pre-colonial era focused on erosion control in combination with water conservation by ridging, mulching, constructing earth bunds and terraces, multiple cropping, fallowing, and planting trees [17, 18, 25, 54]. The colonial administration embarked on extensive development of plantations that served the dual purpose of commercial agricultural

Fig. 4 Increase in gullying due to structures constructed during the reinstatement of the highway section at Km 11, Umunnekwu, Isuikwuato

developments and erosion control measures. The cashew (Anacardium Occidentale), oil palm (Elaeis guineensis), and oil bean (Pentaclethra macrophylla, Benth) trees were used extensively in the Isuikwuato-Okigwe region during this period. The post-colonial period saw a continuation of the deployment of trees with the inclusion of the Gmelina (Gmelina arborea) and the Douglas fir (Pseudotsuga menziesii) trees. Extensive soil conservation schemes including agronomical measures, as well as other techniques like contour plowing, construction of bunds, the construction of concrete (reinforced and unreinforced) structures, and drainage channels in erosion control, were also implemented during the post-colonial period.

However, notwithstanding the early recognition of the wasteful trend in soil loss due to erosion and the several techniques deployed to combat soil erosion in southeastern Nigeria, there has been no clear indication of success from the deployment of these techniques. On the contrary, these efforts seem to have failed, to a large measure, as areas under soil erosion have generally increased [53]. The area exposed to soil erosion (sheet, rill, or gully) in southeastern Nigeria was 1021 km^2 in 1976 while in 2006 it was recorded as 2820 km^2—about a 280% increase [53]. This scale of failure calls for an urgent reassessment of our soil conservation measures/practices with a view to identifying the most appropriate strategy to deal with the problem.

The failure of the soil conservation and agronomical measures implemented in erosion control has always been attributed to farmer apathy or lack of farmer/local cooperation, which culminates in the "bad farming practices" [4, 53]. However, there are several other reasons why soil conservation and agronomical measures implemented in erosion control have failed. These reasons, which generally apply to southeastern Nigeria and are also relevant to the Okigwe-Isuikwuato highway, can be grouped into three broad categories:

(a) Lack of stakeholder participation and/or buy-in: Most medium- to large-scale efforts at soil conservation and erosion control are typically conceived and executed by Governments (Federal, State, and Local) without input and buy-in from the local populace that will be most affected by these programs. Hence, due to this lack of ownership of the program by the local community and the typical ad-hoc nature of most of these programs, the conservation methods are not appropriately understood and generally result in the failure of the program.

(b) Implementation of inappropriate solutions: More often than not, largely due to inadequate understanding of issues relating to erosion control and soil conservation, inappropriate techniques have been deployed in the solution of erosion control and soil conservation problems. In the area of soil conservation, techniques based on the poor/faulty understanding of tropical agriculture have resulted in the use of plants not best suited for conservation. This is particularly evident in the use of cashew trees in erosion-prone areas of southeastern Nigeria, including areas abutting the Okigwe-Isuikwuato Highway. While the leaves of the cashew tree provide cover to the soil, the roots of the tree neither hold the soil together nor encourage the thriving of undergrowth to form a protective cover for the soil [5, 32]. Junge et al. [55] present a summary review of the extensive research (mainly on agronomic measures) conducted on soil conservation in Nigeria that can be implemented to achieve good results. In the area of engineering, the lack of knowledge of hydrology, hydraulics, and geotechnics has resulted in the construction of inappropriately sized, inappropriately located, and inappropriately terminated concrete structures and drainage channels as a means to control or prevent erosion. However, due to the inappropriateness of these structures, they actually exacerbate the problem as the gullies intensify their activities due to the erection of these structures. Several researchers [19, 22, 56] have documented a number of such structures in Imo State, and old Anambra and Imo states, respectively.

(c) Anthropogenic excesses: The lack of effective regulations on the exploitation of the environmental resources with respect to agriculture, mineral exploitation/mining, infrastructural developments, and related activities have also resulted in the failure of soil conservation and erosion control measures. As discussed in the previous sections on the causes of the gullies on the Okigwe-Isuikwuato highway, there is need to implement multifaceted countermeasures as there are factors that lead to the inception and development of gullies. Some of the countermeasures include:

(d) Farming/Agricultural Practice: Generally, farming should be done with conservation in mind. Limits should be placed on the extent and timing of bush burning, which practice should be restricted to the beginning of the dry season. Farming/cultivation and grazing should be restricted around the immediate vicinity of gullies, depending on the location restriction of farming within 0.5–1 km radius of gullies should be implemented in order to give conservation and erosion control measures a chance to succeed. Movement of livestock and grazing should be controlled and carried out with conservation in mind and not indiscriminate.

(e) Infrastructural Development: Infrastructure (roads, buildings, etc.) should be developed in line with environmental best management practices. The removal of vegetation for infrastructure projects should be restricted to the footprint of the project. Considering that the region's geology is highly susceptible to erosion, attention should be paid to drainage from the conception of the schemes. The hydraulics and hydrology of the drainage system should be determined and used in the design of the hydraulic structures/drainage, and these should be updated as changes in the parameters (hydrologic and hydraulic) of the watershed warrant. This has to be done to avoid undersized hydraulic structures which are one of the major causes of gullying on the Okigwe-Isuikwuato highway. All bare surfaces resulting from the development of infrastructure should be reinstated (revegetated) in order to prevent the development of erosion in those bare areas devoid of vegetation.

(f) Mineral Extraction/Mining: There is need to control the current indiscriminate mineral extraction/mining activities. There is need to promulgate and enforce an extensive and thorough full-cycle reclamation regime. Mine owners should be required to submit a reclamation plan before they obtain mining licenses, and they should be made to put money aside for reclamation activities throughout the life of the mine while also taking care of the environment effects of the operation of the mine. Reclamation process should include all activities including slope reduction, leveling, burying of toxic materials, and revegetation which is aimed at the re-establishment of vegetative cover, soil stability, and water conditions at the site such that the site can be used beneficially for fish and wildlife, grazing, forestry, wetlands, and commercial and industrial use as the case may be. The current case whereby mines and quarries are abandoned without any hint of reclamation activities contributes to soil erosion.

From the foregoing, it is readily observable that issues of soil conservation and erosion control are multifaceted and hence will require a multifaceted approach to its solution. Therefore, the most effective approach to adopt in tackling erosion is to consider it as a component of an all-encompassing environmental resource management program. To this end, the most effective solutions are those that are multi-pronged, multifaceted, and sustained through the long run and that harness the efforts of both the local populace and a wide range of professionals including engineers (hydraulic, geotechnical, civil, highway, structural, agricultural, environmental, etc.), environmental scientists, agriculturists (soil scientists, crop scientists, horticulturists, forestry specialists, agricultural extension specialists.), geologists, hydrogeologists, hydrologists, communication specialists, etc.

5 Conclusions

The Isuikwuato-Okigwe region is by the nature of its geology susceptible to erosion (sheet and gully). The development of the Okigwe-Isuikwuato highway in 1976–1978 has led to the development of gullies on the highway. This paper reviewed the causes of these gullies and found that there was interplay of a number of factors at work in this respect. The effect of geology, climate, relief (topography), hydrology, hydraulics, soil type, anthropogenic factors—agricultural practices, infrastructural development, mineral extraction/mining—on the inception and progression of gully development and erosion generally was demonstrated. Considering the scale of the problem, it is clearly evident that strategies for the control and remediation of erosion ravaged areas should involve a serious reforestation program and all other strategies that reduce the rate of flow of surface runoff. Also, special attention should be paid to the hydrology and hydraulics of erosion control structures to ensure that these structures are continuously maintained and upgraded so that they do not become part of the problem that they were erected to solve as is often the case.

Acknowledgements The authors wish to acknowledge the constructive comments we received from Dr. Ibrahim E. Ahmed Amari. Ofure Ehichoya and Mercy Okhiria helped with the preparation of the manuscript.

References

1. Stamp, L.D.: Land utilisation and soil erosion in Nigeria. Geogr. Rev. **28**, 32–45 (1938)
2. Grove, A.T.: Land Use and Soil Conservation in parts of Onitsha and Owerri Provinces. Bulletin No. 21. Geological Survey of Nigeria, Lagos (1951)
3. Grove, A.T.: Soil erosion and population problems in south-east Nigeria. Geogr. J. **117**(3), 291–304 (1951)
4. Ofomata, G.E.K.: Factors of soil erosion in the Enugu area of Nigeria. Niger. Geogr. J. **8**(1), 45–59 (1965)
5. Ofomata, G.E.K.: Soil erosion in Southeastern Nigeria: the views of a geomorphologist. In: Inaugural Lecture Series, vol. 7. University of Nigeria, Nsukka (1987)
6. Ofomata, G.E.K.: The management of soil erosion problems in south eastern Nigeria. In: Proceedings of the International Symposium on Erosion in Southeastern Nigeria, pp.3–12. Federal University of Technology Owerri, Owerri (1988)
7. Nwajide, C.S., Hoque, M.: Gullying processes in south-eastern Nigeria. Niger. Field **44**(2), 63–74 (1979)
8. Ogbukagu, I.N.: Soil erosion in the northern parts of the Awka-Orlu uplands. Niger. J. Min. Geol. **13**, 6–19 (1976)
9. Nwajide, S.C.: A lithostratigraphic analysis of the Nanka sands of southeastern Nigeria. Niger. J. Min. Geol. **16**(2), 103–109 (1979)
10. Nwajide, S.C.: Gullying in the Idemilli river catchment, Anambra State, Nigeria: theory and cure. In: Freeth, S.J., Ofoegbu, C.O., Onuoha, K.M. (eds.) Natural Hazards in West and Central Africa, pp. 149–161. Vieweg, Braunschweig, Wiesbaden (1992)
11. Uma, K.O., Onuoha, K.M.: Groundwater fluxes and gully development in southeastern Nigeria. In: Groundwater and Mineral Resources of Nigeria, Earth Evolution Science Monograph Series, Friedr. Vieweg and John Wiesbaden, Wiesbaden (1987)
12. Egboka, B.C.E., Nwankwor, G.I.: The hydrogeological and geotechnical parameters as causative agents in the generation of erosion in the rain forest belt of Nigeria. J. Africa Earth Sci. **3**(4), 417–425 (1985)
13. Floyd, B.: Soil erosion and deterioration in eastern Nigeria. Niger. Geogr. J. **8**(1), 33–44 (1965)
14. Obi, M.E., Asiegbu, B.O.: The physical properties of some eroded soils of south eastern Nigeria. Soil Sci. **130**(1), 39–45 (1980)
15. Akpokodje, E.G., Olurunfemi, B.N., Etu-Efeotor, J.O.: Geotechnical properties of soils susceptible to erosion in south eastern Nigeria. Niger. J. Appl. Sci. **4**(3), 161–163 (1986)
16. Okagbue, C.O., Ezechi, J.I.: Geotechnical characteristics of soils susceptible to erosion in Eastern Nigeria. Bull. Int. Assoc. Eng. Geol. **38**, 111–118 (1988)

17. Hudec, P.P., Simpson, F., Akpokodje, E.G., Umenweke, M.O.: Anthropogenic contribution to gully initiation and propagation, SE Nigeria. In: Ehlen, J., Haneberg, W.C., Larson, R.A. (eds.) Humans as Geologic Agents, pp. 149–158. Geological Society of America Reviews in Engineering Geology, XVI, Geological Society of America, Boulder, Colorado (2005)
18. Hudec, P.P., Akpokodje, E.G., Umenweke, M.O., Simpson, F., Ondrasik, M.: Study of Causes and Prevention of Gully Erosion in Southeast Nigeria. http://web2.uwindsor.ca/courses/earth_science/hudec/ (2015). Accessed 15 Feb 2015
19. Okagbue, C.O., Uma K.O.: Performance of gully erosion control measures in management. In: Forest Hydrology and Watershed Management, Proceedings of the Vancouver Symposium, August, 1987. International Association of Hydrological Sciences (IAHS) Publication No. 167, pp. 163–172 (1987)
20. Ijioma, M.A.: A conceptual ecosystems models for effective gully erosion management. A case study of the Onu-Igbere gully. In: Proceedings of the International Symposium on Erosion in Southeastern Nigeria. pp. 92–97. Federal University of Technology, Owerri, Owerri (1988)
21. Igbozurike, U.M.: Soil Erosion Prevention and Control Manual. Friedrich Ebert Foundation, Bonn/Lagos (1993)
22. Amangabara, G.T.: Analysis of selected failed gully erosion control works in Imo State. In: Hydrology for Disaster Management: Special Publication of the Nigerian Association of Hydrological Sciences, pp. 279–287. Nigerian Association of Hydrological Sciences, Abuja (2012)
23. Okogbue, C.O., Agbo, J.U.: Gully erosion resulting from anthropogenic-hydrological modifications: case of the Opi-Ugwogo-Abakpa Nike Road, Anambra State, Nigeria. In: Proceeding International Conference Water Resources in Mountainous Regions, Lausanne Switzerland. International Association of Hydrological Sciences (IAHS) Publication No. 194, pp. 407–415 (1990)
24. Akpokodje, E.G., Tse, A.C., Ekeocha, N.: Gully erosion Geohazards in Southeastern Nigeria and management implications. Scientia Africana **9**(1), 20–36 (2010)
25. Hudec, P.P., Simpson, F., Akpokodje, E.G., Umenweke, M.O., Ondrasik, M. Gully erosion of coastal plain sediments of Nigeria. In: Moore, D., Hungr, O. (eds.) Proceedings Eighth International Congress, International Association for Engineering Geology and the Environment, pp. 1835–1841. A.A. Balkema, Rotterdam, Vancouver, Canada, 21–25 Sept 1998
26. Koppen, W.: Das geographisca System der Klimate. In: Koppen, W., Geiger, G.C. (eds.) Handbuch der Klimatologie, pp. 1–44. Gebr, Borntraeger, Berlin (1936)
27. Peel, M.C., Finlayson, B.I., McMahon, T.A.: Updated world map of the Köppen-Geiger climate classification. Hydrol. Earth Syst. Sci. **11**, 1633–1644 (2007)
28. Wigwe, G.A.: Geology and mineral resources. In: Igbozurike, U.M. (ed.) The Isuikwuato-Okigwe Region. Kartopress, Owerri, Nigeria (1986)
29. Igbozurike, U.M.: The Isuikwuato-Okigwe Region. Kartopress, Owerri, Nigeria (1986)
30. Reyment, R.A.: Aspects of the Geology of Nigeria. University of Ibadan Press, Ibadan (1965)
31. Obi, M.E., Salako, F.K., Lal, R.: Relative susceptibility of some southeastern Nigeria soils to erosion. CATENA **16**, 215–225 (1989)
32. Gobin, A.M., Campling, P., Deckers, J.A., Poesen, J., Feyen, J.: Soil erosion assessment at the Udi-Nsukka Cuesta (southeastern Nigeria). Land Degrad. Dev. **10**, 141–160 (1999)
33. Hudec, P.P., Simpson, F., Akpokodje, E.G., Umenweke, M.O.: Termination of gully processes, Southeastern Nigeria. In: Proceedings of the Eight Federal Interagency Sedimentation Conference (8th FISC), pp. 671–679. Reno. NV. USA, 2–6 April 2006
34. Certini, G.: Effects of fire on properties of forest soils: a review. Oecologia **143**(1), 1–10 (2005). https://doi.org/10.1007/s00442-004-1788-8
35. Ziegler, A.D., Giambelluca, T.W.: Importance of rural roads as source areas for runoff in mountainous areas of northern Thailand. J. Hydrol. **196**, 204–229 (1997)
36. MacDonald, L. H., Coe, D.B.R.: Road sediment production and delivery: processes and management. In: Proceedings of the First World Landslide Forum, pp. 385–388. United Nations University, Tokyo, Japan. International Consortium on Landslides, Japan (2008)

37. IMSU.: Environmental Survey Report on the Imo State University Site Okigwe. Imo State University (IMSU) Press, Okigwe, Nigeria (1987)
38. Lal, R.: Soil erosion and sediment transport research in tropical Africa. Hydrol. Sci. J. **30**(2), 239–256 (1985)
39. Pimentel, D.: Soil erosion: a food and environmental threat. Environ. Dev. Sustain. **8**, 119–137 (2006)
40. Zuazo, V.H.D., Pleguezuelo, C.R.R.: Soil erosion and runoff prevention by plant covers: a review. Agron. Sustain. Dev. **28**(1), 65–86 (2008)
41. Fleischner, T.L.: Ecological costs of livestock grazing in Western North America. Conserv. Biol. **8**(3), 629–644 (1994). https://doi.org/10.1046/j.1523-1739.1994.08030629.x
42. Belsky, A.J., Blumenthal, D.M.: Effects of livestock grazing on stand dynamics and soils in upland forests of the interior west. Conserv. Biol. **11**(2), 315–327 (1997). https://doi.org/10.1046/j.1523-1739.1997.95405.x
43. Trimble, S.W., Mendel, A.C.: The cow as a geomorphic agent, a critical review. Geomorphology **13**, 233–253 (1995)
44. Bilotta, G.S., Brazier, R.E., Haygarth, P.M.: The impacts of grazing animals on the quality of soils, vegetation, and surface waters in intensively managed grasslands. Adv. Agron. **94**(6), 237–280 (2007). https://doi.org/10.1016/S0065-2113(06)94006-1
45. Ziegler, A.D., Sutherland, R.A., Giambelluca, T.W.: Runoff generation and sediment transport on unpaved roads, paths, and agricultural land surfaces in northern Thailand. Earth Surf. Proc. Land. **25**, 519–534 (2000)
46. Ziegler, A.D., Sutherland, R.A., Giambelluca, T.W.: Acceleration of Horton overland flow and erosion by footpaths in an agricultural watershed in northern Thailand. Geomorphology **41**, 249–262 (2001)
47. Kuypers, H., Mollema, A., Topper, E.: Erosion control in the tropics, (Agrodok 11), 6th edn. Agromisa Foundation, Wageningen (2005)
48. Vuillaume, G.: The influence of environment parameters on natural erosion in the tropical region of West Africa. Stud. Rep. Hydrol. **32**, 225–251 (1982)
49. Salako, F.K.: Susceptibility of coarse-textured soils to erosion by water in the tropics, (LNS0418030). In: Gabriels, D.M., Ghirardi, G., Nielsen, D.R., Pla Sentis, I., Skidmore, E.L. (eds.) ICTP Lecture Notes: Invited Presentations on College on Soil Physics, pp. 339–362. The Abdus Salam International Centre for Theoretical Physics, Trieste, Italy, 3–21 Mar 2003
50. Lal, R.: Analyses of different processes governing soil erosion by water in the tropics. In: Erosion and Sediment Transport Measurement, Proceedings of the Florence Symposium, June 1981, Publication No. 133, pp. 351–364. International Association of Hydrological Sciences (IAHS) (1981)
51. Obi, M.E., Salako, F.K.: Rainfall parameters influencing erosivity in southeastern Nigeria. CATENA **24**, 275–287 (1995)
52. Lal, R.: Soil erosion on alfisols in Western Nigeria. I. Effects of slope, crop rotation and residue management. Geoderma **16**, 363–373 (1976)
53. World Bank: Nigeria—Erosion and Watershed Management Project. Washington, DC: World Bank. http://documents.worldbank.org/curated/en/2012/04/16232809/nigeria-erosion-watershed-management-project (2012). Accessed 12 May 2015
54. Igbokwe, E.M.: A soil and water conservation system under threat. A visit to Maku, Nigeria. In: Reij, C., Scoones, I., Toulmin, C. (eds.) Sustaining the Soil—Indigenous Soil and WATER CONservation in Africa, pp. 219–243. Earthscan, London (1996)
55. Junge, B., Abaidoo, R., Chikoye, D., Stahr, K.: Soil Conservation in Nigeria: Past and present on-station and on-farm initiatives. Soil and Water Conservation Society, Ankeny, Iowa (2008)
56. Ene, G.E., Okogbue, C.O.: Construction and performance of geo-engineering structures for combating gully erosion in South-Eastern Nigeria. In: Engineering Geology for Society and Territory, vol. 2, pp. 857–864. Springer International Publishing (2015)
57. Wikipedia: States of Nigeria. https://en.wikipedia.org/wiki/Subdivisions_of_Nigeria (2015). Accessed 12 Dec 2015

Assessment of the Causes of Erosion and Gullying Along the Leru–Nkwoagu, Amuda (Isuochi)–Mbala (Isuochi) Highway, Southeastern Nigeria: A Case Study

Site Onyejekwe, Jeremiah C. Obi, and Elamin Ismail

1 Introduction

The nexus between population growth and environmental degradation has been established in published literature. Several studies indicate that population growth is among factors that contribute to environment degradation [1–6]. Population growth leads directly to an increase in the demand for infrastructure required for the general wellbeing of the population. To fulfill this increasing demand, recourse is often made to the natural environment/land resources and involves some form of change in land use.

The expansion of infrastructure to meet the increasing demand of an increasing population requires some form of change in land use which, if not done correctly, ultimately leads to environmental degradation. These include the clearing of virgin forest, reduction of fallow period, expansion into marginal land, large-scale clearance of grounds, creating access road into virgin forests, mining and quarrying operations which lead to deforestation and soil erosion [7]. This is particularly true for the countries of the global South where the function of land as a factor of production is still of prime importance and the implementation of environmental safeguards, where they exist, with respect to infrastructural developments is so lax as to be considered non-existent.

Nigeria is a developing country with a great rural population. Hence, there is a great need to 'open up' these rural areas (i.e., link these rural areas to the major transportation network) so as to provide access to greater social and economic opportunities and thereby improve the general wellbeing of the inhabitants. Successive

S. Onyejekwe (✉)
Road Sector Development Team, Federal Ministry of Works, Abuja, Nigeria

J. C. Obi
Missouri Department of Transportation, Chesterfield, MO 63017, USA

E. Ismail
Bara Geophysical Services, Rolla, MO 65401, USA

I. Pal and S. Kolathayar (eds.), *Sustainable Cities and Resilience*, Lecture Notes in Civil Engineering 183, https://doi.org/10.1007/978-981-16-5543-2_39

political administrations at all levels (federal, state, and local) have had rural road development programs. However, for some reasons, mainly policy shortcomings and notwithstanding the well-documented negative environmental effects of large-scale linear clearances like roads, and utility lines on forests [8], most of these road developments are executed without environmental safeguards. This has resulted in the unintended consequence of accelerated land degradation in terms of soil erosion by water, particularly, sheet and gully erosion. The effect is more prevalent in environmentally sensitive regions such as southeastern Nigeria which has severe issues of soil erosion by water, particularly, sheet and gully erosion [9–16].

Soil erosion has been an issue in southeastern Nigeria for a long while. Several researchers (e.g., [14, 16–29]) have studied the problem of soil erosion in southeastern Nigeria and have attributed the causes of erosion (soil/gully) to factors such as geology/soil type, topography, climate, vegetation, and anthropogenic actions. Almost all severe erosion gullies in southeastern Nigeria are located on moderate to very gently dipping, poorly consolidated sandstones, usually associated with local or regional highlands, among which, the Udi–Orlu and the Okigwe–Ohafia–Arochukwu highlands are the most prominent ([11, 30]. The major highlands, plateau, and their precipitous escarpments are formed by sandstones bedrocks (Ajalli sandstones and Nanka sands) while the lower slopes and plains are underlain by mainly shale units (Imo, Mamu, Nsukka and Bende-Ameki [Ameke] formations) [11, 12, 16, 30]. The gentle slopes of undulating plains are covered by thick and highly sandy overburden. The highest point in the region (343 m) occurs around Okigwe while the southern plains of the region stand at about 61 m above sea level [11, 12, 16, 30]. Gully processes are mostly prevalent in the fine to medium-grained coastal plain sands (Pliocene to recent) and Nanka Sands (Eocene) and the medium- to coarse-grained Nsukka Sandstone and Ajalli Sandstone (Cretaceous) of the Anambra-Imo basin region [12, 16]. The 19 km Leru–Nkwoagu–Mbala highway is situated between Okigwe and Awgu within the scarp slope of the erosion-prone 250 km long north–south trending Arochukwu–Ohafia–Awgu–Udi–Nsukka cuesta in Southeastern Nigeria.

This paper reports on an assessment of the causes of soil erosion and gullying on the 19 km Leru–Nkwoagu–Mbala highway, particularly on the recently upgraded 9 km Leru–Nkwoagu section. Remedial measures for the mitigation of the effect of these environmental issues are also recommended.

2 Physiography and Geology of Study Area

The 19 km Leru–Nkwoagu–Mbala (Isuochi) highway is located within the northern fringes of the old Isuikwuato/Okigwe Local Government Area (LGA) of the old Imo State. The road is now located within the Umunneochi Local Government of Abia State and originates a few meters after the Leru Underpass (6.058433 N, 7.358325 E) on the old Enugu–Port Harcourt highway and ends at the Mmam River (Mbala, Isuochi) (5.915286 N, 7.408039 E) on the Abia/Enugu state border. The study area,

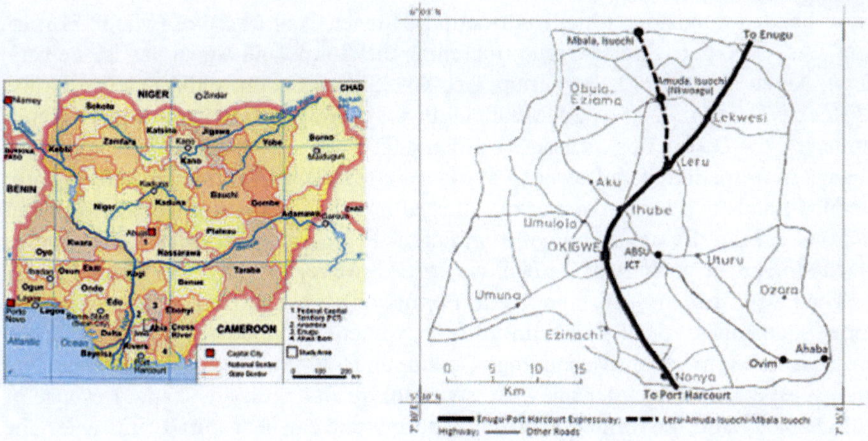

(a) (b)

Fig. 1 **a** Study area (adapted from Wikipedia [48]). **b** Highway alignment

shown in Fig. 1a, is located between Okigwe and Awgu within the scarp slope of the 250 km long north–south trending Arochukwu-Ohafia-Awgu-Udi-Nsukka cuesta. The highway is shown in Fig. 1b. The former Isuikwuato/Okigwe Local Government Area is located within Latitude 5° 40′–6° 03′ N and longitudes 7° 10′–7° 35′ E, which places it within the tropical monsoon (AM) climatic zone based on the Köppen Climate Classification System ([31, 32]. At an average of about 130–200 m above sea level, the region lies on a relatively higher terrain as compared to the other parts of Imo and Abia States of Nigeria.

The Isuikwuato/Okigwe region has two seasons: rainy (March to October) and dry (November to February). Rainfall is typical of tropical rainforest regions and ranges from 2250 to 2500 mm. Annual average monthly maximum and minimum temperature range from 28 to 35 °C (in the dry season) and 19 to 24 °C (in the peak rainy, and the Harmattan seasons), respectively. Mean relative humidity for the dry and rainy seasons ranges from 60 to 70% and 80 to 85%, respectively.

The study area is located within the Anambra Basin (in the southeastern Nigeria sedimentary basin). The Anambra Basin along with the Abakaliki Anticlinorium and the Afikpo Syncline are the major components of the Lower Benue Trough of the Benue Trough of Nigeria [33]. The rocks, rock systems, and geological history of the Isuikwuato/Okigwe region are derived from geological events which span the Cretaceous and Tertiary periods of the Mesozoic and Cenozoic eras, respectively [34]. The oldest rocks in the Isuikwuato-Okigwe region are of lower Cretaceous age, while the youngest ones are still being formed. The following ten main geological formations (in order of stratification, from the youngest to oldest) can be distinguished in the study area [34, 35]: Coastal Plain Sands Formation (Benin Formation); Bende-Ameki (Ameke) Formation; Imo Shale Group; Upper Coal Measure (Nsukka Formation);

False bedded Sandstone (Ajalli Formation); Lower Coal Measure (Mamu Formation); Asata-Nkporo Shales Group (including the Afikpo and Agwu Sandstone variants); Agwu-Ndeaboh Shales Group; Eze-Aku Shales Group; and Asu River Group of Rocks. The Leru–Nkwoagu–Mbala highway falls within the False Bedded Sandstone and the Lower Coal Measure Formations. From the sections of road cuts, the change in formation from Lower Coal Measure for the road section in Leru to False Bedded Sandstone for remainder of the road in Isuochi occurs at about Kilometer (Km) 6. A brief discussion on both formations is presented below. A more detailed discussion on the geology of the Isuikwuato/Okigwe region can be found in [35, 36].

The False Bedded Sandstone (Ajalli Formation) consists of thick, friable, poorly sorted sandstones typically white in color, but sometimes iron-stained. The formation is often marked by repetitive banding of coarse and fine-grained layers [37]. The sand grains, especially the larger ones, are sometimes sub-angular in shape. Because of this characteristic, the formation is highly porous and can be texturally unstable. The formation is highly unstable, being highly porous and weakly consolidated in places. This makes it is very prone to sheet and particularly gully erosion when disturbed mindlessly. The most important structural characteristic is the cross-bedding.

The Lower Coal Measure (Mamu Formation), unlike the Upper Coal Measure, contains a distinctive alternating admixture of sandstones, dark shale, dark blue or gray mudstones and thin bands of sandy shale with thicker lignite and coal materials seams at several horizons. The sandstone is fine to medium-grained and may be white or yellow in color in the lower horizons. They are normally well-bedded, even though cross-bedding is evident in some places [34].

3 Origin of Present Problems

The 19 km Leru–Nkwoagu–Mbala (River Mmam, the Abia/Enugu State border) is a 'legacy' road that has evolved from a footpath to single track earth road then to the present asphaltic concrete/surface-dressed highway. Like most roads in the area, the Leru–Nkwoagu–Mbala highway started as a Native Authority footpath linking Leru to the colonial Native Administration offices at Nkwoagu, Amuda, Isuochi.

The road is of two sections: the 9 km Leru–Nkwoagu, and the 10 km Nkwoagu–Mbala (Enugu/Abia State border). The Nkwoagu–Mbala section was upgraded from a 2-lane earth road to an all-season surface-dressed pavement in 1981 as part of Ihube–Nkwoagu highway construction project. About 4.5 kms of this section from Nkwoagu was paved with asphalt concrete in 2014. From the end of the recent asphaltic concrete overlay at Km 13.5, there are some traces of the 1981 surface-dressed pavement still existing up to Km 15. After this point, the road is essentially an earth road with no trace of the previous surfacing up to the end of the road at Mmam River. The Leru–Nkwoagu Section of the highway on the other hand was upgraded from a 2-lane earth road to an all-season asphaltic concrete pavement in 2013.

The Leru–Nkwoagu–Mbala highway was constructed in the environmentally sensitive southeastern region of Nigeria that, due to the geology of the area, is known to be highly susceptible to soil erosion by water [9–16]. The erosion susceptibility of this area has long been recognized as the colonial government (pre-1960) and successive governments established and maintained cashew plantations as erosion control measures in the area during and after colonial rule [22, 23, 38].

While the opening of roads has well-documented benefits (social, economic, etc.), it similarly has its disadvantages [8, 39]. For environmentally sensitive areas, appropriate safeguards have to be instituted and implemented during and post-construction to mitigate the negative effects of road/highway development. The Leru–Nkwoagu–Mbala highway was constructed without the institution and implementation of these safeguards; this, in combination with factors such as unregulated exploitation of natural resources, poor engineering design, and poor farming practices, have resulted in severe environmental degradation in terms of soil erosion and gullying within some segments of the road corridor. This is particularly so for the Leru–Nkwoagu section which was upgraded in 2013. The Nkwoagu–Mbala section, which was upgraded many years ago in 1981, has inadequately maintained segments that have signs of sheet and gully erosion, particularly the last four kilometers.

The several instances of land degradation, particularly soil erosion and gullying that have occurred as a result of the recent upgrade of the Leru–Nkwoagu section of the highway and other cases on the Nkwoagu–Mbala section are discussed in the following section.

4 Causes of Soil Erosion and Gullying

Several cases of soil erosion and gullying, and other activities unregulated excavations, and quarrying, landslides, slope failures, which ultimately degrade the environment, are present within the road corridor. While the most severe cases are on the recently upgraded Leru–Nkwoagu section of the highway, other cases are present in the Nkwoagu–Mbala section. Following is a discussion of some of the causes of soil erosion and gullying on the highway.

4.1 Highway Drainage and Drainage Structures

Roads and highways are known to induce a concentration of surface runoff, a diversion of concentrated runoff to other catchments, and an increase in catchment size, which enhance gully development after road construction [40]. To limit these risks, roads and highways should be designed in a way that keeps runoff interception, concentration, and deviation minimal [41].

The state-of-the-practice of highway development in Nigeria emphasizes the geometric design aspects of highway development to the exclusion of other aspects, particularly the drainage and its effect on the immediate and wider environment. This is clearly evident on this highway. The is lack of hydrological studies during the design and the upgrading of hydraulic structures with changes in the hydraulic/hydrological regime which is clearly evident has resulted in the deployment of inappropriately sized and sometimes inappropriately located drainage structures which act as trigger points for erosion on the highway.

Instances include: (a) the development gullies in roadside sections without drainage channels or structures and other locations where due to inappropriately sized and located drainage structures a greater percentage of runoff flow outside the drainage channels are shown in Fig. 2a; (b) effect of high-velocity runoff from undersized culverts when running full on the erosion-susceptible geology of the highway corridor is evident at several locations particularly Km 7.0 in Ndiawa, Isuochi (see Fig. 2b) and the worst case is at Km 8.6, near the Local Government Area Headquarters, where a 600 mm diameter twin cell ring culvert is clearly not the appropriate size for the volume of runoff conveyed and has resulted in gullying which breached the highway in the rainy season of 2016. See Fig. 2c, d.

Fig. 2 **a** An instance of a poorly located drainage. **b** Gullying around culvert outlet at Km 7.0. **c** Gullying around culvert outlet at Km 8.6—armor protection still present. **d** Gullying around culvert outlet at Km 8.6—highway breached by gully

Fig. 3 Gully development from poorly reinstated utility (optical fiber) excavation

4.2 *Excavation for Utilities*

Improperly reinstated linear excavations for optical fiber cables on the Isuochi-bound side of the road have developed into erosion channels and gullies in many segments. The excavations which are typically about 250 mm wide and 450 to 600 mm deep have developed into channels of varying proportions, with dimensions that range from 1 to 2 m wide and 3 to 4 m deep in some locations. Figure 3 shows these channels.

4.3 *Excessive Clearing*

There was unnecessary removal of vegetative cover in some segments of the highway, sometimes in excess of 20 m beyond the carriageway shoulders without any form of reinstatement. This is not advised as it exposes the bare soil to erosive forces, thereby exacerbating soil erosion ([42, 43]. Erosion features (splash, rill, sheet, and gully) have developed in the exposed unconsolidated sandy materials in these unnecessarily cleared sections as shown in Fig. 4.

4.4 *Side-Cast Construction*

Shale and lateritic materials from the cut described in §4.6 were side-cast between Km 0.4 and 0.6 on the Isuochi-bound lane side of the roadway; the area beyond the shoulder and the slope of the embankment were neither protected nor was there an effort made to establish vegetative cover for this area, thereby exposing these loose unprotected decomposed shale materials and lateritic materials to the forces

Fig. 4 Excessive site clearing without reinstatement

(a) (b)

Fig. 5 Side-cast construction. **a** Erosion at unprotected verge. **b** Gullying at unprotected side-cast slope

of erosion resulting in the development of sheet erosion, rills, and gully erosion features as shown in Fig. 5. Side-cast construction is one of the methods used for the construction of roads in hilly and mountainous terrains [44]. However, side-cast fills are usually not well-compacted and not draining and are over-steepened which factors make them prone to erosion [43, 44].

4.5 Footpaths and Earth Roads

Footpaths and earth roads are necessary components of rural transportation systems. However, these footpaths and earth roads being neither planned nor designed in terms of their orientation relative to ground contours largely function to promote the concentration of runoff and erosion [14, 45]. When unchecked, these footpaths and

(a) (b)

Fig. 6 **a** Footpath as runoff channel. **b** Earth road as runoff channel

earth roads develop into gullies that continue to erode long after they are no longer in use [46]. The completion of the highway and associated improved access and more footpaths and earth roads have been established in addition to previously existing ones. Figure 6 shows cases where footpaths (Fig. 6a) and earth roads (Fig. 6b) have effectively become drainage channels.

4.6 Slope Failures

Slope failures of varying magnitudes arising from unprotected slope faces and non-provision of earth retention structures where needed can be found along the highway, particularly in the first five kilometers. Three larger slope failures occurred within the first 250 m (Km 0.15–0.4) where the carriageway section was constructed expanding the bench on an existing steep slope at the Leru Hills on the Enugu–Port Harcourt Expressway by cutting into it. This increased both the length and the angle of the slope of the cut and thereby making the formerly stable slope highly susceptible to slope failure [43, 44]. These failures occurred as a result of the effect of water on very steep unprotected cut slopes in an unstable geological formation. This segment of the highway is situated within the Lower Coal Measures (Mamu) geological formation (general lithology—shale with intercalations of sand and clay). In most cases, the unstable soil layer is first eroded thereby putting the slope in disequilibrium. The stable soil (usually lateritic overburden) then fails subsequently. Figure 7 shows a picture (taken from the Enugu-bound carriageway of the Enugu–Port Harcourt Expressway) (Fig. 7a) and an idealized cross section of the segment of the highway (Fig. 7b). Figure 7c shows a larger slope failure, with debris covering part of the Leru-bound lane while Fig. 7d shows alternating layers of stable and unstable soil layers on steep cut slopes.

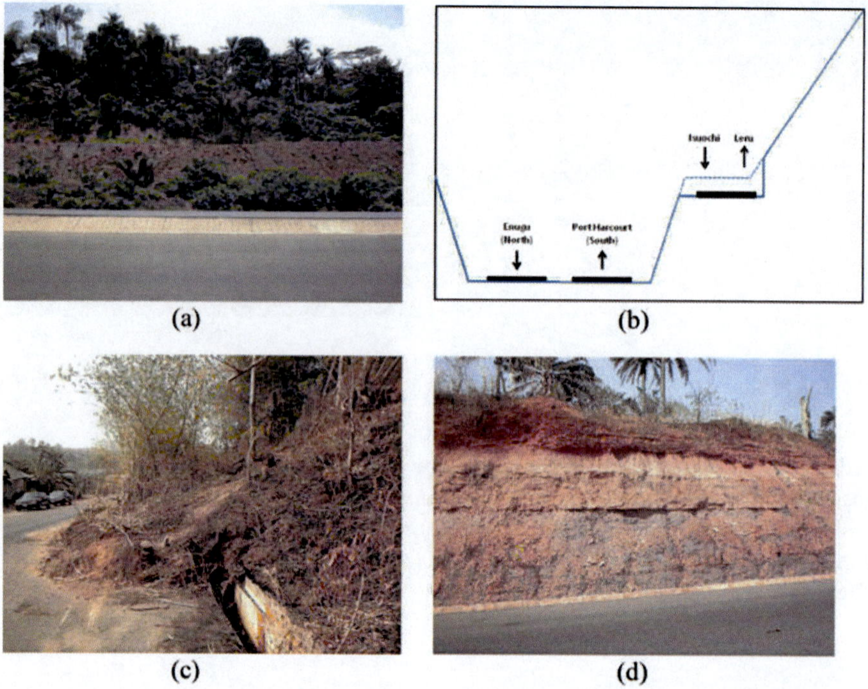

Fig. 7 **a** View of the Leru–Isuochi Highway from the Enugu-bound carriageway of the Enugu–Port Harcourt Expressway. **b** Idealized cross section of the segment of the highway. **c** A larger slope failure, with debris covering part of the Leru-bound lane at Km 0.15. **d** Layers of stable and unstable soil layers on steep cut slopes

5 Recommendations

The requirements of the modern highway and high-speed travel on highways including wider carriageways, wider right-of-way, and the extensive amount of earthworks needed to achieve safe, suitable horizontal and vertical alignments have turned highway construction into a high-risk activity in terms of its propensity to trigger soil erosion by water [47]. This is particularly true for the construction of the highways in environmentally sensitive areas such as areas with known susceptibility to soil erosion by water without the institution of safeguards both during and after construction to prevent soil erosion as in the case of the Leru—Nkwoagu—Mbala highway.

Considering the huge damage wrought by highway-related soil erosion on the environment and the economic cost of the damage, it is of utmost importance and urgency that necessary countermeasures be instituted to check this alarming trend in environmental degradation. To this end, taking into cognizance the scale of the problem at hand, a river basin-wide, intergovernmental, multi-sectorial, multi-disciplinary, and all-stakeholder approach will be required to solve the problems of highway-related soil erosion in the southeastern Nigeria.

In light of the foregoing, the following countermeasures aimed at remediating and preventing highway-related erosion on the Leru—Nkwoagu—Mbala highway and other highways similarly constructed in environmentally sensitive regions are proposed.

(a) Stakeholder participation: Highway development is usually by governments (federal, state, and local) and in this process, public consultations with the communities that will be most affected by highway projects to obtain both their inputs and buy-in are rarely held as should be the case. These communities should be enlightened on the effects of the project on their environment and the measures proposed to remediate these effects. Hence, there is usually a lack of ownership and participation by the communities and other stakeholders. This lack of stakeholder participation has far-reaching effects on the entire highway development process, particularly the implementation of environmental safeguards during and after the execution of highway projects. Absent community and stakeholder participation, the implementation of environmental safeguards is left to government agencies that have all failed in their mandate, as is evident in the ever-increasing area under highway-related gully erosion.

(b) Standard of practice: There is an urgent need to re-calibrate the standard of practice of highway engineering in Nigeria to emphasize other aspects of highway design other than geometric design. Design professionals should be made to pay adequate attention to geological, geotechnical, drainage, and related issues in the design of highway projects. In this regard, capacity-building in hydrology, hydraulics, and geotechnics for highway design professionals is warranted. The lack of capacity in these subject areas has resulted in the construction of inappropriately sized, inappropriately located, and inappropriately terminated concrete structures and drainage channels that act to exacerbate and intensify erosion processes rather than aid the control or prevention of erosion.

(c) Environmental-impact assessment (EIA): Environmental-impact assessments (EIAs) are currently mandated for highway developments. However, they are generally limited in scope and their recommendations are almost never implemented. EIAs should be more detailed and extensive not only covering the direct effects of the road building within the R-O-W but the indirect effects of highway development like mining, development of earth roads and footpaths, bush fires, and deforestation which all lead to land degradation in the larger catchment area. They should also meet all state and local regulations. The recommendations of EIAs should be strictly enforced.

(d) Erosion and Sediment Control (ESC) Plan: An erosion and sediment control (ESC) plan, covering both the construction period and the post-construction period, should be prepared for every highway development project by the lead design engineer in concert with other relevant professionals. The ESC plan should detail best management practices (BMPs) for the mitigation of any erosion and sediment-related issues that may arise during and post-construction. The ESC plan should be made available to all stakeholders—civil society (religious organizations, youth organizations, community development associations [CDAs], professional organizations, traditional councils, etc.) and relevant government (local, state, federal) agencies—who should jointly ensure compliance with the ESC plan during and after construction.

(e) Anthropogenic factors: The lack of effective regulations on the exploitation of the environmental resources with respect to agriculture, mineral exploitation/mining, infrastructural developments, and related activities have also resulted in the failure of soil conservation and erosion control measures. There is a need to bring into force an effective framework for the regulation of these activities at the level of the river basin, the Anambra-Imo River basin in this case.

6 Conclusions

The Leru–Nkwoagu–Mbala highway was constructed without safeguards in an environmentally sensitive region that, due to its geology, is known to be highly susceptible to soil erosion by water. The absence of safeguards, in conjunction with poor design decisions, and other activities associated with improved access like deforestation, illegal excavations, and quarrying have expectedly resulted in several instances of soil erosion, gullying, and general land degradation within the highway. The failure to implement safeguards, poor design considerations, particularly in relation to the aspects of design relating to hydraulics, hydrology, and geotechnics, and the general lack of enforcement of environmental regulations on this highway has been found not to be an outlier. Hence, a river basin-wide, intergovernmental, multi-sectorial, multi-disciplinary, and all-stakeholder approach to the solution of highway-related soil erosion is strongly recommended for adoption.

Acknowledgements The authors wish to acknowledge the constructive comments we received from Dr. Ibrahim E. Ahmed Amari. Ofure Ehichoya and Mercy Okhiria helped with the preparation of the manuscript.

References

1. Mortimore, M.: Population Growth and Land Degradation. Geojournal, vol. 31, no.1, pp. 15–21. Retrieved from http://www.jstor.org/stable/41145902 (1993)
2. Roose, E.: Land husbandry—components and strategy. In: Food and Agricultural Organisation (FAO) Soil Bulletin No. 70. FAO, Rome (1996)
3. Barbier, E.B.: The economic determinants of land degradation in developing countries. Philos. Trans. Royal Soc., Ser. B **352**, 891–899 (1997)
4. Barbier, E.B.: Poverty, development, and environment. Environ. Dev. Econ. **15**(6), 635–660 (2010)

5. Hunter, L.M.: The Environmental Implications of Population Dynamics. RAND, Santa Monica, CA (2000)
6. FAO [Food and Agriculture Organization]: The State of The World's Land and Water Resources for Food and Agriculture: Managing systems at risk. FAO/Earthscan, Rome/London (2011)
7. Titilola, S., Jeje, L.K.: Environmental degradation and its implications for agricultural and rural development: the issue of land Erosion. J. Sustain. Dev. Africa **10**(2), 116–146 (2008). ISSN 1520-5509
8. Laurance, W.F., Goosem, M., Laurance, S.G.W.: Impacts of roads and linear clearings on tropical forests. Trends Ecol. Evol. **24**(12), 659–669 (2009). https://doi.org/10.1016/j.tree.2009.06.009
9. Okagbue, C.O., Ezechi, J.I.: Geotechnical characteristics of soils susceptible to erosion in Eastern Nigeria. Bull. Int. Assoc. Eng. Geol. **38**, 111–118 (1988)
10. Obi, M.E., Salako, F.K., Lal, R.: Relative susceptibility of some south-eastern Nigeria soils to erosion. CATENA **16**(3), 215–225 (1989)
11. Okogbue, C.O., Agbo, J.U.: Gully erosion resulting from anthropogenic-hydrological modifications: case of the Opi-Ugwogo-Abakpa Nike Road, Anambra State, Nigeria. In: Proceeding International Conference Water Resources in Mountainous Regions, Lausanne Switzerland. International Association of Hydrological Sciences (IAHS) Publication No. 194, pp. 407–415 (1990)
12. Hudec, P.P., Simpson, F., Akpokodje, E.G., Umenweke, M.O., Ondrasik, M.: Gully erosion of coastal plain sediments of Nigeria. In: Moore, D., Hungr, O. (eds.) Proceedings Eighth International Congress, International Association for Engineering Geology and the Environment, pp. 1835–1841. A.A. Balkema, Rotterdam, Vancouver, Canada, 21–25 Sept 1998
13. Gobin, A.M., Campling, P., Deckers, J.A., Poesen, J., Feyen, J.: Soil erosion assessment at the Udi-Nsukka Cuesta (southeastern Nigeria). Land Degrad. Dev. **10**(2), 141–160 (1999)
14. Hudec, P.P., Simpson, F., Akpokodje, E.G., Umenweke, M.O.: An-thropogenic contribution to gully initiation and propagation, SE Nigeria. In: Ehlen, J., Haneberg, W.C., Larson, R.A. (eds.) Humans as Geologic Agents, Geological Society of America Reviews in Engineering Geology, XVI, pp. 149–158. Geological Society of America, Boulder, Colorado (2005). https://doi.org/10.1130/2005.4016(13)
15. Hudec, P.P., Simpson, F., Akpokodje, E.G., Umenweke, M.O.: Termination of gully processes, Southeastern Nigeria. In: Proceedings of the Eight Federal Interagency Sedimentation Conference (8th FISC), pp. 671–679. Reno, NV, USA, 2–6 April 2006
16. Hudec, P.P., Akpokodje, E.G., Umenweke, M.O., Simpson, F., Ondrasik, M.: Study of Causes and Prevention of Gully Erosion in Southeast Nigeria. http://web2.uwindsor.ca/courses/earth_science/hudec/ (2015). Accessed 15 Feb 2015
17. Stamp, L.D.: Land utilisation and soil Erosion in Nigeria. Geogr. Rev. **28**, 32–45 (1938)
18. Grove, A.T.: Land use and soil conservation in parts of Onitsha and Owerri provinces. Bulletin No. 21. Geological Survey of Nigeria, Lagos (1951)
19. Grove, A.T.: Soil erosion and population problems in south-east Nigeria. Geogr. J. **117**(3), 291–304 (1951)
20. Ofomata, G.E.K.: Factors of soil erosion in the Enugu area of Nigeria. Niger. Geogr. J. **8**(1), 45–59 (1965)
21. Ofomata, G.E.K.: Soil Erosion in Southeastern Nigeria: the views of a geomorphologist. In: Inaugural Lecture Series, No. 7. University of Nigeria, Nsukka (1987)
22. Ofomata, G.E.K.: The management of soil erosion problems in south eastern Nigeria. In: Proceedings of the International Symposium on EROSION in Southeastern Nigeria. pp. 3–12. Federal University of Technology, Owerri (1988)
23. Floyd, B.: Soil erosion and deterioration in Eastern Nigeria. Niger. Geogr. J. **8**(1), 33–44 (1965)
24. Nwajide, C.S., Hoque, M.: Gullying processes in south-eastern Nigeria. Niger. Field **44**(2), 63–74 (1979)
25. Nwajide, S.C.: A lithostratigraphic analysis of the Nanka sands of south-eastern Nigeria. Niger. J. Min. Geol. **16**(2), 103–109 (1979)

26. Nwajide, S.C.: Gullying in the Idemilli river catchment, Anambra State, Nigeria: theory and cure. In: Freeth, S.J., Ofoegbu, C.O, Onuoha, K.M. (eds.) Natural Hazards in West and Central Africa. pp. 149–161. Vieweg, Wiesbaden (1992)
27. Obi, M.E., Asiegbu, B.O.: The physical properties of some eroded soils of south eastern Nigeria. Soil Sci. **130**(1), 39–45 (1980)
28. Akpokodje, E.G., Olurunfemi, B.N., Etu-Efeotor, J.O.: Geotechnical properties of soils susceptible to erosion in south eastern Nigeria. Niger. J. Appl. Sci. **4**(3), 161–163 (1986)
29. Ene, G.E., Okogbue, C.O.: Construction and performance of geo-engineering structures for combating Gully Erosion in South-Eastern Nigeria. In: Engineering Geology for Society and Territory, vol. 2, pp. 857–864. Springer International Publishing (2015). https://doi.org/10.1007/978-3-319-09057-3_147
30. Akpokodje, E.G., Tse, A.C., Ekeocha, N.: Gully erosion geohazards in southeastern Nigeria and management implications. Scientia Africana **9**(1), 20–36 (2010)
31. Koppen, W.: Das geographisca System der Klimate. In: Koppen, W., Geiger, G.C., (eds.) Handbuch der Klimatologie, pp. 1–44. Gebr, Borntraeger, Berlin (1936)
32. Peel, M.C., Finlayson, B.I., McMahon, T.A.: Updated world map of the Köppen-Geiger climate classification. Hydrol. Earth Syst. Sci. **11**, 1633–1644 (2007). https://doi.org/10.5194/hess-11-1633-2007
33. Burke, K., Dessaugie, T.F.J., Whiteman, A.J.: Geological history of the Benue valley and adjacent areas. In: Dessauvagie, T.F.J., Whiteman, A.J. (eds.), African Geology, Procseeding Conference on African Geology, University of Ibadan, pp. 187–205. University of Ibadan Press, Ibadan, 7–14 Dec 1970
34. Wigwe, G.A.: Geology and mineral resources. In: Igbozurike, U.M. (ed.) The Isuikwuato-Okigwe Region. Kartopress, Owerri, Nigeria (1986)
35. Igbozurike, U.M.: The Isuikwuato-Okigwe Region. Karto-press, Owerri, Nigeria (1986)
36. IMSU [Imo State University]: Environmental Survey Report on the Imo State University Site Okigwe. IMSU Press, Okigwe, Nigeria (1987)
37. Reyment, R.A.: Aspects of the Geology of Nigeria. University of Ibadan Press, Ibadan (1965)
38. CFC.: Common Fund for Commodities (CFC), CFC Technical Paper No. 23, Regional meeting on the development of Cashew nut exports from West Africa, Cotonou, Benin July, 2002. http://common-fund.org/fileadmin/user_upload/Projects/FIGTF/FIGTF_07/Technical_Paper_No._23.pdf (2003). Accessed 12 Jan 2016
39. Caro, T., Dobson, A., Marshall, A.J., Peres, C.A.: Compromise solutions between conservation and road building in the tropics. Curr. Biol. **24**(16), R722–R725 (2014). https://doi.org/10.1016/j.cub.2014.07.007
40. Nyssen, J., Poesen, J., Moeyersons, J., Luyten, E., Veyret-Picot, M., Deckers, J., Mi-tiku, H., Govers, G.: Impact of road building on gully erosion risk: a case study from the northern Ethiopian Highlands. Earth Surf. Proc. Land. **27**(12), 1267–1283 (2002). https://doi.org/10.1002/esp.404
41. Valentin, C., Poesen, J., Li, Y.: Gully erosion: impacts, factors and control. CATENA **63**, 132–153 (2005)
42. Zuazo, V.H.D., Pleguezuelo, C.R.R.: Soil erosion and runoff prevention by plant covers: a review. Agron. Sustain. Dev. **28**(1), 65–86 (2008)
43. Seutloali, K.E., Beckedahl, H.R.: A review of road-related soil erosion: an assessment of causes, evaluation techniques and available control measures. Earth Sci. Res. J. (Geotechnics) **19**(1), 73–80 (2015). http://dx.doi.org/https://doi.org/10.15446/esrj.v19n1.43841
44. Fay, L., Akin, M., and Shi, X.: Cost-effective and sustainable road slope stabilization and erosion control. In: National Cooperative Highway Research Program (NCHRP) Synthesis 430. Transportation Research Board of The National Academies of Sciences, Engineering, and Medicine, Washington, DC (2012)
45. MacDonald, L.H., Coe, D.B.R.: Road sediment production and delivery: processes and management. In: Proceedings of the First World Landslide Forum, United Nations University, Tokyo, Japan, pp. 385–388. International Consortium on Landslides, Japan (2008)

46. Douglas, I., Pietroniro, A.: Predicting road erosion rates in selectively logged tropical rain forests. In: de Boer, D., Froehlich, W., Mizuyama, T. (eds.) Erosion Prediction in Ungauged Basins, Integrating Methods and Techniques. Proceedings of an International Symposium Sapporo, Japan, pp. 199–205. IAHS Press, Wallingford, UK 8–9 July 2003
47. Clyde, C.G., Israelsen, C.E., Packer, P.E.: Erosion Control during Highway Construction: Volume 1. Reports. Paper 536. http://digitalcommons.usu.edu/water_rep/536 (1979)
48. Wikipedia: States of Nigeria. https://en.wikipedia.org/wiki/Subdivisions_of_Nigeria Accessed 12 Dec 2015